Property of
FAMILY OF FAITH
LIBRARY

W9-BIH-748

Family of Faith Library

Linear Algebra with Applications

GARETH WILLIAMS
Stetson University

ᵫcb
Wm. C. Brown Publishers
Dubuque, Iowa

Copyright © 1984 by Allyn and Bacon, Inc.

Copyright © 1989 by Wm. C. Brown Publishers. All rights reserved

Library of Congress Catalog Card Number: 83–13523

ISBN 0–697–06883–8

No part of this publication may be reproduced, stored in a retrieval
system, or transmitted, in any form or by any means, electronic,
mechanical, photocopying, recording, or otherwise, without the prior
written permission of the publisher.

Printed in the United States of America by Wm. C. Brown Publishers
2460 Kerper Boulevard, Dubuque, IA 52001

10 9 8 7 6 5 4 3 2

Contents

* Asterisked chapters, sections, and examples are optional. Many of these areas contain additional applications; some provide further theory. The instructor can use these sections to build around the core material, in order to give the course the desired flavor.

Preface

This book is designed for an introductory course in linear algebra. Its aim is to provide a solid foundation in the mathematics while at the same time teaching the reader how to apply the tools learned. Numerous interesting applications from many fields are presented.

Linear Algebra with Applications has arisen from the second edition of *Computational Linear Algebra with Models*. However, this new book includes more complete discussions of general vector spaces, inner product spaces, isomorphisms, and matrix representations of linear mappings. Furthermore, it starts with a discussion of systems of linear equations.

Calculus is not a prerequisite. Optional sections and examples requiring calculus can be included in classes where students have the necessary background.

Numerous applications, such as those illustrating the role of linear algebra in archaeology, population movement, and weather prediction, have been retained from the previous book. New applications from such fields as coding theory and computer graphics have been added. Discussions involving applications are self-contained, with enough information for the reader to fully appreciate the role played by linear algebra. The instructor should select those applications that will be of most interest to the class. In passing over other applications, readers will still develop a sense of the breadth of the field.

The computer is often used when linear algebra is applied. This text, with an appendix on BASIC and optional exercises on computing, is suitable for a linear algebra course that integrates the computer. Programs are related to sections in the main text where appropriate methods have been developed. A computer diskette, with manual, of programs that can be used with this text is also available from the publishers. Included on this diskette are not only computational programs such as Gauss-Jordan elimination, but also applications such as digraphs and Markov chains.

The book is arranged around a core of twenty-nine sections. These sections incorporate techniques, some interesting general applications, and what I believe is minimal theory. Sections and examples marked with an asterisk contain further numerical techniques, applications, and theory. This framework allows the instructor to cover the core material and then select optional material that gives the course the desired perspective. For example, for a theoretical course with casual coverage of applications, the instructor might select from among the following sections:

· Chapter 1, Sections 1-1 – 1-4
· Chapter 2, Sections 2-1 – 2-5

- Chapter 3, Sections 3-1 – 3-4
- Chapter 4, Sections 4-1 – 4-11
- Chapter 5, Sections 5-1 – 5-3
- Chapter 6, Sections 6-1 – 6-7
- Chapter 7, Sections 7-1, 7-3, 7-4
- Chapter 8, Sections 8-1 – 8-3

For a course oriented toward computation and application, the instructor might select from among the following sections:

- Chapter 1, Sections 1-1 – 1-5
- Chapter 2, Sections 2-1 – 2-8
- Chapter 3, Sections 3-1 – 3-4
- Chapter 4, Sections 4-1 – 4-11
- Chapter 6, Sections 6-1 – 6-5
- Chapter 7, Sections 7-1 – 7-5
- Chapter 8, Sections 8-1 – 8-3
- Chapter 9, Sections 9-1 – 9-3

My hope is that even students in the theoretical course will, through glancing at the many interesting applications included, become aware of the broad spectrum of applications of the subject.

The approach that I have adopted in writing the text is to develop the mathematics first and then provide examples of applications. This, I believe, makes for the clearest text presentation. However, some instructors may prefer to look ahead with the class to an application and then use the application to motivate the mathematics.

Following is a chapter-by-chapter discussion of the text.

Chapter 1. The reader is led step by step from solving systems of two linear equations in two variables through solving square systems of linear equations with a unique solution to solving general systems of linear equations. Starting with the case of unique solutions enables the reader to master the algorithm before complications are introduced. The method of Gauss-Jordan elimination is used rather than that of Gaussian elimination, since I think it is "cleaner" for an introductory course. (Gaussian elimination is, of course, an important *efficient* technique, and it is introduced later in Chapter 8.) The chapter closes with examples of applications of systems of linear equations in electrical networks and traffic flow. The former illustrates a unique solution, the latter many solutions.

Chapter 2. In the first chapter, matrices were used to handle systems of linear equations. This introduction motivates the algebraic development of the theory of matrices in Chapter 2. Instructors who plan to go on to the discussion of general vector spaces in Chapter 5 should stress the algebraic properties that are introduced. Instructors who are more interested in a computation/applications approach can tread gently over the theory, concentrating on computation and applications. Section 2-4 has a beautiful application from archaeology which illustrates the importance of matrix multiplication, transpose, and symmetric matrices. The chapter closes with two optional sections that should have broad appeal. Section 2-7 introduces the reader to Markov chain models in the fields of demography and genetics. Section 2-8 uses graph theory in models in communication and sociology. Instructors who cannot fit these sections into their formal class schedule should encourage their students to browse through them.

Chapter 3. Determinants and their properties are introduced as quickly and painlessly as possible in the first two sections of Chapter 3. Proofs are included for the sake of completeness, but these proofs can be skipped if the instructor so desires. The row reduction numerical technique of evaluating a determinant is also covered.

Chapter 4. The groundwork has been laid for a more theoretical turn in Chapter 4. The structure of the vector space $\mathbf{R}^n$ is developed. Concepts of subspaces, linear dependence, bases, and dimension are introduced. These concepts are given geometrical interpretations whenever possible. Discussions of the dot product, norms, angles, and distances follow. Instructors who intend to go on to general vector spaces and inner product spaces in the following section should bear in mind when teaching this material that it will be generalized later. Pointing out common algebraic properties of, for example, matrices and vectors will make the transition to the general vector space very natural when the time comes. Numerous proofs are included in this chapter. Instructors desiring less rigor can omit proofs and concentrate on examples of computation.

Chapter 5. Chapter 5 on general vector spaces is optional. It can, however, be a highlight of the course and should be included in all but the more computationally oriented courses. Throughout the book, comments on algebraic properties have been leading up to these sections. Vector spaces of matrices and functions are introduced in Section 5-1. Section 5-2 shows how an inner product can be used to control the geometry of a space. This discussion can lead into a self-contained section on special relativity, which provides a splendid example of the power of the theory developed.

Chapter 6. Mappings are introduced geometrically, starting with mappings defined by 2×2 matrices. Coverage of square matrices and rectangular matrices leads up to a treatment of general linear mappings. The concept of a coset is introduced in order to permit discussion of the solution set of $f(v) = u$, $u \neq 0$. These tools are used to introduce geometrical insights concerning the behavior of systems of linear equations. The coset is employed to relate solutions of nonhomogeneous equations to those of homogeneous equations. Section 6-4 shows, for example, how the set of solutions to a system of nonhomogeneous equations can be decomposed into a particular solution and the set of solutions to the corresponding homogeneous system. This decomposition leads to a geometrical interpretation of the solutions. The optional section on differential equations runs parallel to the one on linear equations and provides an excellent example of how mathematical tools developed within a general framework provide "free" information in many seemingly unrelated areas. The section on differential equations can generate elegant insights not usually presented in an undergraduate course. Chapter 6 closes with a section on isomorphisms. The reader will find that all the real finite-dimensional vector spaces that have been studied are isomorphic to some $\mathbf{R}^n$, so in studying $\mathbf{R}^n$ we were in a sense studying all real finite-dimensional vector spaces.

Chapter 7. Eigenvalues and eigenvectors are introduced in Chapter 7, and their role in similarity transformations is discussed. Applications are drawn from sociology, weather prediction, and engineering. The theory of coordinate transformations is presented in Section 7-4. The reader sees the central role coordinate transformations play in computer graphics, since the coordinate system used by the computer may not be the most suitable one for the user.

Chapter 8. Chapter 8 on numerical techniques introduces the pivoting refinement into Gauss-Jordan elimination. Iterative methods are discussed and compared with row reduction methods. The power method for finding the dominant eigenvalue and corresponding eigenvector of a matrix is introduced. Discussion of the eigenvector leads to an illustration of the importance of these techniques to geographers as a means of measuring the relative accessibility of cities.

Chapter 9. Chapter 9 on linear programming takes an in-depth look at an important application of linear algebra. It contains numerous illustrations of ways in which companies are using linear programming. Readers who desire just a brief introduction to linear programming can get it in the first section. Those who want to be able to use the simplex method should include Section 9-2, and those who desire to know why the simplex method works should go on to Section 9-3. The chapter can be covered any time after Section 1-4.

Appendix. The appendix on BASIC can be used simultaneously with the main text or in a block of classes devoted to computing at the end of the course. If the computer is used during the course, the instructor is advised to introduce computing after Section 2-1 on matrices and follow the order presented in the appendix. I feel it is inadvisable to present a long program such as Gauss-Jordan elimination early in the course, concurrently with Chapter 1. Students should have practice in handling matrices first.

It is a pleasure to acknowledge the help that made this book possible. In addition to those who gave input into the first two editions of *Computational Linear Algebra with Models,* I wish to thank my colleague Dennis Kletzing of Stetson University. My thanks go to the following reviewers for their constructive comments: Jean Bevis, David C. Buchthal, G. R. Chapman, Bruce Edwards, Robert Gasper, Vincent Giambalvo, Pat Goeters, Lorraine Keller, Ronald Morash, Rainier Sachs, Don Shriner, and G. P. Styan.

I thank Sally Lifland once again for her editing. It is a pleasure to work with her. I am as usual grateful to my wife Donna for all her mathematical and computing input and for her typing contribution.

G.W.

1

Systems of Linear Equations

Mathematics is, of course, a discipline in its own right. It is, however, more than that—it is also a tool used in many other fields. Linear algebra is a branch of mathematics that plays a central role in modern mathematics and is also of increasing importance to engineers and physical, social, and behavioral scientists. In this course, readers will learn mathematics and also be instructed in the art of applying mathematics; the course is a blend of theory, numerical techniques, and applications. They will develop mathematical models, mathematical "pictures" of certain aspects of situations, and see how such descriptions lead to a deeper understanding of the situation. Actual applications, in fields from the aerospace industry to archaeology, will be discussed. The aim of the course is to develop in students an overall competence in linear algebra as well as an appreciation of the role this subject plays in many other fields.

Many problems in mathematics, engineering, economics, biology, and other sciences reduce to solving systems of linear equations. Historically, linear algebra developed from analyzing methods for solving such equations. This chapter covers numerical techniques for solving systems of linear equations and examples of problems that reduce to solving such equations.

1-1 SYSTEMS OF TWO LINEAR EQUATIONS IN TWO VARIABLES

In this section, we analyze the solving of a system of two linear equations in two variables. This leads to a method for solving larger systems.

Consider the system

$$x - 2y = 1$$
$$4x - 3y = 9$$

We are asked to determine values of x and the corresponding values of y that satisfy both equations simultaneously. Each such pair (x, y) is called a *solution*. As it stands, this is an algebraic problem. However, to gain further insight into the problem, we can present it geometrically.

Each equation represents a straight line. A point (x, y) must lie on both lines in order to satisfy both equations simultaneously. Their intersection will represent the only solution (Figure 1-1).

Let us determine the solution using algebraic techniques. Keep the first equation unchanged but modify the second by adding -4 times the first equation to it. The system becomes

$$x - 2y = 1$$
$$5y = 5$$

This system has the same solution as the original system, but is a simpler system in that the value of y can immediately be seen as 1. Our aim in modifying the second equation in this manner was to eliminate x from it. Thus we have that $y = 1$ at the point of intersection. Substituting back for y into the first equation gives $x = 3$.

Hence the unique solution is $x = 3$, $y = 1$. (Substitute these values back into the original system of equations to see that they are indeed satisfied.)

This discussion raises two questions: Will there always be a solution to such a system? If so, will it always be unique? Let us discuss the existence question first. In the previous example, we interpreted the equations geometrically as straight lines, and the solution was the point of intersection of those lines. Equations of a pair of intersecting straight lines have a solution. However, if

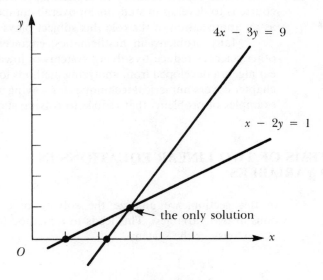

Figure 1-1

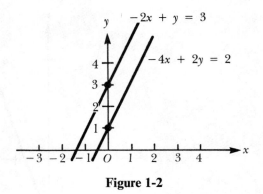

Figure 1-2

the equations represent parallel lines, there are no points in common and a solution does not exist. We would expect that a system such as

$$-2x + y = 3$$
$$-4x + 2y = 2$$

would not have a solution, for each equation represents a straight line of slope 2 (Figure 1-2).

Let us see what happens when we perform the previous type of algebraic manipulation. To eliminate x from the second equation, multiply the first equation by -2 and add it to the second. The system becomes

$$-2x + y = 3$$
$$0y = -4$$

There is no value of y that can be used to satisfy the second equation. A solution does not exist.

We now turn to the possibility of having more than one solution to a system. We have seen that solutions correspond geometrically to points of intersection of a pair of straight lines. At first sight it might appear that two straight lines can intersect in, at most, one point; hence there can be, at most, one solution. However, the case where both equations represent the same geometrical straight line should not be excluded. There would then be an infinite number of solutions; each solution would be represented geometrically by a point on the line. For example,

$$6x - 2y = 4$$
$$9x - 3y = 6$$

would be such a system.

Both equations in Figure 1-3 represent a straight line of slope 3 with y intercept -2. Any point on the straight line is a solution. Isolating y in the first equation, one gets $y = 3x - 2$. The variable x can take on any value, say

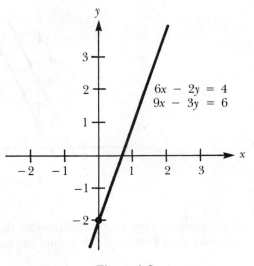

Figure 1-3

r; the corresponding y of the solution is then $y = 3r - 2$. The solutions can be expressed

$$x = r, \qquad y = 3r - 2$$

where r is any real number. Letting r assume various values, we can get specific solutions. For example, when $r = 2$, we get $x = 2$, $y = 4$ as a solution. When $r = -1$, we see that $x = -1$, $y = -5$ is a solution.

In this section we have analyzed solutions to systems of two linear equations in two variables completely. We have found that there may be a unique solution, a solution may not exist, or there may be many solutions.

Our aim is to analyze such systems in many variables. A system of three equations in three variables would be of the form

$$ax + by + cz = p$$
$$dx + ey + fz = q$$
$$gx + hy + iz = r$$

$a, b, \dots, i, p, q, r$ being constants. A solution will be a set of values of x, y, and z that satisfies all the equations simultaneously. Each of the above three equations describes a plane in three-dimensional space. Solutions will correspond to points on all three planes. For such a system we can expect a solution to be unique, nonexistent, or multiple. We illustrate some of the various possibilities in Figure 1-4.

As the number of variables increases, the geometrical interpretation of such a system becomes increasingly complicated and the geometrical discus-

Unique solution :

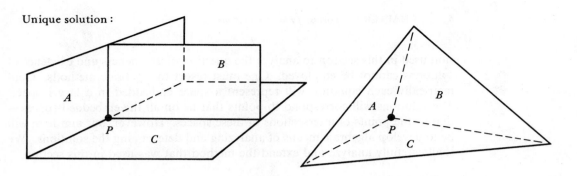

Three planes A, B, C, intersect at a single point P, P corresponds to unique solution

No solutions :

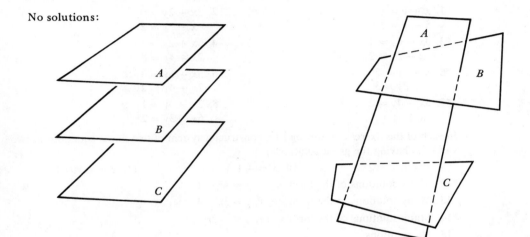

Planes A, B, C have no point of common intersection, no solution

Many solutions :

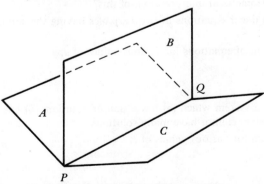

Any point on line PQ will be a solution

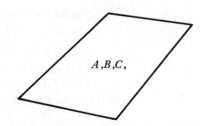

Three equations represent same plane, any point on the plane will be a solution

Figure 1-4

sion used in this section to analyze the question of uniqueness and existence of solutions cannot be employed. One must resort to algebraic methods. Geometrically each equation will represent a space embedded in a larger space. The solutions will correspond to points that lie on all the embedded spaces—that is, the points of intersections of these spaces. However, our aim here will be to develop algebraic means of analyzing and determining the solutions. We shall carefully analyze and extend the method that was used in this section.

EXERCISES

Solve, if possible, the following systems of linear equations. Illustrate your answers.

*1. $x + y = 1$
$\quad x - y = 2$

2. $3x + 2y = 0$
$\quad x - y = 2$

*3. $3x - 6y = 1$
$\quad x - 2y = 2$

4. $x + 2y = 4$
$\quad 2x + 4y = 8$

*5. $x + y = 7$
$\quad x + 2y = 3$

6. $3x + 6y = 3$
$\quad 2x + 4y = 2$

*7. $3x - 4y = 2$
$\quad x + y = 1$

8. $2x - y = 1$
$\quad 6x - 3y = 2$

In each of the Exercises 9 through 15, construct a system of two linear equations in two variables having the given solution(s).

*9. $x = 1, y = 2$ 10. $x = -1, y = 3$ 11. $x = 4, y = 1$

*12. Many solutions of the form $x = r, y = 4r - 1$

13. Many solutions of the form $x = r, y = 3r + 4$

*14. Many solutions of the form $x = r, y = -2r + 1$

15. No solutions

16. Construct three distinct systems of linear equations all having the same solution, $x = 2, y = 3$. Discuss the geometrical significance of this.

*17. Construct a system of three linear equations in two variables having the solution $x = 1, y = 3$. What is the geometrical interpretation of this?

18. Construct a system of three linear equations in two variables having the solution $x = -1, y = 1$.

*19. Consider the following system of equations for various values of d.

$$x + y = 4$$
$$dx + 2y = 2$$

Which values of d result in a system which (a) has a unique solution, (b) has no solution? Prove that the system cannot have many solutions.

20. Consider the following system for various values of c.

$$x + y = 2$$
$$2x + cy = 2c$$

Which values of c result in a system with (a) a unique solution, (b) many solutions? Prove that the system cannot have no solutions.

* Answers to exercises marked with an asterisk are provided in the back of the book.

1-2 MATRICES AND SYSTEMS OF LINEAR EQUATIONS

In this section, we introduce the concept of a matrix and relate matrices to systems of linear equations. Systems of linear equations can be handled conveniently and efficiently in terms of matrices.

Definition 1-1   A *matrix* is a rectangular array of numbers. The numbers in the array are called the *elements* of the matrix.

Examples of matrices in standard notation are

$$\begin{pmatrix} 1 & 2 & 3 \\ 0 & -1 & 1 \end{pmatrix}, \quad \begin{pmatrix} 2 & 3 \\ 1 & 1 \\ 4 & 1 \end{pmatrix}, \quad \begin{pmatrix} 1 & 2 & 3 \\ 4 & 5 & 6 \\ 0 & 1 & 2 \end{pmatrix}$$

Each matrix has a certain number of rows and a certain number of columns. For example, the matrix

$$\begin{pmatrix} 1 & 2 & 3 \\ 0 & -1 & 1 \end{pmatrix}$$

has 2 rows and 3 columns. We describe the *size* of a matrix by specifying the number of rows and columns in the matrix. We call a matrix such as the above one, having 2 rows and 3 columns, a 2×3 *matrix*; the first number indicates the number of rows in the matrix, the second indicates the number of columns. When the number of rows and columns is equal, the matrix is said to be a *square matrix*.

$$\begin{pmatrix} 2 & 3 \\ 1 & 1 \\ 4 & 1 \end{pmatrix} \text{ is a } 3 \times 2 \text{ matrix.}$$

$$\begin{pmatrix} 1 & 2 & 3 \\ 4 & 5 & 6 \\ 0 & 1 & 2 \end{pmatrix} \text{ is a } 3 \times 3 \text{ matrix, a square matrix.}$$

$$\begin{pmatrix} 1 & 2 & 3 \\ 4 & 1 & 0 \end{pmatrix} \text{ and } \begin{pmatrix} -1 & 4 & 3 \\ 2 & 2 & 1 \end{pmatrix} \quad \begin{array}{l} \text{are of the same size;} \\ \text{both are } 2 \times 3 \text{ matrices.} \end{array}$$

Each element of a matrix has a certain location. The location is described in terms of the row and column in which the element lies. Rows are labeled from the top of the matrix, columns from the left. Thus, for example,

$$\begin{array}{c} \\ \text{Row 1} \\ \text{Row 2} \\ \text{Row 3} \end{array} \begin{array}{cccc} \text{Col. 1} & \text{Col. 2} & \text{Col. 3} & \text{Col. 4} \\ \begin{pmatrix} 4 & 6 & -3 & 2 \\ 1 & 5 & ⑨ & -2 \\ 7 & 8 & 3 & 4 \end{pmatrix} \end{array}$$

The circled element 9 is in row 2, column 3 of the matrix. We shall refer to this location as the (2, 3) location; the row is indicated first, followed by the column.

A square $n \times n$ matrix having ones in the (1, 1), (2, 2), (3, 3), ..., (n, n) locations and zeros elsewhere is called a *unit matrix* and denoted I_n. The following matrices are unit matrices.

$$I_2 = \begin{pmatrix} 1 & 0 \\ 0 & 1 \end{pmatrix}, \qquad I_3 = \begin{pmatrix} 1 & 0 & 0 \\ 0 & 1 & 0 \\ 0 & 0 & 1 \end{pmatrix}, \qquad I_4 = \begin{pmatrix} 1 & 0 & 0 & 0 \\ 0 & 1 & 0 & 0 \\ 0 & 0 & 1 & 0 \\ 0 & 0 & 0 & 1 \end{pmatrix}$$

The (1, 1), (2, 2), (3, 3), ... locations of a square matrix are called *diagonal locations*.

Let us now see how matrices relate to systems of linear equations. Consider the system

$$\begin{aligned} 7x_1 + x_2 + x_3 &= 3 \\ 2x_1 + 3x_2 + x_3 &= 5 \\ x_1 - x_2 - 2x_3 &= -5 \end{aligned}$$

We use x_1, x_2, x_3 for the variables here—this is a notation that can be generalized for n variables.

The coefficients of this system form a matrix called the *matrix of coefficients* of the system:

$$\begin{pmatrix} 7 & 1 & 1 \\ 2 & 3 & 1 \\ 1 & -1 & -2 \end{pmatrix}$$

A second matrix that includes the numbers on the right-hand side of the equations is called the *augmented matrix* of the system:

$$\begin{pmatrix} 7 & 1 & 1 & 3 \\ 2 & 3 & 1 & 5 \\ 1 & -1 & -2 & -5 \end{pmatrix}$$

Observe that the augmented matrix completely characterizes the system. With every system of linear equations we can associate an augmented matrix. Conversely, any matrix can be interpreted to be the augmented matrix that defines a system of linear equations.

✦ *Example 1* ──

Consider the matrix

$$\begin{pmatrix} 2 & 3 & -1 & 4 & 6 \\ 5 & 0 & 1 & 2 & 3 \\ 4 & -2 & 5 & 1 & 3 \end{pmatrix}$$

This is the augmented matrix for the system

$$2x_1 + 3x_2 - x_3 + 4x_4 = 6$$
$$5x_1 + x_3 + 2x_4 = 3$$
$$4x_1 - 2x_2 + 5x_3 + x_4 = 3$$

The matrix of coefficients of this system is

$$\begin{pmatrix} 2 & 3 & -1 & 4 \\ 5 & 0 & 1 & 2 \\ 4 & -2 & 5 & 1 \end{pmatrix}$$

Observe that the matrix of coefficients is the *submatrix* of the augmented matrix obtained by leaving out the last column.   ✦

✦ *Example 2* _____

Consider the matrix

$$\begin{pmatrix} 1 & 0 & 0 & -2 \\ 0 & 1 & 0 & 4 \\ 0 & 0 & 1 & 3 \end{pmatrix}$$

This is the augmented matrix of the "straightforward" system

$$x_1 = -2$$
$$ x_2 = 4$$
$$ x_3 = 3$$

The solution to this system is of course $x_1 = -2$, $x_2 = 4$, $x_3 = 3$. The matrix of coefficients of this system is

$$\begin{pmatrix} 1 & 0 & 0 \\ 0 & 1 & 0 \\ 0 & 0 & 1 \end{pmatrix}$$

namely, the unit matrix I_3. Note how the matrix of coefficients appears as a submatrix of the original augmented matrix:

$$\begin{pmatrix} 1 & 0 & 0 & | & -2 \\ 0 & 1 & 0 & | & 4 \\ 0 & 0 & 1 & | & 3 \end{pmatrix}$$
$$\underbrace{}_{I_3}$$

✦

It is in this role that identity matrices will be important in our discussion of linear systems in the next section. There we shall see how systems of linear equations can be conveniently handled in terms of matrices.

EXERCISES

*1. Give the sizes of the following matrices.

(a) $\begin{pmatrix} 1 & 2 \\ -1 & 3 \end{pmatrix}$ (b) $\begin{pmatrix} 1 & 3 \\ -5 & 2 \\ 4 & 3 \end{pmatrix}$

(c) $(0 \quad 1 \quad 2)$ (d) $\begin{pmatrix} 1 & 3 & 4 & 7 & 8 \\ 6 & 7 & 3 & -8 & 4 \end{pmatrix}$

2. Give the sizes of the following matrices.

(a) $\begin{pmatrix} 1 \\ 7 \end{pmatrix}$ (b) $\begin{pmatrix} -5 & 3 & 7 \\ 8 & 4 & -6 \\ 7 & 5 & -7 \end{pmatrix}$

(c) $\begin{pmatrix} 1 & 5 & -9 \\ 8 & 3 & 2 \end{pmatrix}$ (d) $\begin{pmatrix} 7 \\ 8 \\ 4 \end{pmatrix}$

*3. Consider the matrix

$$\begin{pmatrix} 1 & 2 & 3 & 5 \\ -6 & 7 & 0 & 9 \\ 7 & 2 & 1 & 8 \end{pmatrix}$$

Give the (1, 1), (2, 2), (3, 3), (3, 4) elements of this matrix.

4. Consider the matrix

$$\begin{pmatrix} 5 & 4 & 6 & -9 & 3 & -5 \\ 7 & 8 & 0 & 4 & 2 & 8 \\ 9 & -5 & 6 & 6 & -7 & 3 \end{pmatrix}$$

Give the (1, 1), (2, 2), (3, 4), (2, 5) elements of this matrix.

*5. Write out the matrix I_4.

6. Determine the matrix of coefficients and the augmented matrix for each of the following systems of linear equations.

*(a) $x_1 - x_2 = 1$ (b) $x_1 - x_2 + x_3 = 4$ *(c) $x_2 - x_3 = 5$
    $2x_1 + x_2 = 3$ $x_1 + 3x_2 + 4x_3 = 2$ $x_1 + 2x_2 - 4x_3 = 2$
                             $2x_1 - 2x_2 + x_3 = 5$ $2x_1 - x_2 = 4$

(d) $2x_1 + x_2 - x_3 + x_4 = -6$ *(e) $7x_1 - x_2 + x_3 + 2x_4 = 4$
    $3x_1 - x_2 + 2x_3 = 7$ $x_2 - 8x_3 - x_4 = -1$
    $8x_1 + x_2 - 9x_3 + x_4 = 8$ $x_1 + 7x_3 + x_4 = -2$
    $-x_1 - 3x_2 + 8x_3 + x_4 = 9$

7. Any matrix can be interpreted as an augmented matrix defining a system of linear equations. Write out the systems defined by the following matrices.

*(a) $\begin{pmatrix} 2 & 1 & 5 \\ 3 & -2 & 3 \end{pmatrix}$ (b) $\begin{pmatrix} 1 & 3 & 2 & 0 \\ 2 & 4 & -3 & 2 \\ -1 & 6 & 1 & 4 \end{pmatrix}$

$$*(c) \begin{pmatrix} 3 & 5 & -4 & 1 \\ -1 & 2 & 2 & 2 \\ 2 & 3 & 0 & 3 \end{pmatrix} \qquad (d) \begin{pmatrix} 0 & 2 & 1 & 4 & 3 \\ 1 & 3 & 0 & -2 & 5 \\ 4 & 2 & 3 & -7 & 8 \end{pmatrix}$$

$$(e) \begin{pmatrix} 4 & 3 & 0 & 8 & 5 \\ 5 & 6 & 5 & 9 & 1 \\ 0 & 8 & 4 & -2 & 0 \\ -1 & -2 & 7 & 4 & 2 \end{pmatrix}$$

8. Write down the linear systems that have the following augmented matrices:

$$*(a) \begin{pmatrix} 1 & 4 & 3 & 4 & -1 \\ 0 & 1 & -8 & 9 & 0 \\ 0 & 0 & 0 & 1 & 4 \end{pmatrix} \qquad (b) \begin{pmatrix} 1 & 2 & 3 & 4 \\ 0 & 1 & 7 & -2 \\ 0 & 0 & 1 & 3 \end{pmatrix}$$

$$*(c) \begin{pmatrix} 1 & 7 & 5 & -8 \\ 0 & 1 & 8 & 2 \end{pmatrix} \qquad (d) \begin{pmatrix} 1 & 5 & 6 & 4 & 7 \\ 0 & 1 & 3 & -7 & 0 \\ 0 & 0 & 1 & 8 & 3 \end{pmatrix}$$

*9. Give the system of linear equations that has the following augmented matrix:

$$\begin{pmatrix} 1 & 0 & 0 & 2 \\ 0 & 1 & 0 & 1 \\ 0 & 0 & 1 & -3 \end{pmatrix}$$

Give the matrix of coefficients of this system. What is the solution to the system?

10. Give the system of linear equations that has the following augmented matrix:

$$\begin{pmatrix} 1 & 0 & 0 & 0 & 9 \\ 0 & 1 & 0 & 0 & 3 \\ 0 & 0 & 1 & 0 & -6 \\ 0 & 0 & 0 & 1 & 8 \end{pmatrix}$$

Give the matrix of coefficients of this system. What is the solution to the system?

1-3 SYSTEMS OF LINEAR EQUATIONS HAVING UNIQUE SOLUTIONS

In Section 1-1, we manipulated systems of two equations in two variables to obtain solutions. Similar operations can be performed on larger systems of equations without altering the solutions. We call these operations *elementary transformations*, and systems that are related through elementary transformations are called *equivalent systems*. These transformations are of three types:

1. Interchanging two equations.
2. Multiplying an equation throughout by a nonzero constant.
3. Adding a multiple of one equation to another.

We use the notation ≈ for equivalent systems.

Through elementary transformations, a system of linear equations can be transformed into an equivalent simpler system that leads to the solution. The following example illustrates the approach.

✦ Example 1

Solve, if possible, the system

$$x_1 + x_2 + x_3 = 3$$
$$2x_1 + 3x_2 + x_3 = 5$$
$$x_1 - x_2 - 2x_3 = -5$$

We first use elementary transformations to eliminate x_1 from the second and third equations, *using the x_1 of the first equation.*

$$x_1 + x_2 + x_3 = 3 \qquad\qquad x_1 + x_2 + x_3 = 3$$
$$2x_1 + 3x_2 + x_3 = 5 \qquad \approx \qquad x_2 - x_3 = -1$$
$$x_1 - x_2 - 2x_3 = -5 \quad {\scriptstyle \text{Eq. 2}+(-2)\text{Eq. 1} \atop \text{Eq. 3}+(-1)\text{Eq. 1}} \quad -2x_2 - 3x_3 = -8$$

We now eliminate x_2 from the first and third equations, *using the x_2 of the second equation.*

$$x_1 \quad\;\; + 2x_3 = 4$$
$$\approx \qquad x_2 - x_3 = -1$$
$${\scriptstyle \text{Eq. 1}+(-1)\text{Eq. 2} \atop \text{Eq. 3}+(2)\text{Eq. 2}} \quad -5x_3 = -10$$

Multiplying the last equation by $-\frac{1}{5}$ leads to a system that gives x_3.

$$x_1 \quad\;\; + 2x_3 = 4$$
$$\approx \qquad x_2 - x_3 = -1$$
$${\scriptstyle (-\frac{1}{5})\text{Eq. 3}} \qquad x_3 = 2$$

Now we eliminate x_3 from the first and second equations, *using the x_3 of the third equation.*

$$x_1 \qquad\quad = 0$$
$$\approx \qquad x_2 \qquad = 1$$
$${\scriptstyle \text{Eq. 1}+(-2)\text{Eq. 3} \atop \text{Eq. 2}+\text{Eq. 3}} \qquad x_3 = 2$$

The solution to this system and thus also to the original system is $x_1 = 0$, $x_2 = 1$, $x_3 = 2$.

Geometrically, each of the three original equations represents a plane in three-dimensional space, and the solution is the unique point $(0, 1, 2)$ that lies on all three planes.   ✦

That elementary transformations transform systems into other systems having the same solutions is intuitively clear. Obviously the order of the equations does not matter; thus the first transformation is valid. Multiplying one equation through by a nonzero constant does not change the truth of the equality, justifying the second transformation. Adding equal quantities to both sides of an equality result in an equality; this is the equivalent of the third transformation.

Let us now look at the above method in terms of matrices. In solving a system such as the above, one can represent each equivalent system by an augmented matrix. It becomes unnecessary to write down $x_1 \ x_2, \ldots$, since we can work in terms of augmented matrices. Instead of performing elementary transformations on the systems of linear equations, we can perform equivalent transformations, called *elementary matrix transformations*, on the augmented matrices. Elementary matrix transformations are of three classes:

1. Interchanging two rows of the matrix.
2. Multiplying a row by a nonzero constant.
3. Adding a multiple of one row of the matrix to another row.

We say that two matrices are *equivalent* if one is obtained from the other using elementary matrix transformations.

We now illustrate this technique of using elementary matrix transformations. Consider the system of Example 1. On the left is the sequence of equivalent systems of equations that led to the solution, and on the right is the corresponding sequence of equivalent augmented matrices, illustrating the approach we shall be taking from now on.

✦ *Example 2*
_____

Solve the system of equations

$$x_1 + x_2 + x_3 = 3$$
$$2x_1 + 3x_2 + x_3 = 5$$
$$x_1 - x_2 - 2x_3 = -5$$

Equivalent systems of equations:

original
system
$$= \begin{array}{r} x_1 + x_2 + x_3 = 3 \\ 2x_1 + 3x_2 + x_3 = 5 \\ x_1 - x_2 - 2x_3 = -5 \end{array}$$

$\approx$
Eq. 2 + (−2)Eq. 1
Eq. 3 + (−1)Eq. 1
$$\begin{array}{r} x_1 + x_2 + x_3 = 3 \\ x_2 - x_3 = -1 \\ -2x_2 - 3x_3 = -8 \end{array}$$

$\approx$
Eq. 1 + (−1)Eq. 2
Eq. 3 + (2)Eq. 2
$$\begin{array}{r} x_1 \qquad + 2x_3 = 4 \\ x_2 - x_3 = -1 \\ -5x_3 = -10 \end{array}$$

Equivalent augmented matrices (R denotes row):

original
augmented =
matrix
$$\begin{pmatrix} 1 & 1 & 1 & 3 \\ 2 & 3 & 1 & 5 \\ 1 & -1 & -2 & -5 \end{pmatrix}$$

$\approx$
R2 + (−2)R1
R3 + (−1)R1
$$\begin{pmatrix} 1 & 1 & 1 & 3 \\ 0 & 1 & -1 & -1 \\ 0 & -2 & -3 & -8 \end{pmatrix}$$

$\approx$
R1 + (−1)R2
R3 + (2)R2
$$\begin{pmatrix} 1 & 0 & 2 & 4 \\ 0 & 1 & -1 & -1 \\ 0 & 0 & -5 & -10 \end{pmatrix}$$

Equivalent systems of equations:

Equivalent augmented matrices (R denotes row):

$$
\underset{(-\frac{1}{5})\text{Eq. 3}}{\approx}
\begin{array}{rcr}
x_1 & + 2x_3 = & 4 \\
x_2 - & x_3 = & -1 \\
& x_3 = & 2
\end{array}
$$

$$
\underset{(-\frac{1}{5})\text{R3}}{\approx}
\begin{pmatrix}
1 & 0 & 2 & 4 \\
0 & 1 & -1 & -1 \\
0 & 0 & 1 & 2
\end{pmatrix}
$$

$$
\underset{\substack{\text{Eq. 1}+(-2)\text{Eq. 3} \\ \text{Eq. 2}+\text{Eq. 3}}}{\approx}
\begin{array}{rcr}
x_1 & & = 0 \\
& x_2 & = 1 \\
& & x_3 = 2
\end{array}
$$

$$
\underset{\substack{\text{R1}+(-2)\text{R3} \\ \text{R2}+\text{R3}}}{\approx}
\begin{pmatrix}
1 & 0 & 0 & 0 \\
0 & 1 & 0 & 1 \\
0 & 0 & 1 & 2
\end{pmatrix}
$$

The solution is $x_1 = 0$, $x_2 = 1$, $x_3 = 2$.

This is the augmented matrix of the system

$$
\begin{array}{rcl}
x_1 & & = 0 \\
& x_2 & = 1 \\
& & x_3 = 2
\end{array}
$$

The solution is $x_1 = 0$, $x_2 = 1$, $x_3 = 2$.   ◆

Observe that this method of reducing a system of equations to an equivalent simpler system, using matrices, involves creating ones and zeros in certain locations of the matrices. These numbers are created in a systematic manner. The method used is called the *method of Gauss-Jordan elimination.* We now discuss how the method proceeds in the most straightforward situation—a system of n equations in n variables with a unique solution.[†] The matrix of coefficients for such a system will be a square $n \times n$ matrix, and the augmented matrix will be an $n \times (n + 1)$ matrix.

Method of Gauss-Jordan Elimination for n Equations in n Variables Having a Unique Solution

Step 1. Write down the augmented matrix of the system.

Step 2. Interchange rows, if necessary, to ensure that there is a nonzero element in the (1, 1) location. This element is called a *pivot.*

nonzero pivot element here

$$
\begin{pmatrix}
\circledast & * & * & \cdots \\
* & * & * & \cdots \\
* & * & * & \\
\vdots & & &
\end{pmatrix}
$$

[†] We present the algorithm here. The actual proof that it can proceed thus is given in the following section.

Step 3. Create a 1 in the (1, 1) location by multiplying the first row by (1/pivot).

$$\begin{pmatrix} 1 & * & * & \cdots \\ * & * & * \\ * & * & * \\ \vdots \end{pmatrix}$$

Step 4. Add suitable multiples of the first row to each of the following rows to create zeros below the 1 in the first column.

$$\begin{pmatrix} 1 & * & * & \cdots \\ 0 & * & * \\ 0 & * & * \\ \vdots \\ 0 \end{pmatrix}$$

We now have a matrix that corresponds to a system of equations in which x_1 has been eliminated from all equations except the first.

Step 5. Interchange the second row with a *later* row, if necessary, to ensure that there is a nonzero element in the (2, 2) location. This element is now the pivot.

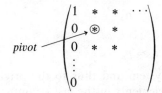

Step 6. Create a 1 in the (2, 2) location by multiplying the second row by (1/pivot).

$$\begin{pmatrix} 1 & * & * & \cdots \\ 0 & 1 & * \\ 0 & * & * \\ \vdots \\ 0 \end{pmatrix}$$

Step 7. Add suitable multiples of the second row to every other row to create zeros above and below the 1 in the (2, 2) location in the second column.

$$\begin{pmatrix} 1 & 0 & * & \cdots \\ 0 & 1 & * \\ 0 & 0 & * \\ \vdots \\ 0 & 0 & * & \cdots \end{pmatrix}$$

We now have a matrix that corresponds to a system of equations in which x_1 has been eliminated from all equations except the first and x_2 from all equations except the second.

Continue this pattern. Select pivots in diagonal locations, create ones in these locations, and then create zeros above and below each 1. Eventually we arrive at the following type of matrix, where $p_1, \ldots, p_n$ are real numbers.

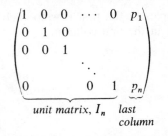

This matrix is called the *reduced echelon form* of the original augmented matrix.

This matrix corresponds to the system of linear equations

$$
\begin{array}{rcl}
x_1 & = & p_1 \\
x_2 & = & p_2 \\
x_3 & = & p_3 \\
& \ddots & \vdots \\
x_n & = & p_n
\end{array}
$$

The solution to this system and thus to the original system is $x_1 = p_1$, $x_2 = p_2, \ldots, x_n = p_n$. The elements in the last column of the reduced echelon form give the solution.

We now present a further example to illustrate the method.

✦ **Example 3**
_____

Let us solve the system of equations

$$
\begin{array}{rcrcrcr}
4x_1 & + & 8x_2 & - & 12x_3 & = & 44 \\
3x_1 & + & 6x_2 & - & 8x_3 & = & 32 \\
-2x_1 & - & x_2 & & & = & -7
\end{array}
$$

We commence with the augmented matrix. The pivots are circled. The ones that are used to create the zeros are also circled.

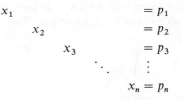

At this stage we want a nonzero element in the $(2, 2)$ location to use as pivot. To achieve this, we interchange the second row with the third row (*a later row*).

$$\underset{R2\leftrightarrow R3}{\approx} \begin{pmatrix} 1 & 2 & -3 & 11 \\ 0 & ③ & -6 & 15 \\ 0 & 0 & 1 & -1 \end{pmatrix} \quad \underset{(\frac{1}{3})R2}{\approx} \begin{pmatrix} 1 & 2 & -3 & 11 \\ 0 & ① & -2 & 5 \\ 0 & 0 & 1 & -1 \end{pmatrix}$$

$$\underset{R1+(-2)R2}{\approx} \begin{pmatrix} 1 & 0 & 1 & 1 \\ 0 & 1 & -2 & 5 \\ 0 & 0 & ① & -1 \end{pmatrix} \quad \underset{\substack{R1+(-1)R3 \\ R2+(2)R3}}{\approx} \begin{pmatrix} 1 & 0 & 0 & 2 \\ 0 & 1 & 0 & 3 \\ 0 & 0 & 1 & -1 \end{pmatrix}$$

We have arrived at the reduced echelon form. The unique solution is $x_1 = 2$, $x_2 = 3$, $x_3 = -1$. ✦

In this section we have limited our discussion to systems of n equations in n variables that have a unique solution. The reduced echelon forms of such systems, as we have seen, assume a particularly straightforward form. In the following section, we shall extend the method to accommodate other types of systems that have a unique solution and also to include systems that have many solutions or no solutions. The above concept of the reduced echelon form will be generalized.

EXERCISES

1. In each of the following questions you are given a matrix followed by a suggested elementary matrix transformation. Determine the matrix that results from performing the transformation in each case.

    *(a) $\begin{pmatrix} 2 & 4 & -4 & 2 \\ 1 & 2 & -1 & 3 \\ -1 & 0 & 4 & 2 \end{pmatrix} \underset{(\frac{1}{2})R1}{\approx}$ (b) $\begin{pmatrix} 0 & -1 & 3 & 4 \\ 2 & 1 & 5 & 2 \\ -3 & 1 & -2 & 6 \end{pmatrix} \underset{R1\leftrightarrow R2}{\approx}$

    *(c) $\begin{pmatrix} 1 & 2 & 3 & -1 \\ 0 & 1 & 2 & 1 \\ 0 & -4 & 5 & -3 \end{pmatrix} \underset{\substack{R1+(-2)R2 \\ R3+(4)R2}}{\approx}$ (d) $\begin{pmatrix} 1 & 0 & 4 & 4 \\ 0 & 1 & -3 & 2 \\ 0 & 0 & 1 & 3 \end{pmatrix} \underset{\substack{R1+(-4)R2 \\ R2+(3)R3}}{\approx}$

    *(e) $\begin{pmatrix} 1 & -2 & 0 & 3 \\ -2 & -1 & 3 & 4 \\ 3 & 1 & 2 & 6 \end{pmatrix} \underset{\substack{R2+(2)R1 \\ R3+(-3)R1}}{\approx}$ (f) $\begin{pmatrix} 1 & 0 & 2 & 3 \\ 0 & 1 & 4 & 1 \\ 0 & 0 & -2 & 6 \end{pmatrix} \underset{(-\frac{1}{2})R3}{\approx}$

2. Interpret each of the following as representing a stage in arriving at the reduced echelon form of an augmented matrix. Why are the indicated transformations selected? What particular aims do they accomplish?

    *(a) $\begin{pmatrix} 1 & 2 & -3 & 4 \\ 0 & 3 & 6 & -9 \\ 0 & 2 & 1 & 4 \end{pmatrix} \underset{(\frac{1}{3})R2}{\approx} \begin{pmatrix} 1 & 2 & -3 & 4 \\ 0 & 1 & 2 & -3 \\ 0 & 2 & 1 & 4 \end{pmatrix}$

    (b) $\begin{pmatrix} 0 & 4 & 2 & 6 \\ 2 & 1 & 3 & 1 \\ -1 & 2 & 4 & 0 \end{pmatrix} \underset{R1\leftrightarrow R2}{\approx} \begin{pmatrix} 2 & 1 & 3 & 1 \\ 0 & 4 & 2 & 6 \\ -1 & 2 & 4 & 0 \end{pmatrix}$

*(c) $\begin{pmatrix} 1 & 3 & 1 & 6 \\ 0 & 0 & 4 & 3 \\ 0 & -2 & 2 & 1 \end{pmatrix} \underset{R2 \leftrightarrow R3}{\approx} \begin{pmatrix} 1 & 3 & 1 & 6 \\ 0 & -2 & 2 & 1 \\ 0 & 0 & 4 & 3 \end{pmatrix}$

(d) $\begin{pmatrix} 1 & 2 & 4 & 0 \\ 0 & 1 & 2 & -3 \\ 0 & -3 & 1 & -2 \end{pmatrix} \underset{\substack{R1 + (-2)R2 \\ R3 + (3)R2}}{\approx} \begin{pmatrix} 1 & 0 & 0 & 6 \\ 0 & 1 & 2 & -3 \\ 0 & 0 & 7 & -11 \end{pmatrix}$

*(e) $\begin{pmatrix} 1 & 0 & 2 & 6 \\ 0 & 1 & -1 & 3 \\ 0 & 0 & 1 & 2 \end{pmatrix} \underset{\substack{R1 + (-2)R3 \\ R2 + R3}}{\approx} \begin{pmatrix} 1 & 0 & 0 & 2 \\ 0 & 1 & 0 & 5 \\ 0 & 0 & 1 & 2 \end{pmatrix}$

The systems of linear equations in Exercises 3–18 all have the same number of equations as variables and all have unique solutions. Solve each system using the method of Gauss-Jordan elimination.

*3. $x_1 - 2x_2 = -8$
$2x_1 - 3x_2 = -11$

4. $2x_1 + 2x_2 = 4$
$3x_1 + 2x_2 = 3$

*5. $x_1 \quad\quad + x_3 = 3$
$2x_2 - 2x_3 = -4$
$- x_2 - 2x_3 = 5$

6. $x_1 + x_2 + 3x_3 = 6$
$x_1 + 2x_2 + 4x_3 = 9$
$2x_1 + x_2 + 6x_3 = 11$

*7. $\quad 2x_2 + 4x_3 = 8$
$2x_1 + 2x_2 \quad\quad = 6$
$x_1 + x_2 + x_3 = 5$

8. $x_1 - 2x_2 - 4x_3 = -9$
$-x_1 + 5x_2 + 10x_3 = 21$
$2x_1 - 3x_2 - 5x_3 = -13$

*9. $x_1 + 2x_2 + 3x_3 = 14$
$2x_1 + 5x_2 + 8x_3 = 36$
$-x_1 - x_2 \quad\quad = -4$

10. $x_1 - x_2 - x_3 = -1$
$-2x_1 + 6x_2 + 10x_3 = 14$
$2x_1 + x_2 + 6x_3 = 9$

*11. $2x_1 + 2x_2 - 4x_3 = 14$
$3x_1 + x_2 + x_3 = 8$
$2x_1 - x_2 + 2x_3 = -1$

12. $x_1 - 2x_2 - 6x_3 = -17$
$2x_1 - 6x_2 - 16x_3 = -46$
$x_1 + 2x_2 - x_3 = -5$

*13. $\frac{3}{2}x_1 \quad\quad + 3x_3 = 15$
$-x_1 + 7x_2 - 9x_3 = -45$
$2x_1 \quad\quad + 5x_3 = 22$

14. $-3x_1 - 6x_2 - 15x_3 = -3$
$x_1 + \frac{3}{2}x_2 + \frac{9}{2}x_3 = \frac{1}{2}$
$-2x_1 - \frac{7}{2}x_2 - \frac{17}{2}x_3 = -2$

*15. $3x_1 + 6x_2 \quad\quad - 3x_4 = 3$
$x_1 + 3x_2 - x_3 - 4x_4 = -12$
$-x_1 - x_2 + x_3 + 2x_4 = 8$
$2x_1 + 3x_2 \quad\quad = 8$

16. $x_1 + 2x_2 + 2x_3 + 5x_4 = 11$
$2x_1 + 4x_2 + 2x_3 + 8x_4 = 14$
$x_1 + 3x_2 + 4x_3 + 8x_4 = 19$
$-x_1 - x_2 + x_3 \quad\quad = 2$

*17. $x_1 + x_2 + 2x_3 + 6x_4 = 11$
$2x_1 + 3x_2 + 6x_3 + 19x_4 = 36$
$3x_2 + 4x_3 + 15x_4 = 28$
$x_1 - x_2 - x_3 - 6x_4 = -12$

18. $\quad x_2 + 2x_3 + 6x_4 = 21$
$-x_1 + x_2 + x_3 + 5x_4 = 12$
$x_1 - x_2 - x_3 - 4x_4 = -9$
$3x_1 - 2x_2 \quad\quad - 6x_4 = -4$

1-4   GAUSS-JORDAN ELIMINATION

In the previous section, we introduced the method of Gauss-Jordan elimination as applicable to systems of *n* equations in *n* variables having a unique solution. We shall now discuss the method in its more general setting, where the number of equations can differ from the number of variables and where there can be a unique solution, many solutions, or no solutions at all. Our approach once again will be to start from the augmented matrix of the given system and to perform a sequence of elementary matrix transformations that will result in a simpler matrix (the reduced echelon form), which leads directly to the solution.

We now present the general definition of reduced echelon form. The reader should observe that the reduced echelon forms discussed in the previous section all conform to this definition.

Definition 1-2   A matrix is in *reduced echelon form* if

1. Any rows consisting entirely of zeros are grouped at the bottom of the matrix.
2. The first nonzero element of each other row is 1. This element is called the *leading* 1.
3. The leading 1 of each row after the first is positioned to the right of the leading 1 of the previous row.
4. All elements directly above and all elements directly below a leading 1 are zeros.

The following matrices are all in reduced echelon form:

$$\begin{pmatrix} 1 & 0 & 0 & 2 \\ 0 & 1 & 0 & 4 \\ 0 & 0 & 1 & -2 \end{pmatrix}, \quad \begin{pmatrix} 1 & 4 & 0 & 0 \\ 0 & 0 & 1 & 0 \\ 0 & 0 & 0 & 1 \end{pmatrix}, \quad \begin{pmatrix} 1 & 2 & 0 & 4 \\ 0 & 0 & 1 & 3 \\ 0 & 0 & 0 & 0 \end{pmatrix},$$

$$\begin{pmatrix} 1 & 4 & 1 & 0 & 4 & 0 & 2 \\ 0 & 0 & 0 & 1 & 3 & 0 & -2 \\ 0 & 0 & 0 & 0 & 0 & 1 & 3 \end{pmatrix},$$

$$\begin{pmatrix} 1 & 0 & 5 & 0 & 4 & 0 & 5 \\ 0 & 1 & 2 & 0 & -3 & 0 & 0 \\ 0 & 0 & 0 & 1 & 2 & 0 & -2 \\ 0 & 0 & 0 & 0 & 0 & 1 & 3 \end{pmatrix}$$

The following matrices are *not* in reduced echelon form:

$$\begin{pmatrix} 1 & 2 & 0 & 4 & 0 \\ 0 & 0 & 0 & 0 & 0 \\ 0 & 0 & 1 & 3 & 0 \\ 0 & 0 & 0 & 0 & 1 \end{pmatrix}, \qquad \begin{pmatrix} 1 & 2 & 0 & 3 & 0 \\ 0 & 0 & 3 & 4 & 0 \\ 0 & 0 & 0 & 0 & 1 \end{pmatrix},$$

There is a row consisting of zeros that is not at the bottom of the matrix.   *The first nonzero element in row 2 is not 1.*

$$\begin{pmatrix} 1 & 0 & 0 & 2 \\ 0 & 0 & 1 & 4 \\ 0 & 1 & 0 & 3 \end{pmatrix}, \qquad \begin{pmatrix} 1 & 2 & 0 & 4 \\ 0 & 1 & 0 & -3 \\ 0 & 0 & 1 & 2 \\ 0 & 0 & 0 & 0 \end{pmatrix}$$

The leading 1 in row 3 is not to the right of the leading 1 in row 2.

The element directly above the leading 1 in row 2 is not 0.

We now illustrate the general method of *Gauss-Jordan elimination*. This is an algorithm whereby a matrix is transformed, using elementary matrix transformations, to reduced echelon form. There are usually many sequences of transformations that can lead to the reduced echelon form. They all, however, lead to the same reduced echelon form. We say that *the reduced echelon form of a given matrix is unique.* (We do not prove this result.) The reader should master the formal Gauss-Jordan algorithm first. This is an important systematic technique that lends itself to implementation on the computer. It will also be used later in other contexts. Once the Gauss-Jordan approach has been learned, one can look for shortcuts that can lead quickly to the reduced echelon form in specific cases.

At the heart of the method is the selection of pivots. Their selection can be a little more involved than for a system of n equations in n variables having a unique solution. However, once they have been selected, their use is the same—to create zeros in columns. The method proceeds as previously once a pivot has been selected. We illustrate the steps involved in the method for a specific matrix.

✦ *Example 1*

Let us reduce the following matrix, A, to reduced echelon form.

$$A = \begin{pmatrix} 0 & 0 & 2 & -2 & 2 \\ 3 & 3 & -3 & 9 & 12 \\ 4 & 4 & -2 & 11 & 12 \end{pmatrix}$$

Step 1. Locate the first column (starting from the left) that does not consist entirely of zeros.

$$\begin{pmatrix} 0 & 0 & 2 & -2 & 2 \\ 3 & 3 & -3 & 9 & 12 \\ 4 & 4 & -2 & 11 & 12 \end{pmatrix}$$

↑
first nonzero column

Step 2. Interchange rows, if necessary, to bring a nonzero element to the top of the first nonzero column. This nonzero element is called a *pivot.*

$$\underset{\underset{R1 \leftrightarrow R2}{\approx}}{pivot \rightarrow} \begin{pmatrix} ③ & 3 & -3 & 9 & 12 \\ 0 & 0 & 2 & -2 & 2 \\ 4 & 4 & -2 & 11 & 12 \end{pmatrix}$$

Step 3. Create a 1 in the pivot location by multiplying the row containing the pivot by (1/pivot).

$$\underset{(\frac{1}{3})R1}{\approx} \begin{pmatrix} 1 & 1 & -1 & 3 & 4 \\ 0 & 0 & 2 & -2 & 2 \\ 4 & 4 & -2 & 11 & 12 \end{pmatrix}$$

Step 4. Create zeros elsewhere in the column of the pivot by adding suitable multiples of the row containing the pivot to all other rows of the matrix.

$$\underset{R3+(-4)R1}{\approx} \begin{pmatrix} 1 & 1 & -1 & 3 & 4 \\ 0 & 0 & 2 & -2 & 2 \\ 0 & 0 & 2 & -1 & -4 \end{pmatrix}$$

Step 5. Cover the row containing the last pivot and all rows above it. Repeat steps 1, 2, and 3 for the remaining submatrix. Repeat step 4 for the whole matrix. Continue thus until the reduced echelon form is reached. We get

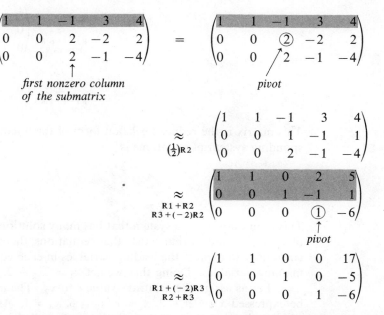

This matrix is in reduced echelon form. It is the reduced echelon form of the given matrix. ✦

We now illustrate how this method is used to solve various systems of equations.

✦ *Example 2*

Solve, if possible, the system

$$2x_1 - 4x_2 + 12x_3 - 10x_4 = 58$$
$$-x_1 + 2x_2 - 3x_3 + 2x_4 = -14$$
$$2x_1 - 4x_2 + 9x_3 - 6x_4 = 44$$

We start with the augmented matrix and reduce it to reduced echelon form. Pivots and leading ones are circled.

$$\begin{pmatrix} ② & -4 & 12 & -10 & 58 \\ -1 & 2 & -3 & 2 & -14 \\ 2 & -4 & 9 & -6 & 44 \end{pmatrix} \underset{(\frac{1}{2})R1}{\approx} \begin{pmatrix} ① & -2 & 6 & -5 & 29 \\ -1 & 2 & -3 & 2 & -14 \\ 2 & -4 & 9 & -6 & 44 \end{pmatrix}$$

$$\underset{\substack{R2+R1 \\ R3+(-2)R1}}{\approx} \begin{pmatrix} 1 & -2 & 6 & -5 & 29 \\ 0 & 0 & ③ & -3 & 15 \\ 0 & 0 & -3 & 4 & -14 \end{pmatrix}$$

$$\underset{(\frac{1}{3})R2}{\approx} \begin{pmatrix} 1 & -2 & 6 & -5 & 29 \\ 0 & 0 & ① & -1 & 5 \\ 0 & 0 & -3 & 4 & -14 \end{pmatrix}$$

$$\underset{\substack{R1+(-6)R2 \\ R3+(3)R2}}{\approx} \begin{pmatrix} 1 & -2 & 0 & 1 & -1 \\ 0 & 0 & 1 & -1 & 5 \\ 0 & 0 & 0 & ① & 1 \end{pmatrix}$$

$$\underset{\substack{R1+(-1)R3 \\ R2+R3}}{\approx} \begin{pmatrix} 1 & -2 & 0 & 0 & -2 \\ 0 & 0 & 1 & 0 & 6 \\ 0 & 0 & 0 & 1 & 1 \end{pmatrix}$$

This matrix is the reduced echelon form of the augmented matrix. The corresponding system of equations is

$$x_1 - 2x_2 = -2$$
$$x_3 = 6$$
$$x_4 = 1$$

This is an example of a system that has many solutions. There are many values of x_1, x_2, x_3, and x_4 that satisfy these equations; there are many solutions. It is customary to express the leading variables in each equation in terms of the remaining variables. Doing this, we get $x_1 = 2x_2 - 2$, $x_3 = 6$, $x_4 = 1$.

Let us assign the arbitrary value r to x_2. The *arbitrary solution* can then be expressed $x_1 = 2r - 2$, $x_2 = r$, $x_3 = 6$, $x_4 = 1$. As r ranges over the set of real numbers, we get all the solutions. r is called a *parameter*. For example, when $r = 1$, we get $x_1 = 0$, $x_2 = 1$, $x_3 = 6$, $x_4 = 1$ as a solution. When $r = -2$, we get $x_1 = -6$, $x_2 = -2$, $x_3 = 6$, $x_4 = 1$ as a solution. ✦

✦ *Example 3*
_____

Solve, if possible, the system

$$x_1 + 2x_2 - x_3 + 3x_4 + x_5 = 2$$
$$2x_1 + 4x_2 - 2x_3 + 6x_4 + 3x_5 = 6$$
$$-x_1 - 2x_2 + x_3 - x_4 + 3x_5 = 4$$

We get

$$\begin{pmatrix} ① & 2 & -1 & 3 & 1 & 2 \\ 2 & 4 & -2 & 6 & 3 & 6 \\ -1 & -2 & 1 & -1 & 3 & 4 \end{pmatrix} \underset{\substack{R2+(-2)R1 \\ R3+R1}}{\approx} \begin{pmatrix} 1 & 2 & -1 & 3 & 1 & 2 \\ 0 & 0 & 0 & 0 & 1 & 2 \\ 0 & 0 & 0 & ② & 4 & 6 \end{pmatrix}$$

$$\underset{R2 \leftrightarrow R3}{\approx} \begin{pmatrix} 1 & 2 & -1 & 3 & 1 & 2 \\ 0 & 0 & 0 & ② & 4 & 6 \\ 0 & 0 & 0 & 0 & 1 & 2 \end{pmatrix}$$

$$\underset{(\frac{1}{2})R2}{\approx} \begin{pmatrix} 1 & 2 & -1 & 3 & 1 & 2 \\ 0 & 0 & 0 & ① & 2 & 3 \\ 0 & 0 & 0 & 0 & 1 & 2 \end{pmatrix}$$

$$\underset{R1+(-3)R2}{\approx} \begin{pmatrix} 1 & 2 & -1 & 0 & -5 & -7 \\ 0 & 0 & 0 & 1 & 2 & 3 \\ 0 & 0 & 0 & 0 & ① & 2 \end{pmatrix}$$

$$\underset{\substack{R1+5R3 \\ R2+(-2)R3}}{\approx} \begin{pmatrix} 1 & 2 & -1 & 0 & 0 & 3 \\ 0 & 0 & 0 & 1 & 0 & -1 \\ 0 & 0 & 0 & 0 & 1 & 2 \end{pmatrix}$$

This matrix is the reduced echelon form of the augmented matrix.

The corresponding system of equations is

$$x_1 + 2x_2 - x_3 \qquad\qquad = 3$$
$$x_4 \qquad = -1$$
$$x_5 = 2$$

Expressing leading variables in terms of remaining variables, we get

$$x_1 = -2x_2 + x_3 + 3, \qquad x_4 = -1, \qquad x_5 = 2$$

Let us assign the arbitrary values r to x_2 and s to x_3. The arbitrary solution is thus

$$x_1 = -2r + s + 3, \qquad x_2 = r, \qquad x_3 = s, \qquad x_4 = -1, \qquad x_5 = 2$$

By letting r and s take on different values, we can get specific solutions. ✦

✦ *Example 4*
_____

This example illustrates a system that has no solutions. Consider the system

$$x_1 - x_2 + 2x_3 = 3$$
$$2x_1 - 2x_2 + 5x_3 = 4$$
$$x_1 + 2x_2 - x_3 = -3$$
$$2x_2 + 2x_3 = 1$$

We get

$$\begin{pmatrix} ① & -1 & 2 & 3 \\ 2 & -2 & 5 & 4 \\ 1 & 2 & -1 & -3 \\ 0 & 2 & 2 & 1 \end{pmatrix} \underset{\substack{R2-(2)R1 \\ R3-R1}}{\approx} \begin{pmatrix} 1 & -1 & 2 & 3 \\ 0 & 0 & 1 & -2 \\ 0 & 3 & -3 & -6 \\ 0 & 2 & 2 & 1 \end{pmatrix} \underset{R2 \leftrightarrow R3}{\approx} \begin{pmatrix} 1 & -1 & 2 & 3 \\ 0 & ③ & -3 & -6 \\ 0 & 0 & 1 & -2 \\ 0 & 2 & 2 & 1 \end{pmatrix}$$

$$\underset{(\frac{1}{3})R2}{\approx} \begin{pmatrix} 1 & -1 & 2 & 3 \\ 0 & ① & -1 & -2 \\ 0 & 0 & 1 & -2 \\ 0 & 2 & 2 & 1 \end{pmatrix} \underset{\substack{R1+R2 \\ R4-(2)R2}}{\approx} \begin{pmatrix} 1 & 0 & 1 & 1 \\ 0 & 1 & -1 & -2 \\ 0 & 0 & ① & -2 \\ 0 & 0 & 4 & 5 \end{pmatrix}$$

$$\underset{\substack{R1-R3 \\ R2+R3 \\ R4-(4)R3}}{\approx} \begin{pmatrix} 1 & 0 & 0 & 3 \\ 0 & 1 & 0 & -4 \\ 0 & 0 & 1 & -2 \\ 0 & 0 & 0 & ⑬ \end{pmatrix} \underset{(\frac{1}{13})R4}{\approx} \begin{pmatrix} 1 & 0 & 0 & 3 \\ 0 & 1 & 0 & -4 \\ 0 & 0 & 1 & -2 \\ 0 & 0 & 0 & ① \end{pmatrix}$$

This matrix is not in reduced echelon form; zeros still have to be created above the 1 in the last row. However, in such a situation, when the last nonzero row is of the form $(0 \quad 0 \quad \cdots \quad 0 \quad 1)$, there is no need to proceed further. The system has no solutions. To see this, let us write down the system that corresponds to the above matrix. We get

$$\begin{aligned} x_1 \qquad\qquad\qquad &= \cdot \; 3 \\ x_2 \qquad\qquad &= -4 \\ x_3 \qquad &= -2 \\ 0 &= \quad 1 \end{aligned}$$

This system cannot be satisfied for any values of x_1, x_2, and x_3; thus the original system has no solutions.   ✦

We complete this section with some general results. The first result confirms the method that was used in the previous section to solve a system of n equations in n variables having a unique solution.

Theorem 1-1   *A system of n equations in n variables has a unique solution if and only if the reduced echelon form of the augmented matrix is of the form*

$$\begin{pmatrix} 1 & 0 & & 0 & p_1 \\ 0 & 1 & & & \vdots \\ & & \ddots & & \\ 0 & & & 1 & p_n \end{pmatrix}$$

$$\underbrace{\phantom{\begin{matrix} 1 & 0 & & 0 \end{matrix}}}_{I_n}$$

Discussion: If the reduced echelon form is of this type, then the given system has the unique solution $x_1 = p_1, \ldots, x_n = p_n$.

Let us now discuss the converse. Suppose a system has a unique solution. If the reduced echelon form is not of the above type, then it must have a last row of zeros or there must be a row in which the leading 1 is not to the immediate right of the leading 1 of the previous row. Both of these situations lead to many solutions. For example,

$$\begin{pmatrix} 1 & 0 & 3 & 1 \\ 0 & 1 & 2 & 4 \\ 0 & 0 & 0 & 0 \end{pmatrix}$$

$\uparrow$
last row of zeros

leads to $x_1 = -3x_3 + 1$, $x_2 = -2x_3 + 4$, and

$$\begin{pmatrix} 1 & 2 & 0 & 0 & 4 \\ 0 & 0 & 1 & 0 & 3 \\ 0 & 0 & 0 & 1 & 5 \end{pmatrix}$$

leading 1 not to immediate
right of leading 1 in row
before

leads to $x_1 = -2x_2 + 4$, $x_3 = 3$, $x_4 = 5$. Thus the reduced echelon form must be of the above type.

The reader is asked to classify all reduced echelon form types of 3×4 matrices in Exercise 18. It should be observed that the only type of 3×4 matrix that corresponds to a unique solution is that having I_3 as the first submatrix.

Observe that in computing the reduced echelon form of the augmented matrix of a system, one at the same time computes the reduced echelon form of the matrix of coefficients. For example, for the system of Example 3, we get

$$\underbrace{\begin{pmatrix} 1 & 2 & -1 & 3 & 1 & | & 2 \\ 2 & 4 & -2 & 6 & 3 & | & 6 \\ -1 & -2 & 1 & -1 & 3 & | & 4 \end{pmatrix}}_{\substack{\textit{matrix of coefficients} \\ \textit{of the system}}} \approx \cdots \approx \begin{pmatrix} 1 & 2 & -1 & 0 & 0 & | & 3 \\ 0 & 0 & 0 & 1 & 0 & | & -1 \\ 0 & 0 & 0 & 0 & 1 & | & 2 \end{pmatrix}$$

augmented matrix (left)

reduced echelon form of augmented matrix (right above)

reduced echelon form of the matrix of coefficients of the system

This observation enables us to rephrase Theorem 1-1 in the following convenient form.

Theorem 1-2   *A system of n equations in n variables has a unique solution if and only if the reduced echelon form of the matrix of coefficients A is the identity matrix I_n.*

Let us now look at a special class of systems of linear equations.

Homogeneous Systems of Linear Equations

A system of linear equations is *homogeneous* if all the constant terms are zeros. The following is an example of a homogeneous system:

$$3x_1 - 2x_2 + \ x_3 = 0$$
$$x_1 + 5x_2 + 2x_3 = 0$$
$$7x_1 - \ x_2 + 3x_3 = 0$$

$$\uparrow$$

All the constant terms are zeros.

Observe that $x_1 = 0$, $x_2 = 0$, $x_3 = 0$ is a solution to this system. *A homogeneous system of linear equations in n variables always has the solution $x_1 = 0$, $x_2 = 0, \ldots, x_n = 0$.* This solution is called the *trivial solution*.

In this section, we have seen examples of systems of linear equations that have a unique solution, many solutions, and no solutions at all. In general, without further analysis, one cannot look at a system and tell which of these possibilities applies. However, when the number of variables is greater than the number of equations *in a homogeneous system*, one knows in advance that there will be many solutions. There will thus be *nontrivial solutions* for such a system. We now derive this result.

Theorem 1-3   *A homogeneous system of linear equations that has more variables than equations has many solutions.*

Proof: Consider a homogeneous system of m linear equations in n variables, with $n > m$. The augmented matrix of this system will have m rows and $n + 1$ columns. Let us look at the reduced echelon form of such an augmented matrix. There will be two distinct types of reduced echelon forms that can arise.

In type 1, the reduction goes "smoothly." Each leading 1 is always to the immediate right of the leading 1 in the previous row. The zeros (if any) are grouped at the bottom of the matrix. Let there be r rows with nonzero elements, $r \leq m$. This matrix gives a system in which the last $n - r$ variables can take on any values. There are many solutions.

In type 2, there will be at least one row in which the leading 1 is not to the immediate right of the leading 1 in the preceding row. We illustrate a typical reduced echelon form of this type above, in which the leading 1 in row 3 is not to the immediate right of the leading 1 in row 2. This example leads to a system of equations in which x_3 can take on any value. In general, for this type of matrix, if the leading 1 in row i is not to the immediate right of the leading 1 in the preceding row, then x_i can take on any value; there are many solutions.

✦ Example 5 ──

Consider the following homogeneous system of three linear equations in five variables. Since the number of variables is greater than the number of equations, the system should have many solutions. Let us solve the system to confirm this.

$$
\begin{aligned}
x_1 + 2x_2 + 2x_3 - x_4 &= 0 \\
2x_1 + 4x_2 + 5x_3 - 3x_4 - x_5 &= 0 \\
-x_1 - 2x_2 - 3x_3 + 3x_4 + 4x_5 &= 0
\end{aligned}
$$

Observe that $x_1 = 0, x_2 = 0, x_3 = 0, x_4 = 0, x_5 = 0$ is a solution. We get, using Gauss-Jordan elimination,

$$
\begin{pmatrix}
1 & 2 & 2 & -1 & 0 & 0 \\
2 & 4 & 5 & -3 & -1 & 0 \\
-1 & -2 & -3 & 3 & 4 & 0
\end{pmatrix}
\underset{\substack{R2+(-2)R1 \\ R3+R1}}{\approx}
\begin{pmatrix}
1 & 2 & 2 & -1 & 0 & 0 \\
0 & 0 & 1 & -1 & -1 & 0 \\
0 & 0 & -1 & 2 & 4 & 0
\end{pmatrix}
$$

$$
\underset{\substack{R1+(-2)R2 \\ R3+R2}}{\approx}
\begin{pmatrix}
1 & 2 & 0 & 1 & 2 & 0 \\
0 & 0 & 1 & -1 & -1 & 0 \\
0 & 0 & 0 & 1 & 3 & 0
\end{pmatrix}
$$

$$
\underset{\substack{R1+(-1)R3 \\ R2+R3}}{\approx}
\begin{pmatrix}
1 & 2 & 0 & 0 & -1 & 0 \\
0 & 0 & 1 & 0 & 2 & 0 \\
0 & 0 & 0 & 1 & 3 & 0
\end{pmatrix}
$$

This reduced echelon form corresponds to the system

$$
\begin{aligned}
x_1 + 2x_2 \quad\quad - x_5 &= 0 \\
x_3 \quad + 2x_5 &= 0 \\
x_4 + 3x_5 &= 0
\end{aligned}
$$

Expressing the leading variables in terms of the remaining variables, we get

$$
x_1 = -2x_2 + x_5, \quad x_3 = -2x_5, \quad x_4 = -3x_5
$$

Let us assign the arbitrary values r to x_2 and s to x_5. There are many solutions:

$$x_1 = -2r + s, \quad x_2 = r, \quad x_3 = -2s, \quad x_4 = -3s, \quad x_5 = s \quad \blacklozenge$$

Note: In this section we have introduced the method of Gauss-Jordan elimination for solving systems of linear equations. There is another popular elimination method for solving systems, called the method of Gaussian elimination. Gaussian elimination is in general more efficient than Gauss-Jordan, in that fewer additions and multiplications are needed to arrive at the solution. This is an important factor when large systems are solved on computers. However, Gauss-Jordan has advantages as an introductory method in that it is cleaner and easier to manage by hand. It is superior for solving small systems. Furthermore, the Gauss-Jordan algorithm is used in a number of other contexts, such as in computation of the inverse of a matrix and in the simplex method in linear programming. The reader will meet these topics later in the course; any readers who are interested in mastering Gaussian elimination at this time and comparing it with Gauss-Jordan can turn to Section 8-2.

EXERCISES

1. State whether or not the following matrices are in reduced echelon form. If a matrix is not in reduced echelon form, explain why it is not.

*(a) $\begin{pmatrix} 1 & 0 & 0 & 0 & 0 \\ 0 & 0 & 1 & 2 & 3 \\ 0 & 0 & 0 & 0 & 0 \end{pmatrix}$ (b) $\begin{pmatrix} 1 & 0 & 0 & 3 \\ 0 & 1 & 0 & 4 \\ 0 & 0 & 2 & 1 \end{pmatrix}$

*(c) $\begin{pmatrix} 1 & 0 & 0 & 3 & 2 \\ 0 & 2 & 0 & 6 & 1 \\ 0 & 0 & 1 & 2 & 3 \end{pmatrix}$ (d) $\begin{pmatrix} 1 & 6 & 0 & 0 & 2 & -1 \\ 0 & 0 & 1 & 0 & 4 & 3 \\ 0 & 0 & 0 & 1 & 3 & 1 \end{pmatrix}$

*(e) $\begin{pmatrix} 1 & 0 & 0 & 2 \\ 0 & 0 & 1 & 3 \\ 0 & 1 & 0 & 4 \end{pmatrix}$ (f) $\begin{pmatrix} 1 & 2 & 0 & 0 & 4 \\ 0 & 0 & 1 & 0 & 6 \\ 0 & 0 & 0 & 1 & 5 \end{pmatrix}$

*(g) $\begin{pmatrix} 1 & 0 & 4 & 2 & 6 \\ 0 & 1 & 2 & 3 & 4 \\ 0 & 0 & 0 & 1 & 2 \\ 0 & 0 & 0 & 0 & 1 \end{pmatrix}$ (h) $\begin{pmatrix} 1 & 5 & 4 & 2 & 1 \\ 0 & 0 & 1 & 5 & 3 \\ 0 & 0 & 0 & 0 & 1 \\ 0 & 0 & 0 & 0 & 0 \end{pmatrix}$

*(i) $\begin{pmatrix} 1 & 5 & 0 & 0 & 0 \\ 0 & 0 & 2 & 0 & 0 \\ 0 & 0 & 0 & 1 & 0 \\ 0 & 0 & 0 & 0 & 1 \end{pmatrix}$

2. Interpret each of the following matrices as being a matrix in the sequence that leads to the reduced echelon form of a given system of linear equations. First select that row for which a pivot is needed at this stage. Then select a suitable nonzero element as pivot for that row, interchanging rows if necessary.

*(a) $\begin{pmatrix} 1 & 0 & 1 & 4 \\ 0 & 1 & 2 & 3 \\ 0 & 0 & 3 & 6 \end{pmatrix}$
(b) $\begin{pmatrix} 1 & 2 & 4 \\ 0 & 3 & 6 \\ 0 & 4 & 2 \end{pmatrix}$

*(c) $\begin{pmatrix} 1 & 3 & 4 & -1 \\ 0 & 0 & 6 & 2 \\ 0 & 4 & 3 & 4 \end{pmatrix}$
(d) $\begin{pmatrix} 1 & 0 & 0 & -1 & 6 \\ 0 & 1 & 3 & 2 & 4 \\ 0 & 0 & 0 & 4 & 2 \\ 0 & 0 & 0 & 3 & 1 \end{pmatrix}$

*(e) $\begin{pmatrix} 1 & 4 & 2 & 4 & 5 \\ 0 & 0 & 0 & 0 & 6 \\ 0 & 0 & 0 & 2 & 3 \\ 0 & 0 & 3 & 1 & 2 \end{pmatrix}$
(f) $\begin{pmatrix} 0 & 4 & -2 \\ 0 & 6 & 4 \\ 2 & 3 & 6 \\ 1 & 1 & 7 \end{pmatrix}$

(g) $\begin{pmatrix} 1 & 0 & 2 & 3 \\ 0 & 1 & 4 & 4 \\ 0 & 0 & 0 & 0 \\ 0 & 0 & 0 & 5 \end{pmatrix}$
(h) $\begin{pmatrix} 1 & 0 & 0 & 2 & 5 & 1 \\ 0 & 1 & 4 & 6 & 1 & 3 \\ 0 & 0 & 0 & 0 & 0 & 2 \\ 0 & 0 & 0 & 0 & 3 & 4 \end{pmatrix}$

3. Each of the following matrices is in reduced echelon form. Write down the system of equations that corresponds to each matrix. Solve each system, if possible.

*(a) $\begin{pmatrix} 1 & 0 & 0 & 0 & 2 \\ 0 & 1 & 2 & 0 & 3 \\ 0 & 0 & 0 & 1 & -1 \end{pmatrix}$
(b) $\begin{pmatrix} 1 & 2 & 0 & 0 & 2 \\ 0 & 0 & 1 & 0 & 3 \\ 0 & 0 & 0 & 1 & 4 \end{pmatrix}$

*(c) $\begin{pmatrix} 1 & 0 & 2 & 0 & 2 & -1 \\ 0 & 1 & 4 & 0 & 3 & 2 \\ 0 & 0 & 0 & 1 & 4 & 1 \end{pmatrix}$
(d) $\begin{pmatrix} 1 & 0 & 2 & 0 & 0 \\ 0 & 1 & 4 & 0 & 0 \\ 0 & 0 & 0 & 1 & 0 \\ 0 & 0 & 0 & 0 & 1 \end{pmatrix}$

*(e) $\begin{pmatrix} 1 & 0 & 2 & 0 & 2 \\ 0 & 1 & -3 & 0 & 1 \\ 0 & 0 & 0 & 1 & 3 \\ 0 & 0 & 0 & 0 & 0 \end{pmatrix}$
(f) $\begin{pmatrix} 1 & 3 & 4 & 0 & 0 & 0 & 1 \\ 0 & 0 & 0 & 1 & 0 & 0 & 2 \\ 0 & 0 & 0 & 0 & 1 & 0 & 4 \\ 0 & 0 & 0 & 0 & 0 & 1 & 3 \end{pmatrix}$

*(g) $\begin{pmatrix} 1 & 2 & 0 & 1 & 0 & 0 & 4 \\ 0 & 0 & 1 & 3 & 0 & 0 & -1 \\ 0 & 0 & 0 & 0 & 1 & 0 & 3 \\ 0 & 0 & 0 & 0 & 0 & 1 & 2 \end{pmatrix}$
(h) $\begin{pmatrix} 1 & 0 & 0 & 0 \\ 0 & 1 & 0 & 0 \\ 0 & 0 & 1 & 0 \\ 0 & 0 & 0 & 1 \end{pmatrix}$

(i) $\begin{pmatrix} 1 & 2 & 0 & 4 & 0 & 1 & 7 \\ 0 & 0 & 1 & 3 & 0 & -1 & 3 \\ 0 & 0 & 0 & 0 & 1 & 4 & 2 \end{pmatrix}$

In Exercises 4–17, solve each system of equations (if possible).

*4. $x_1 + 2x_2 - x_3 - x_4 = 0$
$x_1 + 2x_2 \qquad + x_4 = 4$
$-x_1 - 2x_2 + 2x_3 + 4x_4 = 5$

5. $x_1 + x_2 + x_3 - x_4 = -3$
$2x_1 + 3x_2 + x_3 - 5x_4 = -9$
$x_1 + 3x_2 - x_3 - 6x_4 = -7$
$-x_1 - x_2 - x_3 \qquad = 1$

6. $x_1 - x_2 + x_3 = 3$
$-2x_1 + 3x_2 + x_3 = -8$
$4x_1 - 2x_2 + 10x_3 = 10$

*7. $x_2 + 2x_3 = 7$
$x_1 - 2x_2 - 6x_3 = -18$
$x_1 - x_2 - 2x_3 = -5$
$2x_1 - 5x_2 - 15x_3 = -46$

8. $2x_1 - 4x_2 + 16x_3 - 14x_4 = 10$
$-x_1 + 5x_2 - 17x_3 + 19x_4 = -2$
$x_1 - 3x_2 + 11x_3 - 11x_4 = 4$
$3x_1 - 4x_2 + 18x_3 - 13x_4 = 17$

*9. $x_1 - x_2 + 2x_3 \qquad = 7$
$2x_1 - 2x_2 + 2x_3 - 4x_4 = 12$
$-x_1 + x_2 - x_3 + 2x_4 = -4$
$-3x_1 + x_2 - 8x_3 - 10x_4 = -29$

10. $x_1 + 6x_2 - x_3 - 4x_4 = 0$
$-2x_1 - 12x_2 + 5x_3 + 17x_4 = 0$
$3x_1 + 18x_2 - x_3 - 6x_4 = 0$

*11. $x_1 + 2x_2 + 3x_3 \qquad - x_5 - 2x_6 + x_7 = 2$
$2x_1 + 4x_2 + 6x_3 \qquad - x_5 - 3x_6 + 3x_7 = 7$
$x_1 + 2x_2 + 3x_3 + x_4 \qquad + x_6 + x_7 = 3$
$-3x_1 - 6x_2 - 9x_3 + 2x_4 + 7x_5 + 14x_6 + 2x_7 = 5$

12. $2x_1 - 4x_2 - 14x_3 = 6$
$x_1 - x_2 - 5x_3 = 4$
$2x_1 - 4x_2 - 17x_3 = 9$
$-x_1 + 3x_2 + 10x_3 = -3$
$2x_2 + 2x_3 = 4$

*13. $x_1 + x_2 + 3x_3 + 6x_4 + 24x_5 + 48x_6 = 112$
$2x_1 + 2x_2 + 6x_3 + 13x_4 + 52x_5 + 104x_6 = 241$
$x_3 + 2x_4 + 8x_5 + 16x_6 = 37$
$-x_1 - x_2 - 4x_3 - 7x_4 - 29x_5 - 58x_6 = -136$
$x_1 + 2x_2 + 5x_3 + 11x_4 + 42x_5 + 84x_6 = 197$

14. $x_1 + 2x_2 - 2x_3 + 7x_4 + 7x_5 = 0$
$-x_1 - 2x_2 + 2x_3 - 9x_4 - 11x_5 = 0$
$x_3 - 2x_4 - x_5 = 0$
$2x_1 + 4x_2 - 3x_3 + 12x_4 + 13x_5 = 0$

*15. $x_1 + 2x_2 - 3x_3 + 2x_4 + 5x_5 - x_6 = 0$
$-2x_1 - 4x_2 + 6x_3 - x_4 - 4x_5 + 5x_6 = 0$
$3x_1 + 6x_2 - 9x_3 + 5x_4 + 13x_5 - 4x_6 = 0$

16. $4x_1 + 8x_2 - 12x_3 = 28$
$-x_1 - 2x_2 + 3x_3 = -7$
$2x_1 + 4x_2 - 8x_3 = 16$
$-3x_1 - 6x_2 + 9x_3 = -21$

*17. $x_1 + 3x_2 + 6x_3 - 2x_4 = -7$
$-2x_1 - 5x_2 - 10x_3 + 3x_4 = 10$
$x_1 + 2x_2 + 4x_3 \qquad = 0$
$x_2 + 2x_3 - 3x_4 = -10$

18. Classify the reduced echelon form types of 3×4 matrices. Represent each class by a general matrix having $*$ in possible nonzero locations. The first two classes could be

$$\begin{pmatrix} 1 & 0 & 0 & * \\ 0 & 1 & 0 & * \\ 0 & 0 & 1 & * \end{pmatrix}, \quad \begin{pmatrix} 1 & 0 & * & 0 \\ 0 & 1 & * & 0 \\ 0 & 0 & 0 & 1 \end{pmatrix}$$

19. Theorem 1-3 tells us that a homogeneous system of linear equations that has more variables than equations always has many solutions. Is the analogous result true for nonhomogeneous systems?

1-5*   MODELS INVOLVING SYSTEMS OF LINEAR EQUATIONS

Many models in such diverse fields as electrical engineering, economics, and traffic analysis involve solving systems of linear equations. We now present some of these models.

Electrical Network Analysis

Systems of linear equations are involved in analyzing currents through various circuits in a network. In the planning of electrical networks containing resistances, it is necessary to predict the current load through the various circuits. Two laws that form the foundation of this mathematical analysis are *Kirchhoff's laws*:

1. All the current flowing into a junction must flow out of it.
2. The sum of the *IR* terms (*I* denotes current, *R* resistance) in any direction around a closed path is equal to the total voltage in the path in that direction.

These laws are based on experimental verification in the laboratory.

✦ *Example 1*
_____

Consider the network in Figure 1-5. Let us determine the current through each branch.

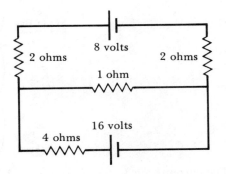

Figure 1-5

The batteries (denoted —|⊢) are 8 volts and 16 volts. The following convention is used to indicate out of which terminal of the battery the current flows: ⊄⊢. The resistances (denoted —⋀⋀—) are one 1-ohm, one 4-ohm, and two 2-ohm. The current entering each battery will be the same as that leaving it.

In Figure 1-6, we label the currents I_1, I_2, and I_3, and the batteries and junctions A, B, C, and D.

Applying Law 1 to each junction,

Junction B, $I_1 + I_2 = I_3$

Junction C, $I_1 + I_2 = I_3$

giving $I_1 + I_2 - I_3 = 0$. Applying Law 2 to various paths,

Path $ABCA$, $2I_1 + 1I_3 + 2I_1 = 8$

Path $DBCD$, $4I_2 + 1I_3 = 16$

It is not necessary to look further at path $ABDCA$. We now have a system of three equations in three unknowns: I_1, I_2, I_3. Path $ABDCA$ leads to an equation that is a combination of the last two; there is no new information.

The problem thus reduces to solving the system of linear equations

$$
\begin{aligned}
I_1 + I_2 - I_3 &= 0 \\
4I_1 + I_3 &= 8 \\
4I_2 + I_3 &= 16
\end{aligned}
$$

Using the method of Gauss-Jordan elimination, we get

$$
\begin{pmatrix} \textcircled{1} & 1 & -1 & 0 \\ 4 & 0 & 1 & 8 \\ 0 & 4 & 1 & 16 \end{pmatrix} \underset{R2+(-4)R1}{\approx} \begin{pmatrix} 1 & 1 & -1 & 0 \\ 0 & \boxed{-4} & 5 & 8 \\ 0 & 4 & 1 & 16 \end{pmatrix}
$$

$$
\underset{(-\frac{1}{4})R2}{\approx} \begin{pmatrix} 1 & 1 & -1 & 0 \\ 0 & \textcircled{1} & -\frac{5}{4} & -2 \\ 0 & 4 & 1 & 16 \end{pmatrix}
$$

$$
\underset{\substack{R1+(-1)R2 \\ R3+(-4)R2}}{\approx} \begin{pmatrix} 1 & 0 & \frac{1}{4} & 2 \\ 0 & 1 & -\frac{5}{4} & -2 \\ 0 & 0 & \textcircled{6} & 24 \end{pmatrix}
$$

$$
\underset{(\frac{1}{6})R3}{\approx} \begin{pmatrix} 1 & 0 & \frac{1}{4} & 2 \\ 0 & 1 & -\frac{5}{4} & -2 \\ 0 & 0 & \textcircled{1} & 4 \end{pmatrix}
$$

$$
\underset{\substack{R1+(-\frac{1}{4})R3 \\ R2+(\frac{5}{4})R3}}{\approx} \begin{pmatrix} 1 & 0 & 0 & 1 \\ 0 & 1 & 0 & 3 \\ 0 & 0 & 1 & 4 \end{pmatrix}
$$

The currents are $I_1 = 1$, $I_2 = 3$, $I_3 = 4$. The units are amps.

The solution is unique, as is to be expected in this physical situation.

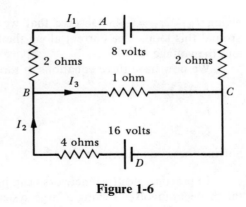

Figure 1-6

✦ *Example 2*
───

Determine the currents through the various branches of the network in Figure 1-7. This example illustrates how one has to be conscious of direction in applying Law 2 for loops.

Law 1 gives

Junction B, $I_1 + I_2 = I_3$
Junction D, $I_3 = I_1 + I_2$

giving $I_1 + I_2 - I_3 = 0$.

Law 2 gives

Path $ABCDA$, $1I_1 + 2I_3 = 12$

and

Path $ABDA$, $1I_1 + 2(-I_2) = 12 + (-16)$

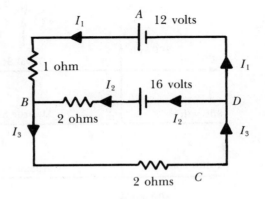

Figure 1-7

giving $I_1 - 2I_2 = -4$. Observe that we have selected the direction $ABDA$ around this loop. The current along the branch BD in this direction is $-I_2$, and the voltage is -16.

We now have three equations in three variables I_1, I_2, and I_3.

$$
\begin{aligned}
I_1 + I_2 - I_3 &= 0 \\
I_1 \qquad + 2I_3 &= 12 \\
I_1 - 2I_2 \qquad &= -4
\end{aligned}
$$

Solving these equations, we get $I_1 = 2$, $I_2 = 3$, $I_3 = 5$ amps.   ✦

In practice, electrical networks can involve many resistances and circuits; the problem involves solving a large system of equations on the computer.

Traffic Flow

Network analysis, as we saw in the previous example, plays an important role in electrical engineering. In recent years, the concepts and tools of network analysis have been found to be very useful in many other fields, such as information theory and the study of transportation systems. The following analysis of traffic flow through a road network during a peak period illustrates how a system of linear equations with many solutions can arise in practice.

Consider the typical road network in Figure 1-8. It represents an area of downtown Jacksonville, Florida. The flow in and out of the network is measured in vehicles per hour (vph). Figures are based on midweek peak traffic hours, 7 A.M. to 9 A.M. and 4 P.M. to 6 P.M. An increase of 2 percent in the overall flow should be allowed for during the Friday evening traffic flow. Let us construct a mathematical model that can be used to analyze this network.

Suppose it becomes necessary to perform road work on the stretch of Adams Street between Laura and Hogan. It is desirable to have as small a

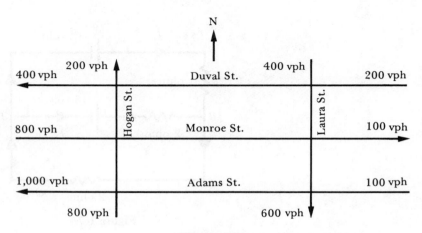

Figure 1-8

flow of traffic as possible along this stretch of road. The flows can be controlled along various branches by means of traffic lights at the junctions. What is the minimum flow possible along Adams that would not lead to traffic congestion? What are the flows along the other branches when this is attained? Our model will enable us to answer these questions.

Let the traffic flows along the various branches be as represented in Figure 1-9. It is reasonable to assume that all traffic entering a junction must leave that junction if there is to be no congestion. This conservation of flow constraint (compare it with the first of Kirchhoff's laws for electrical networks) leads to a system of linear equations.

Consider Junction A. The traffic flowing in is $400 + 200$, and that flowing out is $x_1 + x_5$. This leads to the equation

$$x_1 + x_5 = 600$$

If we look at Junction B, the traffic flowing in is $x_1 + x_6$, and that flowing out is $x_2 + 100$. Thus

$$x_1 + x_6 = x_2 + 100$$

Rearranging this equation, we get

$$x_1 - x_2 + x_6 = 100$$

Carrying out this procedure for each junction and writing the resulting equations in convenient form, we get the following system:

$$
\begin{aligned}
\text{Junction } A, \quad x_1 \qquad\qquad\; + x_5 \qquad\qquad\quad &= \;\; 600 \\
\text{Junction } B, \quad x_1 - x_2 \qquad\qquad + x_6 \qquad\quad &= \;\; 100 \\
\text{Junction } C, \qquad\quad x_2 \qquad\qquad\qquad - x_7 &= \;\; 500 \\
\text{Junction } D, \qquad\qquad\;\; - x_3 \qquad\qquad + x_7 &= \;\; 200 \\
\text{Junction } E, \qquad\qquad\quad x_3 - x_4 \quad - x_6 \quad &= -800 \\
\text{Junction } F, \qquad\qquad\qquad\quad x_4 + x_5 \qquad\qquad &= \;\; 600
\end{aligned}
$$

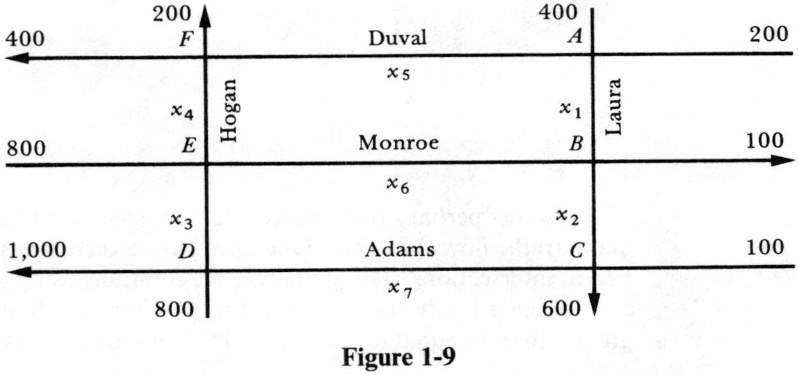

Figure 1-9

The method of Gauss-Jordan elimination is used to solve this system. Observe that the augmented matrix contains many zeros. These zeros greatly reduce the computation involved. We present the reduced echelon form, leaving its derivation as an exercise. In practice, networks are much larger than the one we have illustrated here, and the systems of equations that describe them are thus much larger. The systems are solved on the computer. However the augmented matrices of all such systems contain many zeros. Much research has gone into the development of efficient algorithms for solving such *sparse* systems of equations efficiently. The augmented matrix and reduced echelon form of the above system are as follows:

$$\begin{pmatrix} 1 & 0 & 0 & 0 & 1 & 0 & 0 & 600 \\ 1 & -1 & 0 & 0 & 0 & 1 & 0 & 100 \\ 0 & 1 & 0 & 0 & 0 & 0 & -1 & 500 \\ 0 & 0 & -1 & 0 & 0 & 0 & 1 & 200 \\ 0 & 0 & 1 & -1 & 0 & -1 & 0 & -800 \\ 0 & 0 & 0 & 1 & 1 & 0 & 0 & 600 \end{pmatrix}$$

$$\approx \cdots \approx \begin{pmatrix} 1 & 0 & 0 & 0 & 0 & 1 & -1 & 600 \\ 0 & 1 & 0 & 0 & 0 & 0 & -1 & 500 \\ 0 & 0 & 1 & 0 & 0 & 0 & -1 & -200 \\ 0 & 0 & 0 & 1 & 0 & 1 & -1 & 600 \\ 0 & 0 & 0 & 0 & 1 & -1 & 1 & 0 \\ 0 & 0 & 0 & 0 & 0 & 0 & 0 & 0 \end{pmatrix}$$

The system of equations that corresponds to this reduced echelon form is

$$
\begin{aligned}
x_1 \qquad\qquad\qquad + x_6 - x_7 &= 600 \\
x_2 \qquad\qquad\qquad - x_7 &= 500 \\
x_3 \qquad\qquad - x_7 &= -200 \\
x_4 \quad + x_6 - x_7 &= 600 \\
x_5 - x_6 + x_7 &= 0
\end{aligned}
$$

Expressing the leading variables in terms of the remaining variables, we get

$$
\begin{aligned}
x_1 &= -x_6 + x_7 + 600 \\
x_2 &= x_7 + 500 \\
x_3 &= x_7 - 200 \\
x_4 &= -x_6 + x_7 + 600 \\
x_5 &= x_6 - x_7
\end{aligned}
$$

As was perhaps to be expected, the system has many solutions—there are many traffic flows possible. One does have a certain amount of choice of direction at intersections. Let us analyze the solutions using this system of equations.

We are interested in minimizing the flow x_7. Since all the flows must be greater than or equal to zero, the third equation implies that the minimum flow for x_7 is 200, for otherwise x_3 could become negative. (A negative flow would be interpreted as traffic moving in the direction opposite to the one permitted on the one-way street.) Thus road work must allow for a flow of at least 200 cars per hour on the branch CD in the peak period.

Let us now examine what the flows in the other branches will be when this is attained. $x_7 = 200$ gives

$$x_1 = -x_6 + 800$$
$$x_2 = 700$$
$$x_3 = 0$$
$$x_4 = -x_6 + 800$$
$$x_5 = x_6 - 200$$

Since $x_7 = 200$ implies that $x_3 = 0$ and vice versa, we see that the minimum flow in the branch x_7 can be attained by making $x_3 = 0$—i.e., by closing DE to traffic. The traffic x_2 in BC is then determined uniquely as 700 cars per hour. There is still some freedom in the flows x_1, x_4, x_5, and x_6 along AB, EF, AF, and EB, respectively, since these variables are still not uniquely determined. For example, one possible set of flows in these branches is $x_1 = 500$, $x_4 = 500$, $x_5 = 100$, $x_6 = 300$.

However, since $x_1 \geq 0$, $x_6 \leq 800$; and since $x_5 \geq 0$, $x_6 \geq 200$. Thus $200 \leq x_6 \leq 800$. The flow of traffic along Monroe between Hogan and Laura will be between 200 and 800 vph.

EXERCISES

Determine the currents through the various branches of the electrical networks in Exercises 1—6. (*Hint*: In Exercise 6, it is difficult to decide on the direction of the current along AB. Make a guess. A negative result for that current means that your guess was the wrong one—the current is in the opposite direction. However, the magnitude will still be correct. There is no need to rework the problem.)

*1.   2.

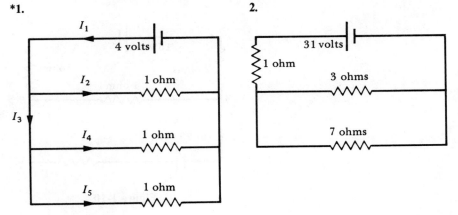

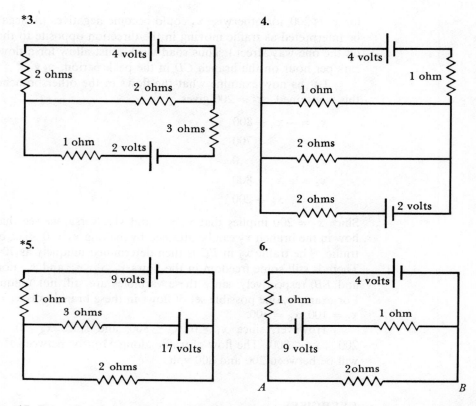

*7. Determine the currents through the various branches of the electrical network in Figure 1–10

(a) when the voltage of battery *C* is 9 volts

(b) when it is 23 volts

Note how the current through the branch *AB* is reversed in (b). What would the voltage of *C* have to be for no current to flow through *AB*?

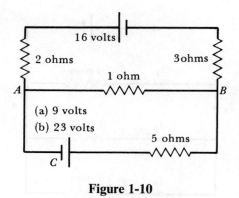

Figure 1-10

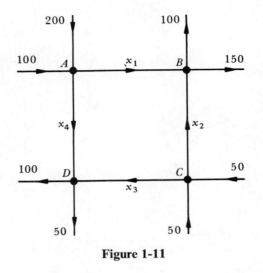

Figure 1-11

*8. Construct a mathematical model that describes the traffic network in Figure 1-11. All streets are one-way in the directions indicated. The units are in vehicles per hour. Give two distinct possible flows of traffic. What is the minimum possible flow that can be expected along *AB*?

9. Figure 1-12 represents the traffic entering and leaving a "roundabout" road junction. Such junctions are very common in Europe. Construct a mathematical model that describes the flow of traffic along various branches. What is the minimum flow theoretically possible along branch *BC*? Is this ever likely to be realized in practice?

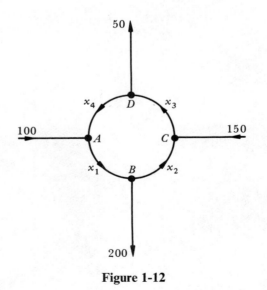

Figure 1-12

Matrices

In the previous chapter, we introduced matrices as convenient tools for handling systems of linear equations. We shall continue to use them in this context. Matrices are, however, useful in many other areas of mathematics and in such diverse fields as archaeology and economics. In this chapter, we shall develop the algebraic theory of matrices and see how this theory is used in some of these areas.

2-1 ADDITION AND SCALAR MULTIPLICATION OF MATRICES

Let us commence the development of an algebraic theory of matrices by defining a concept of equality for matrices.

Definition 2-1 Two matrices are equal if they are of the same size and if their corresponding elements are equal.

This definition will enable us to introduce algebraic equations involving matrices. It immediately allows us to define an operation of addition between certain matrices.

Definition 2-2 Let A and B be matrices of the same size. Their *sum $A + B$* is a matrix obtained by adding together corresponding elements.
The matrix $A + B$ will be of the same size as A and B.

For example,

$$\begin{pmatrix} 1 & 2 & 3 \\ 0 & 4 & 1 \\ 2 & 1 & 0 \end{pmatrix} + \begin{pmatrix} -1 & 0 & 1 \\ 0 & 1 & 1 \\ 2 & 3 & 0 \end{pmatrix} = \begin{pmatrix} 0 & 2 & 4 \\ 0 & 5 & 2 \\ 4 & 4 & 0 \end{pmatrix}$$

These are all 3×3 matrices.

However, we cannot add the matrices

$$\begin{pmatrix} 1 & 2 & 3 \\ 0 & 4 & 1 \\ 2 & 1 & 0 \end{pmatrix} \quad \text{and} \quad \begin{pmatrix} 0 & 1 \\ 0 & -1 \\ 2 & 3 \end{pmatrix}$$

The former is a 3×3 matrix, and the latter is a 3×2 matrix; they are not matrices of the same size. We say that *their sum does not exist.*

In this text, we shall use only the real number system. However, other number systems exist; the other most commonly used system is the complex number system. Most of the theory developed in this course is also appropriate for complex numbers. The term *scalar*, which we shall use, is often used by mathematicians for a number when they do not want to commit themselves to a particular number system. We shall use uppercase letters to denote matrices and lowercase letters to denote scalars. We now introduce a definition of multiplication of a matrix by a scalar.

Definition 2-3   Let A be a matrix and c a scalar. The *scalar multiple* of A by c, denoted cA, is the matrix obtained by multiplying every element of A by c.
    The matrix cA will be of the same size as A.

For example,

$$3\begin{pmatrix} 1 & 3 & -2 \\ 0 & 4 & 5 \end{pmatrix} = \begin{pmatrix} 3 & 9 & -6 \\ 0 & 12 & 15 \end{pmatrix}$$

Subtraction is performed between matrices of the same size by subtracting corresponding elements. For example,

$$\begin{pmatrix} 2 & -3 \\ 4 & 5 \end{pmatrix} - \begin{pmatrix} 1 & 2 \\ 5 & -1 \end{pmatrix} = \begin{pmatrix} 2-1 & -3-2 \\ 4-5 & 5-(-1) \end{pmatrix}$$

$$= \begin{pmatrix} 1 & -5 \\ -1 & 6 \end{pmatrix}$$

Observe that this is equivalent to

$$\begin{pmatrix} 2 & -3 \\ 4 & 5 \end{pmatrix} + \left[(-1)\begin{pmatrix} 1 & 2 \\ 5 & -1 \end{pmatrix} \right] = \begin{pmatrix} 1 & -5 \\ -1 & 6 \end{pmatrix}$$

Thus subtraction is not in effect a new operation; it is a combination of addition and scalar multiplication by -1.

Matrix Notation

A convenient notation has been developed to handle matrices. Every matrix has a certain number of rows and a certain number of columns. Thus when discussing an arbitrary matrix, we talk about an $m \times n$ matrix, a matrix having m rows and n columns.

Every element in a matrix has a certain location, specified by its row and column. If we use the letter A for an $m \times n$ matrix, let us choose the convention a_{ij} to talk about the element in the ith row and the jth column of A:

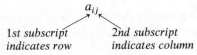

1st subscript
indicates row

2nd subscript
indicates column

Let us refer to a_{ij} as the (i, j)th *element* of A. We can visualize the matrix A as follows:

$$\begin{pmatrix} a_{11} & a_{12} & a_{13} & \cdots & a_{1n} \\ a_{21} & a_{22} & a_{23} & \cdots & a_{2n} \\ a_{31} & a_{32} & a_{33} & \cdots & a_{3n} \\ & & \vdots & & \\ a_{m1} & a_{m2} & a_{m3} & \cdots & a_{mn} \end{pmatrix}$$

For example, the element a_{11} lies in the first row and first column; a_{32} lies in the third row, second column; etc.

✦ *Example 1*

Consider the matrix

$$A = \begin{pmatrix} 1 & -2 & -1 & 3 \\ 3 & 4 & 2 & 1 \\ 5 & 1 & 0 & 6 \end{pmatrix}$$

A is a 3×4 matrix. $a_{11} = 1$, $a_{13} = -1$, $a_{23} = 2$, $a_{34} = 6$, etc. ✦

An arbitrary matrix A can be referred to in terms of an arbitrary element a_{ij}, the element in the ith row and jth column. If one knows what is happening to an arbitrary element of the matrix, one knows what is happening to every element. Thus we write $A = (a_{ij})$ and handle A in the form (a_{ij}).

Let us now look at the form the operations of addition and scalar multiplication of matrices assume in this notation.

Let $A = (a_{ij})$, $B = (b_{ij})$ be arbitrary matrices of the same kind and let c be an arbitrary scalar. The sum $A + B$ is defined to be the matrix obtained by adding corresponding elements. We write

$$A + B = (a_{ij}) + (b_{ij})$$
$$= (a_{ij} + b_{ij})$$

$A + B$ is the matrix whose (i, j)th element is the sum of the (i, j)th elements of A and B.

Scalar multiplication can be expressed

$$cA = c(a_{ij})$$
$$= (ca_{ij})$$

cA is the matrix whose (i, j)th element is the scalar multiple of the (i, j)th element of A by c.

We illustrate the power of this notation in the proofs of the following two important algebraic properties.

Theorem 2-1 *Let A and B be two matrices of the same size. Then*

$$A + B = B + A$$

The order in which two matrices of the same size are added is not important. Matrices of the same size are commutative under addition.

Proof: We shall prove this result first using the longhand notation, then using the compact notation.

Let

$$A = \begin{pmatrix} a_{11} & \cdots & a_{1n} \\ & \vdots & \\ a_{m1} & \cdots & a_{mn} \end{pmatrix} \quad \text{and} \quad B = \begin{pmatrix} b_{11} & \cdots & b_{1n} \\ & \vdots & \\ b_{m1} & \cdots & b_{mn} \end{pmatrix}$$

Then

$$\begin{aligned} A + B &= \begin{pmatrix} a_{11} & \cdots & a_{1n} \\ & \vdots & \\ a_{m1} & \cdots & a_{mn} \end{pmatrix} + \begin{pmatrix} b_{11} & \cdots & b_{1n} \\ & \vdots & \\ b_{m1} & \cdots & b_{mn} \end{pmatrix} \\[2mm] &= \begin{pmatrix} a_{11} + b_{11} & \cdots & a_{1n} + b_{1n} \\ & \vdots & \\ a_{m1} + b_{m1} & \cdots & a_{mn} + b_{mn} \end{pmatrix} \\[2mm] &= \begin{pmatrix} b_{11} + a_{11} & \cdots & b_{1n} + a_{1n} \\ & \vdots & \\ b_{m1} + a_{m1} & \cdots & b_{mn} + a_{mn} \end{pmatrix} \quad \begin{array}{l} \textit{since real numbers are} \\ \textit{commutative under} \\ \textit{addition} \end{array} \\[2mm] &= B + A \end{aligned}$$

This proof is correct, but all the indices make it tedious. The reasoning used in this proof can be applied to arbitrary elements of A and B. Let $A = (a_{ij})$ and $B = (b_{ij})$. Then

$$\begin{aligned} A + B &= (a_{ij}) + (b_{ij}) \\ &= (a_{ij} + b_{ij}) \\ &= (b_{ij} + a_{ij}) \quad \textit{since real numbers are commutative under addition} \\ &= (b_{ij}) + (a_{ij}) \\ &= B + A \end{aligned}$$

✦ *Example 2*

Let $A = \begin{pmatrix} 1 & 2 \\ 3 & -1 \end{pmatrix}$ and $B = \begin{pmatrix} 4 & 5 \\ 6 & 7 \end{pmatrix}$. Then we get

$$A + B = \begin{pmatrix} 1 & 2 \\ 3 & -1 \end{pmatrix} + \begin{pmatrix} 4 & 5 \\ 6 & 7 \end{pmatrix} = \begin{pmatrix} 5 & 7 \\ 9 & 6 \end{pmatrix}$$

$$B + A = \begin{pmatrix} 4 & 5 \\ 6 & 7 \end{pmatrix} + \begin{pmatrix} 1 & 2 \\ 3 & -1 \end{pmatrix} = \begin{pmatrix} 5 & 7 \\ 9 & 6 \end{pmatrix}$$

Observe that

$$A + B = B + A$$

for these specific matrices.   ✦

> **Theorem 2-2**   *Let A, B, and C be matrices of the same size. Then*
>
> $$A + (B + C) = (A + B) + C$$

Matrices of the same size are associative under addition.

Proof: Let $A = (a_{ij})$, $B = (b_{ij})$, and $C = (c_{ij})$. Then

$$\begin{aligned} A + (B + C) &= (a_{ij}) + (b_{ij} + c_{ij}) \\ &= (a_{ij} + b_{ij} + c_{ij}) \\ &= (a_{ij} + b_{ij}) + (c_{ij}) \qquad \text{\textit{by the rule of matrix addition}} \\ &= (A + B) + C \end{aligned}$$

We denote this sum $A + B + C$. Thus

$$A + B + C = (a_{ij} + b_{ij} + c_{ij})$$

This result can be extended to add any finite number of matrices of the same size. The sum is obtained by adding corresponding elements:

$$A + B + C + \cdots + Z = (a_{ij} + b_{ij} + c_{ij} + \cdots + z_{ij})$$

✦ *Example 3*

Let $A = \begin{pmatrix} 1 & 0 \\ -2 & 3 \end{pmatrix}$, $B = \begin{pmatrix} 5 & 1 \\ 2 & 7 \end{pmatrix}$, and $C = \begin{pmatrix} 3 & 4 \\ -1 & 2 \end{pmatrix}$. Then

$$\begin{aligned} A + B + C &= \begin{pmatrix} 1 & 0 \\ -2 & 3 \end{pmatrix} + \begin{pmatrix} 5 & 1 \\ 2 & 7 \end{pmatrix} + \begin{pmatrix} 3 & 4 \\ -1 & 2 \end{pmatrix} \\ &= \begin{pmatrix} 1 + 5 + 3 & 0 + 1 + 4 \\ -2 + 2 - 1 & 3 + 7 + 2 \end{pmatrix} \\ &= \begin{pmatrix} 9 & 5 \\ -1 & 12 \end{pmatrix} \quad ✦ \end{aligned}$$

We now summarize the properties of matrices under addition and scalar multiplication. Here it is assumed that the matrices are of appropriate sizes for adding. *A*, *B*, and *C* are matrices, and *a* and *b* are scalars.

Properties of Matrices under Addition and Scalar Multiplication

1. $A + B = B + A$
2. $A + (B + C) = (A + B) + C$
3. $a(B + C) = aB + aC$
4. $(a + b)C = aC + bC$
5. $(ab)C = a(bC)$

We have already proved and discussed properties 1 and 2. Properties 3, 4, and 5 are derived using similar techniques. The reader is asked to arrive at these results in Exercise 7.

EXERCISES

*1. If

$$A = \begin{pmatrix} 1 & 2 \\ 3 & 0 \end{pmatrix}, \quad B = \begin{pmatrix} -1 & 2 \\ 1 & 1 \end{pmatrix}, \quad C = \begin{pmatrix} 0 & 1 \\ -1 & 4 \end{pmatrix}$$

determine the matrices

(a) $2A$, $3B$, $-2C$
(b) $A + B$, $B + A$, $A + C$, $B + C$
(c) $A + 2B$, $3A + C$, $2A + B - C$ (These are said to be linear combinations of the matrices *A*, *B*, and *C*.)

2. If

$$A = \begin{pmatrix} -1 & 4 & 5 \\ 0 & 1 & 2 \end{pmatrix}, \quad B = \begin{pmatrix} 1 & 1 & 0 \\ 2 & 3 & 1 \end{pmatrix}, \quad C = \begin{pmatrix} 4 & 5 & 0 \\ 2 & -1 & -1 \end{pmatrix}$$

determine

(a) $A - B$, $B + 3C$, $2B + C$
(b) $4A$, $-B$, $3C$
(c) $A + 3B$, $2A - B + C$, $A + 2B - 2C$

*3. If

$$A = \begin{pmatrix} 1 & 2 & -3 & 5 & 7 \\ 0 & 4 & 6 & -1 & 5 \\ -1 & 2 & -7 & 3 & 4 \end{pmatrix}$$

determine a_{11}, a_{24}, a_{34}, and a_{25}.

4. If

$$B = \begin{pmatrix} 2 & 2 & 5 & -7 \\ 4 & 8 & 3 & 2 \\ 6 & 4 & 2 & -1 \\ 3 & 0 & 1 & 9 \end{pmatrix}$$

determine b_{12}, b_{32}, b_{33}, b_{24}, and b_{43}.

*5. If

$$A = \begin{pmatrix} 1 & 2 & -3 \\ 0 & 4 & 5 \end{pmatrix}, \qquad B = \begin{pmatrix} 7 & 3 & 1 \\ 2 & -2 & 4 \end{pmatrix}$$

and $C = A + B$, determine c_{12}, c_{21}, and c_{23}.

6. If

$$A = \begin{pmatrix} 1 & -1 & 2 & 5 \\ 0 & 7 & 1 & 0 \\ 9 & 8 & -3 & 2 \end{pmatrix}, \qquad B = \begin{pmatrix} 1 & 0 & 5 & 2 \\ 6 & -3 & -2 & 4 \\ 2 & 5 & 1 & -3 \end{pmatrix}$$

and $C = 3A - 2B$, determine c_{13}, c_{24}, c_{33}, and c_{34}.

7. Let a and b be scalars and let B and C be matrices of the same size. Prove, using techniques similar to those used in the proofs of Theorems 2-1 and 2-2, that

*(a) $a(B + C) = aB + aC$

(b) $(a + b)C = aC + bC$

(c) $(ab)C = a(bC)$

2-2   MULTIPLICATION OF MATRICES

We have introduced addition and scalar multiplication of matrices; the next natural step in the development of a theory of matrices is the multiplication of matrices. Here again we have complete freedom in defining multiplication. The most natural way may seem to be to multiply matrices in a componentwise manner as follows:

$$\begin{pmatrix} 1 & 2 \\ -1 & 3 \end{pmatrix} \begin{pmatrix} -2 & 4 \\ 5 & -3 \end{pmatrix} = \begin{pmatrix} 1 \times -2 & 2 \times 4 \\ -1 \times 5 & 3 \times -3 \end{pmatrix} = \begin{pmatrix} -2 & 8 \\ -5 & -9 \end{pmatrix}$$

It has been found that this is not the most useful way of multiplying matrices. Mathematicians have introduced a different rule of multiplication. To lead up to this rule, we ask the reader to think about matrices in terms of rows and columns. For example, consider the matrix

$$\begin{pmatrix} 1 & 2 & -1 & 3 \\ 2 & 1 & 4 & 6 \\ 0 & 1 & 3 & 2 \end{pmatrix}$$

The first row is (1 2 −1 3); the second column is $\begin{pmatrix} 2 \\ 1 \\ 1 \end{pmatrix}$; the fourth column

is $\begin{pmatrix} 3 \\ 6 \\ 2 \end{pmatrix}$; and so on.

In the multiplication of two matrices such as

$$\begin{matrix} A & & B \end{matrix}$$
$$\begin{pmatrix} 1 & 2 \\ -1 & 3 \end{pmatrix}\begin{pmatrix} -2 & 4 \\ 5 & -3 \end{pmatrix}$$

we shall interpret the first matrix in terms of its rows and the second in terms of its columns. To get the first row of the product, C, multiply the first row of A by each column of B:

$$\begin{matrix} A & B & & C \end{matrix}$$
$$\begin{pmatrix} 1 & 2 \\ -1 & 3 \end{pmatrix}\begin{pmatrix} -2 & 4 \\ 5 & -3 \end{pmatrix} = \left((1 \ \ 2) \times \begin{pmatrix} -2 \\ 5 \end{pmatrix} \ \bigg| \ (1 \ \ 2) \times \begin{pmatrix} 4 \\ -3 \end{pmatrix} \right)$$

← *first row of C (Dotted lines will be included in examples to clarify components of the matrices.)*

To get the second row of the product, multiply the second row of A by each column of B:

$$\begin{matrix} & C \end{matrix}$$
$$= \left(\begin{array}{c|c} (1 \ \ 2) \times \begin{pmatrix} -2 \\ 5 \end{pmatrix} & (1 \ \ 2) \times \begin{pmatrix} 4 \\ -3 \end{pmatrix} \\ \hline (-1 \ \ 3) \times \begin{pmatrix} -2 \\ 5 \end{pmatrix} & (-1 \ \ 3) \times \begin{pmatrix} 4 \\ -3 \end{pmatrix} \end{array} \right)$$

It remains to define the special multiplication of the rows by the columns: multiply corresponding components and add. In the above matrix C,

$$(1 \ \ 2) \times \begin{pmatrix} -2 \\ 5 \end{pmatrix} = (1 \times (-2)) + (2 \times 5) = 8$$

$$(1 \ \ 2) \times \begin{pmatrix} 4 \\ -3 \end{pmatrix} = (1 \times 4) + (2 \times (-3)) = -2$$

and so on. Thus,

$$\begin{pmatrix} 1 & 2 \\ -1 & 3 \end{pmatrix}\begin{pmatrix} -2 & 4 \\ 5 & -3 \end{pmatrix} = \begin{pmatrix} 8 & -2 \\ 17 & -13 \end{pmatrix}$$

We now illustrate this rule with an example.

✦ *Example 1*

$$\begin{pmatrix} 0 & 1 & 2 \\ -1 & 0 & 3 \\ 0 & 4 & 1 \end{pmatrix} \begin{pmatrix} -1 & 3 \\ 2 & 1 \\ 0 & 1 \end{pmatrix}$$

$$= \begin{pmatrix} (0 \quad 1 \quad 2) \times \begin{pmatrix} -1 \\ 2 \\ 0 \end{pmatrix} & (0 \quad 1 \quad 2) \times \begin{pmatrix} 3 \\ 1 \\ 1 \end{pmatrix} \\ (-1 \quad 0 \quad 3) \times \begin{pmatrix} -1 \\ 2 \\ 0 \end{pmatrix} & (-1 \quad 0 \quad 3) \times \begin{pmatrix} 3 \\ 1 \\ 1 \end{pmatrix} \\ (0 \quad 4 \quad 1) \times \begin{pmatrix} -1 \\ 2 \\ 0 \end{pmatrix} & (0 \quad 4 \quad 1) \times \begin{pmatrix} 3 \\ 1 \\ 1 \end{pmatrix} \end{pmatrix}$$

$$= \begin{pmatrix} (0 \times -1) + (1 \times 2) + (2 \times 0) & (0 \times 3) + (1 \times 1) + (2 \times 1) \\ (-1 \times -1) + (0 \times 2) + (3 \times 0) & (-1 \times 3) + (0 \times 1) + (3 \times 1) \\ (0 \times -1) + (4 \times 2) + (1 \times 0) & (0 \times 3) + (4 \times 1) + (1 \times 1) \end{pmatrix}$$

$$= \begin{pmatrix} 2 & 3 \\ 1 & 0 \\ 8 & 5 \end{pmatrix} \quad ✦$$

Readers will appreciate the usefulness of this rule of multiplication as they proceed in the course. Later in this chapter, for example, we use it in analyzing the behavior of electrical currents, in analyzing the chronology of graves in archaeology, in predicting population distributions, and in analyzing communication networks. Let us at this time proceed to arrive at a deeper understanding of the properties of matrix multiplication so that we shall be in a position to effectively apply this tool. Consider the following three matrices:

$$A = \begin{pmatrix} 1 & 2 & -1 \\ 3 & 1 & 0 \end{pmatrix}, \quad B = \begin{pmatrix} 1 & 0 \\ 2 & 1 \\ 3 & 2 \end{pmatrix}, \quad C = \begin{pmatrix} 1 & -1 & 4 \\ 2 & 0 & 5 \end{pmatrix}$$

On applying the rule for multiplication, we obtain

$$AB = \begin{pmatrix} 1 & 2 & -1 \\ 3 & 1 & 0 \end{pmatrix} \begin{pmatrix} 1 & 0 \\ 2 & 1 \\ 3 & 2 \end{pmatrix} = \begin{pmatrix} 2 & 0 \\ 5 & 1 \end{pmatrix}$$

However, when we attempt to evaluate AC,

$$AC = \begin{pmatrix} 1 & 2 & -1 \\ 3 & 1 & 0 \end{pmatrix} \begin{pmatrix} 1 & -1 & 4 \\ 2 & 0 & 5 \end{pmatrix}$$

$$= \begin{pmatrix} (1 \times 1) + (2 \times 2) + (-1 \times ?) & \end{pmatrix}$$

Observe that we cannot perform the multiplication necessary to get even the first element of the product, since there is no corresponding element in C by which to multiply the -1 in A. It is not possible to multiply A by C according to this matrix multiplication. We say that *the product AC does not exist.*

The product AC does not exist because the number of columns in A is not equal to the number of rows in C. When the number of columns in the first matrix is equal to the number of rows in the second, the product will exist, for there will always be the corresponding elements required to perform the desired multiplication. We thus get the following general result:

The product AB of two matrices A and B exists if and only if the number of columns in A is equal to the number of rows in B.

✦ ***Example 2*** _____

Let

$$A = \begin{pmatrix} 1 & 2 \\ 3 & 4 \end{pmatrix}, \qquad B = \begin{pmatrix} 1 & 2 & 3 \\ 4 & 5 & 6 \end{pmatrix}, \qquad C = \begin{pmatrix} 5 \\ 4 \\ -3 \end{pmatrix}$$

We see that AB exists; A has two columns, and B has two rows. AC does not exist; A has two columns, and C has three rows. BC exists.   ✦

We now investigate the kind of matrix that results from a product. Since the product of two matrices A and B, AB, exists only if the number of columns in A is equal to the number of rows in B, let A be an $m \times n$ matrix and B an $n \times r$ matrix; AB then exists.

The first row of AB is obtained by multiplying the first row of A by each column of B in turn in the appropriate manner. Thus the number of columns in AB is equal to the number of columns in B. The first column of AB results from multiplying the first column of B by each row of A in turn. Thus the number of rows in AB is equal to the number of rows in A. We have the following result for the product of two matrices:

If A is an m × n matrix and B is an n × r matrix, then the product AB will be an m × r matrix.

We can picture this result as follows:

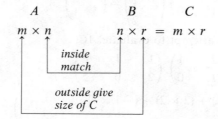

✦ **Example 3**

If A is a 5×6 matrix and B is a 6×7 matrix, what kind of matrix is AB?

First we note that A has six columns and B has six rows; thus AB exists. It will be a 5×7 matrix.

$$
\begin{array}{ccc}
A & B & C \\
5 \times 6 & 6 \times 7 = & 5 \times 7
\end{array}
$$

match

C is
5×7

Any desired element in a product matrix can be computed without calculating all the elements in the product. Let A and B be matrices for which the product AB exists.

To obtain the element in row i and column j of AB, one multiplies the ith row of A with the jth column of B in the appropriate manner.

✦ **Example 4**

If

$$
A = \begin{pmatrix} -1 & 2 \\ 3 & 4 \\ 1 & 2 \end{pmatrix} \quad \text{and} \quad B = \begin{pmatrix} 1 & 2 & 4 & -1 \\ 3 & 6 & 1 & 2 \end{pmatrix}
$$

determine the element in row 2, column 3 of AB.

The product AB will be a 3×4 matrix, since A has three rows and B has four columns.

$$
\begin{pmatrix} -1 & 2 \\ 3 & 4 \\ 1 & 2 \end{pmatrix}\begin{pmatrix} 1 & 2 & 4 & -1 \\ 3 & 6 & 1 & 2 \end{pmatrix} = \begin{pmatrix} x & x & x & x \\ x & x & \otimes & x \\ x & x & x & x \end{pmatrix}
$$

we desire
this element

We take the second row of A and the third column of B, and we get the required element:

$$
(3 \quad 4) \times \begin{pmatrix} 4 \\ 1 \end{pmatrix} = 12 + 4 = 16 \quad ✦
$$

✦ **Example 5**

Let

$$
A = \begin{pmatrix} -1 & 2 & 0 \\ 3 & 4 & 5 \\ 2 & 1 & 1 \end{pmatrix}, \quad B = \begin{pmatrix} 0 & -2 & 1 \\ 1 & 3 & 1 \\ 2 & 4 & 1 \end{pmatrix}
$$

and $C = AB$. Writing

$$C = \begin{pmatrix} c_{11} & c_{12} & c_{13} \\ c_{21} & c_{22} & c_{23} \\ c_{31} & c_{32} & c_{33} \end{pmatrix}$$

determine the elements c_{31} and c_{23}.

To obtain c_{31} we take the third row of A and first column of B:

$$c_{31} = (2 \quad 1 \quad 1) \times \begin{pmatrix} 0 \\ 1 \\ 2 \end{pmatrix} = (2 \times 0) + (1 \times 1) + (1 \times 2) = 3$$

Similarly, c_{23} is obtained using the second row of A and third column of B:

$$c_{23} = (3 \quad 4 \quad 5) \times \begin{pmatrix} 1 \\ 1 \\ 1 \end{pmatrix} = (3 \times 1) + (4 \times 1) + (5 \times 1) = 12 \quad \blacklozenge$$

Partitioning of Matrices*

A matrix may be subdivided into a number of *submatrices*. For example, the following matrix A can be subdivided into the submatrices P, Q, R, and S.

$$A = \begin{pmatrix} 0 & \vdots & -1 & 2 \\ 3 & \vdots & 1 & 4 \\ \cdots & \vdots & \cdots & \cdots \\ -2 & \vdots & 5 & -3 \end{pmatrix} = \begin{pmatrix} P & Q \\ R & S \end{pmatrix}$$

where $P = \begin{pmatrix} 0 \\ 3 \end{pmatrix}$, $Q = \begin{pmatrix} -1 & 2 \\ 1 & 4 \end{pmatrix}$, $R = (-2)$, and $S = (5, -3)$.

Such subdivisions are used both in computations and in theoretical discussions involving matrices.

Provided appropriate rules are followed, the matrix operations previously described can be applied to submatrices as if they were elements of an ordinary matrix.

Let us look at addition. Let A and B be matrices of the same kind. If A and B are partitioned in the same way, their sum is the sum of the corresponding submatrices. For example,

$$A + B = \begin{pmatrix} P & Q \\ R & S \end{pmatrix} + \begin{pmatrix} H & I \\ J & K \end{pmatrix} = \begin{pmatrix} P + H & Q + I \\ R + J & S + K \end{pmatrix}$$

In multiplication, any column partition of the first matrix in a product determines the row partition of the second matrix. For example, let us consider the product AB of the matrices

$$A = \begin{pmatrix} 1 & 2 & -1 \\ 3 & 0 & -2 \\ 4 & -3 & 2 \end{pmatrix} \quad \text{and} \quad B = \begin{pmatrix} -1 & 0 \\ 2 & 1 \\ 5 & 4 \end{pmatrix}$$

Let A be subdivided

$$\left(\begin{array}{cc|c} 1 & 2 & -1 \\ \hline 3 & 0 & -2 \\ 4 & -3 & 2 \end{array}\right) = \begin{pmatrix} P & Q \\ R & S \end{pmatrix}$$

A is interpreted as having two columns in this form. B must be subdivided into a suitable form having two rows for multiplication to be possible.

$$B = \left(\begin{array}{cc} -1 & 0 \\ 2 & 1 \\ \hline 5 & 4 \end{array}\right) = \begin{pmatrix} M \\ N \end{pmatrix}$$

would be a suitable partition for B.

$$AB = \begin{pmatrix} P & Q \\ R & S \end{pmatrix}\begin{pmatrix} M \\ N \end{pmatrix} = \begin{pmatrix} PM + QN \\ RM + SN \end{pmatrix}$$

If A is subdivided into a form having q columns, to obtain a product AB, one must subdivide B into an appropriate form having q rows. There is freedom in the column subdivision of B. If the partition of A is a $p \times q$ form and B is a $q \times s$ form, AB will be partitioned into p rows and s columns.

✦ **Example 6** _____

Let

$$A = \begin{pmatrix} 1 & -1 \\ 3 & 0 \\ 2 & 4 \end{pmatrix} \quad \text{and} \quad B = \begin{pmatrix} 1 & 2 & -1 \\ 1 & 3 & 1 \end{pmatrix}$$

Consider the partition of A

$$\left(\begin{array}{c|c} 1 & -1 \\ 3 & 0 \\ \hline 2 & 4 \end{array}\right) = \begin{pmatrix} P & Q \\ R & S \end{pmatrix}$$

A under this partition is interpreted as a 2×2 matrix. For the product to exist, B must be appropriately partitioned into a matrix having 2 rows, $\begin{pmatrix} H \\ J \end{pmatrix}$. The number of columns of P and Q determine the number of rows of H and J, since PH and QJ must exist:

$$AB = \begin{pmatrix} P & Q \\ R & S \end{pmatrix}\begin{pmatrix} H \\ J \end{pmatrix} = \begin{pmatrix} PH + QJ \\ RH + SJ \end{pmatrix}$$

Thus an appropriate partition for B is $\left(\begin{array}{c|cc} 1 & 2 & -1 \\ \hline 1 & 3 & 1 \end{array}\right)$, where B is interpreted as a 2×1 matrix.

Under these partitions, A is 2×2, B is 2×1, and AB is 2×1. There are other possible partitions of B; can you give them, and also the corresponding partitions of AB in each case?   ✦

These techniques with submatrices are particularly appropriate for work involving large matrices that has to be carried out on a computer. Computers can only handle matrices containing up to a certain number of elements. Larger matrices can, however, be accommodated in terms of submatrices. For example, the PET (Commodore 2001 Series microcomputer) can handle, in matrix form, matrices consisting of up to 255 elements. Suppose we want to multiply matrices A and B, where A is a 10×5 matrix (50 elements) and B is a 5×40 matrix (200 elements). The computer will accept A and B; however, their product AB will be a 10×40 matrix (400 elements) and the computer cannot calculate the product as it stands. Let us use the following suitable subdivisions of A and B:

$$A = \begin{pmatrix} P & | & Q \\ 10 \times 2 & & 10 \times 3 \end{pmatrix}, \quad B = \begin{pmatrix} H & | & I \\ 2 \times 20 & | & 2 \times 20 \\ \hline J & | & K \\ 3 \times 20 & | & 3 \times 20 \end{pmatrix}$$

Then

$$AB = (P \mid Q)\left(\frac{H \mid I}{J \mid K}\right) = \begin{pmatrix} PH + QJ & | & PI + QK \\ 10 \times 20 & 10 \times 20 & 10 \times 20 & 10 \times 20 \end{pmatrix}$$

Both matrices $PH + QJ$ and $PI + QK$ are 10×20 matrices consisting of 200 elements each. These can be calculated on the computer separately, to give the product AB.

EXERCISES

*1. Let $A = \begin{pmatrix} 0 & 1 \\ 2 & 3 \end{pmatrix}$, $B = \begin{pmatrix} 4 \\ 5 \end{pmatrix}$, $C = \begin{pmatrix} -1 & 3 \\ 2 & 1 \end{pmatrix}$, and $D = \begin{pmatrix} 1 & 3 & -1 \\ 2 & 4 & 1 \end{pmatrix}$. Determine the following products if they exist: AB, AC, AD, BA, DB, BD, A^2, C^3, $AC + CA$. ($A^2 = AA$, etc.) Predict the size of matrix that will result, beforehand, in each case.

2. Let $A = \begin{pmatrix} -1 & 0 \\ 2 & 1 \\ 3 & 4 \end{pmatrix}$, $B = \begin{pmatrix} 1 \\ 2 \\ 3 \end{pmatrix}$, $C = (-1, 2, 3)$, and $D = (1, 3)$. Determine the following products if they exist: AB, BC, CB, AD, DA, CA, BD. Predict the size of matrix that will result, beforehand, in each case.

*3. If $A = \begin{pmatrix} 0 & 1 \\ 2 & 4 \end{pmatrix}$, $B = \begin{pmatrix} -1 & 3 \\ 2 & 1 \end{pmatrix}$, $C = \begin{pmatrix} 1 \\ 3 \end{pmatrix}$, and $D = \begin{pmatrix} 1 & 2 & -1 \\ 0 & 1 & 3 \end{pmatrix}$, compute the following if they exist:

(a) $AB + BA$ (b) $AC + BC$ (c) $AD - 3(BD)$
(d) $3(DC) + (4A)C$ (e) $AC + 2(BD)$

*4. If $A = \begin{pmatrix} -1 & 0 & 2 \\ 3 & 4 & 5 \\ 1 & 2 & 0 \end{pmatrix}$ and $B = \begin{pmatrix} 1 & -1 \\ 3 & -2 \\ 3 & -2 \end{pmatrix}$, determine the elements

(a) in the third row and first column of the product matrix AB

(b) in the second row and second column of the product matrix AB

5. If $A = \begin{pmatrix} 1 & 3 & 0 \\ 2 & 2 & 1 \\ 3 & 1 & 0 \end{pmatrix}$, $B = \begin{pmatrix} -1 & 0 & -1 \\ -2 & 1 & 1 \\ 0 & 1 & 2 \end{pmatrix}$, and

$$C = \begin{pmatrix} c_{11} & c_{12} & c_{13} \\ c_{21} & c_{22} & c_{23} \\ c_{31} & c_{32} & c_{33} \end{pmatrix} = AB$$

determine c_{21}, c_{22}, and c_{33} without evaluating the whole product matrix.

6. Let A be a 3×5 matrix, B a 5×2 matrix, C a 3×4 matrix, D a 4×2 matrix, and E a 4×5 matrix. Determine which of the following exist and give the size of the resulting matrix when it does exist.

*(a) $AB + C$ (b) $AB + CD$ *(c) $3(EB) + 4D$

(d) $CD - 2(CE)B$ *(e) $2(EB) + DA$

7. (a) Let A be an $n \times n$ matrix. Prove that A^2 is $n \times n$.

(b) Let A be an $m \times n$ matrix with $m \neq n$. Prove that A^2 does not exist. Thus one can only talk about powers of a square matrix.

8. Consider two matrices P and Q such that $PQ = 0$, the zero matrix of the appropriate kind. (A zero matrix is one whose elements are all zeros.) Does this imply that either $P = 0$ or $Q = 0$? (*Hint*: Construct an example.)

*9. Consider three matrices A, B, and C such that $AC = BC$. Does this imply that $A = B$?

10. Let $A = \begin{pmatrix} 1 & 2 & 3 \\ -1 & 1 & 4 \\ 0 & 1 & 2 \end{pmatrix}$ and $B = \begin{pmatrix} -1 & -2 \\ 0 & 3 \\ 4 & 1 \end{pmatrix}$. For each partition of A given below find all partitions of B that can be used for calculating AB. Give the corresponding dimensions of AB in each case.

*(a) $\left(\begin{array}{c|cc} 1 & 2 & 3 \\ \hline -1 & 1 & 4 \\ 0 & 1 & 2 \end{array} \right)$ (b) $\left(\begin{array}{c|cc} 1 & 2 & 3 \\ -1 & 1 & 4 \\ \hline 0 & 1 & 2 \end{array} \right)$ (c) $\left(\begin{array}{c|c|c} 1 & 2 & 3 \\ -1 & 1 & 4 \\ \hline 0 & 1 & 2 \end{array} \right)$

11. Let A and B be $n \times n$ matrices.

*(a) Partition A into n row matrices $A_1, \ldots, A_n$, keeping B as it is. Determine the product AB in terms of submatrices.

(b) Partition B into n column matrices $B^1, \ldots, B^n$, keeping A unchanged. Determine AB in terms of submatrices.

*(c) Partition A into n row matrices $A_1, \ldots, A_n$ and B into n column matrices $B^1, \ldots, B^n$. Determine AB.

(d) Partition A into n column matrices $A^1, \ldots, A^n$ and B into n row matrices $B_1, \ldots, B_n$. Does AB exist for these partitions?

Later, we find that partitions of matrices into row and column matrices are very useful.

2-3 PROPERTIES OF MATRICES UNDER MULTIPLICATION

We have found that matrices are both commutative and associative under addition. We shall now discuss these concepts under multiplication.

Consider the matrices

$$A = \begin{pmatrix} 1 & 2 \\ -1 & 0 \end{pmatrix} \quad \text{and} \quad B = \begin{pmatrix} 3 & 1 \\ 1 & 2 \end{pmatrix}$$

We compute AB and BA.

$$AB = \begin{pmatrix} 1 & 2 \\ -1 & 0 \end{pmatrix}\begin{pmatrix} 3 & 1 \\ 1 & 2 \end{pmatrix} = \begin{pmatrix} 5 & 5 \\ -3 & -1 \end{pmatrix}$$

$$BA = \begin{pmatrix} 3 & 1 \\ 1 & 2 \end{pmatrix}\begin{pmatrix} 1 & 2 \\ -1 & 0 \end{pmatrix} = \begin{pmatrix} 2 & 6 \\ -1 & 2 \end{pmatrix}$$

We find that $AB \neq BA$. Thus:

Matrices are not commutative under multiplication.

The order of multiplication is important for matrices, unlike for real numbers. There will be special cases of matrices A and B where $AB = BA$, but in general this result does not hold.

We now give the result for the associative property of matrices under multiplication, without proof.

Theorem 2-3   *If A, B, and C are matrices of suitable sizes so that the indicated products exist,*

$$A(BC) = (AB)C$$

Matrices are associative under multiplication.

A significance of this result is that an expression ABC (without brackets) can have meaning. We define the product of three matrices by $ABC = A(BC) = (AB)C$. In such a product it does not matter which multiplication is carried out first, BC or AB. We can picture the matching of the rows and columns of the various matrices as follows. Let A be an $m \times n$ matrix, B $n \times r$ and C $r \times p$.

This result can be extended to the product of any finite number of matrices, $ABCD \cdots$; this product is well defined without brackets indicating the order of multiplication.

✦ Example 1

Let $A = \begin{pmatrix} 1 & 2 \\ 0 & 1 \end{pmatrix}$, $B = \begin{pmatrix} 1 & 3 \\ -1 & 2 \end{pmatrix}$, $C = \begin{pmatrix} 0 & 1 \\ -1 & 3 \end{pmatrix}$, and $D = \begin{pmatrix} 2 \\ 3 \end{pmatrix}$. Determine $ABCD$.

We see that rows and columns match in the following way; the product exists and is a 2×1 matrix.

$$
\begin{array}{cccc}
A & B & C & D \\
2 \times 2 & 2 \times 2 & 2 \times 2 & 2 \times 1
\end{array}
$$

product will be a 2×1 matrix

We get

$$AB = \begin{pmatrix} 1 & 2 \\ 0 & 1 \end{pmatrix}\begin{pmatrix} 1 & 3 \\ -1 & 2 \end{pmatrix} = \begin{pmatrix} -1 & 7 \\ -1 & 2 \end{pmatrix}$$

$$ABC = \begin{pmatrix} -1 & 7 \\ -1 & 2 \end{pmatrix}\begin{pmatrix} 0 & 1 \\ -1 & 3 \end{pmatrix} = \begin{pmatrix} -7 & 20 \\ -2 & 5 \end{pmatrix}$$

$$ABCD = \begin{pmatrix} -7 & 20 \\ -2 & 5 \end{pmatrix}\begin{pmatrix} 2 \\ 3 \end{pmatrix} = \begin{pmatrix} 46 \\ 11 \end{pmatrix}$$

The matrices in such a product can be grouped together in any manner for the actual computation, as long as the order is maintained. For example, in this product, we could have computed BC first to get $A(BC)D$, and so on.

✦

We now introduce the *sigma notation*, a tool that is useful in theoretical work involving matrix multiplication. Let $f(k)$ be an algebraic expression involving the variable k. For values of $k = 1, 2, 3, \ldots$, $f(k)$ assumes the values $f(1), f(2), f(3), \ldots$. The notation $\sum_{k=1}^{n} f(k)$ is used to denote the sum of the first n positive integer values of $f(k)$:

$$\sum_{k=1}^{n} f(k) = f(1) + f(2) + f(3) + \cdots + f(n)$$

Very often, in theoretical work, as we shall write $\sum_{k} f(k)$ to signify summation over the relevant values of k. The letter k here is a *dummy variable* that can be replaced by any letter we like. We now give some examples illustrating the use of this notation.

$$\sum_{k=1}^{5} k^2 = 1^2 + 2^2 + 3^2 + 4^2 + 5^2$$

$$\sum_{k=1}^{6} a_{2k} = a_{21} + a_{22} + a_{23} + a_{24} + a_{25} + a_{26}$$

(This last expression could be interpreted as the sum of the elements in the second row of a matrix A which has six columns.)

$$\sum_{j=1}^{3} (a_j + b_j) = a_1 + b_1 + a_2 + b_2 + a_3 + b_3$$

Let us now use the sigma notation in a convenient way to express matrix multiplication. Let A be an $m \times n$ matrix and B an $n \times r$ matrix. Then we know that the product $C = AB$ with be an $m \times r$ matrix. The general element c_{ij} of C is obtained by taking the product of the ith row of A and the jth column of B.

$$c_{ij} = (a_{i1} \quad a_{i2} \quad \cdots \quad a_{in}) \times \begin{pmatrix} b_{1j} \\ b_{2j} \\ \vdots \\ b_{nj} \end{pmatrix} = a_{i1}b_{1j} + a_{i2}b_{2j} + \cdots + a_{in}b_{nj}$$

Observe that in this sum each term is of the form $a_{ik}b_{kj}$, with k varying from 1 to n. We can express c_{ij} in the compact form

$$c_{ij} = \sum_{k=1}^{n} a_{ik}b_{kj}$$

The (i, j)th element of the product AB is $\sum_{k=1}^{n} a_{ik}b_{kj}$. We can thus express the product of two matrices in terms of an arbitrary element in the following way:

$$AB = \left(\sum_{k=1}^{n} a_{ik}b_{kj} \right)$$

✦ *Example 2*

If A is a 5×6 matrix and B is a 6×7 matrix, give an expression in sigma notation for the element in the fourth row and second column of AB.

The element in the ith row and jth column of AB is $\sum_{k=1}^{6} a_{ik}b_{kj}$. The summation here is from 1 to 6, as this is the number of columns in A and rows in B. For the element in the fourth row and second column of AB, $i = 4, j = 2$. Hence this element is $\sum_{k=1}^{6} a_{4k}b_{k2}$.   ✦

The following theorem tells us how the operations of addition and multiplication interact for matrices.

Theorem 2-4   *If A, B, and C are matrices of suitable sizes, so that the indicated operations can be performed, then*

$$A(B + C) = AB + AC$$
$$(B + C)A = BA + CA$$

This property of matrices is called the distributive property.

Proof: Let $A = (a_{ij})$, $B = (b_{ij})$, and $C = (c_{ij})$. Then $B + C = (b_{ij} + c_{ij})$.

$$A(B + C) = \left(\sum_k a_{ik}(b_{kj} + c_{kj}) \right)$$

$$= \left(\sum_k (a_{ik}b_{kj} + a_{ik}c_{kj}) \right) \quad \text{since real numbers are distributive}$$

$$= \left(\sum_k a_{ik}b_{kj} \right) + \left(\sum_k a_{ik}c_{kj} \right)$$

$$= AB + AC$$

The proof of the second part of the theorem is similar.

The third and final result involving matrix multiplication is now given.

Theorem 2-5   *Let a be a scalar and let B and C be matrices such that the product BC exists. Then*

$$a(BC) = (aB)C = B(aC)$$

Proof: We leave the proof of this theorem as an exercise for the reader.

We express $a(BC)$ as aBC, since it does not matter whether we perform the scalar multiplication or the matrix multiplication first.

✦ *Example 3*

Compute $3\begin{pmatrix} 1 & 2 \\ 4 & 7 \end{pmatrix}\begin{pmatrix} 7 & -1 \\ 0 & 2 \end{pmatrix}$ in two different ways.

We get

$$3\left[\begin{pmatrix} 1 & 2 \\ 4 & 7 \end{pmatrix}\begin{pmatrix} 7 & -1 \\ 0 & 2 \end{pmatrix} \right] = 3\begin{pmatrix} 7 & 3 \\ 28 & 10 \end{pmatrix} = \begin{pmatrix} 21 & 9 \\ 84 & 30 \end{pmatrix}$$

and

$$\left[3\begin{pmatrix} 1 & 2 \\ 4 & 7 \end{pmatrix} \right]\begin{pmatrix} 7 & -1 \\ 0 & 2 \end{pmatrix} = \begin{pmatrix} 3 & 6 \\ 12 & 21 \end{pmatrix}\begin{pmatrix} 7 & -1 \\ 0 & 2 \end{pmatrix} = \begin{pmatrix} 21 & 9 \\ 84 & 30 \end{pmatrix} \quad ✦$$

Let us for convenience summarize these algebraic properties of matrices under multiplication.

Properties of Matrices under Multiplication

1. $A(BC) = (AB)C$
2. $A(B + C) = AB + AC$ and $(B + C)A = BA + CA$
3. $a(BC) = (aB)C = B(aC)$

We now give an example which illustrates the importance of matrix algebra in the theory of electrical circuits. Furthermore, this example gives us a physical interpretation for the associativity and noncommutativity of matrices under multiplication.

✦ *Example 4*

Many networks are designed to accept signals at certain points and to deliver a modified version of the signals. The usual arrangement at such locations is illustrated in Figure 2-1. A current i_1 at voltage V_1 is delivered into the black box, and it in some way determines the current i_2 at voltage V_2. The term *black box* is used to imply that very often one is ignorant of the actual structure of the interior of the two-port. The two-port may, for example, be a transistor. The only assumption we shall make concerning the terminal is that it behaves in a linear manner—the input and output currents and voltages are related by the expression

$$\begin{pmatrix} V_2 \\ i_2 \end{pmatrix} = \begin{pmatrix} a_{11} & a_{12} \\ a_{21} & a_{22} \end{pmatrix} \begin{pmatrix} V_1 \\ i_1 \end{pmatrix}$$

This is very often the case in practice.

The matrix $\begin{pmatrix} a_{11} & a_{12} \\ a_{21} & a_{22} \end{pmatrix}$ is called the *transmission matrix.*

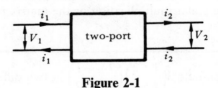

Figure 2-1

Figure 2-2 is an example of a two-port. The interior consists of a resistance R, connected as shown. We shall illustrate that the currents and voltages do indeed behave in the linear manner described above. *Ohm's law* describes the behavior of currents in such circuits; it tells us that the voltage drop across any portion of a circuit is equal to the current multiplied by the resistance of that part. $V_1 = V_2$, since the terminals are connected directly. Applying Ohm's law to the part AB of the circuit, we see that the voltage drop across AB must be V_1 (voltage drops occur across resistances). The current through AB is $i_1 - i_2$ in the direction A to B. Thus $V_1 = (i_1 - i_2)\, R$.

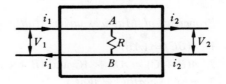

Figure 2-2

We have the two equations

$$V_1 = V_2$$
$$V_1 = (i_1 - i_2) R$$

These two equations can be rearranged in the form

$$V_2 = V_1 + 0i_1$$

$$i_2 = -\frac{1}{R} V_1 + i_1$$

to give the matrix equation

$$\begin{pmatrix} V_2 \\ i_2 \end{pmatrix} = \begin{pmatrix} 1 & 0 \\ -\dfrac{1}{R} & 1 \end{pmatrix} \begin{pmatrix} V_1 \\ i_1 \end{pmatrix}$$

Thus the transmission matrix is $\begin{pmatrix} 1 & 0 \\ -\dfrac{1}{R} & 1 \end{pmatrix}$.

We now give illustrations of three other common two-ports and their transmission matrices. These matrices can be arrived at by applying Ohm's law in a manner similar to that used in the previous illustration. The matrices are

$$\begin{pmatrix} 1 & -R \\ 0 & 1 \end{pmatrix} \quad \textit{for Figure 2-3}$$

Figure 2-3

$$\begin{pmatrix} 1 + (R_1/R_2) & -R_1 \\ -1/R_2 & 1 \end{pmatrix} \quad \textit{for Figure 2-4}$$

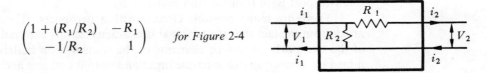

Figure 2-4

$$\begin{pmatrix} 1 & -R_1 \\ -1/R_2 & 1 + (R_1/R_2) \end{pmatrix} \quad \textit{for Figure 2-5}$$

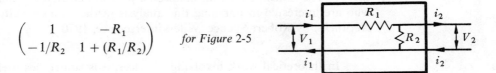

Figure 2-5

If a number of two-ports are placed in series as in Figure 2-6, matrix operations make the dependence of the output current and voltage on the input immediately apparent.

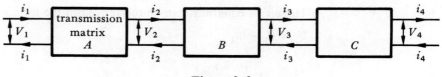

Figure 2-6

Let the transmission matrices of the two-ports be A, B, and C. Considering each port separately, we have

$$\begin{pmatrix} V_2 \\ i_2 \end{pmatrix} = A \begin{pmatrix} V_1 \\ i_1 \end{pmatrix}, \qquad \begin{pmatrix} V_3 \\ i_3 \end{pmatrix} = B \begin{pmatrix} V_2 \\ i_2 \end{pmatrix}, \qquad \begin{pmatrix} V_4 \\ i_4 \end{pmatrix} = C \begin{pmatrix} V_3 \\ i_3 \end{pmatrix}$$

Combining the first two equations by substituting the value of $\begin{pmatrix} V_2 \\ i_2 \end{pmatrix}$ from the first equation into the second equation gives

$$\begin{pmatrix} V_3 \\ i_3 \end{pmatrix} = BA \begin{pmatrix} V_1 \\ i_1 \end{pmatrix}$$

Then substituting this value of $\begin{pmatrix} V_3 \\ i_3 \end{pmatrix}$ into the third equation gives

$$\begin{pmatrix} V_4 \\ i_4 \end{pmatrix} = CBA \begin{pmatrix} V_1 \\ i_1 \end{pmatrix}$$

Note that the order of matrix multiplication is important, for matrices are not commutative under multiplication. The physical interpretation of this is that, in general, the order of the two-ports in the circuit is important. Since CBA is a matrix, we see that the three individual two-ports have been combined to give one two-port with transmission matrix CBA.

There are many possible circuits within two-ports. The name *two-port* arises from the fact that the terminal is characterized by external measurements of the two ports. The four elements of the transmission matrix can be determined by measuring two separate input and output currents and voltages. (The reader is asked to determine the transmission matrix from such data in the exercises.) Three-ports and, in general, *n*-ports are analyzed similarly in network theory. The transmission matrix for an *n*-port will be an $n \times n$ matrix. Readers who are interested in pursuing this analysis further can consult *Linear Active Networks*, by Robert Spence, Wiley-Interscience, 1970. ✦

In theoretical work involving matrices, it is sometimes useful to express a matrix in terms of its columns. Let $A_1, \ldots, A_p$ be the column *submatrices* of a matrix A. For example, the column submatrices of the matrix

$$A = \begin{pmatrix} 1 & 2 & 3 & 5 \\ -1 & 4 & 6 & -3 \\ 1 & 2 & 0 & 2 \end{pmatrix}$$

are

$$A_1 = \begin{pmatrix} 1 \\ -1 \\ 1 \end{pmatrix}, \qquad A_2 = \begin{pmatrix} 2 \\ 4 \\ 2 \end{pmatrix}, \qquad A_3 = \begin{pmatrix} 3 \\ 6 \\ 0 \end{pmatrix}, \qquad A_4 = \begin{pmatrix} 5 \\ -3 \\ 2 \end{pmatrix}$$

We write $A = (A_1 \quad A_2 \quad A_3 \quad A_4)$.

There are a number of equivalent ways of multiplying matrices. The following is a useful way of expressing the product of two matrices in terms of the column submatrices of the second matrix. We shall, for example, use this result in arriving at a method for computing the inverse of a matrix later in the course.

> **Theorem 2-6**   Let B be an $n \times r$ matrix with column submatrices $B_1, \ldots, B_r$. If A is an $m \times n$ matrix, then the product AB can be expressed
>
> $$AB = (AB_1 \quad \cdots \quad AB_r)$$
>
> That is, the columns of the product matrix are AB_1, AB_2, etc.

Proof: Let us compute the jth column of AB. The first element in the jth column of AB is the product of the first row of A and the jth column of B. It is $\sum_k a_{1k}b_{kj}$. The second element in the jth column of AB is the product of the second row of A and the jth column of B; $\sum_k a_{2k}b_{kj}$. We find that the jth column of AB is

$$\begin{pmatrix} \sum_k a_{1k}b_{kj} \\ \sum_k a_{2k}b_{kj} \\ \vdots \\ \sum_k a_{nk}b_{kj} \end{pmatrix}$$

This is the matrix AB_j, where B_j is the jth column submatrix of B. Thus

$$AB = (AB_1 \quad \cdots \quad AB_r)$$

✦ Example 5 ──

Let us illustrate the theorem for

$$A = \begin{pmatrix} 1 & 2 & -1 \\ 3 & 0 & 4 \end{pmatrix} \quad \text{and} \quad B = \begin{pmatrix} 1 & 0 & 4 \\ 2 & 3 & 1 \\ -1 & 2 & -1 \end{pmatrix}$$

We get

$$AB = \begin{pmatrix} 1 & 2 & -1 \\ 3 & 0 & 4 \end{pmatrix} \begin{pmatrix} 1 & 0 & 4 \\ 2 & 3 & 1 \\ -1 & 2 & -1 \end{pmatrix}$$

$$= \left(\begin{pmatrix} 1 & 2 & -1 \\ 3 & 0 & 4 \end{pmatrix} \begin{pmatrix} 1 \\ 2 \\ -1 \end{pmatrix} \quad \begin{pmatrix} 1 & 2 & -1 \\ 3 & 0 & 4 \end{pmatrix} \begin{pmatrix} 0 \\ 3 \\ 2 \end{pmatrix} \quad \begin{pmatrix} 1 & 2 & -1 \\ 3 & 0 & 4 \end{pmatrix} \begin{pmatrix} 4 \\ 1 \\ -1 \end{pmatrix} \right)$$

$$= \begin{pmatrix} 6 & 4 & 7 \\ -1 & 8 & 8 \end{pmatrix}$$

The reader can verify that the product of these two matrices, computed in the standard elementwise manner, is indeed this matrix.   ✦

EXERCISES

*1. Compute the products AB and BA for the matrices

$$A = \begin{pmatrix} 1 & 3 \\ -1 & 0 \end{pmatrix} \quad \text{and} \quad B = \begin{pmatrix} 2 & 4 \\ 5 & 1 \end{pmatrix}$$

Observe that $AB \neq BA$, illustrating that matrices are not commutative under multiplication.

2. (a) Compute the products AB and BA if they exist for the matrices

$$A = \begin{pmatrix} 1 & 2 \\ 3 & -1 \end{pmatrix} \quad \text{and} \quad B = \begin{pmatrix} 2 & 0 & 4 \\ -2 & 5 & 1 \end{pmatrix}$$

Observe again that $AB \neq BA$.

(b) Let A be an $m \times n$ matrix. Prove that AB and BA will both exist only if B is an $n \times m$ matrix.

*3. Compute the products AB and BA for the matrices

$$A = \begin{pmatrix} 1 & 2 \\ -1 & 0 \end{pmatrix} \quad \text{and} \quad B = \begin{pmatrix} 2 & -2 \\ 1 & 3 \end{pmatrix}$$

Observe that the products are equal. In certain cases we have $AB = BA$, but this is the exception rather than the rule.

4. Compute the products $A(BC)$ and $(AB)C$ for the matrices

$$A = \begin{pmatrix} 1 & 2 \\ -1 & 0 \\ 1 & 1 \end{pmatrix}, \quad B = \begin{pmatrix} 2 & 4 \\ -2 & 3 \end{pmatrix}, \quad C = \begin{pmatrix} 1 \\ 2 \end{pmatrix}$$

Observe that these products are equal, illustrating the associative property of matrices under multiplication.

*5. Compute the product ABC for the following matrices in two distinct ways:

$$A = \begin{pmatrix} 1 & 2 \\ -1 & 3 \end{pmatrix}, \qquad B = \begin{pmatrix} 3 & 1 & 2 \\ 4 & 3 & 1 \end{pmatrix}, \qquad C = \begin{pmatrix} 1 & 2 \\ 3 & 4 \\ 1 & 0 \end{pmatrix}$$

6. Given that A is 4×2, B is 2×6, C is 3×4, and D is 6×3, determine the sizes of the following products, if they exist.

    *(a) ABC **(b)** ABD ***(c)** CAB **(d)** $DCAB$ ***(e)** $ABDC$

7. If P is 3×2, Q is 2×1, R is 1×3, S is 3×1, and T is 3×3, predict the sizes of the following products and sums, if they exist.

    (a) PQR **(b)** $PQ + TPQ$ **(c)** $5QR + 2TPR$

    (d) $4SPQ + 3PQ$ **(e)** $QRSR + QR$

8. Write out the following in full, and interpret in terms of matrix products.

    *(a) $\sum\limits_{k=1}^{4} a_{4k}b_{k3}$ **(b)** $\sum\limits_{k=1}^{5} a_{2k}b_{k6}$ ***(c)** $\sum\limits_{k=1}^{3} a_{3k}b_{k1}$

    (d) $\sum\limits_{j=1}^{4} p_{2j}q_{j5}$ ***(e)** $\sum\limits_{s=1}^{3} p_{1s}q_{s4}$ **(f)** $\sum\limits_{k=1}^{2} a_{6k}b_{k2}$

9. If A and B are both 5×5 matrices, use sigma notation to express the elements in

    *(a) the first row and first column of AB

    (b) the second row and third column of AB

    *(c) the fourth row and fifth column of AB

    (d) the fifth row and second column of AB

10. If

$$A = \begin{pmatrix} 0 & 2 & -1 \\ 3 & 1 & 4 \end{pmatrix} \quad \text{and} \quad B = \begin{pmatrix} 1 & 2 & 3 & 0 \\ -1 & 1 & -2 & 0 \\ 0 & 3 & 1 & 4 \end{pmatrix}$$

compute $AB_1, \ldots, AB_4$ and show that $AB = (AB_1 \quad \cdots \quad AB_4)$.

11. Let A and B be square matrices of the same kind. Prove that in general

$$(A + B)^2 \neq A^2 + 2AB + B^2$$

Under what condition is $(A + B)^2 = A^2 + 2AB + B^2$?

12. If A and B are square matrices of the same kind and are such that $AB = BA$, prove that

$$(AB)^2 = A^2B^2$$

By constructing an example, show that this identity does not hold in general for all square matrices of the same kind.

13. Let A be an arbitrary square matrix. Prove that $A^r A^s = A^{r+s}$ and $(A^r)^s = A^{rs}$ for all nonnegative integer values of r and s.

14. If P and Q are 4×6 and 6×7 matrices respectively, use sigma notation to express the elements in

    *(a) the second row and third column of PQ

    (b) the fourth row and sixth column of PQ

*(c) the first row and seventh column of PQ

(d) the second row and first column of PQ

15. Determine the transmission matrices for the composite two-ports given in Figure 2-7. They are combinations of two-ports discussed in Example 4 of this section. Note that the transmission matrices differ—this is a physical interpretation of the fact that matrices are not commutative under multiplication. (The transmission matrices of the individual two-ports are given in Example 4.)

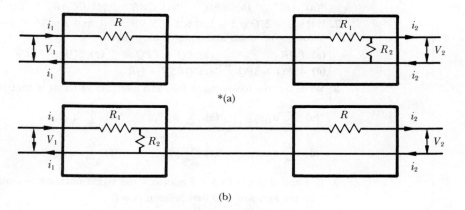

*(a)

(b)

Figure 2-7

16. The two-port in Figure 2-8 consists of three two-ports placed in a series. The transmission matrices are indicated. What is the transmission matrix of the composite two-port? If the input voltage is 2 volts and current is 3 amps, determine the output voltage and current.

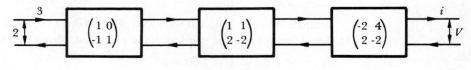

Figure 2-8

†17. Write a computer program that can be used for evaluating the transmission matrix of a composite two-port such as the one in Exercise 16 and for determining output voltage and current given certain inputs.

18. The internal circuits of the two-port in Figure 2-9 are not known. External measurements of certain currents and voltages are

*(a)

V_1 volts	i_1 amps	V_2 volts	i_2 amps
3	2	3	1
6	5	6	3

† Exercises on the computer have been included. These are all optional.

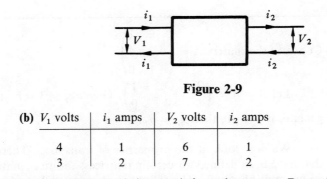

Figure 2-9

(b)

V_1 volts	i_1 amps	V_2 volts	i_2 amps
4	1	6	1
3	2	7	2

Determine the transmission matrix for each two-port. Determine the output voltage and currents for each two-port when the inputs are $V_1 = 1$ volt, $i_1 = 3$ amps.

19. Write a computer program that can be used to determine transmission matrices when certain input and output voltages and currents are known, as in Exercise 18. Using your program, check the results of Exercise 18.

2-4 SPECIAL CLASSES OF MATRICES AND AN APPLICATION IN ARCHAEOLOGY

We now examine certain relations between matrices and define special types of matrices.

If A is an $m \times n$ matrix, the *transpose* of A, denoted by A^t, is the $n \times m$ matrix obtained by writing the rows of A as columns. The first row of A becomes the first column of A^t, the second row of A becomes the second column of A^t, etc.

✦ Example 1

Determine the transposes of the following matrices, A, B, and C:

$$A = \begin{pmatrix} 0 & -2 \\ 1 & 3 \end{pmatrix}, \quad B = \begin{pmatrix} 1 & 2 & 1 \\ 3 & -4 & 1 \end{pmatrix}, \quad C = \begin{pmatrix} 1 \\ -2 \\ 3 \end{pmatrix}$$

A is a 2×2 matrix; A^t is the 2×2 matrix $\begin{pmatrix} 0 & 1 \\ -2 & 3 \end{pmatrix}$. B is a 2×3 matrix; B^t is the 3×2 matrix $\begin{pmatrix} 1 & 3 \\ 2 & -4 \\ 1 & 1 \end{pmatrix}$. C is a 3×1 matrix; C^t is the 1×3 matrix $(1 \quad -2 \quad 3)$. ✦

In determination of the transpose of a matrix A, the element that was in row i and column j of A goes to row j and column i of A^t. We illustrate this property, which is used in theoretical work involving transposes, with the following example.

✦ *Example 2*

Consider the matrix $A = \begin{pmatrix} 2 & 1 \\ 3 & 4 \\ -1 & 0 \end{pmatrix}$.

Let $B = A^t = \begin{pmatrix} 2 & 3 & -1 \\ 1 & 4 & 0 \end{pmatrix}$. Then $a_{21} = 3 = b_{12}$; $a_{32} = 0 = b_{23}$, etc. In general, if $B = A^t$, then $a_{ij} = b_{ji}$. ✦

We now look at the properties of transpose. There are three operations that have been defined for certain matrices: addition, matrix multiplication, and scalar multiplication. The following theorem tells us how transpose works in conjunction with these operations. Assume that the sizes of the matrices are such that the operations can be performed.

Theorem 2-7

(1) $(A + B)^t = A^t + B^t$

(2) $(AB)^t = B^t A^t$ (*Note the reversal of the order of the matrices on the right.*)

(3) $(cA)^t = cA^t$, *for any scalar c*

(4) $(A^t)^t = A$ (*The transpose of the transposed matrix is the original matrix.*)

Proof: For part (1), let $A = (a_{ij})$ and $B = (b_{ij})$. After the necessary additions and transpositions have been carried out, $(A + B)^t$ and $A^t + B^t$ will both be matrices. We prove that these matrices are equal by proving that corresponding arbitrary elements are equal.

The (i, j)th element of $(A + B)^t$ = (j, i)th element of $A + B$

$$= a_{ji} + b_{ji}$$

The (i, j)th element of $A^t + B^t$ = (i, j)th element of A^t
$$+ (i, j)\text{th element of } B^t$$
$$= (j, i)\text{th element of } A$$
$$+ (j, i)\text{th element of } B$$
$$= a_{ji} + b_{ji}$$

Thus corresponding arbitrary elements of $(A + B)^t$ and $A^t + B^t$ are equal, proving that these matrices are equal.

For part (2),

The (i, j)th element of $(AB)^t$ = (j, i)th element of AB

$$= \sum_k a_{jk} b_{ki}$$

The (i, j)th element of $B^t A^t$ = product of ith row of B^t and jth column of A^t
$$= \text{product of } i\text{th column of } B \text{ and } j\text{th row of } A$$
$$= \sum_k a_{jk} b_{ki}$$

Corresponding elements of $(AB)^t$ and $B^t A^t$ are equal, proving the equality of these matrices.

The reader is asked to prove parts (3) and (4) in Exercise 7.

The *main diagonal* of a square matrix A is the set of elements a_{11}, $a_{22}, \ldots, a_{nn}$, where the two subscripts are equal. These elements make up the diagonal going from the top left to the bottom right of the matrix, as the following example illustrates:

the main diagonal

The main diagonal is the set $\{3, 2, -1\}$.

A square matrix A is said to be *symmetric* if it is equal to its transpose; that is, if $A = A^t$. The symmetry in such a matrix occurs about its main diagonal. Elements above the main diagonal match those below the main diagonal in pairs, as the following example illustrates:

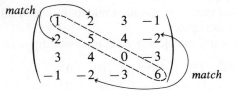

For the above matrix, $a_{21} = a_{12} = 2$, $a_{42} = a_{24} = -2$, etc; all the elements above the main diagonal are duplicated below the main diagonal. In general, a matrix A is symmetric if $a_{ij} = a_{ji}$ for every element a_{ij}.

The following are further examples of symmetric matrices:

$$\begin{pmatrix} 1 & 2 \\ 2 & 0 \end{pmatrix}, \quad \begin{pmatrix} 3 & -2 & 5 \\ -2 & 4 & 0 \\ 5 & 0 & 1 \end{pmatrix}, \quad \begin{pmatrix} 1 & 3 & 2 & 4 \\ 3 & 5 & -3 & 6 \\ 2 & -3 & 4 & 7 \\ 4 & 6 & 7 & 0 \end{pmatrix}$$

Symmetric matrices form an important class. They arise frequently in applications, as we shall see, and they have special properties. They have been studied in great depth by mathematicians. We shall look at their special properties from time to time.

✦ ***Example 3***

Consider the following matrix, representing the distances between various U.S. cities.

	1	2	3	4	5	6
1. Atlanta	0	702	1,411	2,215	667	862
2. Chicago	702	0	1,013	2,092	1,374	841
3. Denver	1,411	1,013	0	1,157	2,074	1,866
4. Los Angeles	2,215	2,092	1,157	0	2,733	2,797
5. Miami	667	1,374	2,074	2,733	0	1,336
6. New York	862	841	1,866	2,797	1,336	0

To determine the mileage between any two of the cities listed, one looks along the row of one city and the column of the other. The element that lies at the intersection of the row and the column is the required mileage. For example, the distance between Miami and Denver is 2,074 miles.

Notice that we can look up the mileage between any two cities in two distinct ways. For example, we can look up Miami in the row and Denver in the column to arrive at 2,074. We arrive at the same number if we look up Denver in the row and Miami in the column. Each number appears twice in the matrix; there is a symmetry about the matrix. It is, in fact, symmetric.

Every such mileage matrix will be symmetric.   ✦

A *diagonal matrix* is a square matrix in which all the elements not on the main diagonal are zero.

The following matrices are examples of diagonal matrices:

$$\begin{pmatrix} 1 & 0 \\ 0 & 2 \end{pmatrix}, \qquad \begin{pmatrix} 3 & 0 & 0 \\ 0 & 5 & 0 \\ 0 & 0 & 8 \end{pmatrix}, \qquad \begin{pmatrix} 0 & 0 & 0 \\ 0 & 7 & 0 \\ 0 & 0 & -6 \end{pmatrix}$$

The unit matrix was introduced earlier in the discussion of systems of linear equations. We can now interpret it as a special diagonal matrix. A *unit matrix* is a diagonal matrix in which every diagonal element is 1. The $n \times n$ unit matrix is denoted I_n. Thus

$$I_2 = \begin{pmatrix} 1 & 0 \\ 0 & 1 \end{pmatrix}, \qquad I_3 = \begin{pmatrix} 1 & 0 & 0 \\ 0 & 1 & 0 \\ 0 & 0 & 1 \end{pmatrix}, \qquad \text{etc.}$$

Theorem 2-8   *If A is an $n \times n$ matrix, then $I_nA = AI_n = A$.*

Thus I_n plays a role in the multiplication of $n \times n$ matrices similar to the role of 1 in the multiplication of real numbers—hence the terminology *unit* $n \times n$ *matrix*.

Proof: To prove this result we employ a very useful tool in the manipulation of matrices called the *Kronecker delta*. It is defined as follows:

$$\delta_{ab} \begin{cases} = 1 & \text{if} \quad a = b \\ = 0 & \text{if} \quad a \neq b \end{cases}$$

Hence, $\delta_{11} = 1$, $\delta_{12} = 0$, $\delta_{22} = 1$, $\delta_{24} = 0$, etc.

I_n may be represented by (δ_{ij}). This gives

$$I_n A = \left(\sum_{k=1}^{n} \delta_{ik} a_{kj} \right)$$

$$= (\delta_{i1} a_{1j} + \delta_{i2} a_{2j} + \cdots + \delta_{ii} a_{ij} + \ldots + \delta_{in} a_{nj})$$

The only nonzero term in this expansion is the term $\delta_{ii} a_{ij}$, or a_{ij}. Hence,

$$I_n A = (a_{ij})$$
$$= A$$

The proof of $AI_n = A$ is similar; we leave it as an exercise for the reader.

A matrix all of whose elements are zero is called a *zero matrix*. The following are examples of zero matrices:

$$\begin{pmatrix} 0 & 0 \\ 0 & 0 \end{pmatrix}, \quad \begin{pmatrix} 0 & 0 & 0 \\ 0 & 0 & 0 \end{pmatrix}, \quad \begin{pmatrix} 0 & 0 & 0 \\ 0 & 0 & 0 \\ 0 & 0 & 0 \end{pmatrix}$$

If O is the zero matrix of the same size as a matrix A,

$$A + O = A$$

and

$$A - A = O$$

If o is the zero scalar,

$$oA = O$$

A zero matrix plays a role in matrix theory similar to the role of o in the case of real numbers.

Our final definition in this section is that of the *trace* of a square matrix. If $A = (a_{ij})$ is a $n \times n$ matrix, the trace of A, denoted tr(A), is the sum of the diagonal elements:

$$\text{tr}(A) = a_{11} + a_{22} + \cdots + a_{nn}$$

$$= \sum_{i=1}^{n} a_{ii}$$

✦ Example 4 _____

Determine the trace of the matrix $A = \begin{pmatrix} 0 & 1 & 2 & 3 \\ -1 & 4 & 2 & 1 \\ 3 & 4 & -1 & 2 \\ 0 & 1 & 2 & -4 \end{pmatrix}$.

$$\text{tr}(A) = 0 + 4 - 1 - 4 = -1 \qquad ✦$$

The trace of a matrix plays an important role in matrix theory and matrix applications because of its properties and the ease with which it can be evaluated. It is important in such fields as statistical mechanics, general relativity, and quantum mechanics, where it has physical significance.

Results analogous to those for the transpose exist for the trace of the sum of two matrices, the trace of the product of two matrices, and the trace of the scalar multiple of a matrix.

Theorem 2-9   *If A and B are two n × n matrices, then*

(1) $\text{tr}(A + B) = \text{tr}(A) + \text{tr}(B)$
(2) $\text{tr}(AB) = \text{tr}(BA)$
(3) $\text{tr}(cA) = c\text{tr}(A)$, *for any scalar c*

Proof: For part (1),

$$\text{tr}(A + B) = \sum_{i=1}^{n} (a_{ii} + b_{ii})$$

$$= \sum_{i=1}^{n} a_{ii} + \sum_{i=1}^{n} b_{ii}$$

$$= \text{tr}(A) + \text{tr}(B)$$

The reader is asked to prove parts (2) and (3) in Exercise 8.

The following example illustrates how scientists are using some of these techniques of linear algebra in archaeology.

✦ Example 5* _____

A problem confronting archaeologists is that of placing sites and artifacts in proper chronological order. Let us look at this general problem in terms of graves and varieties of pottery found in graves. This approach to *sequence dating* or *seriation* in archaeology began with the work of Flinders Petrie in the late nineteenth century. Petrie studied graves in the cemeteries of Nagada, Ballas, Abadiyeh, and Hu, all located in what was prehistoric Egypt. (Recent radiocarbon dating shows that the graves ranged from 6000 B.C. to 2500 B.C.) Petrie used the data from approximately 900 graves to order them and assign a time period or sequence to each type of pottery found.

An assumption usually made in archaeology is that two graves that lie close together in temporal order are more likely to have similar contents than are two graves that lie further apart. The model we now construct leads to information concerning the relative common contents of graves. We construct a matrix A, all of whose elements are either 1 or 0, that relates graves to pottery. Label the graves 1, 2, ... and the types of pottery 1, 2, ... Let the matrix A be defined by

$$a_{ij} = \begin{cases} 1 & \text{if grave } i \text{ contains pottery type } j \\ 0 & \text{if grave } i \text{ does not contain pottery } j \end{cases}$$

The matrix A contains all the information about the pottery content of the various graves. The following result tells us how information is now extracted from A.

The element g_{ij} of the matrix $G = AA^t$ is equal to the number of varieties of pottery found in both graves i and j. Let us verify this result:

$$g_{ij} = \sum_k a_{ik} b_{kj}$$

where $(b_{kj}) = A^t$. But $b_{kj} = a_{jk}$. Thus

$$g_{ij} = \sum_k a_{ik} a_{jk}$$

$$g_{ij} = a_{i1} a_{j1} + a_{i2} a_{j2} + a_{i3} a_{j3} + \cdots + a_{in} a_{jn}$$

Each term in this sum, such as $a_{i3} a_{j3}$, will be either 0 or 1. It will be 1 if and only if a_{i3} and a_{j3} are both 1—that is, if and only if variety 3 pottery is common to both grave i and grave j. Thus g_{ij} will give the number of varieties of pottery found in both graves. The relative magnitudes of the elements of the matrix G thus lead scientists to the relative temporal proximity of the various graves.

The matrices A and A^t also lead to information concerning the sequence dating of the varieties of pottery. The element v_{ij} of the matrix $V = A^t A$ is equal to the number of graves in which the ith and jth varieties of pottery both appear. The assumption is made that the larger the number of graves in which two varieties of pottery appear, the closer they are chronologically. The reader is asked to verify this result in Exercise 21.

Furthermore, the matrices $G(= AA^t)$ and $V(= A^t A)$ are symmetric. (See Exercise 21.)

We illustrate the method for a situation involving four graves and three types of pottery. Let the following matrix A represent the pottery contents of the various graves:

$$A = \begin{pmatrix} 1 & 0 & 1 \\ 1 & 0 & 0 \\ 0 & 1 & 1 \\ 0 & 1 & 0 \end{pmatrix}$$

Thus, for example, $a_{32} = 1$ implies that grave 3 contains pottery type 2; $a_{43} = 0$ implies that grave 4 does not contain pottery type 3. G is determined:

$$G = AA^t = \begin{pmatrix} 1 & 0 & 1 \\ 1 & 0 & 0 \\ 0 & 1 & 1 \\ 0 & 1 & 0 \end{pmatrix} \begin{pmatrix} 1 & 1 & 0 & 0 \\ 0 & 0 & 1 & 1 \\ 1 & 0 & 1 & 0 \end{pmatrix} = \begin{pmatrix} 2 & 1 & 1 & 0 \\ 1 & 1 & 0 & 0 \\ 1 & 0 & 2 & 1 \\ 0 & 0 & 1 & 1 \end{pmatrix}$$

Since G is symmetric, the information contained in the elements above the main diagonal is duplicated in the elements below it. We systematically look at the elements above the diagonal.

$g_{12} = 1 \Rightarrow$ graves 1 and 2 have one type of pottery in common.

$g_{13} = 1 \Rightarrow$ graves 1 and 3 have one type of pottery in common.

$g_{14} = 0 \Rightarrow$ graves 1 and 4 do not have common content.

$g_{23} = 0 \Rightarrow$ graves 2 and 3 do not have common content.

$g_{24} = 0 \Rightarrow$ graves 2 and 4 do not have common content.

$g_{34} = 1 \Rightarrow$ graves 3 and 4 have one type of pottery in common.

Thus graves 1 and 2 are close together in time, having common content, and so are graves 1 and 3, 3 and 4. The other graves, having no common content, are further apart. Let us start with graves 1 and 2 and write

1—2

Since grave 3 is close to 1 but not close to 2, we next get

3—1—2

Finally, grave 4 is close to 3 but not to 1 or 2. Thus,

4—3—1—2

The information available does not tell us which way time flows. There are two possibilities:

$$4 \rightarrow 3 \rightarrow 1 \rightarrow 2 \quad \text{or} \quad 4 \leftarrow 3 \leftarrow 1 \leftarrow 2$$

The archaeologist usually knows which of the extreme graves—4 or 2 in our case—came first, so that the chronological order is then known.

The matrices G and V, through the relative magnitudes of their elements, contain information about the chronological order of the graves and pottery. These matrices are in practice large, and the information cannot be extracted as easily as it was in the above illustration. For example, Petrie examined 900 graves; his matrix G would be a 900×900 matrix. Special mathematical techniques have been developed for extracting information from these matrices, these methods being executed on the computer. Readers who are interested in pursuing this topic further should consult "Some Problems and Methods in Statistical Archaeology," by David G. Kendall, in *World Archaeology*, **1**, 61–76, 1969.   ✦

We now discuss another area of application of these techniques—the analysis of stresses within bodies in engineering.

✦ **Example 6**
_____

Consider a body subject to external forces. The external forces cause forces distributed throughout the body. These *internal forces* transmit the effects of the external forces. Knowledge of their distribution is necessary in order that the body may be designed so that the forces do not exceed its strength. This example illustrates how the analysis of such forces is carried out in terms of matrices.

Let O be an arbitrary point in the body. Consider a cross section of the body through O. The material on one side of this section exerts forces on the material on the other side. It is customary to analyze the effect of such forces relative to the plane at the point in terms of *stresses* or forces per unit area at the point. One examines the stress exerted by the portion on one side of the plane on the portion on the other side of the plane. A convention is that pull is positive and push is negative.

Consider the body in Figure 2-10, under the action of external forces.

Let O be the point of interest in the body, and let AB be a cross section through O. Consider the stress exerted on the side C. Let it be P. This stress

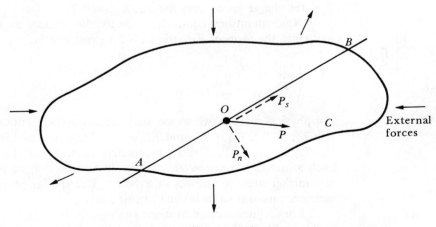

Figure 2-10

is analyzed by decomposing it into two components: one perpendicular to the plane, P_n, called *normal stress*, and one parallel to the plane, P_s, called *shearing stress* ($P = P_n + P_s$). The normal stress tends to elongate the body, and the shearing stress tends to cause the plane to slide on an adjacent plane.

To completely specify the forces at a point O, the stresses for three mutually perpendicular planes passing through the point must be given. The approach taken is to construct rectangular coordinate axes at the point (the relevant point becomes the origin) and to let the three perpendicular planes be the xy, xz, and yz planes. We now introduce the notation that is usually used. Consider the xz plane. The normal stress is denoted T_y (it is in the direction of the y axis). The shearing stress is considered in terms of its two components in the x and z directions, denoted T_{yx} and T_{yz}, respectively (see Figure 2-11). Thus, this stress is $T_{yx}i + T_yj + T_{yz}k$.

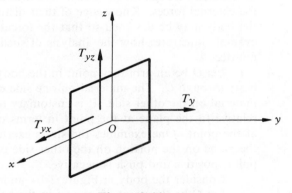

Figure 2-11

Similarly, the components of the stress on the xy plane are T_z, T_{zx}, and T_{zy}. On the yz plane, they are T_x, T_{xy}, and T_{xz}.

Nine quantities completely specify the stresses at O; they are usually written in the form of a matrix called a *stress matrix*:

$$\begin{pmatrix} T_x & T_{xy} & T_{xz} \\ T_{yx} & T_y & T_{yz} \\ T_{zx} & T_{zy} & T_z \end{pmatrix}$$

For physical reasons (which we shall not go into here), this stress matrix is such that $T_{yx} = T_{xy}$, $T_{zx} = T_{xz}$, and $T_{zy} = T_{yz}$. Thus, in actual fact, the six quantities T_x, T_y, T_z, T_{xy}, T_{xz}, and T_{yz} completely characterize the stresses at a point. Such a matrix is a *symmetric matrix*. In Section 7-4, we return to this model and further analyze stresses in a body. The special properties of symmetric matrices turn out to be crucial to this analysis.

For an introduction to stress analysis, see *Introduction to Solid Mechanics*, by James McD. Baxter Brown, John Wiley & Sons, 1973.   ✦

EXERCISES

1. Determine the transpose of each of the following matrices. Indicate whether or not the matrix is symmetric.

    *(a) $\begin{pmatrix} -1 & 2 \\ 2 & -3 \end{pmatrix}$ (b) $\begin{pmatrix} 1 & 2 \\ 0 & 3 \end{pmatrix}$ *(c) $\begin{pmatrix} 3 & -1 \\ 2 & 4 \end{pmatrix}$

    (d) $\begin{pmatrix} 4 & 5 & 6 \\ -1 & 2 & 3 \\ 0 & 1 & 2 \end{pmatrix}$ *(e) $\begin{pmatrix} 2 & -1 & 3 \\ 4 & 5 & 6 \\ 7 & 8 & 9 \end{pmatrix}$ (f) $\begin{pmatrix} 1 & 2 \\ 3 & 4 \\ 1 & 0 \end{pmatrix}$

    *(g) $\begin{pmatrix} 1 & -1 & 3 \\ -1 & 2 & 0 \\ 3 & 0 & 4 \end{pmatrix}$ (h) $\begin{pmatrix} 3 & 1 & 4 & 5 \\ 2 & 1 & 0 & -1 \end{pmatrix}$ *(i) $\begin{pmatrix} 1 \\ 2 \\ 3 \end{pmatrix}$

    (j) $(-1 \quad 4 \quad 7 \quad 0)$

2. Each of the following matrices is to be symmetric. Determine the elements indicated with a *.

    *(a) $\begin{pmatrix} 1 & 2 & 4 \\ * & 6 & * \\ 4 & 5 & 2 \end{pmatrix}$ (b) $\begin{pmatrix} 3 & 5 & * \\ * & 8 & 4 \\ -3 & * & 3 \end{pmatrix}$ (c) $\begin{pmatrix} -3 & * & 8 & 9 \\ -4 & 7 & * & 7 \\ * & 2 & 6 & 4 \\ * & 7 & * & 9 \end{pmatrix}$

3. (a) If A is an $n \times n$ matrix, prove that $AI_n = A$. (This is the second part of Theorem 2-8.)

    (b) If A is an $m \times n$ matrix, prove that $AI_n = A$.

4. Determine the traces of the following matrices:

    *(a) $\begin{pmatrix} 1 & 2 \\ -1 & 3 \end{pmatrix}$ (b) $\begin{pmatrix} 0 & -1 \\ 2 & 3 \end{pmatrix}$

    (c) $\begin{pmatrix} 1 & 2 & 3 \\ 0 & -1 & 1 \\ 4 & 1 & 0 \end{pmatrix}$ *(d) $\begin{pmatrix} -1 & 2 & 3 & 4 \\ 0 & 0 & 0 & 0 \\ 1 & 2 & 3 & 4 \\ 1 & 2 & 3 & 1 \end{pmatrix}$

5. (a) Consider two matrices of the same kind, A and B, such that $A^t = B^t$. Prove that $A = B$.

    (b) Consider two matrices of the same kind, A and B, such that $\text{tr}(A) = \text{tr}(B)$. Does this imply that $A = B$?

6. Let A be a diagonal matrix. Prove that $A = A^t$.

7. Prove the following results for transpose:

    (a) $(cA)^t = cA^t$, c being a scalar, A a matrix

    (b) $(A^t)^t = A$

    *(c) $(A + B + C)^t = A^t + B^t + C^t$

    (d) $(ABC)^t = C^t B^t A^t$

8. Prove the following results for trace:

    *(a) $\mathrm{tr}(cA) = c\mathrm{tr}(A)$, c being a scalar, A a square matrix

    (b) $\mathrm{tr}(AB) = \mathrm{tr}(BA)$, A and B being square matrices of the same size

    (c) $\mathrm{tr}(A) = \mathrm{tr}(A^t)$, A being a square matrix

    (d) $\mathrm{tr}(A + B + C) = \mathrm{tr}(A) + \mathrm{tr}(B) + \mathrm{tr}(C)$, A, B, and C being square matrices of the same size

9. Prove that the product of two diagonal matrices of the same size is a diagonal matrix.

10. *(a) Prove that the sum of two symmetric matrices of the same size is symmetric.

    (b) Prove that the scalar multiple of a symmetric matrix is symmetric.

11. Prove that a symmetric matrix is necessarily square.

12. A matrix A is said to be antisymmetric if $A = -A^t$. Give an example of an antisymmetric matrix. Prove that such a matrix is a square matrix having diagonal elements all zero. Prove that the sum of two antisymmetric matrices is itself antisymmetric. Further, prove that a scalar multiple of an antisymmetric matrix is antisymmetric.

13. If A is a square matrix, prove that

    *(a) $A + A^t$ is symmetric

    (b) $A - A^t$ is antisymmetric

14. Prove that any square matrix A can be decomposed into the sum of a symmetric matrix B and an antisymmetric matrix C: $A = B + C$. (*Hint*: Use Exercise 13.)

15. Prove that $(I_3)^2 = (I_3)^3 = I_3$. Generalize this result by proving that $(I_n)^m = I_n$, where m is any positive integer.

16. An *upper triangular matrix* is a square matrix of the type

$$\begin{pmatrix} a_{11} & a_{12} & \cdots & a_{1n} \\ & a_{22} & & \\ 0 & & \ddots & \\ & & & a_{nn} \end{pmatrix}$$

having zeros below the main diagonal. Prove that the product of two upper triangular matrices of the same kind is an upper triangular matrix. (A lower triangular matrix is a square matrix having zeros above the main diagonal.)

17. *(a) Let A be a 4×4 matrix defined by $a_{ij} = \delta_{i\,j-1}$. Write down A.

    (b) B is a 5×5 matrix defined by $b_{ij} = \delta_{i+1\,j-1}$. Write down B.

18. Expand the sums

    *(a) $\displaystyle\sum_{k=1}^{4} \delta_{2k} a_{k4}$ (b) $\displaystyle\sum_{k=1}^{3} b_{2k} \delta_{k3}$

19. Let A be an arbitrary matrix. Prove that the products AA^t and A^tA always exist. (This result has been assumed in the archaeology model.)

20. The following matrices describe the pottery contents of various graves. For each situation determine possible chronological orderings of first the graves and then the pottery types.

*(a) $A = \begin{pmatrix} 1 & 0 \\ 0 & 1 \\ 1 & 1 \end{pmatrix}$ (b) $A = \begin{pmatrix} 0 & 0 & 1 \\ 1 & 1 & 0 \\ 1 & 0 & 1 \\ 0 & 1 & 0 \end{pmatrix}$ *(c) $A = \begin{pmatrix} 1 & 0 & 1 & 0 \\ 0 & 1 & 1 & 1 \\ 1 & 1 & 1 & 1 \end{pmatrix}$

(d) $A = \begin{pmatrix} 0 & 0 & 0 & 1 \\ 1 & 1 & 0 & 0 \\ 0 & 0 & 1 & 1 \\ 1 & 0 & 1 & 0 \end{pmatrix}$ (e) $A = \begin{pmatrix} 1 & 0 & 1 & 0 \\ 1 & 0 & 0 & 0 \\ 0 & 1 & 0 & 1 \\ 0 & 1 & 1 & 0 \end{pmatrix}$

21. Let A be a matrix of ones and zeros that defines the relationship between pottery varieties and graves, as in Example 5. Let $V = A^t A$. Prove that v_{ij} gives the number of graves in which the ith and jth varieties of pottery are both found. This leads to information concerning the temporal proximities of various varieties of pottery: the assumption is made that the larger the number of graves in which two varieties of pottery appear, the closer they are chronologically. Prove that $G(= AA^t)$ and $V(= A^t A)$ are symmetric matrices.

22. Let A be the matrix that defines the relationship between pottery and graves in archaeology. Write a computer program for determining $G(= AA^t)$ and $V(= A^t A)$.

23. The model used here in archaeology can be used in sociology to analyze relationships within a group of people. Let us look at the relationship of friendship in a group of people. Let use assume that all friendships are mutual. Label the people 1 to n and define a matrix A:

$$a_{ii} = 0 \quad \text{for all } i \text{ (diagonal element zero)}$$

$$a_{ij} = \begin{cases} 1 & \text{if } i \text{ and } j \text{ are friends} \\ 0 & \text{if } i \text{ and } j \text{ are not friends} \end{cases}$$

(a) Prove that if $G = AA^t$, then g_{ij} is the number of friends that i and j have in common.

(b) Suppose we drop the condition of mutual friendship, i.e., it is possible that even if i interprets j as a friend j might not consider i a friend. How might this affect the model?

2-5 THE INVERSE OF A MATRIX

Let us now use our newly developed matrix algebra to further our understanding of and control over systems of linear equations. This discussion will lead very naturally into a numerical technique for computing the inverse of a matrix.

Consider a system of m equations in n variables. The matrix of coefficients A will be an $m \times n$ matrix. Let us now use the standard matrix notation for this matrix. The system can thus be written

$$a_{11}x_1 + \cdots + a_{1n}x_n = y_1$$
$$\vdots$$
$$a_{m1}x_1 + \cdots + a_{mn}x_n = y_m$$

Represent this system in column matrix form:

$$\begin{pmatrix} a_{11}x_1 + \cdots + a_{1n}x_n \\ \vdots \\ a_{m1}x_1 + \cdots + a_{mn}x_n \end{pmatrix} = \begin{pmatrix} y_1 \\ \vdots \\ y_m \end{pmatrix}$$

This representation enables us to express the system in the following convenient matrix product form:

$$\begin{pmatrix} a_{11} & \cdots & a_{1n} \\ & \vdots & \\ a_{ml} & \cdots & a_{mn} \end{pmatrix} \begin{pmatrix} x_1 \\ \vdots \\ x_n \end{pmatrix} = \begin{pmatrix} y_1 \\ \vdots \\ y_m \end{pmatrix}$$

Writing the matrix of coefficients as A and the column matrices as X and Y, we get

$$AX = Y$$

We shall use this way of expressing a system of linear equations a great deal henceforth.

✦ **Example 1** ———————————————————————————

Consider the system of equations

$$2x_1 + 3x_2 - x_3 = 4$$
$$x_1 - x_2 + x_3 = 1$$
$$3x_1 + x_2 - x_3 = 2$$

The matrix of coefficients is

$$\begin{pmatrix} 2 & 3 & -1 \\ 1 & -1 & 1 \\ 3 & 1 & -1 \end{pmatrix}$$

The system can be expressed in terms of the single matrix equation

$$\begin{pmatrix} 2 & 3 & -1 \\ 1 & -1 & 1 \\ 3 & 1 & -1 \end{pmatrix} \begin{pmatrix} x_1 \\ x_2 \\ x_3 \end{pmatrix} = \begin{pmatrix} 4 \\ 1 \\ 2 \end{pmatrix} \qquad ✦$$

Consider now a number of systems $AX = Y_1$, $AX = Y_2, \ldots, AX = Y_k$, all having the *same square* matrix of coefficients A and each having a *unique solution*. The Y's are given and the X's have to be determined. One could of course go through the method of Gauss-Jordan elimination for each system, solving for the X's. Let us write the augmented matrices as $(A \mid Y_1), (A \mid Y_2), \ldots, (A \mid Y_k)$. This procedure would lead to reduced echelon forms $(I_n \mid X_1), (I_n \mid X_2), \ldots, (I_n \mid X_k)$, and the solutions would be $X_1, X_2, \ldots, X_k$. However, the reduction of A to I_n would be repeated for each system; this involves a great deal of unnecessary work. The systems can be represented by one large augmented matrix $(A \mid Y_1 \quad Y_2 \quad \cdots \quad Y_k)$, and the Gauss-Jordan method can be performed on this one matrix. We would get

$$(A \mid Y_1 \quad Y_2 \quad \cdots \quad Y_k) \approx \cdots \approx (I_n \mid X_1 \quad X_2 \quad \cdots \quad X_k)$$

leading to the solutions $X_1, X_2, \ldots, X_k$.

✦ *Example 2*

Solve the following three systems of linear equations:

$$
\begin{aligned}
x_1 - x_2 + 3x_3 &= b_1 \\
2x_1 - x_2 + 4x_3 &= b_2 \\
-x_1 + 2x_2 - 4x_3 &= b_3
\end{aligned}
\quad \text{for} \quad
\begin{pmatrix} b_1 \\ b_2 \\ b_3 \end{pmatrix} =
\begin{pmatrix} 8 \\ 11 \\ -11 \end{pmatrix},
\begin{pmatrix} 0 \\ 1 \\ 2 \end{pmatrix},
\begin{pmatrix} 3 \\ 3 \\ -4 \end{pmatrix}
\text{ in turn}
$$

The augmented matrix that represents all three systems is

$$
\begin{pmatrix}
1 & -1 & 3 & \mid & 8 & 0 & 3 \\
2 & -1 & 4 & \mid & 11 & 1 & 3 \\
-1 & 2 & -4 & \mid & -11 & 2 & -4
\end{pmatrix}
$$

Performing Gauss-Jordan elimination, we get

$$
\begin{pmatrix}
① & -1 & 3 & \mid & 8 & 0 & 3 \\
2 & -1 & 4 & \mid & 11 & 1 & 3 \\
-1 & 2 & -4 & \mid & -11 & 2 & -4
\end{pmatrix}
\underset{\substack{R2+(-2)R1 \\ R3+R1}}{\approx}
\begin{pmatrix}
1 & -1 & 3 & \mid & 8 & 0 & 3 \\
0 & ① & -2 & \mid & -5 & 1 & -3 \\
0 & 1 & -1 & \mid & -3 & 2 & -1
\end{pmatrix}
$$

$$
\underset{\substack{R1+R2 \\ R3+(-1)R1}}{\approx}
\begin{pmatrix}
1 & 0 & 1 & \mid & 3 & 1 & 0 \\
0 & 1 & -2 & \mid & -5 & 1 & -3 \\
0 & 0 & ① & \mid & 2 & 1 & 2
\end{pmatrix}
$$

$$
\underset{\substack{R1+(-1)R3 \\ R2+(2)R3}}{\approx}
\begin{pmatrix}
1 & 0 & 0 & \mid & 1 & 0 & -2 \\
0 & 1 & 0 & \mid & -1 & 3 & 1 \\
0 & 0 & 1 & \mid & 2 & 1 & 2
\end{pmatrix}
$$

The solutions to the three systems are $x_1 = 1$, $x_2 = -1$, $x_3 = 2$; $x_1 = 0$, $x_2 = 3$, $x_3 = 1$; and $x_1 = -2$, $x_2 = 1$, $x_3 = 2$.   ✦

We now commence our discussion of the inverse of a matrix.

Definition 2-4 An $n \times n$ matrix A is *invertible* if and only if there exists a matrix B such that

$$AB = BA = I_n$$

Here I_n is the unit $n \times n$ matrix. B is called the *multiplicative inverse* of A and is denoted A^{-1}.

Theorem 2-10 *The multiplicative inverse of an invertible matrix A is unique.*

Proof: Let B and C be multiplicative inverses of A. Thus $AB = I_n$. Multiplying both sides by C,

$$C(AB) = CI_n$$
$$(CA)B = C$$
$$I_n B = C$$
$$B = C$$

Thus an invertible matrix A has only one inverse.

✦ **Example 3** _____

Prove that the matrix A, $\begin{pmatrix} 1 & 2 \\ 3 & 4 \end{pmatrix}$, has inverse B, $\begin{pmatrix} -2 & 1 \\ \frac{3}{2} & -\frac{1}{2} \end{pmatrix}$

We have that

$$AB = \begin{pmatrix} 1 & 2 \\ 3 & 4 \end{pmatrix}\begin{pmatrix} -2 & 1 \\ \frac{3}{2} & -\frac{1}{2} \end{pmatrix} = \begin{pmatrix} 1 & 0 \\ 0 & 1 \end{pmatrix}$$

and

$$BA = \begin{pmatrix} -2 & 1 \\ \frac{3}{2} & -\frac{1}{2} \end{pmatrix}\begin{pmatrix} 1 & 2 \\ 3 & 4 \end{pmatrix} = \begin{pmatrix} 1 & 0 \\ 0 & 1 \end{pmatrix}$$

Thus the inverse of A is the matrix B.   ✦

Note that the definition of inverse is limited to square matrices. For both the products AB and BA to give I_n, A and B must both be $n \times n$ matrices. Not every square matrix has an inverse. The reader will see a square matrix without an inverse in a following example.

A^{-1} plays a role in the theoretical discussion of certain systems of linear equations. The following is a key result in such discussions.

Theorem 2-11   *Let $AX = Y$ be a system of n equations in n variables. If A^{-1} exists, the solution is unique and is given by $X = A^{-1}Y$.*

Proof: Let us first of all prove that a solution exists and that $X = A^{-1}Y$ is a solution. For this value of X,

$$AX = AA^{-1}Y = I_n Y = Y$$

Thus $X = A^{-1}Y$ is a solution.

We now prove the uniqueness. Let X_1 and X_2 be solutions. Thus $AX_1 = Y$ and $AX_2 = Y$. Multiplying both equations by A^{-1} gives

$$A^{-1}AX_1 = A^{-1}Y \quad \text{and} \quad A^{-1}AX_2 = A^{-1}Y$$
$$I_n X_1 = A^{-1}Y \quad \text{and} \quad I_n X_2 = A^{-1}Y$$
$$X_1 = A^{-1}Y \quad \text{and} \quad X_2 = A^{-1}Y$$

Thus $X_1 = X_2$; the solution is unique.

The above result is of theoretical importance. It enables us to *represent* the solution to $AX = Y$ as $A^{-1}Y$, if A^{-1} exists. It is not usually used in practice to numerically solve the system $AX = Y$, as the method of Gauss-Jordan elimination is more efficient. More will be said about this later in the section.

We now use the uniqueness result of this theorem in deriving a method for computing the inverse of a matrix. This method is based on the previous Gauss-Jordan technique for solving a number of systems of linear equations using one large augmented matrix.

Let A be an invertible matrix. Then $AA^{-1} = I_n$.

Denote the columns of A^{-1} as $X_1, X_2, \ldots, X_n$ and the columns of I_n as $Y_1, Y_2, \ldots, Y_n$. Thus,

$$A^{-1} = (X_1 \quad X_2 \quad \cdots \quad X_n) \qquad \text{and} \qquad I_n = (Y_1 \quad Y_2 \quad \cdots \quad Y_n)$$

The matrix equation $AA^{-1} = I_n$ can now be expressed

$$A(X_1 \quad X_2 \quad \cdots \quad X_n) = (Y_1 \quad Y_2 \quad \cdots \quad Y_n)$$

Matrix multiplication gives[†]

$$(AX_1 \quad AX_2 \quad \cdots \quad AX_n) = (Y_1 \quad Y_2 \quad \cdots \quad Y_n)$$

Equating the columns of these two matrices,

$$AX_1 = Y_1, AX_2 = Y_2, \ldots, AX_n = Y_n$$

Thus $X_1, X_2, \ldots, X_n$, the columns of A^{-1}, are solutions to the following systems of equations:

$$AX = Y_1, AX = Y_2, \ldots, AX = Y_n$$

Since A^{-1} exists, the previous theorem implies that each of these systems has a unique solution. Thus the previous technique for solving a number of systems can be used to compute $X_1, X_2, \ldots, X_n$. We perform the Gauss-Jordan method on $(A \mid Y_1 \quad Y_2 \quad \cdots \quad Y_n)$,

$$(A \mid Y_1 \quad Y_2 \quad \cdots \quad Y_n) \approx \cdots \approx (I_n \mid X_1 \quad X_2 \quad \cdots \quad X_n)$$

Since $(Y_1 \quad Y_2 \quad \cdots \quad Y_n) = I_n$ and $(X_1 \quad X_2 \quad \cdots \quad X_n) = A^{-1}$, this may be expressed

$$(A \mid I_n) \approx \cdots \approx (I_n \mid A^{-1})$$

Thus, if we compute the reduced echelon form of $(A \mid I_n)$, we get a matrix of the form $(I_n \mid B)$, where B will be the inverse of A.

On the other hand, if we compute the reduced echelon form of $(A \mid I_n)$ for a given matrix A and find that it is not of the form $(I_n \mid B)$, then the only assumption that we made in the above derivation must be false: A cannot be an invertible matrix.

[†] Theorem 2-6, Section 2-3.

We now summarize the results of this discussion.

Gauss–Jordan Elimination Method for Finding the Inverse of a Matrix

Let A be an $n \times n$ matrix. Adjoin the identity $n \times n$ matrix I_n to A to form the matrix $(A \mid I_n)$. Compute the reduced echelon form of $(A \mid I_n)$.

1. If the reduced echelon form is of the type $(I_n \mid B)$, then B will be the inverse of A.
2. If the reduced echelon form is not of the type $(I_n \mid B)$—that is, if the first n columns do not form I_n—the inverse of A does not exist.

The following examples illustrate the method.

✦ Example 4

Determine the inverse of the matrix

$$A = \begin{pmatrix} 1 & 2 & 0 \\ 2 & 1 & -1 \\ 3 & 1 & 1 \end{pmatrix}$$

We get

$$(A \mid I_3) = \begin{pmatrix} ① & 2 & 0 & \vdots & 1 & 0 & 0 \\ 2 & 1 & -1 & \vdots & 0 & 1 & 0 \\ 3 & 1 & 1 & \vdots & 0 & 0 & 1 \end{pmatrix} \underset{\substack{R2+(-2)R1 \\ R3+(-3)R1}}{\approx} \begin{pmatrix} 1 & 2 & 0 & \vdots & 1 & 0 & 0 \\ 0 & ⃝{-3} & -1 & \vdots & -2 & 1 & 0 \\ 0 & -5 & 1 & \vdots & -3 & 0 & 1 \end{pmatrix}$$

$$\underset{(-\frac{1}{3})R2}{\approx} \begin{pmatrix} 1 & 2 & 0 & \vdots & 1 & 0 & 0 \\ 0 & ① & \frac{1}{3} & \vdots & \frac{2}{3} & -\frac{1}{3} & 0 \\ 0 & -5 & 1 & \vdots & -3 & 0 & 1 \end{pmatrix}$$

$$\underset{\substack{R1+(-2)R2 \\ R3+(5)R2}}{\approx} \begin{pmatrix} 1 & 0 & -\frac{2}{3} & \vdots & -\frac{1}{3} & \frac{2}{3} & 0 \\ 0 & 1 & \frac{1}{3} & \vdots & \frac{2}{3} & -\frac{1}{3} & 0 \\ 0 & 0 & ⃝{\frac{8}{3}} & \vdots & \frac{1}{3} & -\frac{5}{3} & 1 \end{pmatrix}$$

$$\underset{(\frac{3}{8})R3}{\approx} \begin{pmatrix} 1 & 0 & -\frac{2}{3} & \vdots & -\frac{1}{3} & \frac{2}{3} & 0 \\ 0 & 1 & \frac{1}{3} & \vdots & \frac{2}{3} & -\frac{1}{3} & 0 \\ 0 & 0 & ① & \vdots & \frac{1}{8} & -\frac{5}{8} & \frac{3}{8} \end{pmatrix}$$

$$\underset{\substack{R1+(\frac{2}{3})R3 \\ R2+(-\frac{1}{3})R3}}{\approx} \begin{pmatrix} 1 & 0 & 0 & \vdots & -\frac{1}{4} & \frac{1}{4} & \frac{1}{4} \\ 0 & 1 & 0 & \vdots & \frac{5}{8} & -\frac{1}{8} & -\frac{1}{8} \\ 0 & 0 & 1 & \vdots & \frac{1}{8} & -\frac{5}{8} & \frac{3}{8} \end{pmatrix}$$

Thus

$$A^{-1} = \begin{pmatrix} -\frac{1}{4} & \frac{1}{4} & \frac{1}{4} \\ \frac{5}{8} & -\frac{1}{8} & -\frac{1}{8} \\ \frac{1}{8} & -\frac{5}{8} & \frac{3}{8} \end{pmatrix} \qquad ✦$$

The following example illustrates the application of the method for a matrix that does not have an inverse.

✦ *Example 5*

Determine the inverse, if it exists, of the matrix

$$A = \begin{pmatrix} 1 & -1 & 0 & 2 \\ 3 & 1 & 5 & -1 \\ -1 & 1 & 0 & 0 \\ 2 & 2 & 0 & 1 \end{pmatrix}$$

We get

$$(A \mid I_4) = \begin{pmatrix} ① & -1 & 0 & 2 & \mid & 1 & 0 & 0 & 0 \\ 3 & 1 & 5 & -1 & \mid & 0 & 1 & 0 & 0 \\ -1 & 1 & 0 & 0 & \mid & 0 & 0 & 1 & 0 \\ 2 & 2 & 0 & 1 & \mid & 0 & 0 & 0 & 1 \end{pmatrix}$$

$$\underset{\substack{R2+(-3)R1 \\ R3+R1 \\ R4+(-2)R1}}{\approx} \begin{pmatrix} 1 & -1 & 0 & 2 & \mid & 1 & 0 & 0 & 0 \\ 0 & ④ & 5 & -7 & \mid & -3 & 1 & 0 & 0 \\ 0 & 0 & 0 & 2 & \mid & 1 & 0 & 1 & 0 \\ 0 & 0 & 0 & -3 & \mid & -2 & 0 & 0 & 1 \end{pmatrix}$$

$$\underset{(\frac{1}{4})R2}{\approx} \begin{pmatrix} 1 & -1 & 0 & 2 & \mid & 1 & 0 & 0 & 0 \\ 0 & ① & \frac{5}{4} & -\frac{7}{4} & \mid & -\frac{3}{4} & \frac{1}{4} & 0 & 0 \\ 0 & 0 & 0 & 2 & \mid & 1 & 0 & 1 & 0 \\ 0 & 0 & 0 & -3 & \mid & -2 & 0 & 0 & 1 \end{pmatrix}$$

$$\underset{R1+R2}{\approx} \begin{pmatrix} 1 & 0 & \frac{5}{4} & \frac{1}{4} & \mid & \frac{1}{4} & \frac{1}{4} & 0 & 0 \\ 0 & 1 & \frac{5}{4} & -\frac{7}{4} & \mid & -\frac{3}{4} & \frac{1}{4} & 0 & 0 \\ 0 & 0 & 0 & ② & \mid & 1 & 0 & 1 & 0 \\ 0 & 0 & 0 & -3 & \mid & -2 & 0 & 0 & 1 \end{pmatrix}$$

The pivot is now the 2 (circled). The reduced echelon form will not have a 1 in the (3, 3) diagonal location. The reduced echelon form cannot thus be of the form $(I_4 \mid B)$. We need not proceed further; A^{-1} does not exist.   ✦

We now illustrate how a matrix inverse can be used to solve a system of equations if the matrix of coefficients is invertible. This example is followed by a discussion of the method.

✦ *Example 6*

Solve the system of equations

$$\begin{aligned} x_1 + 2x_2 \quad\quad &= \quad 4 \\ 2x_1 + x_2 - x_3 &= \quad 2 \\ 3x_1 + x_2 + x_3 &= -2 \end{aligned}$$

This system can be written in the following matrix form:

$$\begin{pmatrix} 1 & 2 & 0 \\ 2 & 1 & -1 \\ 3 & 1 & 1 \end{pmatrix} \begin{pmatrix} x_1 \\ x_2 \\ x_3 \end{pmatrix} = \begin{pmatrix} 4 \\ 2 \\ -2 \end{pmatrix}$$

If the matrix of coefficients is invertible, the unique solution is then

$$\begin{pmatrix} x_1 \\ x_2 \\ x_3 \end{pmatrix} = \begin{pmatrix} 1 & 2 & 0 \\ 2 & 1 & -1 \\ 3 & 1 & 1 \end{pmatrix}^{-1} \begin{pmatrix} 4 \\ 2 \\ -2 \end{pmatrix}$$

This inverse does, in fact, exist and was computed in Example 4. We get

$$\begin{pmatrix} x_1 \\ x_2 \\ x_3 \end{pmatrix} = \begin{pmatrix} -\frac{1}{4} & \frac{1}{4} & \frac{1}{4} \\ \frac{5}{8} & -\frac{1}{8} & -\frac{1}{8} \\ \frac{1}{8} & -\frac{5}{8} & \frac{3}{8} \end{pmatrix} \begin{pmatrix} 4 \\ 2 \\ -2 \end{pmatrix} = \begin{pmatrix} -1 \\ \frac{5}{2} \\ -\frac{3}{2} \end{pmatrix}$$

The unique solution is $x_1 = -1$, $x_2 = \frac{5}{2}$, $x_3 = -\frac{3}{2}$.   ✦

Most systems of linear equations of appreciable size are solved on the computer. Two factors are important when one is using a computer—efficiency and accuracy. To solve the system, the matrix inverse method requires roughly twice the number of arithmetic operations (addition, subtraction, multiplication, division) required by Gauss-Jordan elimination.[†] Furthermore, the more operations that are performed, the larger the possible error. Thus Gauss-Jordan is in general more efficient and accurate than the matrix inverse method.

One may be tempted to assume that given a number of systems $AX = Y_1, \ldots, AX = Y_k$, each having the same invertible matrix of coefficients, it would be efficient to calculate A^{-1} and then compute the solutions using $X_1 = A^{-1}Y_1, \ldots, X_k = A^{-1}Y_k$. This approach would involve one computation of A^{-1} and then a number of multiplications. In general, however, it is more efficient and accurate to solve the systems using a large augmented matrix that represents all the systems and an elimination method, as described earlier in this section.

There are certain instances when the matrix inverse method is used. In the following section, we illustrate one such occasion which involves solving a number of systems. By their nature, the elements of A in that example lend themselves to an efficient algorithm for computing the inverse, thus making the matrix inverse method suitable.

The concept of the inverse of a matrix is of importance in areas other than discussions of linear systems of equations. We shall, for example, see its central role in the discussion of coordinate transformations later in this course.

[†] See, for example, *Computational Matrix Algebra*, by David I. Steinberg, McGraw-Hill Kogakusha, Ltd., 1974, page 103.

The following example illustrates how matrices having inverses are used in coding theory.

✦ **Example 7**

Governments use sophisticated methods of coding and decoding messages. One type of code used that is extremely difficult to break makes use of a large matrix to encode a message. The receiver of the message decodes it using the inverse of the matrix. We illustrate the method with a 3×3 matrix.

Let the message be

PREPARE TO ATTACK

and the matrix be

$$\begin{pmatrix} -3 & -3 & -4 \\ 0 & 1 & 1 \\ 4 & 3 & 4 \end{pmatrix}$$

We assign a number to each letter of the alphabet. For convenience, let us associate each letter with its position in the alphabet. A is associated with 1, B with 2, etc. Let a space between words be denoted by the number 27. Thus the message becomes

P	R	E	P	A	R	E	*	T	O	*	A	T	T	A	C	K
16	18	5	16	1	18	5	27	20	15	27	1	20	20	1	3	11

Since we are going to use a 3×3 matrix to encode, we break the enumerated message up into a sequence of 3×1 matrices as follows.

$$\begin{pmatrix} 16 \\ 18 \\ 5 \end{pmatrix}, \quad \begin{pmatrix} 16 \\ 1 \\ 18 \end{pmatrix}, \quad \begin{pmatrix} 5 \\ 27 \\ 20 \end{pmatrix}, \quad \begin{pmatrix} 15 \\ 27 \\ 1 \end{pmatrix}, \quad \begin{pmatrix} 20 \\ 20 \\ 1 \end{pmatrix}, \quad \begin{pmatrix} 3 \\ 11 \\ 27 \end{pmatrix}$$

Observe that it was necessary to add a space at the end of the message in order to complete the last matrix. We now put the message into code by multiplying each of the above column matrices by the encoding matrix:

$$\begin{pmatrix} -3 & -3 & -4 \\ 0 & 1 & 1 \\ 4 & 3 & 4 \end{pmatrix}\begin{pmatrix} 16 \\ 18 \\ 5 \end{pmatrix} = \begin{pmatrix} -122 \\ 23 \\ 138 \end{pmatrix}, \quad \begin{pmatrix} -3 & -3 & -4 \\ 0 & 1 & 1 \\ 4 & 3 & 4 \end{pmatrix}\begin{pmatrix} 16 \\ 1 \\ 8 \end{pmatrix} = \begin{pmatrix} -83 \\ 9 \\ 99 \end{pmatrix}$$

and so on. The column matrices obtained are

$$\begin{pmatrix} -122 \\ 23 \\ 138 \end{pmatrix}, \quad \begin{pmatrix} -83 \\ 9 \\ 99 \end{pmatrix}, \quad \begin{pmatrix} -176 \\ 47 \\ 181 \end{pmatrix}, \quad \begin{pmatrix} -130 \\ 28 \\ 145 \end{pmatrix}, \quad \begin{pmatrix} -124 \\ 21 \\ 144 \end{pmatrix}, \quad \begin{pmatrix} -150 \\ 38 \\ 153 \end{pmatrix}$$

The coded message is transmitted in the following linear form.

$-122, 23, 138, -83, 9, 99, -176, 47, 181, -130, 28, 145, -124, 21, 144,$
$-150, 38, 153$

To decode the message, the receiver writes this string as a sequence of column matrices and repeats the technique using the inverse of the original matrix. The inverse in this case is

$$\begin{pmatrix} 1 & 0 & 1 \\ 4 & 4 & 3 \\ -4 & -3 & -3 \end{pmatrix}$$

Thus, for example, using the first three numbers in the coded message,

$$\begin{pmatrix} 1 & 0 & 1 \\ 4 & 4 & 3 \\ -4 & -3 & -3 \end{pmatrix} \begin{pmatrix} -122 \\ 23 \\ 138 \end{pmatrix} = \begin{pmatrix} 16 \\ 18 \\ 5 \end{pmatrix}$$

leading to 16, 18, 5 and PRE, the first three letters in the original message. ✦

EXERCISES

1. Express each of the following systems as a single matrix equation $AX = Y$.

   *(a) $x_1 + 3x_2 = 5$
   $2x_1 - x_2 = 6$

   (b) $2x_1 - 3x_2 + x_3 = 4$
   $4x_1 - x_2 + 2x_3 = -1$
   $x_1 + x_2 - x_3 = 2$

   (c) $x_1 + 2x_2 - 3x_3 + 4x_4 = 0$
   $x_1 + x_2 + x_4 = 5$
   $3x_1 + 2x_2 + x_3 + 2x_4 = 4$

In Exercises 2–6, solve the given systems using the Gauss-Jordan method of elimination.

*2. $x_1 + 2x_2 = b_1$
   $3x_1 + 5x_2 = b_2$ for $\begin{pmatrix} b_1 \\ b_2 \end{pmatrix} = \begin{pmatrix} 3 \\ 8 \end{pmatrix}, \begin{pmatrix} 4 \\ 9 \end{pmatrix}, \begin{pmatrix} 3 \\ 7 \end{pmatrix}$ in turn

3. $x_1 + x_2 = b_1$
   $2x_1 + 3x_2 = b_2$ for $\begin{pmatrix} b_1 \\ b_2 \end{pmatrix} = \begin{pmatrix} 0 \\ 1 \end{pmatrix}, \begin{pmatrix} 5 \\ 13 \end{pmatrix}, \begin{pmatrix} 1 \\ 2 \end{pmatrix}$ in turn

*4. $x_1 - 2x_2 + 3x_3 = b_1$
   $x_1 - x_2 + 2x_3 = b_2$
   $2x_1 - 3x_2 + 6x_3 = b_3$ for $\begin{pmatrix} b_1 \\ b_2 \\ b_3 \end{pmatrix} = \begin{pmatrix} 6 \\ 5 \\ 14 \end{pmatrix}, \begin{pmatrix} -5 \\ -3 \\ -8 \end{pmatrix}, \begin{pmatrix} 4 \\ 3 \\ 9 \end{pmatrix}$ in turn

5. $x_1 + 2x_2 - x_3 = b_1$
   $-x_1 - x_2 + x_3 = b_2$
   $3x_1 + 7x_2 - x_3 = b_3$ for $\begin{pmatrix} b_1 \\ b_2 \\ b_3 \end{pmatrix} = \begin{pmatrix} -1 \\ 1 \\ -1 \end{pmatrix}, \begin{pmatrix} 6 \\ -4 \\ 18 \end{pmatrix}, \begin{pmatrix} 0 \\ -2 \\ -4 \end{pmatrix}$ in turn

*6. $x_1 + x_2 + 5x_3 = b_1$
   $x_1 + 2x_2 + 8x_3 = b_2$
   $2x_1 + 4x_2 + 16x_3 = b_3$ for $\begin{pmatrix} b_1 \\ b_2 \\ b_3 \end{pmatrix} = \begin{pmatrix} 2 \\ 5 \\ 10 \end{pmatrix}, \begin{pmatrix} 3 \\ 2 \\ 4 \end{pmatrix}$ in turn

7. Consider the systems $AX = Y_1$, $AX = Y_2$, both having the same 3×3 matrix of coefficients A.

   (a) Is it possible for $AX = Y_1$ to have a unique solution and $AX = Y_2$ to have many solutions?

   (b) Is it possible for $AX = Y_1$ to have a unique solution and $AX = Y_2$ to have no solutions?

   (c) Is it possible for $AX = Y_1$ to have many solutions and $AX = Y_2$ to have no solutions?

   (d) Solve the system

$$
\begin{aligned}
x_1 + 2x_2 + 4x_3 &= b_1 \\
x_1 + x_2 + 2x_3 &= b_2 \\
2x_1 + 3x_2 + 6x_3 &= b_3
\end{aligned}
\quad \text{for} \quad
\begin{pmatrix} b_1 \\ b_2 \\ b_3 \end{pmatrix} =
\begin{pmatrix} 8 \\ 5 \\ 13 \end{pmatrix},
\begin{pmatrix} 5 \\ 3 \\ 11 \end{pmatrix}
\text{ in turn}
$$

Determine the inverses (if they exist) of the following matrices using the Gauss-Jordan elimination method.

*8. $\begin{pmatrix} 1 & 0 \\ 2 & 1 \end{pmatrix}$ **9.** $\begin{pmatrix} 2 & 1 \\ 4 & 3 \end{pmatrix}$ ***10.** $\begin{pmatrix} 0 & 2 \\ -\frac{1}{3} & \frac{1}{3} \end{pmatrix}$

11. $\begin{pmatrix} 1 & 2 & 3 \\ 0 & 1 & 2 \\ 4 & 5 & 3 \end{pmatrix}$ ***12.** $\begin{pmatrix} 2 & 0 & 4 \\ -1 & 3 & 1 \\ 0 & 1 & 2 \end{pmatrix}$ **13.** $\begin{pmatrix} 0 & 3 & 3 \\ 1 & 2 & 3 \\ 1 & 4 & 6 \end{pmatrix}$

*14. $\begin{pmatrix} 1 & 2 & -1 \\ 3 & -1 & 0 \\ 2 & -3 & 1 \end{pmatrix}$ **15.** $\begin{pmatrix} 1 & 2 & 3 \\ 2 & -1 & 4 \\ 0 & -1 & 1 \end{pmatrix}$ ***16.** $\begin{pmatrix} 1 & 2 & -1 \\ 2 & 4 & -3 \\ 1 & -2 & 0 \end{pmatrix}$

17. $\begin{pmatrix} -3 & -1 & 1 & -2 \\ -1 & 3 & 2 & 1 \\ 1 & 2 & 3 & -1 \\ -2 & 1 & -1 & -3 \end{pmatrix}$ ***18.** $\begin{pmatrix} 1 & 1 & 0 & 0 \\ 0 & 1 & 1 & 0 \\ 1 & 0 & 0 & 1 \\ 0 & 0 & 1 & 1 \end{pmatrix}$

Solve the following systems of equations by determining the inverse of the matrix of coefficients and then using matrix multiplication.

*19.
$$
\begin{aligned}
x_1 + 3x_2 &= 5 \\
2x_1 + x_2 &= 10
\end{aligned}
$$

20.
$$
\begin{aligned}
x_1 + 2x_2 - x_3 &= 2 \\
x_1 + x_2 + 2x_3 &= 0 \\
x_1 - x_2 - x_3 &= 1
\end{aligned}
$$

*21.
$$
\begin{aligned}
x_1 - x_2 &= 1 \\
x_1 + x_2 + 2x_3 &= 2 \\
x_1 + 2x_2 + x_3 &= 0
\end{aligned}
$$

22.
$$
\begin{aligned}
x_1 + x_2 + 2x_3 + x_4 &= 5 \\
2x_1 + 2x_3 + x_4 &= 6 \\
x_2 + 3x_3 - x_4 &= 1 \\
3x_1 + 2x_2 + 2x_4 &= 7
\end{aligned}
$$

23. Using the matrix inverse method, solve the following system of equations

$$
\begin{aligned}
x + 2y - z &= b_1 \\
x + y + 2z &= b_2 \\
x - y - z &= b_3
\end{aligned}
\quad \text{for} \quad
\begin{pmatrix} b_1 \\ b_2 \\ b_3 \end{pmatrix} =
\begin{pmatrix} 1 \\ 2 \\ 3 \end{pmatrix},
\begin{pmatrix} 0 \\ 1 \\ 4 \end{pmatrix},
\begin{pmatrix} 5 \\ 2 \\ 3 \end{pmatrix}
\text{ in turn}
$$

Exercises 24–27 are on coding theory. Associate each letter with its position in the alphabet, as in Example 7.

*24. Encode the message RETREAT using the matrix $\begin{pmatrix} 4 & -3 \\ 3 & -2 \end{pmatrix}$.

25. Encode the message ATTACK AT DAWN using the matrix $\begin{pmatrix} 1 & 2 & 1 \\ 2 & 3 & 1 \\ -2 & 0 & 1 \end{pmatrix}$.

*26. Decode the message 49, 38, -5, -3, -61, -39, which was encoded using the matrix of Exercise 24.

27. Decode the message 71, 100, -1, 28, 43, -5, 84, 122, -11, 77, 112, -13, 61, 89, -8, 71, 104, -13, which was encoded using the matrix of Exercise 25.

28. Prove that a diagonal matrix is invertible if and only if all its diagonal elements are nonzero. Find a rule that can be used for determining the inverse of an invertible diagonal matrix.

29. An *upper triangular matrix* is a square matrix of the type

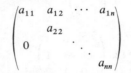

Prove that an upper triangular matrix is invertible if and only if all its diagonal elements are nonzero. Prove that the inverse of an invertible upper triangular matrix is itself an upper triangular matrix.

30. Let A be an invertible matrix and c a nonzero scalar. Prove that

(a) A^{-1} is invertible with $(A^{-1})^{-1} = A$

(b) $(cA)^{-1} = \dfrac{1}{c} A^{-1}$

(c) A^n is invertible with $(A^n)^{-1} = (A^{-1})^n$ for $n = 1, 2, 3, \ldots$

(d) If B and C are matrices such that $AB = AC$, then $B = C$.

*31. Let A be an invertible matrix having inverse $\begin{pmatrix} 2 & 1 \\ 4 & 3 \end{pmatrix}$. Determine the matrix A.

[*Hint:* Use Exercise 30(a).]

32. Consider the matrix $A = \begin{pmatrix} 1 & 2 \\ 3 & 4 \end{pmatrix}$ having inverse $\begin{pmatrix} -2 & 1 \\ 1.5 & -0.5 \end{pmatrix}$. Determine

(a) $(3A)^{-1}$,   (b) $(A^3)^{-1}$.

[*Hint:* Use Exercises 30(b) and (c).]

33. Check your answers to Exercises 8–18 using the computer (Appendix G).

34. Write a program for solving a system of linear equations using the inverse of the matrix of coefficients. Use your program to solve the system

$$\begin{aligned} x_1 + 2x_2 \quad\quad &= \quad 4 \\ 2x_1 + x_2 - x_3 &= \quad 2 \\ 3x_1 + x_2 + x_3 &= -2 \end{aligned}$$

The answer is $x_1 = -1$, $x_2 = 2.5$, $x_3 = -1.5$. Use your program to check your answers to Exercises 19–22.

35. Write a program that can be used for Exercise 23. Check your answers.

2-6* LEONTIEF INPUT-OUTPUT MODEL IN ECONOMICS

In this section, we introduce the Leontief input-output model that is used to analyze the interdependence of industries or sectors of an economic situation. Wassily Leontief received the Nobel Prize in Economic Science in 1973 for his work in this area. The practical applications of this model have proliferated; it has now become a standard tool for investigating economic structures ranging from cities and corporations to states and countries. The compilation of input-output tables has become part of the statistical program of all the developed countries and many of the less developed ones.

Consider an economic situation involving n interdependent industries. The output of any one industry is needed as input by other industries, and even possibly by the industry itself. We shall see how a mathematical model involving a system of linear equations can be constructed to analyze such a situation. Let us assume, for the sake of simplicity, that each industry produces one commodity. Let a_{ij} denote the amount of input of a certain commodity i to produce unit output of commodity j. The first subscript refers to input, the second to output. In our model let the amounts of input and output be in dollars. Thus, for example, $a_{34} = 0.45$ means that 45 cents' worth of commodity 3 is required to produce 1 dollar's worth of commodity 4.

The elements a_{ij}, called *input coefficients*, define a matrix $A = (a_{ij})$ called the *input-output matrix* which describes the interdependence of the industries involved. Note that each column of A specifies the input requirements for the production of one unit of the output of a particular industry. For example, if the system involves three industries and

$$A = \begin{pmatrix} a_{11} & a_{12} & a_{13} \\ a_{21} & a_{22} & a_{23} \\ a_{31} & a_{32} & a_{33} \end{pmatrix} = \begin{pmatrix} 0.25 & 0.40 & 0.50 \\ 0.35 & 0.10 & 0.20 \\ 0.20 & 0.30 & 0.10 \end{pmatrix}$$

then the elements in the second column give the inputs required from each industry to produce one dollar's worth of the commodity produced by the second industry.

Let us make two observations about input-output matrices:

1. The sum of the elements in each column corresponds to the total input cost of producing 1 dollar's worth of output. Let us assume that such a matrix describes an economically feasible situation; therefore, the sum of the elements in each column will be less than unity.

$$\sum_{i=1}^{n} a_{ij} < 1 \qquad \text{for } j = 1, 2, \ldots, n$$

2. It follows that each element of the matrix is less than unity, and by the interpretation given to these elements, each one must be greater than or equal to zero.

$$0 \le a_{ij} < 1 \qquad \text{for } i, j = 1, 2, \ldots, n$$

✦ ***Example 1***

A national input-output matrix describes interindustry relations that constitute the economic fabric of a country. We now display part of the matrix that describes the interindustry structure of the U.S. economy for 1972. The structure is actually described in terms of the flow among 79 producing sectors; the matrix is thus a 79 × 79 matrix. We cannot, of course, display the whole matrix because of lack of space. We list the first 10 sectors to give the reader a feel for the categories involved.

1. Livestock and livestock products
2. Agricultural crops
3. Forestry and fishery products
4. Agricultural, forestry, and fishery services
5. Iron and ferroalloy ores mining
6. Nonferrous metal ores mining
7. Coal mining
8. Crude petroleum and natural gas
9. Stone and clay mining and quarrying
10. Chemical and fertilizer mineral mining

The matrix based on these sectors is

	1	2	3	4	⋯
1	.26110	.02481	0	.05278	⋯
2	.23277	.03218	0	.01444	
3	0	0	.00467	.00294	
4	.02821	.03673	.02502	.02959	
5	0	0	0	0	
6	0	0	0	0	
7	0	.00002	0	0	
8	0	0	0	0	
9	.00001	.00251	0	.00034	
10	0	.00130	0	0	
⋮	⋮				

Thus, for example, $a_{72} = .00002$ implies that \$0.00002 from the coal mining sector (sector 7) goes into producing \$1 from the agricultural crops sector (sector 2).   ✦

We now extend the model to include an *open sector*. The products of the industries involved may go not only into the other producing industries but also into other nonproducing sectors of the economy such as consumers and governments. We group all these nonproducing classes together into what is called the *open sector*. The open sector in the 1972 model of the U.S. economy referred to above included, for example, both federal government purchases and state and local government purchases. Let d_i be the demand of the open sector from the ith industry. Let x_i be the total output of industry i required to meet the demands of all n industries and the open sector.

Since a_{i1} is the amount from industry i required to produce unit output in industry 1, $a_{i1}x_1$ will be the amount required to produce x_1 units of output. Similarly, $a_{i2}x_2$ will be required to produce x_2 units in industry 2, and so on. We find that

$$\underset{\substack{\uparrow \\ \text{\itshape total} \\ \text{\itshape output of} \\ \text{\itshape industry i}}}{x_i} = \underset{\substack{\uparrow \\ \text{\itshape demand of} \\ \text{\itshape industry 1}}}{a_{i1}x_1} + \underset{\substack{\uparrow \\ \text{\itshape demand of} \\ \text{\itshape industry 2}}}{a_{i2}x_2} + \cdots + \underset{\substack{\uparrow \\ \text{\itshape demand of} \\ \text{\itshape industry n}}}{a_{in}x_n} + \underset{\substack{\uparrow \\ \text{\itshape demand of} \\ \text{\itshape the open sector}}}{d_i}$$

Thus the output levels required of the entire set of n industries in order to meet these demands are given by the system of n linear equations

$$x_1 = a_{11}x_1 + a_{12}x_2 + \cdots + a_{1n}x_n + d_1$$
$$x_2 = a_{21}x_1 + a_{22}x_2 + \cdots + a_{2n}x_n + d_2$$
$$\vdots$$
$$x_n = a_{n1}x_1 + a_{n2}x_2 + \cdots + a_{nn}x_n + d_n$$

This system of n equations can be written in matrix form

$$\begin{pmatrix} x_1 \\ x_2 \\ \vdots \\ x_n \end{pmatrix} = \begin{pmatrix} a_{11} & a_{12} & \cdots & a_{1n} \\ a_{21} & a_{22} & \cdots & a_{2n} \\ \vdots & & & \\ a_{n1} & a_{n2} & \cdots & a_{nn} \end{pmatrix} \begin{pmatrix} x_1 \\ x_2 \\ \vdots \\ x_n \end{pmatrix} + \begin{pmatrix} d_1 \\ d_2 \\ \vdots \\ d_n \end{pmatrix}$$

Let us introduce the following notation:

$$X = \begin{pmatrix} x_1 \\ x_2 \\ \vdots \\ x_n \end{pmatrix}, \text{ the } output\ matrix, \quad \text{and} \quad D = \begin{pmatrix} d_1 \\ d_2 \\ \vdots \\ d_n \end{pmatrix}, \text{ the } demand\ matrix$$

The system of equations can now be written as a single matrix equation with the terms having the indicated significance:

$$\underset{\substack{\uparrow \\ \text{\itshape total} \\ \text{\itshape output}}}{X} = \underset{\substack{\uparrow \\ \text{\itshape interindustry} \\ \text{\itshape portion of} \\ \text{\itshape the output}}}{AX} + \underset{\substack{\uparrow \\ \text{\itshape open sector} \\ \text{\itshape portion of} \\ \text{\itshape the output}}}{D}$$

When the model is applied to the economy of a country, X gives the total output of each of the producing sectors of the economy and AX describes the contributions made by the various sectors to fulfilling the intersectional input requirements of the economy. D is equal to $(X - AX)$, or the difference between total output X and industry transaction AX; D represents the economy's GNP.

The equation $X = AX + D$ is applied in practice in a variety of ways depending on which variables are known and which are not known. For example, an analyst seeking to determine the implications of a change in government purchases or consumer's demand on the economy described by A might assign values to D and solve the equation for X—in other words, project the amount of output from each sector needed to attain various GNPs. (Example 2 will illustrate this application of the model.) On the other hand, an economist knowing the limited production capacity of an economic system described by A might want to consider X as given and solve the equation for D to project the maximum GNP the system can achieve. (Exercises 7, 8, and 9 illustrate this application of the model.)

✦ *Example 2*

Consider an economy consisting of three industries having the following input-output matrix A. Compute the output levels required of the industries to meet the demands of the other industries and of the open sector in each case.

$$A = \begin{pmatrix} \frac{1}{5} & \frac{1}{5} & \frac{3}{10} \\ \frac{1}{2} & \frac{1}{2} & 0 \\ 0 & 0 & \frac{1}{5} \end{pmatrix}, \qquad D = \begin{pmatrix} 9 \\ 12 \\ 16 \end{pmatrix}, \begin{pmatrix} 6 \\ 9 \\ 8 \end{pmatrix}, \begin{pmatrix} 12 \\ 18 \\ 32 \end{pmatrix}$$

The units of D are millions of dollars. We wish to compute the X's that correspond to the various D's.

X is given by $X = AX + D$. Let us rearrange the equation to bring the X's together:

$$X - AX = D$$
$$(I - A)X = D$$

To solve this system of linear equations for X, we can use either Gauss-Jordan elimination or the matrix inverse method. In practice the matrix inverse method is used here—a discussion of the merits of this approach is given below. We get

$$X = (I - A)^{-1}D$$

This is the equation that is used to determine X when A and D are given. For our matrix A,

$$I - A = \begin{pmatrix} 1 & 0 & 0 \\ 0 & 1 & 0 \\ 0 & 0 & 1 \end{pmatrix} - \begin{pmatrix} \frac{1}{5} & \frac{1}{5} & \frac{3}{10} \\ \frac{1}{2} & \frac{1}{2} & 0 \\ 0 & 0 & \frac{1}{5} \end{pmatrix} = \begin{pmatrix} \frac{4}{5} & -\frac{1}{5} & -\frac{3}{10} \\ -\frac{1}{2} & \frac{1}{2} & 0 \\ 0 & 0 & \frac{4}{5} \end{pmatrix}$$

$(I - A)^{-1}$ is computed using Gauss-Jordan elimination:

$$(I - A)^{-1} = \begin{pmatrix} \frac{5}{3} & \frac{2}{3} & \frac{5}{8} \\ \frac{5}{3} & \frac{8}{3} & \frac{5}{8} \\ 0 & 0 & \frac{5}{4} \end{pmatrix}$$

We can efficiently compute $X = (I - A)^{-1}D$ for each of the three values of D by forming a matrix having the various values of D as columns:

$$X = \underset{\substack{\uparrow \\ (I-A)^{-1}}}{\begin{pmatrix} \frac{5}{3} & \frac{2}{3} & \frac{5}{8} \\ \frac{5}{3} & \frac{8}{3} & \frac{5}{8} \\ 0 & 0 & \frac{5}{4} \end{pmatrix}} \underset{\substack{\uparrow \quad \uparrow \quad \uparrow \\ \textit{various values} \\ \textit{of } D}}{\begin{pmatrix} 9 & 6 & 12 \\ 12 & 9 & 18 \\ 16 & 8 & 32 \end{pmatrix}} = \underset{\substack{\uparrow \quad \uparrow \quad \uparrow \\ \textit{corresponding} \\ \textit{outputs}}}{\begin{pmatrix} 33 & 21 & 52 \\ 57 & 39 & 88 \\ 20 & 10 & 40 \end{pmatrix}}$$

The output levels of the three industries to meet the demands

$$\begin{pmatrix} 9 \\ 12 \\ 16 \end{pmatrix}, \quad \begin{pmatrix} 6 \\ 9 \\ 8 \end{pmatrix}, \quad \text{and} \quad \begin{pmatrix} 12 \\ 18 \\ 32 \end{pmatrix}$$

would have to be

$$\begin{pmatrix} 33 \\ 57 \\ 20 \end{pmatrix}, \quad \begin{pmatrix} 21 \\ 39 \\ 10 \end{pmatrix}, \quad \text{and} \quad \begin{pmatrix} 52 \\ 88 \\ 40 \end{pmatrix}$$

respectively, the units being millions of dollars.   ✦

In practice, as we saw in the case of the U.S. economy, analyses of this type often involve many sectors, implying large input-output matrices. There is usually a great deal of computation involved in implementing the model. Thus computers are used. However, computer time is expensive and efficient methods are needed. A great deal of research goes into discovering efficient algorithms for performing various matrix computations. In this model, we have observed that the elements of A are such that $0 \leq a_{ij} < 1$; in fact, most of the elements of A in practice are of small or zero magnitude, as in the illustration of the U.S. economy. This characteristic of A has led to an appropriate numerical method for computing $(I - A)^{-1}$ that makes the matrix inverse method more efficient than the Gauss-Jordan elimination method for solving $(I - A)X = D$ for X. We now describe this method of computing $(I - A)^{-1}$. Consider the following matrix multiplication for any positive integer m:

$$\begin{aligned}
(I - A)(I &+ A + A^2 + \cdots + A^m) \\
&= I(I + A + A^2 + \cdots + A^m) - A(I + A + A^2 + \cdots + A^m) \\
&= (I + A + A^2 + \cdots + A^m) - (A + A^2 + A^3 + \cdots + A^{m+1}) \\
&= I - A^{m+1}
\end{aligned}$$

The elements of successive powers of A become small rapidly and A^{m+1} approaches the zero matrix. Thus, for an appropriate m,

$$(I - A)(I + A + \cdots + A^m) = I$$

This implies that

$$(I - A)^{-1} = I + A + \cdots + A^m$$

This expression is used on the computer, in practice, to determine $(I - A)^{-1}$ for this model.

Today the concept of a world economy has become a tangible reality. A dramatic demonstration of the extent of global interdependence is provided by the recent "oil crisis," whose direct and indirect effects were felt in the furthest corners of five continents. We mentioned earlier how the input-ouput model has become a standard tool for investigating economic interdependences of systems ranging from cities to the U.S. economy. In 1973, the United Nations commissioned an input-ouput model of the world economy. This model was developed with special financial support from the Netherlands. The aim of the model was to transform the vast collection of economic facts that describe the world economy into an organized system from which economic projections could and have been made. These projections, for the years 1980, 1990, and 2000, became available in 1978.

In the model, the world is divided into 15 distinct geographic regions, each one described by an individual input-output matrix. The regions are then linked by a larger matrix which is used in an input-output model. Overall more than 200 variables enter into the model, and naturally it is implemented on the computer. By feeding in projected values for certain variables, researchers use the model in a variety of ways to create scenarios of future world economic possibilities. Readers who are interested in pursuing this world model further are encouraged to read "The World Economy of the Year 2000," by Wassily W. Leontief, in *Scientific American*, September, 1980, page 166.

EXERCISES

1. Consider the following input-output matrix that defines the interdependency of the given five industries.

	1	2	3	4	5
1. Auto	0.15	0.10	0.05	0.05	0.10
2. Steel	0.40	0.20	0.10	0.10	0.10
3. Electricity	0.10	0.25	0.20	0.10	0.20
4. Coal	0.10	0.20	0.30	0.15	0.10
5. Chemical	0.05	0.10	0.05	0.02	0.05

Determine

*(a) the amount of electricity consumed in producing $1 worth of steel

(b) the amount of steel consumed in producing $1 worth in the auto industry

*(c) the largest consumer of coal

(d) the largest consumer of electricity

*(e) On which industry is the auto industry most dependent?

Consider the following economies of either two or three industries. Determine the output levels required of each industry in each situation to meet the demands of the other industries and of the open sector. The units are millions of dollars.

2. $A = \begin{pmatrix} 0.20 & 0.60 \\ 0.40 & 0.10 \end{pmatrix}$, $D = \begin{pmatrix} 24 \\ 12 \end{pmatrix}$, $\begin{pmatrix} 8 \\ 6 \end{pmatrix}$, $\begin{pmatrix} 0 \\ 12 \end{pmatrix}$ in turn

*3. $A = \begin{pmatrix} 0.10 & 0.40 \\ 0.30 & 0.20 \end{pmatrix}$, $D = \begin{pmatrix} 6 \\ 12 \end{pmatrix}$, $\begin{pmatrix} 18 \\ 6 \end{pmatrix}$, $\begin{pmatrix} 24 \\ 12 \end{pmatrix}$ in turn

4. $A = \begin{pmatrix} 0.30 & 0.60 \\ 0.35 & 0.10 \end{pmatrix}$, $D = \begin{pmatrix} 42 \\ 84 \end{pmatrix}$, $\begin{pmatrix} 0 \\ 10 \end{pmatrix}$, $\begin{pmatrix} 14 \\ 7 \end{pmatrix}$, $\begin{pmatrix} 42 \\ 42 \end{pmatrix}$ in turn

*5. $A = \begin{pmatrix} 0.20 & 0.20 & 0.10 \\ 0 & 0.40 & 0.20 \\ 0 & 0.20 & 0.60 \end{pmatrix}$, $D = \begin{pmatrix} 4 \\ 8 \\ 8 \end{pmatrix}$, $\begin{pmatrix} 0 \\ 8 \\ 16 \end{pmatrix}$, $\begin{pmatrix} 8 \\ 24 \\ 8 \end{pmatrix}$ in turn

6. $A = \begin{pmatrix} 0.20 & 0.20 & 0 \\ 0.40 & 0.40 & 0.60 \\ 0.40 & 0.10 & 0.40 \end{pmatrix}$, $D = \begin{pmatrix} 36 \\ 72 \\ 36 \end{pmatrix}$, $\begin{pmatrix} 36 \\ 0 \\ 18 \end{pmatrix}$, $\begin{pmatrix} 36 \\ 0 \\ 0 \end{pmatrix}$, $\begin{pmatrix} 0 \\ 18 \\ 18 \end{pmatrix}$ in turn

Consider the following economies of either two or three industries. The output level of each industry is given. Determine the amounts available for the open sector from each industry.

*7. $A = \begin{pmatrix} 0.20 & 0.40 \\ 0.50 & 0.10 \end{pmatrix}$, $X = \begin{pmatrix} 8 \\ 10 \end{pmatrix}$

8. $A = \begin{pmatrix} 0.10 & 0.20 & 0.30 \\ 0 & 0.10 & 0.40 \\ 0.50 & 0.40 & 0.20 \end{pmatrix}$, $X = \begin{pmatrix} 10 \\ 10 \\ 20 \end{pmatrix}$

*9. $A = \begin{pmatrix} 0.10 & 0.10 & 0.20 \\ 0.20 & 0.10 & 0.30 \\ 0.40 & 0.30 & 0.15 \end{pmatrix}$, $X = \begin{pmatrix} 6 \\ 4 \\ 5 \end{pmatrix}$

10. Write a computer program that can be used to analyze the situations of Exercises 2–6. Let the program use the built-in matrix inverse function MAT B = INV (C) of BASIC to compute $(I - A)^{-1}$. Check your answers to Exercises 2–6.

11. Let A be a matrix that describes the interdependence of industries; all its elements will be nonnegative and each column will add up to less than 1. Write a computer program for determining output levels that incorporates the method $(I - A)^{-1} = I + A + A^2 + \cdots + A^m$ for determining the inverse of $I - A$.

2-7*   STOCHASTIC MATRICES: A POPULATION MOVEMENT MODEL

Certain types of matrices, called *stochastic matrices*, are of great use in the study of random phenomena where the exact outcome is not known but probabilities can be determined. In this section, we introduce stochastic matrices, derive some of their properties, and give examples of their application. One example is an analysis of population movement between a city and its suburbs. The trend from the city to suburbia is one that should be of concern not only to sociologists and politicians, but to all of us. This example shows how a mathematical model can be constructed to analyze this important problem.

If the outcome of an event is certain to occur, we say that its probability is 1. On the other hand, if it will not occur, its probability is said to be 0. Other probabilities can be represented by fractions between 0 and 1: the larger the fraction, the greater the probability p of that outcome occurring. Thus we have the restriction $0 \le p \le 1$ for a probability p.

If any one of n completely independent outcomes is equally likely and if m of these are of interest to us, then the probability p that one of these outcomes will occur is defined to be the fraction m/n.

As an example, let us work out the probability that, when a coin is flipped, it will land heads. There are two possible outcomes, one of which is of interest (landing heads). The probability p of this occurrence is $\frac{1}{2}$.

As another example, consider the event of drawing a single card from a deck of 52 playing cards. What is the probability that the card will be either an ace or a king? First of all, we see that there are 52 possible outcomes. There are 4 aces and 4 kings in the pack. Since there are 8 outcomes of interest to us, the probability is $\frac{8}{52}$, or $\frac{2}{13}$.

Let us now discuss special matrices whose elements are probabilities.

Definition 2-5   A *stochastic matrix* is a square matrix whose elements are probabilities and whose rows each add up to 1.

Thus, if P is a stochastic matrix, then P is square and

$$0 \le p_{ij} \le 1 \qquad \text{and} \qquad \sum_j p_{ij} = 1$$

The following matrices are stochastic matrices:

$$\begin{pmatrix} \frac{1}{2} & \frac{1}{2} \\ \frac{1}{3} & \frac{2}{3} \end{pmatrix} \qquad \begin{pmatrix} 0 & 1 \\ \frac{3}{4} & \frac{1}{4} \end{pmatrix} \qquad \begin{pmatrix} 1 & 0 & 0 \\ 0 & \frac{1}{2} & \frac{1}{2} \\ \frac{3}{4} & \frac{1}{8} & \frac{1}{8} \end{pmatrix}$$

The matrix $\begin{pmatrix} \frac{1}{2} & 1 \\ \frac{3}{4} & \frac{1}{4} \end{pmatrix}$ would not be a stochastic matrix—its first row is not a stochastic vector because the sum of its elements is not 1.

A general 2×2 stochastic matrix can be represented by

$$\begin{pmatrix} x & 1 - x \\ y & 1 - y \end{pmatrix}$$

where $0 \le x \le 1$, $0 \le y \le 1$.

Stochastic matrices have certain properties. The following theorem illustrates one of these properties.

Theorem 2-12   *If A and B are stochastic matrices of the same kind, then AB is a stochastic matrix.*

Proof: We shall prove this theorem for 2×2 stochastic matrices. Let

$$\begin{pmatrix} x & 1 - x \\ y & 1 - y \end{pmatrix} \quad \text{and} \quad \begin{pmatrix} z & 1 - z \\ m & 1 - m \end{pmatrix}$$

represent two 2×2 arbitrary stochastic matrices. We know that their product is

$$\begin{pmatrix} xz + (1 - x)m & x(1 - z) + (1 - x)(1 - m) \\ yz + (1 - y)m & y(1 - z) + (1 - y)(1 - m) \end{pmatrix}$$

Consider the first row. The sum of its components is

$$xz + (1 - x)m + x(1 - z) + (1 - x)(1 - m)$$
$$= xz + m - xm + x - xz + 1 - x - m + xm = 1$$

Next consider the $(1, 1)$ element, $xz + (1 - x)m$. $x \ge 0$, $z \ge 0$, and $(1 - x)m \ge 0$, since $x \le 1$ and $m \ge 0$. Hence this element is ≥ 0. Similarly, the $(1, 2)$ element $x(1 - z) + (1 - x)(1 - m)$ is ≥ 0. The elements must both be ≤ 1 since their sum is 1. Hence the first row satisfies the requirements of a stochastic matrix.

It can be shown in a similar manner that the second row satisfies the requirements, proving that the product matrix is a stochastic matrix.

It follows from this result that if A is a stochastic matrix, then so are A^2 and all A^n, where n is any positive integer.

Now let us look at applications of stochastic matrices in the social sciences.

✦ *Example 1* ————————————————————————————————

This example gives an analysis of land use succession using available statistics for center-city Toronto for the period 1952–1962.

The data are represented in the form of a stochastic matrix (p_{ij}), where each element p_{ij} represents the probability of succession from use i to use j. In this context, the stochastic matrix is called a *matrix of transition probabilities*. In this model, change is defined to include all new construction, demolitions, or major structural adjustments in the building stock. The matrix in Table 2-1 is stochastic; each element satisfies $0 \leq p_{ij} \leq 1$, and the sum of the elements in each row is 1.

Table 2-1. *Land Use Succession and Transition Probabilities for the Total City, 1952–1962*

1952 *Existing use*	Terminal use									
	1	2	3	4	5	6	7	8	9	10
1. Low density residential	0.13	0.34	0.10	0.04	0.04	0.22	0.03	0.02	0.00	0.08
2. High-density residential	0.02	0.41	0.05	0.04	0.00	0.04	0.00	0.00	0.00	0.44
3. Office	0.00	0.07	0.43	0.05	0.01	0.28	0.14	0.00	0.00	0.02
4. General commercial	0.02	0.01	0.09	0.30	0.09	0.27	0.05	0.08	0.01	0.08
5. Auto commercial	0.00	0.00	0.11	0.07	0.70	0.06	0.00	0.01	0.00	0.05
6. Parking	0.08	0.05	0.14	0.08	0.12	0.39	0.04	0.00	0.01	0.09
7. Warehousing	0.01	0.03	0.02	0.12	0.03	0.11	0.38	0.21	0.01	0.08
8. Industry	0.01	0.02	0.02	0.03	0.03	0.08	0.18	0.61	0.00	0.02
9. Transportation	0.01	0.18	0.14	0.04	0.10	0.39	0.03	0.03	0.08	0.00
10. Vacant	0.25	0.08	0.03	0.03	0.05	0.15	0.22	0.13	0.00	0.06

Let us now interpret some of the information contained in this matrix.

For example, $p_{36} = 0.28$. This represents the probability that what was an office in 1952 had become a parking space by 1962. The sixth column gives the probabilities that various areas had become parking areas by 1962. These relatively large figures reveal the increasingly dominant role of parking in land use. Relatively high proportions of all types of land were converted into parking.

The diagonal elements represent the probabilities that land use remained in the same category. For example, $p_{77} = 0.38$ represents the probability that warehousing land remained warehousing land. The relatively high figures of these diagonal numbers reveal the marked tendency for land to remain in the same broad category. The exceptions are land used for transportation and vacant land.

It is interesting that $p_{2\ 10}$ is equal to 0.44. This is the probability that area that was high-density residential in 1952 had become vacant by 1962. Note that this is an area considered to be in center-city Toronto.

The reader who is interested in following up this type of analysis is encouraged to read the article from which these data have been abstracted: Larry S. Bourne, "Physical Adjustment Processes and Land Use Succession: Toronto," in *Economic Geography*, **47**, No. 1, January, 1971, pages 1–15. Another article by the same author which outlines the use of such matrices is "A Spatial Allocation—Land Use Conversion Model of Urban Growth," in *Journal of Regional Science*, 9, August, 1969, pages 261–272. References to other similar works are found in both articles. ✦

✦ *Example 2*

This example is an analysis of the population movement between cities and their surrounding suburbs in the United States. The numbers given are based on statistics in *Statistical Abstracts of the U.S.A.*, 1969–1971.

The number of people (in thousands of persons one year old or over) who lived in cities in the U.S. during 1971 was 57,633. The number of people who lived in the surrounding suburbs was 71,549. We represent this information by the matrix $X_0 = (57{,}633 \quad 71{,}549)$.

Consider the population flow from city centers to suburbs. During the period 1969–1971, the average probability per year of a person staying in the city was 0.96. Thus the probability of moving to the suburbs was 0.04 (assuming that all those moving went to the suburbs).

Consider now the reverse population flow from suburbia to city. The probability of a person moving to the city was 0.01; the probability of a person remaining in suburbia was 0.99.

These probabilities can be written in the form of a stochastic matrix

$$P = (p_{ij}) = \begin{array}{c} \\ Initial \end{array} \begin{array}{cc} & Final \\ & \begin{array}{cc} City & Suburb \end{array} \\ \begin{array}{c} City \\ Suburb \end{array} & \begin{pmatrix} 0.96 & 0.04 \\ 0.01 & 0.99 \end{pmatrix} \end{array}$$

To determine the probability of moving from location A to location B, look up row A and column B.

Consider the population of the city centers in 1972—one year later. 0.96 is the probability of a person living in the city staying there; it represents the fraction of those living in the city who stay there. We have that

City population in 1972 = people who remained from 1971
+ people who moved in from the suburbs
$$= (57{,}633 \times 0.96) + (71{,}549 \times 0.01)$$
$$= 56{,}043.2$$

Similarly,

Suburban population in 1972 = people who moved in from cities
+ people who stayed, from 1971
$$= (57{,}633 \times 0.04) + (71{,}549 \times 0.99)$$
$$= 73{,}138.8$$

Note that we can arrive at these two numbers using matrix multiplication.

$$(57{,}633 \quad 71{,}549) \begin{pmatrix} 0.96 & 0.04 \\ 0.01 & 0.99 \end{pmatrix} = (56{,}043.2 \quad 73{,}139.8)$$

or

$$X_1 = X_0 P$$

where X_0 is the population matrix in 1971, P is the stochastic matrix giving probability distributions, and X_1 is the population matrix one year later.

Assuming that the population flow represented by the matrix P is unchanged, the population distribution X_2 after 2 years is given by

$$X_2 = X_1 P$$

that is,

$$X_2 = X_0 P^2$$

After 3 years the population distribution is given by

$$X_3 = X_2 P$$
$$= X_0 P^3$$

Assuming that the matrix of transition probabilities does not vary, we can predict the distribution any number of years later from the general result

$$X_n = X_0 P^n$$

Observe that the matrix P^n is a stochastic matrix and that it does "take X_0 into X_n," n stages later. This result can be generalized to

$$X_{i+n} = X_i P^n$$

That is, P^n can be used to predict the distribution n stages later from any given distribution. P^n is called the *n-step transition matrix*. Furthermore, the (i, j)th element of P^n gives the probability of going from state S_i to state S_j in n steps.

The predictions of this model are

	City	Suburb
$X_1 =$	(56043.2	73138.8)
$X_2 =$	(54532.8	74649.2)
$X_3 =$	(53098	76084)
$X_4 =$	(51734.9	77447.1)
$X_5 =$	(50440	78742)

etc.

Observe how the city population is decreasing annually while that of the suburbs is increasing. We return to this model in Section 7-2. There we find that the sequence $X_1, X_2, X_3, \ldots$ approaches (25836.4   103345.6).

The probabilities in this model depend only on the current status of a person—whether he or she is living in the city or suburbia. This type of model, where the probability of going from one state to another depends only on the current state rather than on a more complete historical description, is called a *Markov chain*.

These concepts can be extended to Markov processes involving more than two states. The reader meets a model involving three states in a following example.

A further modification of the model would give an improved estimate of future population distributions. Let us take into account the fact that the population of the United States increased by 1% per year during the period 1969–1971. We assume that the population will increase by the same amount annually during the years immediately following 1971.

Thus, if the population was b during any year, the population the following year would be $b + 1\%$ of b.

$$b + b/100 = \tfrac{101}{100}b$$

Our model now becomes

$$X_1 = \tfrac{101}{100}X_0 P$$

Note that since $\tfrac{101}{100}$ is a scalar, we are multiplying a matrix by a scalar and so the multiplying sequence does not matter.

$$X_1 = \tfrac{101}{100}(57{,}633 \quad 71{,}549)\begin{pmatrix} 0.96 & 0.04 \\ 0.01 & 0.99 \end{pmatrix}$$

$$= (56603.632 \quad 73870.188)$$

The population of city centers is now 56603.632 thousand and that of suburbia is 73870.188 thousand.

For 1973, furture population distribution based on a uniform 1% annual population increase would be

$$X_2 = \left(\tfrac{101}{100}\right)^2 X_0 P^2$$

After n years, it would be

$$X_n = \left(\tfrac{101}{100}\right)^n X_0 P^n$$

Note that in this model we have made the assumption that the increase in population is 1% in both cities and suburbia since it is 1% nationally. The increase might be slightly higher than 1% in the cities and correspondingly less than 1% in suburbia. A more accurate model would incorporate a breakdown of this growth statistic. The reader is asked to carry out this modification in Exercise 10. ✦

✦ *Example 3*
_____

Markov chains are useful tools for scientists in many fields. We now discuss the role of Markov chains in the field of *genetics*.

Genetics is the branch of biology that deals with heredity. It is the study of units called genes which determine the characteristics living things inherit from their parents. The inheritance of such traits as sex, height, eye color, and hair color of human beings, and such traits as petal color and leaf shape of plants, are governed by genes. Because many diseases are inherited, genetics is important in medicine. In agriculture, breeding methods based on genetic principles led to important advances in both plant and animal breeding. High-yield hybrid corn ranks as one of the most important contributions of genetics to increasing food production. We shall consider a model developed for analyzing the behavior of simple traits involving a pair of genes. We illustrate the concepts involved in terms of the crossing of guinea pigs.

The traits that we shall study in the guinea pigs are the traits of long hair and short hair. The length of hair is governed by a pair of genes which we shall denote A and a. A guinea pig may have any one of the combinations AA, Aa (which is genetically the same as aA), or aa. Each of these classes is called a *genotype*. The AA type guinea pig is indistinguishable in appearance from the Aa type—both have long hair—while the aa type has short hair. We say that the A gene *dominates* the a genes. An animal is called *dominant* if it has AA genes, *hybrid* with Aa genes, and *recessive* with aa genes.

When two guinea pigs are crossed, the offspring inherits one gene from each parent in a random manner. Given the genotypes of the parents, we can determine the probabilities of the genotype of the offspring. Consider a given population of guinea pigs. Let us perform a series of experiments in which *we keep crossing offspring with dominant animals only*. Thus we keep crossing AA, Aa, and aa with AA. What are the probabilities of the offspring being AA, Aa, or aa in each of these cases?

Consider the crossing of AA with AA. The offspring will have one gene from each parent, so it will be of type AA. Thus the probabilities of AA, Aa, and aa resulting are 1, 0, and 0, respectively. All offspring will have long hair.

Next consider the crossing of Aa with AA. Taking one gene from each parent, we have the possibilities AA, AA (taking the A from the first parent with each A in turn from the second parent), aA, and aA (taking the a from the first parent with each A in turn from the second parent). Thus the probabilities of AA, Aa, and aa resulting are $\frac{1}{2}$, $\frac{1}{2}$, and 0, respectively. All offspring again have long hair.

Finally, on crossing aa with AA, we get the possibilities aA, aA (taking the first a in aa with each A in AA in turn), aA, and aA. Thus the probabilities of AA, Aa, and aa are 0, 1, and 0, respectively.

All offspring resulting from these experiments have long hair. This series of experiments is a Markov chain having transition matrix

$$P = \begin{array}{c} \\ AA \\ Aa \\ aa \end{array} \begin{array}{ccc} AA & Aa & aa \\ \begin{pmatrix} 1 & 0 & 0 \\ \frac{1}{2} & \frac{1}{2} & 0 \\ 0 & 1 & 0 \end{pmatrix} \end{array}$$

Consider an initial population of guinea pigs made up of an equal number of each genotype. Let the initial matrix X_0 be $(\frac{1}{3} \quad \frac{1}{3} \quad \frac{1}{3})$, representing the fraction of guinea pigs of each type initially. The components of $X_1, X_2, X_3, \ldots$

will give the fractions of following generations that are of each genotype. We get

$$
\begin{array}{cccc}
 & AA & Aa & aa \\
X_0 = & (\ \frac{1}{3} & \frac{1}{3} & \frac{1}{3} \ ) \\
X_1 = X_0 P & = (\ \frac{1}{2} & \frac{1}{2} & 0 \ )
\end{array}
$$
first generation; distribution is $\frac{1}{2}$ type AA, $\frac{1}{2}$ type Aa

$$
X_2 = X_0 P^2 = (\ \tfrac{3}{4} \quad \tfrac{1}{4} \quad 0 \ )
$$
second generation; distribution is $\frac{3}{4}$ type AA, $\frac{1}{4}$ type Aa

$$
X_3 = X_0 P^3 = (\ \tfrac{7}{8} \quad \tfrac{1}{8} \quad 0 \ )
$$
$$
X_4 = X_0 P^4 = (\ \tfrac{15}{16} \quad \tfrac{1}{16} \quad 0 \ )
$$

etc.

Observe that the *aa* type disappears after the initial generation and that the *Aa* type becomes a smaller and smaller fraction for each successive generation. The sequence in fact approaches the matrix

$$
\begin{array}{ccc}
AA & Aa & aa
\end{array}
$$
$$
X = (\ 1 \quad 0 \quad 0)
$$

The genotype *AA* in this model is called an *absorbing state*.

Here we have considered the case of crossing offspring with a dominant animal. The reader is asked to construct a similar model that describes the crossing of offspring with a hybrid in Exercise 12. Some of the offspring will have long hair and some short hair in that series of experiments. ◆

EXERCISES

*1. State which of the following matrices are stochastic and which are not.

(a) $\begin{pmatrix} \frac{1}{4} & \frac{3}{4} \\ 0 & 1 \end{pmatrix}$ (b) $\begin{pmatrix} \frac{1}{2} & \frac{1}{2} \\ -1 & 0 \end{pmatrix}$ (c) $\begin{pmatrix} \frac{1}{2} & \frac{1}{4} & \frac{1}{4} \\ 1 & 0 & 0 \\ \frac{1}{8} & \frac{1}{8} & \frac{3}{4} \end{pmatrix}$

2. A stochastic matrix the sum of whose columns is 1 is called *doubly stochastic*. Give an example of a doubly stochastic matrix. Is the product of two doubly stochastic matrices of the same kind doubly stochastic?

*3. Use the matrix of transition probabilities in Example 1 to answer the following questions:

(a) What is the probability that land used for industry in 1952 was used for offices in 1962?

(b) What is the probability that land used for parking in 1952 was in a high-density residential area in 1962?

(c) Vacant land in 1952 had the highest probability of becoming what kind of land by 1962?

(d) Which was the most stable usage of land over the period 1952–1962?

4. Assuming the same matrix of transition probabilities for land usage in Toronto for the period 1962–1972 as for the period 1952–1962, determine the matrix of transition probabilities for the period 1952–1972. Generalize this result.

*5. In the original model in Example 2, determine

   (a) the probability of moving from center city to the suburbs in two years

   (b) the probability of moving from the suburbs to center city in three years.

6. Draw up a model of population flow between metropolitan areas and nonmetropolitan areas given that their respective populations in 1971 were 129,182 and 69,723 (in thousands of persons one year old or over). The probabilities are given by the matrix

$$\text{Initial} \quad \begin{array}{c} \\ \text{Metro} \\ \text{Nonmetro} \end{array} \overset{\displaystyle \begin{array}{c} \textit{Final} \\ \text{Metro} \quad \text{Nonmetro} \end{array}}{\begin{pmatrix} 0.99 & 0.01 \\ 0.02 & 0.98 \end{pmatrix}}$$

7. Construct a model of population flow between cities, suburbs, and nonmetropolitan areas. Their respective populations in 1971 were 57,633, 71,549, and 69,723 (in thousands of persons one year old or over). The stochastic matrix giving the probabilities of the moves is

$$\text{Initial} \quad \begin{array}{c} \\ \text{City} \\ \text{Suburb} \\ \text{Nonmetro} \end{array} \overset{\displaystyle \begin{array}{c} \textit{Final} \\ \text{City} \quad \text{Suburb} \quad \text{Nonmetro} \end{array}}{\begin{pmatrix} 0.96 & 0.03 & 0.01 \\ 0.01 & 0.98 & 0.01 \\ 0.015 & 0.005 & 0.98 \end{pmatrix}}$$

This model is a refinement on the model of Exercise 6 in that the metro population is broken down into city and suburb. It is also a more complete model than that of Example 2 of this section, leading to more accurate predictions than that model.

*8. The following stochastic matrix gives occupational transition probabilities.

$$\begin{array}{c} \textit{Initial} \\ \textit{Generation} \end{array} \quad \begin{array}{c} \\ \text{White collar} \\ \text{Manual} \end{array} \overset{\displaystyle \begin{array}{c} \textit{Following Generation} \\ \text{White collar} \quad \text{Manual} \end{array}}{\begin{pmatrix} 1 & 0 \\ 0.2 & 0.8 \end{pmatrix}}$$

   (a) If the father is a manual worker, what is the probability that the son will be a white-collar worker?

   (b) If there are 10,000 in the white-collar category and 20,000 in the manual category, what will be the distribution one generation later? Three generations later?

9. The following stochastic matrix gives occupational transition probabilities.

$$\begin{array}{c} \textit{Initial} \\ \textit{Generation} \end{array} \quad \begin{array}{c} \\ \text{Nonfarming} \\ \text{Farming} \end{array} \overset{\displaystyle \begin{array}{c} \textit{Following Generation} \\ \text{Nonfarming} \quad \text{Farming} \end{array}}{\begin{pmatrix} 1 & 0 \\ 0.4 & 0.6 \end{pmatrix}}$$

(a) If the father is a farmer, what is the probability that the son will be a farmer?

(b) If there are 10,000 in the nonfarming category and 1,000 in the farming category at a certain time, what will the distribution be one generation later? Four generations later?

(c) If the father is a farmer, what is the probability that the grandson (two generations later) will be a farmer?

10. Carry out the proposed modification of the model in Example 2. Assume that the population increase (that allows for births and deaths) per year in the cities is 1.2% and in suburbia is 0.8%. [*Hint*: Represent this information in the form of a 2×2 matrix A. Then $X_1 = X_0 PA$, $X_2 = X_0 (PA)(PA)$, ..., $X_n = X_0 (PA)^n$.] Note that the order of multiplication does matter here, since matrices are not commutative (unlike the situation in the previous model, where the scalar $\frac{101}{100}$ was used).

11. A certain society is made up of two groups, a large majority group and a small minority group. During a certain period in the history of the society, it is fashionable to become part of the majority group by marrying into this group, by adopting the culture of the group, etc. Initially, the populations of a single generation of the two groups are 50 million and $\frac{1}{2}$ million. The following is a matrix of transition probabilities.

		Following Generation	
		Majority group	Minority group
Initial	Majority group	1	0
Generation	Minority group	0.2	0.8

The matrix represents the probabilities of a following generation being in a certain group.

(a) Assuming that the total population of each successive group remains the same, construct a model that gives the populations after n generations.

(b) Allow for a 1% uniform population increase in each successive generation.

(c) Assume that in each generation there is a 2% increase in population of the majority group and a 3% decrease in population of the minority group due to the births and deaths. Allow for these figures in your model.

A state is said to be an *absorbing state* if it is not possible to leave it. The majority group is an absorbing state in all the models of this exercise, the probability of leaving the majority group being 0. The minority group is being absorbed into the majority group. The minority culture will be lost to future generations unless the society acts.

12. Construct a model, similar to that of Example 3, that describes the crossing of offspring of guinea pigs with hybrids only. (We shall pursue this model further in Exercise 4, Section 7-2.) There is no absorbing state here.

13. (a) Write a computer program for the model of Example 2 that gives the city and suburb populations annually from 1971 to 1981.

(b) Modify the program to allow for a uniform 1% annual increase in population. Use your program to observe the effect of varying the uniform increase. This increase is called a *parameter* of the model.

(c) Modify the above program to allow for a 1.2% annual increase in city population and an 0.8% annual increase in suburbia population (Exercise 10) due to births and deaths. Use your program to observe the effect of varying these parameters.

*14. A market analysis group studying car purchasing trends in a certain region has concluded that a new car is purchased once every 3 years on an average.

The buying patterns are described by the matrix

$$P = \begin{matrix} & \text{Small} & \text{Large} \\ \text{Small} & \\ \text{Large} & \end{matrix} \begin{pmatrix} 80\% & 20\% \\ 40\% & 60\% \end{pmatrix}$$

The elements of P are to be interpreted as follows. The first row indicates that of the current small cars, 80% will be replaced with a small car, 20% with a large car. The second row implies that 40% of the current large cars will be replaced with small cars while 60% will be replaced by large cars.

Write the elements of P as follows:

$$\begin{pmatrix} \frac{80}{100} & \frac{20}{100} \\ \frac{40}{100} & \frac{60}{100} \end{pmatrix} = \begin{pmatrix} 0.8 & 0.2 \\ 0.4 & 0.6 \end{pmatrix}$$

P is thus a stochastic matrix; it defines a Markov chain model of the buying trends.

If there are currently 40,000 small cars and 50,000 large cars in the region, what is your prediction of the distribution in 12 years' time?

15. The conclusion of an analysis of voting trends in a certain state is that the voting patterns of successive generations are described by the following matrix P.

$$P = \begin{matrix} & \text{Dem.} & \text{Rep.} & \text{Ind.} \\ \text{Democrat} & \\ \text{Republican} & \\ \text{Independent} & \end{matrix} \begin{pmatrix} 80\% & 15\% & 5\% \\ 20\% & 70\% & 10\% \\ 60\% & 30\% & 10\% \end{pmatrix}$$

Among the Democrats of one generation, 80% of their next generation are Democrats, 15% are Republicans, 5% are independents, etc. Express P as a stochastic matrix that defines a Markov chain model of these voting patterns.

If there are 2.5 million registered Democrats, 1.5 million registered Republicans, and 0.25 million registered independents at a certain period, what is the distribution likely to be in the next generation?

2-8 COMMUNICATION MODEL AND GROUP RELATIONSHIPS

Many branches of the social sciences and business use models from an area of mathematics called *graph theory* to analyze relationships among elements of a group. The group could consist of people within a certain company; for example, the techniques of graph theory can be used to analyze the hierarchy within a company. Sociologists apply the results of graph theory to analyze the relations among people that make up a group. Principles of graph theory are also used to improve the efficiency of computer networks. In this section, we introduce the reader to this increasingly important field which uses matrix algebra. To motivate the concepts involved, let us look at the following communication system.

Consider a communication system involving five stations labeled $P_1, \ldots,$ P_5. The communication links could be roads, phone lines, etc. Certain stations are linked by two-way communication links, others by one-way links. Still others have only indirect communication by way of intermediate stations. Suppose the network of interest is described by the diagram in Figure 2-12. Curves joining stations represent links; the arrows give the directions of those links. Stations P_1 and P_2 have two-way direct communication. Station P_3 can send a message to station P_5 directly, whereas station P_5 can only send a message to P_3 by way of P_4. Station P_4 can send a message to P_1 by way of stations P_3 and P_2 or by way of P_5 and P_2.

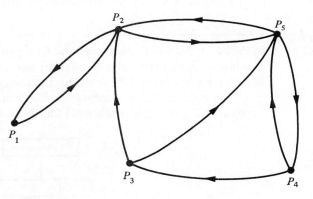

Figure 2-12

Such communication networks can be vast, involving many stations. The diagrams of vast systems become complicated and difficult to interpret in this geometrical manner. The mathematical vocabulary and theory we now present can be used to analyze such networks. This theory can be used, for example, to give the minimum number of links needed to send a message from one station to another.

A *graph*† is a finite collection of vertices $P_1, P_2, \ldots, P_n$ together with a finite number of edges joining certain pairs of vertices. Figure 2-13 gives some examples of graphs. Observe that it is possible to have an edge from a vertex to itself (P_1 in the second example).

In the communication network there is a sense of direction in the edges. This leads to a special class of graphs called digraphs. We now focus on this class.

A *digraph* (for "directed graph") is a graph in which each edge carries a direction, with not more than one edge in any one direction between any pair

† In this context the term graph has a meaning different from that of the graph of a function.

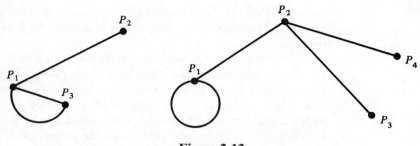

Figure 2-13

of vertices. The graph of the communication network is a digraph. Graphs having more than a single edge in a direction between two vertices are called *multidigraphs.* We shall not be concerned with such graphs here.

✦ *Example 1* ──

In 1921 E. I. duPont de Nemours and Company Ltd. became the first company to establish committees as a permanent part of its top management structure. The digraph in Figure 2-14 describes the current internal structure of DuPont.

The industrial departments are Elastomer Chemicals, Electrochemicals, Explosives, Fabrics and Finishes, Film, Industrial and Biochemicals, Interna-

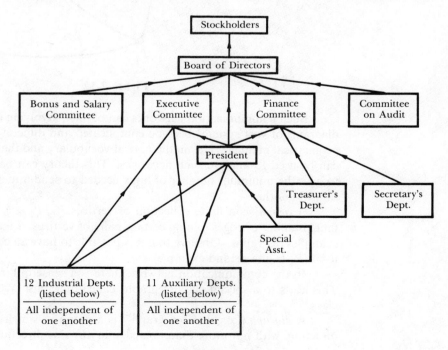

Figure 2-14

tional, Organic Chemicals, Photo Products, Pigments, Polychemicals, and Textile Fibres. Each of these departments is under a General Manager.

The auxiliary departments are Advertising, Central Research, Development, Employee Relations, General Services, Legal, Public Relations, Purchasing, Traffic (each of which is under a Director), Engineering (under a Chief Engineer), and Economist.

Readers who desire further discussion of the DuPont structure should consult *Management and Organization*, 2nd ed., by Henry L. Sisk, South-Western Publishing Co., 1973, from which Figure 2-14 is reproduced by special permission. The text contains further references to the internal organization of DuPont and other companies.   ✦

It is convenient to represent digraphs by matrices. These representations, as we shall see, can be used to analyze digraphs. The *adjacency matrix A* of a digraph consists of ones and zeros and is defined by

$$a_{ij} = \begin{cases} 1 & \text{if there is an edge from vertex } P_i \text{ to vertex } P_j \\ 0 & \text{otherwise} \end{cases}$$

The adjacency matrix of the communication digraph is

$$A = \begin{pmatrix} 0 & 1 & 0 & 0 & 0 \\ 1 & 0 & 0 & 0 & 1 \\ 0 & 1 & 0 & 0 & 1 \\ 0 & 0 & 1 & 0 & 1 \\ 0 & 1 & 0 & 1 & 0 \end{pmatrix}$$

For example, $a_{32} = 1$, since there is an edge from P_3 to P_2; $a_{34} = 0$, since there is no edge from P_3 to P_4.

✦ *Example 2*
___

The network in Figure 2-15 describes a system of one-way streets in a downtown area. Let us interpret the network as a digraph and give the adjacency matrix of this network.

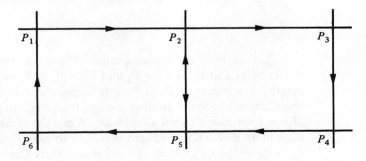

Figure 2-15

Label the intersections $P_1, \ldots, P_6$; these are the vertices of the digraph. The adjacency matrix is

$$A = \begin{pmatrix} 0 & 1 & 0 & 0 & 0 & 0 \\ 0 & 0 & 1 & 0 & 1 & 0 \\ 0 & 0 & 0 & 1 & 0 & 0 \\ 0 & 0 & 0 & 0 & 1 & 0 \\ 0 & 1 & 0 & 0 & 0 & 1 \\ 1 & 0 & 0 & 0 & 0 & 0 \end{pmatrix} \quad \blacklozenge$$

A (*chain* or *path*) between two vertices of a digraph is a sequence of directed edges that allows one to proceed in a continuous manner from one vertex to the other. For example, in the communication digraph (digraph and adjacency matrix in Figure 2-16), there is a chain from P_1 to P_4; this is denoted $P_1 \to P_2 \to P_5 \to P_4$. The adjacency matrix is

$$A = \begin{pmatrix} 0 & 1 & 0 & 0 & 0 \\ 1 & 0 & 0 & 0 & 1 \\ 0 & 1 & 0 & 0 & 1 \\ 0 & 0 & 1 & 0 & 1 \\ 0 & 1 & 0 & 1 & 0 \end{pmatrix}$$

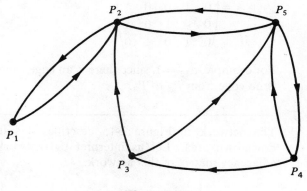

Figure 2-16

Let us look at certain chains. There are various chains from P_3 to P_1; let us look at $P_3 \to P_2 \to P_1$ and $P_3 \to P_5 \to P_2 \to P_1$. These chains differ in length. We call $P_3 \to P_2 \to P_1$ a two-chain and $P_3 \to P_5 \to P_2 \to P_1$ a three-chain, according to the number of edges traversed. In general, a chain consisting of $n + 1$ vertices is called an *n-chain*. When the same vertex occurs more than once in an *n*-chain, the *n*-chain is said to be *redundant*. In the above network, $P_4 \to P_5 \to P_4 \to P_3$ would be a redundant three-chain. We are primarily interested in *n*-chains that are not redundant.

In many instances, one is interested in the shortest chains (if any exist) between two vertices. In the communication network, for example, the routes $P_4 \rightarrow P_5 \rightarrow P_2 \rightarrow P_1$ and $P_4 \rightarrow P_3 \rightarrow P_2 \rightarrow P_1$ would be of interest, whereas the $P_4 \rightarrow P_3 \rightarrow P_5 \rightarrow P_2 \rightarrow P_1$ route would not be of interest.

The following theorem from graph theory gives information about chains.

Theorem 2-13   *Let A be an adjacency matrix representing a digraph. The (i, j)th element of A^n gives the number of n-chains from P_i to P_j.*

Thus, in a given graph, to find the number of four-chains from vertex P_2 to P_5, for example, one would compute A^4; the $(2, 5)$ element of this matrix would give the number of four-chains. In particular, we shall see that this theorem can be used to reveal the number of links in the shortest chains between a given pair of vertices.

Proof: a_{ik} will be the number of one-chains from P_i to P_k, and a_{kj} will be the number of one-chains from P_k to P_j. Thus $a_{ik}a_{kj}$ is the number of distinct two-chains from P_i to P_j passing through P_k. $\sum_k a_{ik}a_{kj}$ will give the total number of distinct two-chains from P_i to P_j. $\sum_k a_{ik}a_{kj}$ is the (i, j)th element of A^2. Thus the elements of A^2 give the numbers of two-chains between vertices.

$a_{ik}a_{ks}a_{sj}$ gives the number of three-chains from P_i to P_j by way of P_k and P_s. The total number of three-chains from P_i to P_j is given by

$$\sum_k \sum_s a_{ik}a_{ks}a_{sj}$$

This is the (i, j)th element of A^3. In this manner, the elements of A^3 give the numbers of three-chains between vertices.

We can extend these concepts to n-chains. $a_{ik}a_{ks} \cdots a_{rj}$ would give the number of distinct n-chains from P_i to P_j by way of $P_k, P_s, \ldots, P_r$. $\sum_k \sum_s \cdots \sum_r a_{ik}a_{ks} \cdots a_{rj}$ would give the total number of n-chains between P_i and P_j. This is the (i, j)th element of A^n. Thus A^n represents the n-chains between vertices, proving the theorem.

Let us use the notation $a_{ij}^{(n)}$ for the (i, j)th element of A^n. We have seen that A^n represents the n-chains between vertices. These chains can include redundant chains. However, if $a_{ij}^{(n)} = c \ (\neq 0)$ and $a_{ij}^{(m)} = 0$ for all $m < n$, then these c distinct n-chains must be chains of minimum length linking P_i to P_j—there can be no shorter chains. Furthermore, there can be no redundant chains among these, for then they would not be the chains of minimum length linking P_i to P_j.

Let us check the theorem and illustrate its application with our communication network. In particular, we know that there are two three-chains linking station P_4 to station P_1 and that these are the chains of minimum length. Successive powers of the adjacency matrix A are determined.

$$A = \begin{pmatrix} 0 & 1 & 0 & 0 & 0 \\ 1 & 0 & 0 & 0 & 1 \\ 0 & 1 & 0 & 0 & 1 \\ 0 & 0 & 1 & 0 & 1 \\ 0 & 1 & 0 & 1 & 0 \end{pmatrix}, \quad A^2 = \begin{pmatrix} 1 & 0 & 0 & 0 & 1 \\ 0 & 2 & 0 & 1 & 0 \\ 1 & 1 & 0 & 1 & 1 \\ 0 & 2 & 0 & 1 & 1 \\ 1 & 0 & 1 & 0 & 2 \end{pmatrix},$$

$$A^3 = \begin{pmatrix} 0 & 2 & 0 & 1 & 0 \\ 2 & 0 & 1 & 0 & 3 \\ 1 & 2 & 1 & 1 & 2 \\ 2 & 1 & 1 & 1 & 3 \\ 0 & 4 & 0 & 2 & 1 \end{pmatrix}, \quad A^4 = \begin{pmatrix} 2 & 0 & 1 & 0 & 3 \\ 0 & 6 & 0 & 3 & 1 \\ 2 & 4 & 1 & 2 & 4 \\ 1 & 6 & 1 & 3 & 3 \\ 4 & 1 & 2 & 1 & 6 \end{pmatrix}, \quad A^5 = \cdots$$

Let us focus our attention on communication between the two stations of interest—namely, communication from station P_4 to station P_1.

A gives $a_{41} = 0$; no direct link

A^2 gives $a_{41}^{(2)} = 0$; no two-chain communication

A^3 gives $a_{41}^{(3)} = 2$; two distinct three-chains from P_4 to P_1

These are the chains of minimum length. The matrix approach is confirmed for this test case.

As a second example, let us determine the minimum number of links for communication from station P_1 to station P_3.

A gives $a_{13} = 0$; no direct link

A^2 gives $a_{13}^{(2)} = 0$; no two-chain communication

A^3 gives $a_{13}^{(3)} = 0$; no three-chain communication

A^4 gives $a_{13}^{(4)} = 1$; a single four-chain from P_1 to P_3

This result is confirmed when we examine the graph. The quickest way of sending a message from station P_1 to station P_3 is by using the four-chain $P_1 \rightarrow P_2 \rightarrow P_5 \rightarrow P_4 \rightarrow P_3$. The model that we have discussed is somewhat limited; although it gives us the shortest linkage possible, it does not tell us which intermediate stations to use. A more complete model that would yield this information has yet to be developed by mathematicians.

✦ *Example 3* ────────────────────────────────────

Graph theory is used by sociologists to analyze group relationships.

Consider a group of six people. A sociologist is interested in determining which one of the six has most influence over, or dominates, the others. The group is asked to fill out the following questionnaire:

· Your name _____

· Person whose opinion you value most _____

These answers are then tabulated. Let us, for convenience, label the group members $M_1, M_2, \ldots, M_6$. The results are given in Table 2-2. The sociologist at this point makes the assumption that the person whose opinion a member values most is the person who influences that member most. This seems to be a natural assumption to make. The question the group was asked is more tactful than "Who influences you most?"

Table 2-2

Group member	Person whose opinion he/she values most
M_1	M_5
M_2	M_4
M_3	Own
M_4	M_3
M_5	M_3
M_6	M_5

Let us represent these results by means of the digraph in Figure 2-17. The edges correspond to direct influence; the arrows indicate direction of influence. M_1 values the opinion of M_5 most; thus M_1 is influenced by M_5. The arrows go from right to left in the table.

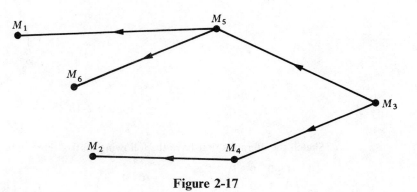

Figure 2-17

In this type of graph, one-chains correspond to direct influence. Chains involving two, three, or more links correspond to indirect influences; the longer the chain, the more remote and presumably the smaller the influence. From Figure 2-17 it becomes apparent that M_3 is the person who exerts most influence over the group.

As groups become larger, the analysis becomes more complex. A matrix description of the graph is used, its powers revealing the indirect influences.

Readers who are interested in further applications of this theory in sociology are referred to "A Method of Matrix Analysis of Group Structure," by R. Duncan Luce and Albert D. Perry, in *Readings in Mathematical Social Science*, edited by Paul F. Lazarsfeld and Neil W. Henry, MIT Press, 1968.

✦

EXERCISES

*1. Determine the matrix that represents each of the digraphs in Figure 2-18.

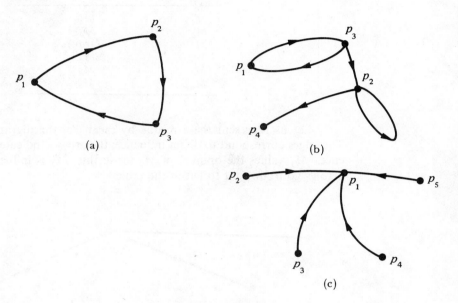

(a)

(b)

(c)

Figure 2-18

Sketch the digraphs that have the following matrix representations.

*2. $\begin{pmatrix} 1 & 0 & 0 & 1 \\ 0 & 1 & 0 & 0 \\ 0 & 0 & 0 & 1 \\ 1 & 0 & 0 & 0 \end{pmatrix}$
   3. $\begin{pmatrix} 0 & 1 & 1 \\ 1 & 1 & 0 \\ 0 & 1 & 0 \end{pmatrix}$
   *4. $\begin{pmatrix} 1 & 1 & 0 & 0 \\ 0 & 1 & 1 & 0 \\ 0 & 0 & 1 & 1 \\ 1 & 0 & 0 & 1 \end{pmatrix}$

5. Prove that the matrix representation of a digraph is necessarily square.

6. What would be the characteristics of a digraph whose matrix representation contained only zeros as diagonal elements?

7. A digraph is said to be *connected* if there is a chain from any one vertex to any other. Give an example of

   (a) a connected graph   **(b)** a graph that is not connected

*8. The following matrix defines a communication network:

$$\begin{pmatrix} 0 & 1 & 0 & 1 \\ 1 & 0 & 0 & 0 \\ 0 & 0 & 0 & 1 \\ 1 & 0 & 1 & 0 \end{pmatrix}$$

Sketch the network. How many links are needed for communication from

(a) station P_1 to station P_3 **(b)** station P_2 to station P_3

In each of Exercises 9–13, the matrix A describes a communication network. Draw the network. Powers of A are given. Interpret the elements that have been circled.

*9. $A = \begin{pmatrix} 0 & 1 & 0 & 0 \\ 1 & 0 & 0 & 0 \\ 0 & 1 & 0 & 0 \\ 0 & 0 & 1 & 0 \end{pmatrix}$, $A^2 = \begin{pmatrix} 1 & 0 & 0 & 0 \\ 0 & ① & 0 & ⓪ \\ ① & 0 & 0 & 0 \\ 0 & ① & 0 & 0 \end{pmatrix}$, $A^3 = \begin{pmatrix} 0 & ① & 0 & 0 \\ 1 & 0 & 0 & ⓪ \\ 0 & ① & 0 & 0 \\ ① & 0 & 0 & 0 \end{pmatrix}$

10. $A = \begin{pmatrix} 0 & 0 & 1 & 0 \\ 0 & 0 & 1 & 0 \\ 0 & 1 & 0 & 0 \\ 0 & 0 & 1 & 0 \end{pmatrix}$, $A^2 = \begin{pmatrix} 0 & 1 & 0 & 0 \\ 0 & 1 & 0 & ⓪ \\ 0 & 0 & ① & 0 \\ 0 & ① & 0 & 0 \end{pmatrix}$, $A^3 = \begin{pmatrix} 0 & 0 & ① & 0 \\ 0 & 0 & 1 & 0 \\ 0 & ① & 0 & ⓪ \\ 0 & 0 & 1 & 0 \end{pmatrix}$

*11. $A = \begin{pmatrix} 0 & 1 & 0 & 1 \\ 0 & 0 & 1 & 0 \\ 1 & 0 & 0 & 0 \\ 0 & 0 & 1 & 0 \end{pmatrix}$, $A^2 = \begin{pmatrix} 0 & 0 & ② & 0 \\ ① & 0 & 0 & 0 \\ 0 & 1 & 0 & ① \\ 1 & 0 & 0 & 0 \end{pmatrix}$, $A^3 = \begin{pmatrix} ② & 0 & 0 & 0 \\ 0 & 1 & 0 & ① \\ 0 & 0 & ② & 0 \\ 0 & ① & 0 & 1 \end{pmatrix}$

12. $A = \begin{pmatrix} 0 & 1 & 0 & 0 & 1 \\ 0 & 0 & 1 & 1 & 0 \\ 0 & 0 & 0 & 0 & 0 \\ 0 & 0 & 1 & 0 & 0 \\ 0 & 0 & 0 & 1 & 0 \end{pmatrix}$, $A^2 = \begin{pmatrix} 0 & 0 & 1 & ② & 0 \\ 0 & 0 & ① & 0 & 0 \\ 0 & 0 & 0 & 0 & 0 \\ 0 & 0 & 0 & 0 & 0 \\ 0 & 0 & ① & 0 & 0 \end{pmatrix}$, $A^3 = \begin{pmatrix} 0 & 0 & ② & 0 & 0 \\ 0 & 0 & 0 & 0 & 0 \\ ⓪ & 0 & 0 & 0 & 0 \\ 0 & 0 & 0 & 0 & 0 \\ 0 & 0 & 0 & 0 & 0 \end{pmatrix}$

13. $A = \begin{pmatrix} 0 & 0 & 0 & 0 & 0 \\ 1 & 0 & 0 & 1 & 0 \\ 0 & 1 & 0 & 1 & 0 \\ 0 & 0 & 1 & 0 & 1 \\ 1 & 1 & 0 & 0 & 0 \end{pmatrix}$, $A^2 = \begin{pmatrix} 0 & 0 & 0 & 0 & 0 \\ 0 & 0 & 1 & 0 & ① \\ ① & 0 & 1 & 1 & 1 \\ 1 & ② & 0 & 1 & 0 \\ ① & 0 & ⓪ & ① & 0 \end{pmatrix}$, $A^3 = \begin{pmatrix} 0 & 0 & 0 & 0 & 0 \\ ① & ② & 0 & 1 & 0 \\ 1 & ② & 1 & ① & ① \\ ② & 0 & 1 & 2 & 1 \\ 0 & 0 & ① & 0 & 1 \end{pmatrix}$

14. Let A be the adjacency matrix of a digraph. What do you know about the digraph in each of the following cases?

*(a) The third row of A consists entirely of zeros.

(b) The fourth column of A consists entirely of zeros.

*(c) The sum of the elements in the fifth row of A is 3.

(d) The sum of elements in the second column of A is 2.

*(e) The second row of A^3 consists entirely of zeros.

(f) The third column of A^4 consists entirely of zeros.

15. Graph theory was invented in 1736 by a mathematician named Euler while he was in the process of analyzing the famous Königsberg bridge problem.[†] The city of Königsberg had seven bridges linking islands in the River Pregel to the banks and to each other as shown in Figure 2-19. The residents of Königsberg wanted to know if it was possible, starting at any point, to walk across each bridge exactly once and then return to the starting point. Translate the Königsberg bridge problem into a graph problem and solve it. (*Hint*: Let the banks be the vertices of a digraph and the bridges be arcs.)

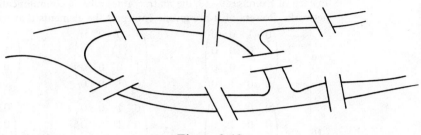

Figure 2-19

16. The digraphs in (a)–(l) have adjacency matrices with the stated characteristics. What can you tell about each digraph?
 *(a) The second row consists entirely of zeros.
 (b) The third column consists entirely of zeros.
 (c) The fourth row has ones in all locations except the diagonal location.
 (d) The fifth column has ones in all locations except the diagonal location.
 *(e) The sum of the elements in the third row is 5.
 (f) The sum of the elements in the second column is 4.
 *(g) The number of ones in the matrix is 5.
 (h) The number of ones in the matrix is 7.
 *(i) The sum of the elements in the second row of the fourth power of the adjacency matrix is 3.
 (j) The sum of the elements in the third column of the fifth power of the adjacency matrix is 2.
 *(k) The fourth row of the square of the adjacency matrix consists entirely of zeros.
 (l) The third column of the fourth power of the adjacency matrix consists entirely of zeros.

The following tables represent information obtained from questionaires given to groups of people. In each case construct a graph that describes the leadership structure within the group and the matrix that represents this information. Determine the person(s) exerting most influence within each group.

[†] At the time of Euler, Königsberg was part of Prussia. It is now called Kaliningrad and is a part of the Soviet Union.

*17. Group member	Person whose opinion he/she values most
M_1	M_2
M_2	M_4
M_3	M_2
M_4	Own

18. Group member	Person whose opinion he/she values most
M_1	M_4
M_2	Own
M_3	M_5
M_4	M_3
M_5	Own

19. Group member	Person whose opinion he/she values most
M_1	M_2
M_2	M_3
M_3	M_1
M_4	M_2
M_5	M_3

The following matrices describe the relationship "friendship" between people of a group. $a_{ij} = 1$ if M_i is a friend of M_j; $a_{ij} = 0$ if he or she is not. Draw the digraphs that represent this relationship for the various groups. Note that in this case the matrices that describe the relationship are symmetric. What is the significance of this? Can such a relationship be defined by a nonsymmetric matrix? What would the implication then be?

*20.
$$\begin{pmatrix} 0 & 1 & 0 & 0 & 0 \\ 1 & 0 & 0 & 0 & 1 \\ 0 & 0 & 0 & 1 & 0 \\ 0 & 0 & 1 & 0 & 0 \\ 0 & 1 & 0 & 0 & 0 \end{pmatrix}$$

21.
$$\begin{pmatrix} 0 & 0 & 0 & 0 & 0 & 1 \\ 0 & 0 & 1 & 0 & 0 & 0 \\ 0 & 1 & 0 & 0 & 1 & 0 \\ 0 & 0 & 0 & 0 & 0 & 1 \\ 0 & 0 & 1 & 0 & 0 & 0 \\ 1 & 0 & 0 & 1 & 0 & 0 \end{pmatrix}$$

22. What would be the characteristics of a digraph that had a symmetric matrix representation? Give an example (other than the previous one) of a real situation that would be described by such a graph.

23. A structure in a digraph of interest particularly to social scientists is that of a clique. A *clique* is defined to be a maximal (largest) subset of a digraph consisting of three or more vertices, each pair of which is mutually related. The application of this concept to the relationship "friendship" is immediate: three or more people form a clique if they are all friends and if they do not have any mutual friendships with anyone outside that set. Give an example of a digraph that contains a clique.

3

Determinants

Associated with every square matrix is a number called its *determinant*. The determinant is a tool used in many branches of mathematics, science, and engineering. In this chapter, the determinant is defined, and its properties are developed. The reader will see its use in developing a formula for the inverse of a matrix. Engineering readers will see how it is employed to determine principal stresses and directions at a point in a body.

3-1 INTRODUCTION TO DETERMINANTS

Consider a 2×2 matrix $\begin{pmatrix} a_{11} & a_{12} \\ a_{21} & a_{22} \end{pmatrix}$. The determinant of such a matrix is defined to be the number $a_{11}a_{22} - a_{12}a_{21}$. The determinant of this matrix is usually denoted $\begin{vmatrix} a_{11} & a_{12} \\ a_{21} & a_{22} \end{vmatrix}$.

✦ *Example 1* ─────────────────────────────

Find the determinant of the matrix $\begin{pmatrix} -1 & 2 \\ 0 & 1 \end{pmatrix}$.

$$\begin{vmatrix} -1 & 2 \\ 0 & 1 \end{vmatrix} = (-1 \times 1) - (2 \times 0) = -1 \quad ✦$$

The definition of the determinant of a 3×3 matrix is given in terms of the determinants of 2×2 matrices, that of a 4×4 matrix in terms of the determinants of 3×3 matrices, etc. For this definition we need the concept of a *minor.* Let A be an $n \times n$ matrix. Associated with an arbitrary element a_{ij} of A is an $n - 1 \times n - 1$ matrix obtained by deleting the ith row and jth column. The determinant of this submatrix is called the minor of a_{ij}. There will exist such a minor for every element of the matrix. Denote the minor of the element a_{ij} by A_{ij}.

✦ **Example 2**

If $A = \begin{pmatrix} 1 & 0 & 3 \\ 4 & -1 & 2 \\ 0 & 1 & 1 \end{pmatrix}$, then the minor of the element a_{11}, the element in

the first row and first column, is $A_{11} = \begin{vmatrix} -1 & 2 \\ 1 & 1 \end{vmatrix} = -3$. The minors of the

other first row elements are

$$A_{12} = \begin{vmatrix} 4 & 2 \\ 0 & 1 \end{vmatrix} = 4, \qquad A_{13} = \begin{vmatrix} 4 & -1 \\ 0 & 1 \end{vmatrix} = 4$$

The minors of the second row elements are

$$A_{21} = \begin{vmatrix} 0 & 3 \\ 1 & 1 \end{vmatrix} = -3, \qquad A_{22} = \begin{vmatrix} 1 & 3 \\ 0 & 1 \end{vmatrix} = 1, \qquad A_{23} = \begin{vmatrix} 1 & 0 \\ 0 & 1 \end{vmatrix} = 1$$

The third row minors are

$$A_{31} = \begin{vmatrix} 0 & 3 \\ -1 & 2 \end{vmatrix} = 3, \qquad A_{32} = \begin{vmatrix} 1 & 3 \\ 4 & 2 \end{vmatrix} = -10, \qquad A_{33} = \begin{vmatrix} 1 & 0 \\ 4 & -1 \end{vmatrix} = -1$$

✦

We now define the determinant of a 3×3 matrix

$$A = \begin{pmatrix} a_{11} & a_{12} & a_{13} \\ a_{21} & a_{22} & a_{23} \\ a_{31} & a_{32} & a_{33} \end{pmatrix}$$

using the first row. The elements of the first row are multiplied by their corresponding minors and summed as follows:

$$\begin{vmatrix} a_{11} & a_{12} & a_{13} \\ a_{21} & a_{22} & a_{23} \\ a_{31} & a_{32} & a_{33} \end{vmatrix} = a_{11}A_{11} - a_{12}A_{12} + a_{13}A_{13}$$

Note that the signs alternate.

✦ *Example 3*

Find the determinant of the matrix $\begin{pmatrix} 1 & 2 & -1 \\ 3 & 0 & 1 \\ 4 & 2 & 1 \end{pmatrix}$.

We know that $a_{11} = 1$, $a_{12} = 2$, and $a_{13} = -1$.

$$A_{11} = \begin{vmatrix} 0 & 1 \\ 2 & 1 \end{vmatrix}, \qquad A_{12} = \begin{vmatrix} 3 & 1 \\ 4 & 1 \end{vmatrix}, \qquad \text{and} \qquad A_{13} = \begin{vmatrix} 3 & 0 \\ 4 & 2 \end{vmatrix}$$

Hence

$$\begin{vmatrix} 1 & 2 & -1 \\ 3 & 0 & 1 \\ 4 & 2 & 1 \end{vmatrix} = 1 \begin{vmatrix} 0 & 1 \\ 2 & 1 \end{vmatrix} - 2 \begin{vmatrix} 3 & 1 \\ 4 & 1 \end{vmatrix} + (-1) \begin{vmatrix} 3 & 0 \\ 4 & 2 \end{vmatrix}$$

$$= 1[(0 \times 1) - (1 \times 2)] - 2[(3 \times 1) - (1 \times 4)]$$
$$\quad - [(3 \times 2) - (0 \times 4)]$$
$$= 1(0 - 2) - 2(3 - 4) - (6 - 0)$$
$$= -2 + 2 - 6$$
$$= -6 \qquad ✦$$

The determinant of the 3×3 matrix A can be written using sigma notation:

$$\begin{vmatrix} a_{11} & a_{12} & a_{13} \\ a_{21} & a_{22} & a_{23} \\ a_{31} & a_{32} & a_{33} \end{vmatrix} = \sum_{j=1}^{3} (-1)^{1+j} a_{1j} A_{1j}$$

The $(-1)^{1+j}$ ensures that each term has the proper sign.

The determinants of larger matrices are defined similarly in terms of the first row and the minors of the first row. For example, the determinant of a 4×4 matrix is

$$\begin{vmatrix} a_{11} & a_{12} & a_{13} & a_{14} \\ a_{21} & a_{22} & a_{23} & a_{24} \\ a_{31} & a_{32} & a_{33} & a_{34} \\ a_{41} & a_{42} & a_{43} & a_{44} \end{vmatrix} = a_{11}A_{11} - a_{12}A_{12} + a_{13}A_{13} - a_{14}A_{14}$$

Note that the signs always alternate.

✦ *Example 4*

$$
\begin{vmatrix} 3 & 0 & -1 & 2 \\ 4 & 1 & 3 & 2 \\ 6 & -1 & 2 & 0 \\ 1 & 2 & 4 & 1 \end{vmatrix} = 3 \begin{vmatrix} 1 & 3 & 2 \\ -1 & 2 & 0 \\ 2 & 4 & 1 \end{vmatrix} - 0 \begin{vmatrix} 4 & 3 & 2 \\ 6 & 2 & 0 \\ 1 & 4 & 1 \end{vmatrix}
$$

$$
+ (-1) \begin{vmatrix} 4 & 1 & 2 \\ 6 & -1 & 0 \\ 1 & 2 & 1 \end{vmatrix} - 2 \begin{vmatrix} 4 & 1 & 3 \\ 6 & -1 & 2 \\ 1 & 2 & 4 \end{vmatrix}
$$

$$
= 3 \left[1 \begin{vmatrix} 2 & 0 \\ 4 & 1 \end{vmatrix} - 3 \begin{vmatrix} -1 & 0 \\ 2 & 1 \end{vmatrix} + 2 \begin{vmatrix} -1 & 2 \\ 2 & 4 \end{vmatrix} \right]
$$

$$
- \left[4 \begin{vmatrix} -1 & 0 \\ 2 & 1 \end{vmatrix} - 1 \begin{vmatrix} 6 & 0 \\ 1 & 1 \end{vmatrix} + 2 \begin{vmatrix} 6 & -1 \\ 1 & 2 \end{vmatrix} \right]
$$

$$
- 2 \left[4 \begin{vmatrix} -1 & 2 \\ 2 & 4 \end{vmatrix} - 1 \begin{vmatrix} 6 & 2 \\ 1 & 4 \end{vmatrix} + 3 \begin{vmatrix} 6 & -1 \\ 1 & 2 \end{vmatrix} \right]
$$

$$
= 3(2 + 3 - 16) - (-4 - 6 + 26) - 2(-32 - 22 + 39)
$$

$$
= 3(-11) - (16) - 2(-15) = -33 - 16 + 30 = -19 \qquad ✦
$$

In terms of sigma notation, the determinant of a 4×4 matrix is

$$
\begin{vmatrix} a_{11} & a_{12} & a_{13} & a_{14} \\ a_{21} & a_{22} & a_{23} & a_{24} \\ a_{31} & a_{32} & a_{33} & a_{34} \\ a_{41} & a_{42} & a_{43} & a_{44} \end{vmatrix} = \sum_{j=1}^{4} (-1)^{1+j} a_{1j} A_{1j}
$$

The determinant of the $n \times n$ matrix $\begin{pmatrix} a_{11} & \cdots & a_{1n} \\ & \vdots & \\ a_{n1} & \cdots & a_{nn} \end{pmatrix}$ is defined as

$$
\begin{vmatrix} a_{11} & \cdots & a_{1n} \\ & \vdots & \\ a_{n1} & \cdots & a_{nn} \end{vmatrix} = a_{11}A_{11} - a_{12}A_{12} + a_{13}A_{13} - \cdots (-1)^{1+n} a_{1n}A_{1n}
$$

Using sigma notation, this is written

$$
\begin{vmatrix} a_{11} & \cdots & a_{1n} \\ & \vdots & \\ a_{n1} & \cdots & a_{nn} \end{vmatrix} = \sum_{j=1}^{n} (-1)^{1+j} a_{1j} A_{1j}
$$

We defined the determinant of a 3×3 matrix in terms of determinants of 2×2 matrices, and the determinant of a 4×4 matrix in terms of determinants of 3×3 matrices. The determinant of an $n \times n$ matrix is defined in terms of the determinants of $(n - 1) \times (n - 1)$ matrices. Since we gave a definition of the determinant of a 2×2 matrix, the determinants of all square matrices of finite order are thus defined. We call such a definition an *inductive definition*. Since a number can be regarded as a 1×1 matrix, we complete this definition by defining the determinant of a 1×1 matrix to be that number.

A square matrix is said to be *singular* if it has a zero determinant; otherwise it is *nonsingular*. These concepts are important in discussing solutions to systems of linear equations, as the following example illustrates.

✦ *Example 5*
_____

Consider the following system of linear equations:

$$x - 2y + 3z + \ w = 0$$
$$2x - 3y + 2z - \ w = 0$$
$$3x - 5y + 5z \qquad = 0$$
$$x - \ y - \ z - 2w = 0$$

It can be shown that there are many solutions to this system. One solution, for example, is $x = 10$, $y = 7$, $z = 1$, $w = 1$; another solution is $x = 0$, $y = 1$, $z = 1$, $w = -1$.

The matrix of coefficients is

$$\begin{pmatrix} 1 & -2 & 3 & 1 \\ 2 & -3 & 2 & -1 \\ 3 & -5 & 5 & 0 \\ 1 & -1 & -1 & -2 \end{pmatrix}$$

It can be verified that the determinant of this matrix is zero:

$$\begin{vmatrix} 1 & -2 & 3 & 1 \\ 2 & -3 & 2 & -1 \\ 3 & -5 & 5 & 0 \\ 1 & -1 & -1 & -2 \end{vmatrix} = 0$$

and therefore the matrix of coefficients is singular.

Later we shall see that the determinant of the matrix of coefficients of certain systems of linear equations is related to the existence and uniqueness of solutions to those equations. For example, if the matrix of coefficients is singular, as is the case here, then the system cannot have a single solution; it has either many solutions or none at all.   ✦

We have defined determinants in terms of the first row; however, the determinant of a matrix can be found using any row or column.

For example, the determinant of a 3×3 matrix can be found using the third row rather than the first row. The rule is

$$\begin{vmatrix} a_{11} & a_{12} & a_{13} \\ a_{21} & a_{22} & a_{23} \\ a_{31} & a_{32} & a_{33} \end{vmatrix} = a_{31}A_{31} - a_{32}A_{32} + a_{33}A_{33}$$

The elements of the third row play a role in this expansion analogous to the one played by the elements of the first row in the previous expansion.

✦ Example 6

Find the determinant of the matrix $\begin{pmatrix} 1 & 2 & -1 \\ 3 & 0 & 1 \\ 4 & 2 & 1 \end{pmatrix}$ using the third row.

$$\begin{vmatrix} 1 & 2 & -1 \\ 3 & 0 & 1 \\ 4 & 2 & 1 \end{vmatrix} = 4 \begin{vmatrix} 2 & -1 \\ 0 & 1 \end{vmatrix} - 2 \begin{vmatrix} 1 & -1 \\ 3 & 1 \end{vmatrix} + 1 \begin{vmatrix} 1 & 2 \\ 3 & 0 \end{vmatrix}$$

$$= 8 - 8 - 6 = -6$$

Note that we have already evaluated this determinant in terms of the first row in Example 3 of this section and that we got the same number, -6.   ✦

Using sigma notation, the expansion of the determinant in terms of the third row is

$$\begin{vmatrix} a_{11} & a_{12} & a_{13} \\ a_{21} & a_{22} & a_{23} \\ a_{31} & a_{32} & a_{33} \end{vmatrix} = \sum_{j=1}^{3} (-1)^{3+j} a_{3j} A_{3j}$$

There is a useful rule that can be memorized for expanding a determinant using any row or column. This rule gives the signs in the expansion—expansions using certain rows and columns begin with a minus sign rather than a plus sign. In any case, they then alternate. The rule is illustrated below.

$$\begin{pmatrix} + & - & + & - & \cdots \\ - & + & - & + & \cdots \\ + & - & + & - & \cdots \\ & & \vdots & & \end{pmatrix}$$

A certain sign, plus or minus, is attached to every element in the matrix. If one expands using the second row or the fourth row, then the signs in the expansion go $-$, $+$, $-$, etc.

✦ Example 7

Find the determinant of the matrix $\begin{pmatrix} 1 & 2 & -1 \\ 3 & 0 & 1 \\ 4 & 2 & 1 \end{pmatrix}$ using the second column.

Using the above rule, the signs involved are

$$\begin{pmatrix} + & - & + \\ - & + & - \\ + & - & + \end{pmatrix}$$

Hence, using the second column, the signs of the terms in the expansion will go $-$, $+$, $-$. Thus

$$\begin{vmatrix} 1 & 2 & -1 \\ 3 & 0 & 1 \\ 4 & 2 & 1 \end{vmatrix} = -2 \begin{vmatrix} 3 & 1 \\ 4 & 1 \end{vmatrix} + 0 \begin{vmatrix} 1 & -1 \\ 4 & 1 \end{vmatrix} - 2 \begin{vmatrix} 1 & -1 \\ 3 & 1 \end{vmatrix}$$

$$= 2 + 0 - 8 = -6$$

Note that this matrix is the one in Example 6 and that we get the same value for the determinant. ✦

Let us now give the formal definition of the determinant of a square matrix first in terms of an arbitrary row and then in terms of an arbitrary column.

The determinant of the matrix

$$\begin{pmatrix} a_{11} & \cdots & a_{1n} \\ & \vdots & \\ a_{n1} & \cdots & a_{nn} \end{pmatrix}$$

in terms of the kth row is

$$\begin{vmatrix} a_{11} & \cdots & a_{1n} \\ & \vdots & \\ a_{n1} & \cdots & a_{nn} \end{vmatrix} = \sum_{j=1}^{n} (-1)^{k+j} a_{kj} A_{kj}$$

In terms of the hth column, it is

$$\sum_{i=1}^{n} (-1)^{i+h} a_{ih} A_{ih}$$

The minor A_{kj} multiplied by $(-1)^{k+j}$ is called the *cofactor* of a_{kj}. Denote this cofactor C_{kj}. Thus

$$C_{kj} = (-1)^{k+j} A_{kj}$$

This method of finding the determinant of a matrix by making use of the definition is not the most efficient way. We present a numerical method in Section 3-3 which is more suitable for programming on a computer. However, as we shall see later, this method is of theoretical importance.

EXERCISES

1. Evaluate the following determinants:

*(a) $\begin{vmatrix} 1 & 2 \\ 0 & 1 \end{vmatrix}$ (b) $\begin{vmatrix} -2 & -1 \\ 3 & 1 \end{vmatrix}$ *(c) $\begin{vmatrix} -1 & 2 & 3 \\ 0 & 1 & 4 \\ 1 & 1 & 2 \end{vmatrix}$

(d) $\begin{vmatrix} 0 & 0 & 1 \\ 2 & 3 & 0 \\ 1 & -1 & 3 \end{vmatrix}$ ***(e)** $\begin{vmatrix} -1 & 2 & 3 \\ 0 & 1 & 4 \\ 1 & 2 & 1 \end{vmatrix}$ **(f)** $\begin{vmatrix} 1 & 2 & 3 & -1 \\ 1 & -1 & 0 & 2 \\ 0 & 1 & 0 & 1 \\ 0 & 0 & -1 & 2 \end{vmatrix}$

2. Find out whether the following matrices are singular or nonsingular.

*(a) $\begin{pmatrix} 1 & 2 \\ 1 & 2 \end{pmatrix}$ **(b)** $\begin{pmatrix} 1 & 2 \\ 0 & 0 \end{pmatrix}$

*(c) $\begin{pmatrix} 3 & 4 \\ 1 & 2 \end{pmatrix}$ **(d)** $\begin{pmatrix} 1 & 2 & 3 \\ 2 & 4 & 6 \\ 0 & 1 & 2 \end{pmatrix}$

*(e) $\begin{pmatrix} -1 & 0 & 2 \\ 2 & 1 & -4 \\ 3 & 4 & -6 \end{pmatrix}$ **(f)** $\begin{pmatrix} 1 & 2 & 3 & 4 \\ 0 & -1 & 2 & 3 \\ -1 & -2 & -3 & -4 \\ 0 & 1 & 0 & 1 \end{pmatrix}$

3. Write in sigma notation the determinant of
    *(a) an arbitrary 5×5 matrix   **(b)** an arbitrary 6×6 matrix

*4. Evaluate the following determinants in terms of their first columns, then in terms of their second columns.

(a) $\begin{vmatrix} -1 & 2 & 3 \\ 0 & 1 & 2 \\ -1 & 1 & 1 \end{vmatrix}$ **(b)** $\begin{vmatrix} 1 & 3 & 0 \\ 4 & 1 & 2 \\ 5 & 0 & 1 \end{vmatrix}$ **(c)** $\begin{vmatrix} -1 & 2 & 5 & 4 \\ 0 & 3 & 2 & 0 \\ 0 & 4 & 1 & -1 \\ 0 & 1 & 1 & 3 \end{vmatrix}$

*5. Evaluate the following determinants in terms of their third rows, then in terms of their second columns. Notice that the arithmetic is simplest when one expands the determinant in terms of the row or column containing the most zeros.

(a) $\begin{vmatrix} 1 & 0 & 2 \\ 3 & 1 & 4 \\ 5 & 6 & 1 \end{vmatrix}$ **(b)** $\begin{vmatrix} 2 & 1 & 3 \\ 1 & 1 & 1 \\ 0 & 0 & 1 \end{vmatrix}$ **(c)** $\begin{vmatrix} 3 & 4 & 1 \\ 2 & 1 & 1 \\ 1 & 1 & 1 \end{vmatrix}$

(d) $\begin{vmatrix} 2 & 0 & 1 \\ 3 & 0 & 2 \\ 4 & 1 & 1 \end{vmatrix}$ **(e)** $\begin{vmatrix} 1 & 0 & 1 & 2 \\ 4 & 0 & 3 & 4 \\ 5 & 1 & 2 & 3 \\ 0 & 0 & -1 & 2 \end{vmatrix}$

6. Evaluate the determinant $\begin{vmatrix} a & a^2 & a^3 \\ b & b^2 & b^3 \\ c & c^2 & c^3 \end{vmatrix}$ by using the second column.

*7. Write out the expansion of the determinant of a general 3×3 matrix in terms of the third column, then express your answer in sigma notation.

8. Write out the expansion of the determinant of a general 4×4 matrix in terms of the fourth column, then express your answer in sigma notation.

9. If A is the diagonal matrix

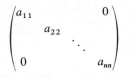

prove that $|A| = a_{11}a_{22} \cdots a_{nn}$.

  Hence, for such a matrix A, $|A| = 0$ if and only if one of the diagonal elements is zero.

10. If A is an upper triangular matrix, a matrix of the type

$$\begin{pmatrix} a_{11} & a_{12} & \cdots & a_{1n} \\ & a_{22} & & \\ & & \ddots & \\ 0 & & & a_{nn} \end{pmatrix}$$

prove that $|A| = a_{11}a_{22} \cdots a_{nn}$.

  Hence, for such a matrix A, $|A| = 0$ if and only if one of the diagonal elements is zero.

3-2 PROPERTIES OF DETERMINANTS

The determinant of a square matrix is a number associated with that matrix. In this section, we develop properties of determinants and find that the determinants of various matrices are related.

Theorem 3-1   *If all the elements of one row or column of a square matrix are zero, then the determinant of that matrix is zero.*

Proof: Let the elements of the kth row of A be zero. If we expand the determinant of A in terms of the kth row,

$$|A| = \sum_{j=1}^{n} (-1)^{k+j} a_{kj} A_{kj}$$

Since the elements a_{kj}, where $j = 1, \ldots, n$, are zero, $|A| = 0$. Similarly, if all the elements in one column are zero, expanding in terms of that column shows the determinant to be zero.

  In the following theorems, we see how elementary matrix transformations affect determinants.

Theorem 3-2   *If a matrix B is obtained from A by multiplying the elements of a row or column by a nonzero scalar c, then $|B| = c|A|$.*

Proof: Suppose B is obtained from A by multiplying its kth row by c. Hence the kth row of B is ca_{kj}, where $j = 1, \ldots, n$, and all other elements of

B are the same as the corresponding elements of A. Expand the determinant of B in terms of the kth row:

$$B = \sum_{j=1}^{n} (-1)^{k+j} b_{kj} B_{kj}$$

$$= \sum_{j=1}^{n} (-1)^{k+j} c a_{kj} A_{kj} = c \sum_{j=1}^{n} (-1)^{k+j} a_{kj} A_{kj}$$

$$= c|A|$$

The proof for columns is similar.

✦ *Example 1* ─────────────────────────────────────

If $A = \begin{pmatrix} 1 & 2 & 3 \\ 0 & 1 & 1 \\ 4 & -1 & 0 \end{pmatrix}$ and $B = \begin{pmatrix} 1 & 6 & 3 \\ 0 & 3 & 1 \\ 4 & -3 & 0 \end{pmatrix}$, then the second column of B is three times the second column of A. Evaluating the determinants, we get $|A| = -3$ and $|B| = -9$. Hence $|B| = 3|A|$, illustrating Theorem 3-2.   ✦

Theorem 3-3   *If B is obtained from A by interchanging two rows (columns), then* $|B| = -|A|$.

Proof: The proof is by induction. We shall prove the result for row interchanges; the proof for column interchanges is similar.

Let A be a 2×2 matrix and B be obtained from it by interchanging rows.

$$|A| = \begin{vmatrix} a_{11} & a_{12} \\ a_{21} & a_{22} \end{vmatrix} = a_{11}a_{22} - a_{12}a_{21}$$

$$|B| = \begin{vmatrix} a_{21} & a_{22} \\ a_{11} & a_{12} \end{vmatrix} = a_{21}a_{12} - a_{22}a_{11} = -A$$

Thus the result of the theorem holds for 2×2 matrices. Let us assume that the result holds for $n \times n$ matrices. We shall show that it then holds for $(n + 1) \times (n + 1)$ matrices, thereby proving by induction that it holds for all square matrices.

Let A be an $(n + 1) \times (n + 1)$ matrix and B be obtained from it by interchanging two rows. Expand B in terms of a row that is not one of those interchanged, such as the kth row. Then

$$|B| = \sum_{j=1}^{n+1} (-1)^{k+j} b_{kj} B_{kj}$$

Each b_{kj}, where $j = 1, \ldots, n + 1$, is identical to the corresponding a_{kj}. Each B_{kj}, where $j = 1, \ldots, n + 1$, is obtained from the corresponding A_{kj}

by interchanging two rows. Hence $b_{kj} = a_{kj}$, and $B_{kj} = -A_{kj}$, where $j = 1, \ldots, n + 1$. Thus

$$|B| = -\sum_{j=1}^{n+1} (-1)^{k+j} a_{kj} A_{kj} = -|A|$$

By induction the theorem holds for all square matrices.

✦ *Example 2*

If $A = \begin{pmatrix} 1 & 2 & -1 \\ 3 & 0 & 1 \\ 4 & 2 & 1 \end{pmatrix}$ and $B = \begin{pmatrix} -1 & 2 & 1 \\ 1 & 0 & 3 \\ 1 & 2 & 4 \end{pmatrix}$, then B can be obtained from A by interchanging the first and third columns. Evaluating the determinants, we find that $|A| = -6$ and $|B| = 6$, illustrating Theorem 3-3.   ✦

Theorem 3-4   *If two rows (columns) of a square matrix A are proportional, then* $|A| = 0$.

Proof: Let the kth row (column) of a matrix A be c times the hth, c being a nonzero scalar. Let B be the matrix obtained on multiplying every element of the hth row (column) of A by c. Thus $|B| = c|A|$. The matrix B has identical kth and hth rows (columns). When these identical rows (columns) are interchanged, the determinant remains unchanged. However, according to Theorem 3-3, the sign should be reversed. This is only possible if $|B| = 0$. Hence $|A| = 0$.

Note that this proof demonstrates a special case of this theorem: a matrix with two identical rows (columns) has a zero determinant.

✦ *Example 3*

The matrix $A = \begin{pmatrix} 1 & 2 & 3 \\ 0 & 1 & 2 \\ 2 & 4 & 6 \end{pmatrix}$ is such that the third row is twice the first. Evaluating, we find that $|A| = 0$, illustrating Theorem 3-4.   ✦

Theorem 3-5   *If a matrix B is obtained from a square matrix A by adding a multiple of one row (column) to another row (column), then* $|B| = |A|$.

Observe that this is an important elementary matrix transformation.

Proof: Let a multiple of the kth row of A be added to the ith row. The element b_{ij} of the ith row of B is of the form $a_{ij} + ca_{kj}$. All elements in rows other than the ith row of B are identical to the corresponding elements in A. Expanding the determinant of B in terms of the ith row, we have

$$|B| = \sum_{j=1}^{n} (-1)^{i+j} b_{ij} B_{ij}$$

$$= \sum_{j=1}^{n} (-1)^{i+j} (a_{ij} + ca_{kj}) A_{ij}$$

$$= \sum_{j=1}^{n} (-1)^{i+j} a_{ij} A_{ij} + c \sum_{j=1}^{n} (-1)^{i+j} a_{kj} A_{ij}$$

$$= |A| + c|A'|$$

where A' is obtained from A by replacing the ith row by the kth row. But since the ith row and the kth row of A' are identical, $|A'| = 0$. Hence $|B| = |A|$.

The proof involving columns is similar.

✦ **Example 4**

Evaluate the following determinant:

$$|A| = \begin{vmatrix} 0 & 1 & 3 & 0 \\ 4 & 5 & 7 & 2 \\ -2 & -1 & -3 & 0 \\ 4 & 2 & 5 & 0 \end{vmatrix}$$

Multiply the third row by 2 and add the result to the fourth row to get

$$|A| = \begin{vmatrix} 0 & 1 & 3 & 0 \\ 4 & 5 & 7 & 2 \\ -2 & -1 & -3 & 0 \\ 0 & 0 & -1 & 0 \end{vmatrix}$$

Now expand in terms of the last row to get

$$-(-1) \begin{vmatrix} 0 & 1 & 0 \\ 4 & 5 & 2 \\ -2 & -1 & 0 \end{vmatrix} = -1 \begin{vmatrix} 4 & 2 \\ -2 & 0 \end{vmatrix} = -4 \quad ✦$$

In this manner, Theorem 3-5 can be used to create zeros in certain locations of the determinant in order to make evaluation easier.

Theorem 3-6 *If A and B are square matrices, B being the transpose of A, then $|A| = |B|$.*

Proof: We shall proceed by induction. The theorem is true for 1×1 matrices. Assume that the theorem is true for all square matrices of order *n*. We shall show that it then holds for matrices of order $n + 1$ and hence, by induction, that it holds for all square matrices.

Let A and B be $(n + 1) \times (n + 1)$ matrices. Expand $|A|$ in terms of the first row to get

$$|A| = \sum_{j=1}^{n+1} (-1)^{1+j} a_{1j} A_{1j} \tag{1}$$

Expand $|B|$ in terms of the first column to get

$$|B| = \sum_{i=1}^{n+1} (-1)^{i+1} b_{i1} B_{i1} \tag{2}$$

But since $B = A^t$, the first row of A is identical to the first column of B; $a_{1j} = b_{j1}$, where $j = 1, \ldots, n + 1$. A_{1j} and B_{j1} are both determinants of $n \times n$ matrices. Since $B = A^t$, the one $n \times n$ matrix is the transpose of the other. Hence, by the hypothesis involving determinants of square matrices of order n, $A_{1j} = B_{j1}$ for all $j = 1, \ldots, n + 1$.

The two right-hand sides of identities (1) and (2) are therefore equal, $|A| = |B|$. By induction, the theorem holds for all square matrices.

Theorem 3-7 *The determinant of the product of two matrices is equal to the product of their determinants*:

$$|AB| = |A||B|$$

We state this theorem without proof.

We shall now give an alternative definition of determinants that is used mainly for theoretical purposes.[†] Both definitions are equivalent.

The second definition is in terms of the *permutation symbol*. The permutation symbol is defined as follows:

$$\varepsilon_{i_1 i_2 \cdots i_n} = \begin{cases} +1 & \text{if } i_1, i_2, \ldots, i_n \text{ is an even permutation of } 1, 2, \ldots, n \\ -1 & \text{if } i_1, i_2, \ldots, i_n \text{ is an odd permutation of } 1, 2, \ldots, n \\ 0 & \text{otherwise} \end{cases}$$

$i_1, i_2, \ldots, i_n$ is an even permutation of $1, 2, \ldots, n$ if it can be rearranged in the form $1, 2, \ldots, n$ through an even number of interchanges. Otherwise it is an odd permutation.

♦ Example 5

$\varepsilon_{13425} = +1$, because 13425 can be rearranged in the form 12345 in two interchanges:

$$13425 \rightarrow 13245 \rightarrow 12345 \qquad ♦$$

[†] The remainder of this section is optional.

◆ *Example 6*
_____

$\varepsilon_{13452} = -1$, because

$$13452 \to 13425 \to 13245 \to 12345$$

An odd number of interchanges is required. ◆

◆ *Example 7*
_____

$\varepsilon_{13243} = 0$, since two of the indices are identical. ◆

The alternative definition of the determinant in terms of this symbol is

$$|A| = \sum_{p=1}^{n} \cdots \sum_{q=1}^{n} \varepsilon_{p \cdots q} a_{1p} \cdots a_{nq}$$

It can be seen that this definition is equivalent to the previous one for a 2×2 matrix $A = \begin{pmatrix} a_{11} & a_{12} \\ a_{21} & a_{22} \end{pmatrix}$:

$$|A| = \sum_{p=1}^{2} \sum_{q=1}^{2} \varepsilon_{pq} a_{1p} a_{2q} = \varepsilon_{12} a_{11} a_{22} + \varepsilon_{21} a_{12} a_{21}$$

$$= a_{11} a_{22} - a_{12} a_{21}$$

EXERCISES

*1. If $A = \begin{pmatrix} 1 & 2 & 3 \\ 0 & 1 & 1 \\ 4 & -1 & 2 \end{pmatrix}$ and $B = \begin{pmatrix} 1 & 2 & -3 \\ 0 & 1 & -1 \\ 4 & -1 & -2 \end{pmatrix}$, prove that $|B| = -|A|$ by expanding the determinants. This illustrates Theorem 3-2.

*2. If $A = \begin{pmatrix} 1 & 2 & 3 \\ 1 & 1 & 1 \\ 4 & 1 & 2 \end{pmatrix}$ and $B = \begin{pmatrix} 1 & 2 & 3 \\ 4 & 1 & 2 \\ 1 & 1 & 1 \end{pmatrix}$, prove that $|B| = -|A|$ by expanding the determinants. This illustrates Theorem 3-3.

3. If $A = \begin{pmatrix} 1 & 3 & 1 \\ 2 & 6 & 2 \\ 4 & 1 & 2 \end{pmatrix}$, prove that $|A| = 0$ by expanding the determinant. This illustrates Theorem 3-4.

*4. Simplify the following determinants by creating zeros using Theorem 3-5. Then evaluate.

(a) $\begin{vmatrix} 1 & 2 & 3 \\ 2 & 4 & 1 \\ 1 & 1 & 1 \end{vmatrix}$ (b) $\begin{vmatrix} 0 & 1 & 5 \\ 1 & 1 & 6 \\ 2 & 2 & 7 \end{vmatrix}$

(c) $\begin{vmatrix} 2 & 1 & -1 \\ 3 & -1 & 1 \\ 1 & 4 & -4 \end{vmatrix}$ (d) $\begin{vmatrix} 3 & -1 & 0 \\ 4 & 2 & 1 \\ 1 & 1 & 2 \end{vmatrix}$

5. If $A = \begin{pmatrix} 1 & 2 & 3 \\ 0 & 1 & 1 \\ 4 & -1 & 2 \end{pmatrix}$ and $B = \begin{pmatrix} 1 & 0 & 4 \\ 2 & 1 & -1 \\ 3 & 1 & 2 \end{pmatrix}$, prove that $|A| = |B|$ by expanding the determinants. Why would you expect this result?

*6. We have computed the same determinant in two different ways below and obtained two different answers. One approach is correct, the other incorrect. Which is the correct answer?

$$\begin{vmatrix} 1 & 2 & 3 \\ -1 & 2 & 4 \\ 2 & 4 & 7 \end{vmatrix} \underset{(2)R1-R3}{=} \begin{vmatrix} 0 & 0 & -1 \\ -1 & 2 & 4 \\ 2 & 4 & 7 \end{vmatrix} = (-1)\begin{vmatrix} -1 & 2 \\ 2 & 4 \end{vmatrix} = (-1)(-4-4) = 8$$

$$\begin{vmatrix} 1 & 2 & 3 \\ -1 & 2 & 4 \\ 2 & 4 & 7 \end{vmatrix} \underset{R3-(2)R1}{=} \begin{vmatrix} 1 & 2 & 3 \\ -1 & 2 & 4 \\ 0 & 0 & 1 \end{vmatrix} = (1)\begin{vmatrix} 1 & 2 \\ -1 & 2 \end{vmatrix} = (1)(2+2) = 4$$

The mistake illustrated by this exercise is a very common one made in evaluating determinants.

7. Let A be a square matrix whose columns are $A_1, \ldots, A_i + A_i', \ldots, A_n$. The ith column is interpreted as being the sum of two column matrices A_i and A_i'. Prove that

$$|A| = |A_1 \quad \cdots \quad A_i + A_i' \quad \cdots \quad A_n|$$
$$= |A_1 \quad \cdots \quad A_i \quad \cdots \quad A_n| + |A_1 \quad \cdots \quad A_i' \quad \cdots \quad A_n|$$

Illustrate this result by showing that

$$\begin{vmatrix} 1 & 3 & -1 \\ 2 & 5 & 2 \\ 3 & 5 & 4 \end{vmatrix} = \begin{vmatrix} 1 & 1 & -1 \\ 2 & 4 & 2 \\ 3 & 3 & 4 \end{vmatrix} + \begin{vmatrix} 1 & 2 & -1 \\ 2 & 1 & 2 \\ 3 & 2 & 4 \end{vmatrix} = 9$$
$$\uparrow\uparrow\uparrow$$
$$A_i + A_i'A_iA_i'$$

(*Hint*: Expand the determinant of A using the ith column.)

*8. Evaluate the following: ε_{12345}, ε_{32415}, ε_{324151}.

9. Give the definition for the determinant of an arbitrary 3×3 matrix in terms of the permutation symbol. Show that this definition is equivalent to the original definition given in terms of the minors.

3-3   THE EVALUATION OF A DETERMINANT

We have introduced the concept of the determinant and have given an inductive, theoretical method for its evaluation. Here we give a numerical method that is suitable for programming on the computer.

In the method of Gauss-Jordan elimination, to solve a system of linear equations we used certain allowable transformations to change the initial system into another system which was easier to solve. We use a similar approach to compute a determinant. In the previous section, we saw that adding a multiple of one row to another left a determinant unchanged and that interchanging two rows negated a determinant. We use these operations to transform a determinant into an upper triangular form. The following theorem tells us that the determinant of such a matrix is the product of its diagonal elements.

Theorem 3-8   *If A is an upper triangular matrix, then its determinant is the product of the diagonal elements.*

Proof: A is an upper triangular matrix—that is, a matrix of the type

$$\begin{pmatrix} a_{11}a_{22} & \cdots & a_{1n} \\ & a_{22} & \cdots & a_{2n} \\ & & \ddots & \\ 0 & & & a_{nn} \end{pmatrix}$$

All the elements below the diagonal are zero. In evaluating $|A|$, we expand each determinant below in terms of the first column:

$$|A| = \begin{vmatrix} a_{11}a_{12} & \cdots & a_{1n} \\ & a_{22} & \cdots & a_{2n} \\ & & \ddots & \\ 0 & & & a_{nn} \end{vmatrix} = a_{11} \begin{vmatrix} a_{22}a_{23} & \cdots & a_{2n} \\ & a_{33} & \cdots & a_{3n} \\ & & \ddots & \\ 0 & & & a_{nn} \end{vmatrix}$$

$$= a_{11}a_{22} \begin{vmatrix} a_{33}a_{34} & \cdots & a_{3n} \\ & a_{44} & \cdots & a_{4n} \\ & & \ddots & \\ 0 & & & a_{nn} \end{vmatrix} = \cdots = (a_{11}a_{22} \cdots a_{nn})$$

proving the theorem.

We now summarize the method.

Numerical Evaluation of a Determinant

Transform the given determinant into an upper triangular determinant using two types of transformations:

1. Adding a multiple of one row to another row. This transformation leaves the determinant unchanged.
2. Interchanging two rows of the determinant. This transformation negates the determinant.

The zeros *below* the main diagonal are created systematically in the columns, proceeding from left to right.

The final determinant, in upper triangular form, is the product of its diagonal elements. The given determinant will be either this number or its negative, depending on the number of row interchanges performed.

The following examples illustrate the method.

✦ *Example 1*

Evaluate

$$\begin{vmatrix} 1 & 0 & 2 & 1 \\ 2 & -1 & 1 & 0 \\ 1 & 0 & 0 & 3 \\ -1 & 0 & 2 & 1 \end{vmatrix}$$

Using the above transformations,

$$\begin{vmatrix} 1 & 0 & 2 & 1 \\ 2 & -1 & 1 & 0 \\ 1 & 0 & 0 & 3 \\ -1 & 0 & 2 & 1 \end{vmatrix} \underset{\substack{R2+(-2)R1 \\ R3+(-1)R1 \\ R4+R1}}{=} \begin{vmatrix} 1 & 0 & 2 & 1 \\ 0 & -1 & -3 & -2 \\ 0 & 0 & -2 & 2 \\ 0 & 0 & 4 & 2 \end{vmatrix}$$

$$\underset{R4+(2)R3}{=} \begin{vmatrix} 1 & 0 & 2 & 1 \\ 0 & -1 & -3 & -2 \\ 0 & 0 & -2 & 2 \\ 0 & 0 & 0 & 6 \end{vmatrix}$$

Thus the determinant is $1 \times -1 \times -2 \times 6 = 12$.   ✦

It sometimes becomes necessary to interchange rows to obtain the upper triangular matrix, as the following example illustrates. Whenever rows are interchanged, remember to negate the determinant.

✦ *Example 2*

Evaluate

$$\begin{vmatrix} 1 & 2 & 3 & 1 \\ 2 & 4 & 3 & 1 \\ 1 & 3 & 4 & 2 \\ 2 & 5 & 6 & 4 \end{vmatrix}$$

We have

$$\begin{vmatrix} 1 & 2 & 3 & 1 \\ 2 & 4 & 3 & 1 \\ 1 & 3 & 4 & 2 \\ 2 & 5 & 6 & 4 \end{vmatrix} \underset{\substack{R2+(-2)R1 \\ R3+(-1)R1 \\ R4+(-2)R1}}{=} \begin{vmatrix} 1 & 2 & 3 & 1 \\ 0 & 0 & -3 & -1 \\ 0 & 1 & 1 & 1 \\ 0 & 1 & 0 & 2 \end{vmatrix}$$

At this point we normally use the element in the (2, 2) location to create the necessary zeros in the second column. We cannot do this here, since that element is 0. We must interchange the second and third rows:

$$\underset{R2 \leftrightarrow R3}{=} - \begin{vmatrix} 1 & 2 & 3 & 1 \\ 0 & 1 & 1 & 1 \\ 0 & 0 & -3 & -1 \\ 0 & 1 & 0 & 2 \end{vmatrix}$$

$$\underset{R4 + (-1)R2}{=} - \begin{vmatrix} 1 & 2 & 3 & 1 \\ 0 & 1 & 1 & 1 \\ 0 & 0 & -3 & -1 \\ 0 & 0 & -1 & 1 \end{vmatrix}$$

$$\underset{R4(-\frac{1}{3})R3}{=} - \begin{vmatrix} 1 & 2 & 3 & 1 \\ 0 & 1 & 1 & 1 \\ 0 & 0 & -3 & -1 \\ 0 & 0 & 0 & \frac{4}{3} \end{vmatrix}$$

$$= -(1 \times 1 \times -3 \times \tfrac{4}{3}) = 4 \qquad \blacklozenge$$

If, at any stage, the diagonal element is zero and all elements below it in that column are also zero, then the value of the determinant is zero; it is not necessary to continue to determine the upper triangular matrix.

The following example illustrates this.

◆ **Example 3** ————————————————————————————————

Evaluate

$$\begin{vmatrix} 1 & -1 & 0 & 2 \\ -1 & 1 & 2 & 3 \\ 2 & -2 & 3 & 4 \\ 6 & -6 & 5 & 1 \end{vmatrix}$$

We have

$$\begin{vmatrix} 1 & -1 & 0 & 2 \\ -1 & 1 & 2 & 3 \\ 2 & -2 & 3 & 4 \\ 6 & -6 & 5 & 1 \end{vmatrix} = \begin{vmatrix} 1 & -1 & 0 & 2 \\ 0 & 0 & 2 & 5 \\ 0 & 0 & 3 & 0 \\ 0 & 0 & 5 & -11 \end{vmatrix} = 0$$

diagonal element   all elements below the zero
is zero   diagonal element are zero

We know that the determinant is zero at this point because the zero diagonal element will appear as a diagonal element in the eventual upper triangular matrix; hence on multiplying the diagonal elements of this upper triangular matrix we will get zero.   ◆

EXERCISES

Evaluate the following determinants using elementary matrix transformations.

*1.
$$\begin{vmatrix} 1 & 0 & -1 \\ 2 & 1 & 2 \\ -1 & 1 & 1 \end{vmatrix}$$

2.
$$\begin{vmatrix} 2 & 1 & 1 \\ 4 & 0 & 1 \\ -1 & 2 & 0 \end{vmatrix}$$

*3.
$$\begin{vmatrix} 1 & -2 & 3 \\ -1 & 2 & 1 \\ 2 & 1 & 3 \end{vmatrix}$$

4.
$$\begin{vmatrix} 2 & -1 & 3 \\ 4 & -2 & 1 \\ 1 & -\frac{1}{2} & 1 \end{vmatrix}$$

*5.
$$\begin{vmatrix} -1 & 2 & 0 & 1 \\ 1 & 1 & -1 & 0 \\ 2 & 1 & 1 & 0 \\ -1 & -1 & 0 & 1 \end{vmatrix}$$

6.
$$\begin{vmatrix} 1 & 0 & -2 & 1 \\ 2 & 1 & 0 & 2 \\ -1 & 1 & -2 & 1 \\ 3 & 1 & -1 & 0 \end{vmatrix}$$

*7.
$$\begin{vmatrix} -1 & 1 & 2 & 1 \\ 1 & -1 & 3 & -1 \\ 2 & -2 & 3 & 1 \\ 1 & -1 & 0 & 1 \end{vmatrix}$$

8.
$$\begin{vmatrix} 1 & -1 & 0 & 2 \\ -1 & 1 & 0 & 0 \\ 2 & -2 & 0 & 1 \\ 3 & 1 & 5 & -1 \end{vmatrix}$$

3-4 CRAMER'S RULE

In this section, we use determinants in the analysis of certain systems of linear equations.

Theorem 3-9   *A system of n equations in n variables has a unique solution if and only if the matrix of coefficients is nonsingular. If the matrix is singular, solutions may or may not exist.*

Proof: Let A be the matrix of coefficients for a system of n equations in n variables. We know that the system has a unique solution if and only if the reduced echelon form of A is I_n (Theorem 1-2, Section 1-4).

Let the reduced echelon form of A be E. We have a sequence of elementary matrix transformations that leads from A to E:

$$A \approx \cdots \approx E$$

Modify these transformations so that when a row is divided by a pivot the matrix is multiplied by that pivot. With this modified sequence of transformations, the determinant is preserved to within a sign. Let the pivots be $p_1, \ldots, p_n$. We get

$$|A| = \pm p_1 p_2 \cdots p_n |E|$$

Since the pivots are all nonzero, this implies that $|A| \neq 0$ if and only if $|E| \neq 0$.

If the reduced echelon form E is I_n, then $|E| \neq 0$; if E is not of the form I_n, then it necessarily has a zero diagonal element and $|E| = 0$. Thus $|A| \neq 0$ if and only if $E = I_n$. Since E being I_n is equivalent to the system having a unique solution, we see that A is nonsingular if and only if the system has a unique solution.

The possibilities of solutions existing or not existing if the matrix of coefficients is singular are demonstrated by the following two examples.

✦ *Example 1* ───────────────────────────────

$$x_1 - 2x_2 + 3x_3 = 1$$
$$3x_1 - 4x_2 + 5x_3 = 3$$
$$2x_1 - 3x_2 + 4x_3 = 2$$

The determinant of the matrix of coefficients is 0. Solutions can be shown to be $x_1 = r + 1$, $x_2 = 2r$, $x_3 = r$.   ✦

✦ *Example 2* ───────────────────────────────

$$x_1 + 2x_2 + 3x_3 = 3$$
$$2x_1 + x_2 + 3x_3 = 3$$
$$x_1 + x_2 + 2x_3 = 0$$

The determinant of the matrix of coefficients is 0. It can be shown that there are no solutions to this system.   ✦

Consider the following system of equations:

$$a_{11}x_1 + \cdots + a_{1n}x_n = y_1$$
$$\vdots$$
$$a_{m1}x_1 + \cdots + a_{mn}x_n = y_n$$

The system can be expressed in terms of column matrices as

$$\begin{pmatrix} a_{11}x_1 \\ \vdots \\ a_{m1}x_1 \end{pmatrix} + \cdots + \begin{pmatrix} a_{1n}x_n \\ \vdots \\ a_{mn}x_n \end{pmatrix} = \begin{pmatrix} y_1 \\ \vdots \\ y_n \end{pmatrix}$$

giving

$$x_1 \begin{pmatrix} a_{11} \\ \vdots \\ a_{m1} \end{pmatrix} + \cdots + x_n \begin{pmatrix} a_{1n} \\ \vdots \\ a_{mn} \end{pmatrix} = \begin{pmatrix} y_1 \\ \vdots \\ y_n \end{pmatrix}$$

Denote the column matrices of the matrix of coefficients $A_1, \ldots, A_n$ and the column of constants Y. The system can now be expressed

$$x_1 A_1 + \cdots + x_n A_n = Y$$

We shall find it convenient to use this representation of the system in the proof of the following method, called *Cramer's rule*, for solving a system of n equations in n variables that has a unique solution.

Cramer's Rule *Consider the following system of n equations in n variables:*

$$a_{11}x_1 + \cdots + a_{1n}x_n = y_1$$
$$\vdots$$
$$a_{n1}x_1 + \cdots + a_{nn}x_n = y_n$$

If the matrix of coefficients is nonsingular, this system has a unique solution given by

$$x_k = \frac{1}{|A|} \begin{vmatrix} a_{11} & \cdots & a_{1(k-1)} & y_1 & a_{1(k+1)} & \cdots & a_{1n} \\ \vdots & & & & & & \vdots \\ a_{n1} & \cdots & a_{n(k-1)} & y_n & a_{n(k+1)} & \cdots & a_{nn} \end{vmatrix}$$

the kth column of the matrix of coefficients
is replaced by $y_1, \ldots, y_n$

where $k = 1, 2, \ldots, n.$

Proof: The determinant on the right-hand side is obtained by replacing the kth column of A by Y. Expressing this determinant in terms of its columns, we get, using the properties of determinants,

$$\begin{vmatrix} A_1 & \cdots & Y & \cdots & A_n \end{vmatrix} = \begin{vmatrix} A_1 & \cdots & (x_1A_1 + \cdots + x_nA_n) & \cdots & A_n \end{vmatrix}$$
$$= \begin{vmatrix} A_1 & \cdots & x_1A_1 & \cdots & A_n \end{vmatrix} + \cdots + \begin{vmatrix} A_1 & \cdots & x_nA_n & \cdots & A_n \end{vmatrix}$$

(see Exercise 24)

$$= x_1\begin{vmatrix} A_1 & \cdots & A_1 & \cdots & A_n \end{vmatrix} + \cdots + x_n\begin{vmatrix} A_1 & \cdots & A_n & \cdots & A_n \end{vmatrix}$$

The only nonzero determinant in this sum is $\begin{vmatrix} A_1 & \cdots & A_k & \cdots & A_n \end{vmatrix}$; all the other determinants have two identical columns. Thus

$$\begin{vmatrix} A_1 & \cdots & Y & \cdots & A_n \end{vmatrix} = x_k\begin{vmatrix} A_1 & \cdots & A_k & \cdots & A_n \end{vmatrix}$$
$$= x_k|A|$$

This proves that the x_k given above define a solution to the system if solutions exist. Since the matrix of coefficients is nonsingular, we know from the previous theorem that a unique solution exists. Hence this must be it.

Cramer's rule is useful in theoretical work involving systems of linear equations because it gives an explicit formula for the solution. However, it is not usually used to obtain numerical solutions to large systems of equations, because the work involved in evaluating large determinants is enormous. We illustrate the rule for a system of three equations in three variables.

◆ *Example 3* ───────────────────────────────────────

Solve the following system of equations using Cramer's rule:

$$
\begin{aligned}
x_1 - x_2 + x_3 &= 2 \\
x_1 + 2x_2 &= 1 \\
x_1 - x_3 &= 4
\end{aligned}
$$

The matrix of coefficients is

$$
A = \begin{pmatrix} 1 & -1 & 1 \\ 1 & 2 & 0 \\ 1 & 0 & -1 \end{pmatrix} \quad \text{and} \quad \begin{pmatrix} y_1 \\ y_2 \\ y_3 \end{pmatrix} = \begin{pmatrix} 2 \\ 1 \\ 4 \end{pmatrix}
$$

$|A| = -5$. Thus a unique solution exists. By Cramer's rule,

$$
x_1 = -\tfrac{1}{5} \begin{vmatrix} 2 & -1 & 1 \\ 1 & 2 & 0 \\ 4 & 0 & -1 \end{vmatrix} = \tfrac{13}{5}
$$

$$
x_2 = -\tfrac{1}{5} \begin{vmatrix} 1 & 2 & 1 \\ 1 & 1 & 0 \\ 1 & 4 & -1 \end{vmatrix} = -\tfrac{4}{5}
$$

$$
x_3 = -\tfrac{1}{5} \begin{vmatrix} 1 & -1 & 2 \\ 1 & 2 & 1 \\ 1 & 0 & 4 \end{vmatrix} = -\tfrac{7}{5}
$$

$x_1 = \tfrac{13}{5}, x_2 = -\tfrac{4}{5}, x_3 = -\tfrac{7}{5}$ is the unique solution.   ◆

Previously we gave a numerical method for determining the inverse of a matrix. Here we derive a formula for the inverse. As in the case of Cramer's rule, this result is of theoretical importance, but it is not a useful method for actually determining the inverse of a matrix. It gives an explicit formula for the inverse of an arbitrary square matrix. We obtain the formula for the inverse in proving the following useful existence theorem.

Theorem 3-10   *A square $n \times n$ matrix is invertible if and only if it is nonsingular.*

Proof: First of all, assume that A is invertible. Hence the inverse of A, A^{-1}, exists, and $AA^{-1} = I_n$. Taking the determinant, we have $|AA^{-1}| = 1$. Thus $|A||A^{-1}| = 1$, proving that $|A| \neq 0$ and that A is nonsingular.

To prove the converse, we associate the nonsingular matrix A with a system of linear equations $AX = Y$, where A is its matrix of coefficients, and we apply Cramer's rule. Here Y is a column matrix with n arbitrary components. The system has a unique solution given by

$$x_1 = \frac{1}{|A|} |Y \quad A_2 \quad \cdots \quad A_n|, \ldots, x_n = \frac{1}{|A|} |A_1 \quad \cdots \quad A_{n-1} \quad Y|$$

Expanding $|Y \quad A_2 \quad \cdots \quad A_n|$ in terms of the first column, expanding $|A_1 \quad Y \quad \cdots \quad A_n|$ in terms of the second column, etc., we can express the solution as

$$x_1 = \frac{1}{|A|} [y_1 A_{11} + \cdots + (-1)^{n+1} y_n A_{n1}], \ldots,$$

$$x_n = \frac{1}{|A|} [(-1)^{1+n} y_1 A_{1n} + \cdots + (-1)^{n+n} y_n A_{nn}]$$

where A_{ij} is the minor of the element a_{ij} of A.

This solution may be expressed as a single matrix equation

$$X = BY$$

where

$$B = \frac{1}{|A|} \begin{pmatrix} A_{11} & \cdots & (-1)^{n+1} A_{n1} \\ & \vdots & \\ (-1)^{1+n} A_{1n} & \cdots & (-1)^{n+n} A_{nn} \end{pmatrix} \tag{1}$$

Combining equations $AX = Y$ and $X = BY$ by inserting the expression for X from the latter into the former, we get

$$ABY = Y$$

Similarly, by substituting for Y from the first equation into the second, we get

$$BAX = X$$

Since X and Y can take all values, these results imply that $AB = BA = I_n$, the unit $n \times n$ matrix, and hence that $B = A^{-1}$.

The proof of this theorem gives a formula for the inverse of a matrix. The inverse of A is given by B in equation (1). It remains to express this equation in a more meaningful way. To this end we define the *adjoint* of a square matrix A, denoted adj(A).

$$\text{adj}(A) = \begin{pmatrix} +A_{11} & -A_{12} & +A_{13} & \cdots & (-1)^{1+n} A_{1n} \\ -A_{21} & +A_{22} & -A_{23} & & \\ +A_{31} & -A_{32} & +A_{33} & & \vdots \\ \vdots & & & & \\ (-1)^{n+1} A_{n1} & \cdots & & & (-1)^{n+n} A_{nn} \end{pmatrix}^t$$

Observe that adj(A) is the transpose of the matrix of signed minors—it is the transpose of the matrix consisting of minors, each with the appropriate sign. Using this definition and equation (1), we get the following expression for the inverse of a matrix.

$$A^{-1} = \frac{1}{|A|} \, \text{adj}(A)$$

We now reinforce our understanding of this formula by using it to compute a number of inverses. However, we again emphasize the fact that the Gauss-Jordan method is a more efficient numerical method for computing an inverse. The above is an expression for the inverse of a matrix that is used in theoretical work. The Gauss-Jordan method does not lead to a general expression for an inverse—it leads to the inverse of a specific matrix.

✦ Example 4 ───────────────────────────────

Consider the matrix $A = \begin{pmatrix} 1 & 2 \\ 3 & 4 \end{pmatrix}$.

$|A| = 4 - 6 = -2 \neq 0$. Hence A^{-1} exists. We know that $A_{11} = 4$, $A_{12} = 3$, $A_{21} = 2$, and $A_{22} = 1$. The matrix of signed minors is

$$\begin{pmatrix} 4 & -3 \\ -2 & 1 \end{pmatrix}$$

and the adjoint matrix is the transpose of this matrix:

$$\text{adj}(A) = \begin{pmatrix} 4 & -2 \\ -3 & 1 \end{pmatrix}$$

The inverse of A is thus

$$A^{-1} = -\tfrac{1}{2}\begin{pmatrix} 4 & -2 \\ -3 & 1 \end{pmatrix}$$

As we check, we confirm that

$$-\tfrac{1}{2}\begin{pmatrix} 1 & 2 \\ 3 & 4 \end{pmatrix}\begin{pmatrix} 4 & -2 \\ -3 & 1 \end{pmatrix} = -\tfrac{1}{2}\begin{pmatrix} 4 & -2 \\ -3 & 1 \end{pmatrix}\begin{pmatrix} 1 & 2 \\ 3 & 4 \end{pmatrix} = \begin{pmatrix} 1 & 0 \\ 0 & 1 \end{pmatrix} \quad ✦$$

✦ Example 5 ───────────────────────────────

Consider $A = \begin{pmatrix} 1 & 2 & 0 \\ 2 & 1 & -1 \\ 3 & 1 & 1 \end{pmatrix}$.

$|A| = -8$. Hence A^{-1} exists.

$$\text{adj}(A) = \begin{pmatrix} \begin{vmatrix} 1 & -1 \\ 1 & 1 \end{vmatrix} & -\begin{vmatrix} 2 & -1 \\ 3 & 1 \end{vmatrix} & \begin{vmatrix} 2 & 1 \\ 3 & 1 \end{vmatrix} \\ -\begin{vmatrix} 2 & 0 \\ 1 & 1 \end{vmatrix} & \begin{vmatrix} 1 & 0 \\ 3 & 1 \end{vmatrix} & -\begin{vmatrix} 1 & 2 \\ 3 & 1 \end{vmatrix} \\ \begin{vmatrix} 2 & 0 \\ 1 & -1 \end{vmatrix} & -\begin{vmatrix} 1 & 0 \\ 2 & -1 \end{vmatrix} & \begin{vmatrix} 1 & 2 \\ 2 & 1 \end{vmatrix} \end{pmatrix}^t$$

$$= \begin{pmatrix} 2 & -5 & -1 \\ -2 & 1 & 5 \\ -2 & 1 & -3 \end{pmatrix}^t = \begin{pmatrix} 2 & -2 & -2 \\ -5 & 1 & 1 \\ -1 & 5 & -3 \end{pmatrix}$$

Hence $A^{-1} = -\frac{1}{8} \begin{pmatrix} 2 & -2 & -2 \\ -5 & 1 & 1 \\ -1 & 5 & -3 \end{pmatrix}$. ◆

EXERCISES

In Exercises 1–6, solve the systems of linear equations using Cramer's rule.

*1. $x_1 + 2x_2 = 8$
$2x_1 + 5x_2 = 19$

2. $x_1 + 2x_2 = 3$
$3x_1 + x_2 = -1$

*3. $x_1 + x_2 + x_3 = 0$
$2x_1 - 5x_2 - 3x_3 = 10$
$4x_1 + 8x_2 + 2x_3 = 4$

4. $2x_1 - x_2 + 2x_3 = 11$
$x_1 + 2x_2 - x_3 = -3$
$3x_1 - 2x_2 - 3x_3 = -1$

*5. $x_1 - x_2 + x_3 = 1$
$2x_1 - 2x_2 + 5x_3 = 0$
$-x_1 + 3x_2 - 4x_3 = 3$

6. $x_1 + 3x_2 - x_3 = 1$
$2x_1 + x_2 + x_3 = 4$
$3x_1 + 4x_2 + 2x_3 = -1$

In Exercises 7–14, determine whether or not each matrix has an inverse. If a matrix has an inverse, find the inverse using the formula of this section.

*7. $\begin{pmatrix} 1 & 4 \\ 3 & 2 \end{pmatrix}$

8. $\begin{pmatrix} -2 & -1 \\ 7 & 3 \end{pmatrix}$

*9. $\begin{pmatrix} 1 & 2 \\ 2 & 4 \end{pmatrix}$

10. $\begin{pmatrix} 2 & 1 \\ 4 & 3 \end{pmatrix}$

*11. $\begin{pmatrix} 1 & 2 & 3 \\ 0 & 1 & 2 \\ 4 & 5 & 3 \end{pmatrix}$

12. $\begin{pmatrix} 0 & 3 & 3 \\ 1 & 2 & 3 \\ 1 & 4 & 6 \end{pmatrix}$

*13. $\begin{pmatrix} 0 & 3 & 3 \\ 1 & 2 & 3 \\ 2 & 4 & 6 \end{pmatrix}$

14. $\begin{pmatrix} 1 & 2 & -1 \\ 2 & 4 & -3 \\ 1 & -2 & 0 \end{pmatrix}$

*15. Apply Theorem 3-9 to determine values of λ for which the system of equations

$$(1 - \lambda)x_1 + 6x_2 = 0$$
$$5x_1 + (2 - \lambda)x_2 = 0$$

can have many solutions. Such λ's are called *eigenvalues* of a matrix of coefficients. Eigenvalues are important in applications, as the reader will see in following sections. Determine the solutions for each λ.

16. Determine values of λ for which the system of equations

$$
\begin{aligned}
(5 - \lambda)x_1 + \quad 4x_2 + \quad 2x_3 &= 0 \\
4x_1 + (5 - \lambda)x_2 + \quad 2x_3 &= 0 \\
2x_1 + \quad 2x_2 + (2 - \lambda)x_3 &= 0
\end{aligned}
$$

has many solutions.

17. Prove that the system of equations $AX = \lambda X$ has the possibility of more than one solution if and only if $|A - \lambda I| = 0$. Here I is the appropriate unit matrix.

18. If C is a nonsingular matrix, prove that $|C||C^{-1}| = 1$.

19. (a) Let A and B be invertible matrices of the same kind. Prove that the product matrix AB is invertible with the inverse defined by

$$(AB)^{-1} = B^{-1}A^{-1}$$

(b) Let A, B, and C be invertible matrices of the same kind. Prove that ABC is invertible with the inverse $(ABC)^{-1} = C^{-1}B^{-1}A^{-1}$.

20. Is the sum of two invertible matrices always invertible?

21. Prove that if A is an invertible matrix then $(A^t)^{-1} = (A^{-1})^t$. Thus the inverse of the transpose of a matrix is the transpose of its inverse.

22. Prove that if A is a symmetric invertible matrix then A^{-1} is also symmetric.

23. Let A be invertible. If B and C are matrices such that $AB = AC$, prove that $B = C$.

24. The following property of determinants was used in the proof of Cramer's rule.

$$
\begin{vmatrix} A_1 & \cdots & (x_1A_1 + \cdots + x_nA_n) & \cdots & A_n \end{vmatrix}
$$
$$
\uparrow
$$
$$
(kth\ column)
$$

$$
= \begin{vmatrix} A_1 & \cdots & x_1A_1 & \cdots & A_n \end{vmatrix} + \cdots + \begin{vmatrix} A_1 & \cdots & x_nA_n & \cdots & A_n \end{vmatrix}
$$

Prove this result by expanding the first determinant using the kth column.

4

The Vector Space $\mathbf{R}^n$

At this time the course becomes geometrical in nature. We shall develop the mathematical tools for discussing two- and three-dimensional spaces, leading up to n-dimensional space. Matrices and systems of linear equations will be useful tools in this development.

4-1 INTRODUCTION TO VECTORS

The relative locations of points in a plane are usually discussed in terms of a coordinate system. For example, in Figure 4-1 the location of each point in a plane can be described relative to a rectangular coordinate system. The point A in Figure 4-1 is the point $(2, 1)$.

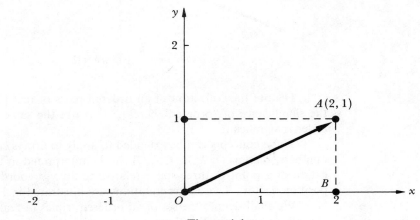

Figure 4-1

Furthermore, A is a certain distance in a certain direction from $(0, 0)$; the distance and direction are characterized by the length and direction of the line segment from O to A. We call such a directed line segment a *position vector* and denote it $\overline{OA}$. We represent the position vector geometrically by a line having an arrow at its head. There are thus two ways of interpreting $(2, 1)$; it defines the location of a point A in the plane, and it also defines the position vector $\overline{OA}$.

✦ *Example 1*

Sketch the position vectors $\overline{OA} = (4, 1)$, $\overline{OB} = (-3, 5)$, and $\overline{OC} = (-5, -2)$.
We get Figure 4-2.

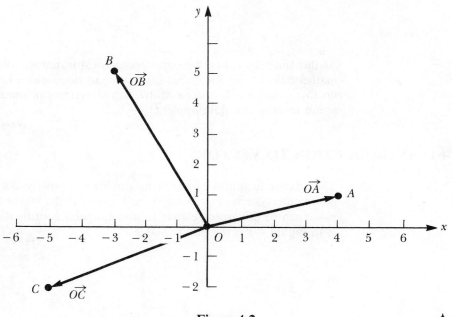

Figure 4-2   ✦

Denote the collection of all ordered pairs of real numbers by **R**^{2}. Note the significance of "ordered" here; $(2, 1)$ is not the same vector as $(1, 2)$. The order is significant.

These concepts can be extended to apply to arrays consisting of three real numbers, such as $(1, 2, 3)$. $(1, 2, 3)$ can be interpreted in two ways—as the location of a point in three-space relative to an xyz coordinate system, or as a position vector. These interpretations are illustrated in Figure 4-3.

We shall denote the set of all ordered triples of real numbers by **R**^{3}.

We now generalize these concepts with the following definition.

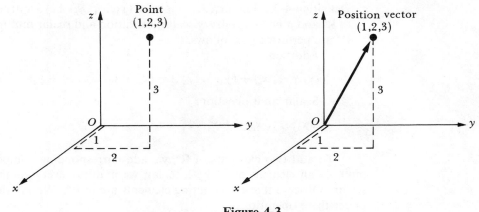

Figure 4-3

Definition 4-1   Let $(x_1, \ldots, x_n)$ be a sequence of n real numbers. The set of all such sequences is called *n-space* and is denoted $\mathbf{R}^n$. $(1, 2, 3, \ldots, n)$, for example, is an element of $\mathbf{R}^n$.

Many of the results and techniques that we develop for $\mathbf{R}^n$ with $n > 3$ will be useful mathematical tools without direct geometrical significance. The elements of $\mathbf{R}^n$ can, however, be interpreted as locations of points in *n*-space, or as position vectors in *n*-space. It is difficult to visualize an *n*-space for $n > 3$, but the reader is encouraged to try to form an intuitive picture. A geometrical "feel" for what is taking place often makes an algebraic discussion easier to follow. The mathematical structures that we shall place on $\mathbf{R}^n$ will be motivated by structures on $\mathbf{R}^2$ and $\mathbf{R}^3$ which have direct geometrical interpretation.
    In keeping with this scheme, let us denote the set of real numbers by $\mathbf{R}$.

✦ *Example 2*
_____

$\mathbf{R}^4$ is the collection of all sets of four ordered real numbers. For example, $(1, 2, 3, 4)$ and $(-1, 0, \frac{1}{2}, 3)$ are elements of $\mathbf{R}^4$.
    $\mathbf{R}^5$ is the collection of all sets of five ordered real numbers. For example, $(-1, 0, 3, 4, 2)$ and $(\frac{1}{2}, \frac{1}{2}, 2, 0, 3)$ are in this set.   ✦

Definition 4-2   Two elements of $\mathbf{R}^n$, $(x_1, \ldots, x_n)$ and $(y_1, \ldots, y_n)$, are said to be *equal* if $x_1 = y_1, \ldots, x_n = y_n$. Thus two elements of $\mathbf{R}^n$ are equal if their corresponding *components* are equal.

Let us now develop an algebraic structure for $\mathbf{R}^n$. The following definitions of the operations of addition and scalar multiplication on $\mathbf{R}^n$ are similar to their definitions for matrices. We shall also find that the algebraic properties of $\mathbf{R}^n$ under these operations are similar to those of matrices.

Definition 4-3   Let $(x_1, \ldots, x_n)$ and $(y_1, \ldots, y_n)$ be arbitrary elements of **R**^{n} and a be an arbitrary scalar. Addition and scalar multiplication on **R**^{n} are performed as follows.

Addition:

$$(x_1, \ldots, x_n) + (y_1, \ldots, y_n) = (x_1 + y_1, \ldots, x_n + y_n)$$

Scalar multiplication:

$$a(x_1, \ldots, x_n) = (ax_1, \ldots, ax_n)$$

To add two elements of **R**^{n}, we add corresponding components, and to multiply an element of **R**^{n} by a scalar, we multiply every component by that scalar. Observe that the resulting elements are in **R**^{n}. We say that **R**^{n} is *closed* under these operations.

R^n with these two operations is called a *vector space*, and its elements are called *vectors*.

We shall henceforth in this text interpret **R**^{n} to be a vector space.

✦ *Example 3* ───────────────────────────────

$(-1, 2, 3, 4)$ and $(0, -3, 2, 3)$ are elements of **R**^{4}. Adding these vectors, we get

$$(-1, 2, 3, 4) + (0, -3, 2, 3) = (-1, -1, 5, 7)$$

To illustrate scalar multiplication, we multiply $(-1, 2, 3, 4)$ by 4:

$$4(-1, 2, 3, 4) = (-4, 8, 12, 16)$$

Note that the resulting vector under each operation is in the original vector space, **R**^{4}.   ✦

The vector $(0, \ldots, 0)$ having n zero components is called the *zero vector* of **R**^{n}. For example, $(0, 0, 0)$ is the zero vector of **R**^{3}. As we proceed, we shall find that zero vectors play a central role in the development of vector spaces.

We now give examples to illustrate the geometrical interpretations of these operations and some applications.

✦ *Example 4* ───────────────────────────────

This example illustrates a geometrical interpretation of vector addition.

Consider the two vectors $(3, 1)$ and $(2, 5)$. On adding,

$$(3, 1) + (2, 5) = (5, 6)$$

In Figure 4-4, we interpret the vectors as position vectors.

Construct the parallelogram $OABC$ having the vectors $(3, 1)$ and $(2, 5)$ as adjacent sides. Then the vector $(5, 6)$, the sum, will be the diagonal of this parallelogram.

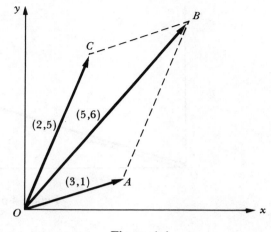

Figure 4-4   ✦

In general, if **u** and **v** are vectors in the same vector space, then we can geometrically interpret **u** + **v** as the diagonal of the parallelogram defined by **u** and **v**. (See Figure 4-5.)

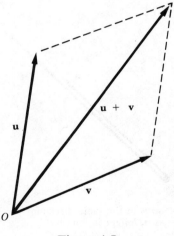

Figure 4-5

This way of visualizing vector addition is useful in all vector spaces.

✦ *Example 5*
_____

This example gives us a geometrical interpretation of scalar multiplication.

Consider the position vector **u** = (2, 1) in Figure 4-6. Let us multiply this vector by 3 to get the vector **v** = (6, 3). Observe that this vector is in the same direction as **u** but 3 times its length.

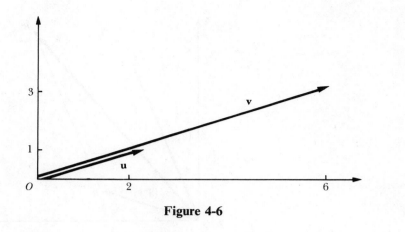

Figure 4-6   ✦

In general, if **u** is an arbitrary vector and a is a scalar, there are the following geometrical interpretations for a**u**, depending on a. The length of the vector a**u** will always be a times the length of **u**. The direction of a**u** will be either the same as that of **u** or the opposite of that of **u**. (See Figures 4-7 and 4-8.)

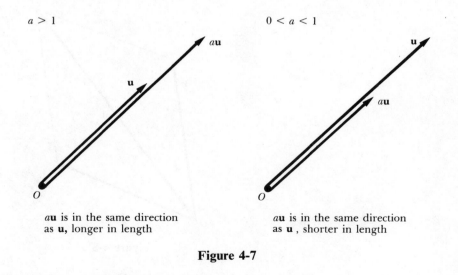

$a > 1$

a**u** is in the same direction as **u,** longer in length

$0 < a < 1$

a**u** is in the same direction as **u** , shorter in length

Figure 4-7

✦ *Example 6* ───

Physical quantities such as velocities and forces that have both magnitude and direction are called vectors. They can be represented mathematically by elements of **R**^{2} or **R**^{3}, as this example illustrates.

A force of two pounds tends to pull a body in a direction $N\ 60°\ E$. We shall now see how we can represent this force by an element of **R**^{2}. Let the

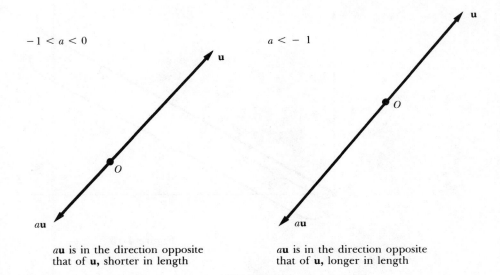

$-1 < a < 0$

$a < -1$

$a\mathbf{u}$ is in the direction opposite that of $\mathbf{u}$, shorter in length

$a\mathbf{u}$ is in the direction opposite that of $\mathbf{u}$, longer in length

Figure 4-8

body be at O, with axes OE and ON representing easterly and northerly directions from O, respectively (Figure 4-9). To obtain the direction $N\ 60°\ E$, one turns $60°$ in an easterly direction from N. This gives the direction OA. The length of the vector $\overline{OA}$ is 2; this represents the magnitude of the force. The lengths of OB and AB will then be $\sqrt{3}$ and 1. Hence the force is represented by the vector $\overline{OA}(\sqrt{3}, 1)$, an element of $\mathbf{R}^2$.

Using this same example, we can see how multiplication of a vector by a scalar has physical interpretation. The vector $\overline{OB} = 3(\sqrt{3}, 1)$—that is, the vector $(3\sqrt{3}, 3)$—represents a force with three times the magnitude of $\overline{OA}$,

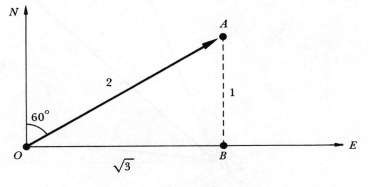

Figure 4-9

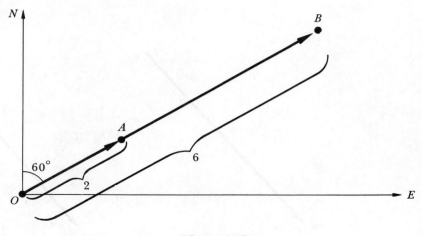

Figure 4-10

acting in the same direction as $\overline{OA}$. It is thus a force of six pounds acting in the same direction as $\overline{OA}$ (Figure 4-10).   ✦

✦ *Example 7* _____

We now look at a physical interpretation of vector addition. Two forces acting simultaneously on a body are equivalent to a single force, called the *resultant force*. Experiments show that if two such forces are represented by vectors in **R**² or **R**³, the resultant force is represented by the sum of these vectors. This law is called the *parallelogram of forces*.

Let (1, 3) and (2, 1) represent two forces acting on a body situated at O in Figure 4-11. On adding (1, 3) and (2, 1), we find that the resultant force is

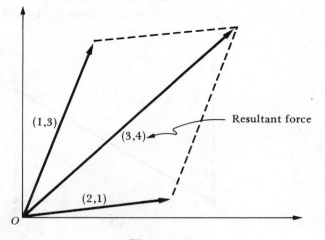

Figure 4-11

(3, 4). From our previous discussion we know that the resultant force will be defined by the diagonal of the parallelogram having the given vectors as adjacent sides—hence the term parallelogram of forces.   ✦

We now give a practical example of addition and scalar multiplication of the elements of $\mathbf{R}^4$.

✦ *Example 8* _____

Consider the following system of equations:

$$x - 2y + 3z + w = 0$$
$$2x - 3y + 2z - w = 0$$
$$3x - 5y + 5z = 0$$
$$x - y - z - 2w = 0$$

There are four equations in four variables x, y, z, and w. We can interpret a solution $x = x_1$, $y = y_1$, $z = z_1$, $w = w_1$ as a vector (x_1, y_1, z_1, w_1), an element of $\mathbf{R}^4$. For example, $x_1 = 10$, $y_1 = 7$, $z_1 = 1$, $w_1 = 1$ is a solution that can be written as the vector $(10, 7, 1, 1)$. (Substitute these values into the equations to see that every equation is satisfied.)

Another solution is $(0, 1, 1, -1)$. Such a set of equations may have only a single solution or many solutions. Note that $(0, 0, 0, 0)$, called the *zero vector*, will always be a solution to such a system.

Let (x_1, y_1, z_1, w_1) and (x_2, y_2, z_2, w_2) represent two arbitrary solutions. Thus

$$x_1 - 2y_1 + 3z_1 + w_1 = 0$$
$$2x_1 - 3y_1 + 2z_1 - w_1 = 0$$
$$3x_1 - 5y_1 + 5z_1 = 0$$
$$x_1 - y_1 - z_1 - 2w_1 = 0$$

and

$$x_2 - 2y_2 + 3z_2 + w_2 = 0$$
$$2x_2 - 3y_2 + 2z_2 - w_2 = 0$$
$$3x_2 - 5y_2 + 5z_2 = 0$$
$$x_2 - y_2 - z_2 - 2w_2 = 0$$

Adding the two first equations, the two second equations, etc., we get

$$(x_1 + x_2) - 2(y_1 + y_2) + 3(z_1 + z_2) + (w_1 + w_2) = 0$$
$$2(x_1 + x_2) - 3(y_1 + y_2) + 2(z_1 + z_2) - (w_1 + w_2) = 0$$
$$3(x_1 + x_2) - 5(y_1 + y_2) + 5(z_1 + z_2) = 0$$
$$(x_1 + x_2) - (y_1 + y_2) - (z_1 + z_2) - 2(w_1 + w_2) = 0$$

implying that $(x_1 + x_2, y_1 + y_2, z_1 + z_2, w_1 + w_2)$ also represents a solution; that is, $(x_1, y_1, z_1, w_1) + (x_2, y_2, z_2, w_2)$, the vector sum of the solutions, is a solution.

We have seen that $(10, 7, 1, 1)$ and $(0, 1, 1, -1)$ are solutions. According to our theory, $(10, 7, 1, 1) + (0, 1, 1, -1)$, or $(10, 8, 2, 0)$, should also be a solution. Substitute $x = 10$, $y = 8$, $z = 2$, and $w = 0$ into the equations to see that they are satisfied, proving that this is a solution.

If c is any scalar, then $c(x_1, y_1, z_1, w_1)$ will also be a solution. (The proof is left as an exercise.) ✦

We call a system of equations like the above, with the constant terms being zero, a *homogeneous linear system*. We shall later prove that if any two solutions to such a system are written in vector form then their vector sum will be a solution. The scalar multiple of any vector solution will also be a solution. New solutions can be generated in this manner from known solutions. Furthermore, the properties that we develop for the vector space **R***ⁿ* will enable us to further our understanding of the behavior of systems of linear equations.

Subtraction is performed on elements of **R***ⁿ* by subtracting corresponding components. For example, in **R**⁴,

$$(5, 3, -6, 1) - (2, 1, 3, -4) = (3, 2, -9, 5)$$

Observe that this is equivalent to

$$(5, 3, -6, 1) + [(-1)(2, 1, 3, -4)] = (3, 2, -9, 5)$$

Thus subtraction is in effect not a new operation on **R***ⁿ*, but a combination of addition and scalar multiplication by -1. It is important to realize that we have only *two* independent operations on **R***ⁿ*—namely, addition and scalar multiplication.

Column Vectors

Up to this point we have defined only *row vectors*; that is, the components of a vector were written in row form. We shall find that it is more suitable sometimes to use vectors in the form of columns, both in theoretical discussions and in applications. We define addition and scalar multiplication in **R***ⁿ*, when the elements are written in column form, again in a componentwise manner:

$$\begin{pmatrix} x_1 \\ \vdots \\ x_n \end{pmatrix} + \begin{pmatrix} y_1 \\ \vdots \\ y_n \end{pmatrix} = \begin{pmatrix} x_1 + y_1 \\ \vdots \\ x_n + y_n \end{pmatrix} \quad \text{and} \quad a\begin{pmatrix} x_1 \\ \vdots \\ x_n \end{pmatrix} = \begin{pmatrix} ax_1 \\ \vdots \\ ax_n \end{pmatrix}$$

For example, in **R**²,

$$\begin{pmatrix} 1 \\ 2 \end{pmatrix} + \begin{pmatrix} 0 \\ 1 \end{pmatrix} = \begin{pmatrix} 1 \\ 3 \end{pmatrix} \quad \text{and} \quad 4\begin{pmatrix} 1 \\ 2 \end{pmatrix} = \begin{pmatrix} 4 \\ 8 \end{pmatrix}$$

EXERCISES

*1. Sketch the position vectors $(1, 0)$ and $(0, 1)$ in $\mathbf{R}^2$. The notation $\mathbf{i}$ is often used for $(1, 0)$ and $\mathbf{j}$ for $(0, 1)$ in science.

2. Sketch the position vectors $(1, 0, 0)$, $(0, 1, 0)$, and $(0, 0, 1)$ in $\mathbf{R}^3$. The notations $\mathbf{i}$, $\mathbf{j}$, and $\mathbf{k}$ are often used for these three vectors, respectively.

3. Sketch the following position vectors in $\mathbf{R}^2$.
   *(a) $\overrightarrow{OA} = (5, 6)$, $\overrightarrow{OB} = (-3, 2)$, $\overrightarrow{OC} = (1, -3)$
   (b) $\overrightarrow{OP} = (2, 4)$, $\overrightarrow{OQ} = (-4, -5)$, $\overrightarrow{OR} = (3, -3)$
   (c) $\mathbf{u} = (1, 1)$, $\mathbf{v} = (-1, -4)$, $\mathbf{w} = (5, 3)$
   (We shall sometimes find it convenient to use letters such as $\mathbf{u}$, $\mathbf{v}$, and $\mathbf{w}$ for position vectors.)

4. Sketch the following position vectors in $\mathbf{R}^3$.
   *(a) $\overrightarrow{OA} = (2, 3, 1)$, $\overrightarrow{OB} = (0, 5, -1)$, $\overrightarrow{OC} = (-1, 3, 4)$
   (b) $\mathbf{u} = (1, 1, 1)$, $\mathbf{v} = (-1, -2, -4)$, $\mathbf{w} = (0, 0, -3)$

5. Multiply the following vectors by the given scalars. Interpret the results for (a)–(e) geometrically.
   *(a) $(1, 4)$ by 3   (b) $(-1, 3)$ by -2
   *(c) $(2, 6)$ by $\frac{1}{2}$ (d) $(2, 4, 2)$ by $-\frac{1}{2}$
   *(e) $(-1, 2, 3)$ by 3   (f) $(-1, 2, 3, -2)$ by 4
   (g) $(1, -4, 3, -2, 5)$ by -5 (h) $(3, 0, 1, 4, 2, -1)$ by 3

6. Compute the following vector expressions for $\mathbf{u} = (1, 2)$, $\mathbf{v} = (4, -1)$, and $\mathbf{w} = (-3, 5)$. Interpret your results geometrically.
   *(a) $\mathbf{u} + \mathbf{v}$ *(b) $\mathbf{u} + 3\mathbf{v}$ (c) $\mathbf{v} + \mathbf{w}$
   (d) $2\mathbf{u} + \mathbf{w}$ *(e) $-\mathbf{u} + 2\mathbf{v}$ (f) $2\mathbf{v} - 3\mathbf{w}$

7. Compute the following vector expressions for $\mathbf{u} = (2, 1, 3)$, $\mathbf{v} = (-1, 3, 2)$, and $\mathbf{w} = (2, 4, -2)$.
   *(a) $\mathbf{u} + \mathbf{v}$ (b) $2\mathbf{u} + \mathbf{v}$ *(c) $\mathbf{u} + 3\mathbf{w}$
   (d) $-2\mathbf{v} + \mathbf{w}$ *(e) $3\mathbf{v} - 4\mathbf{w}$ (f) $-4\mathbf{u} + 2\mathbf{v}$

*8. Consider the following system of homogeneous equations:

$$\begin{aligned} x + 2y - z - 2w &= 0 \\ 2x + 5y \quad\;\; - 2w &= 0 \\ 4x + 9y - 2z - 6w &= 0 \\ x + 3y + z \quad\;\; &= 0 \end{aligned}$$

You are given two solutions, $(5, -2, 1, 0)$ and $(11, -4, 1, 1)$. Using the operations of $\mathbf{R}^4$, generate five other solutions.

9. $(0, -1, 1, -1)$ and $(3, 7, 2, 1)$ are solutions of the system

$$\begin{aligned} x - y + z + 2w &= 0 \\ 3x - 2y + z + 3w &= 0 \\ 5x - 4y + 3z + 7w &= 0 \\ 2x - y \quad\;\; + w &= 0 \end{aligned}$$

Use the operations of $\mathbf{R}^4$ to generate five further solutions.

10. If **a**, **b**, and **c** are the column vectors $\begin{pmatrix} 1 \\ 2 \\ -1 \end{pmatrix}$, $\begin{pmatrix} 3 \\ 0 \\ 1 \end{pmatrix}$, and $\begin{pmatrix} -1 \\ 0 \\ 5 \end{pmatrix}$, respectively, all elements of **R**^{3}, determine

    *(a) **a** + **b**   (b) **b** + 3**c**

    *(c) 3**a**   (d) 2**a** + 8**c**

11. Which of the following physical quantities are scalars and which are vectors?

    *(a) temperature   *(b) acceleration   (c) pressure

    *(d) frequency   (e) gravity   (f) position

    (g) time   (h) sound   (i) cost

12. Determine the resultant of the following forces in vectorial form. Draw a diagram of the forces and their resultant.

    *(a) the forces (3, 2) and (5, −5)

    (b) the forces (−1, 3) and (−2, −3)

    (c) a force of two pounds in a northerly direction and a force of three pounds in a westerly direction

4-2 ALGEBRAIC PROPERTIES OF VECTORS

Having defined operations of addition and of scalar multiplication on **R**^{n} and having glimpsed the significance of these operations, we now turn our attention to examining their algebraic properties. Knowledge of these properties will broaden the usefulness of the mathematical tools in applications and will also guide us in the further development of the field. As for matrices, we now find that vectors in **R**^{n} are commutative and associative under addition.

Theorem 4-1   *Let* **u**, **v**, *and* **w** *be vectors in* **R**^{n}. *Then*

(1) *Vectors are commutative under addition:*

$$\mathbf{u} + \mathbf{v} = \mathbf{v} + \mathbf{u}$$

(2) *Vectors are associative under addition:*

$$\mathbf{u} + (\mathbf{v} + \mathbf{w}) = (\mathbf{u} + \mathbf{v}) + \mathbf{w}$$

Proof: For part (1), let $\mathbf{u} = (x_1, \ldots, x_n)$ and $\mathbf{v} = (y_1, \ldots, y_n)$. Then

$$
\begin{aligned}
\mathbf{u} + \mathbf{v} &= (x_1, \ldots, x_n) + (y_1, \ldots, y_n) \\
&= (x_1 + y_1, \ldots, x_n + y_n) \\
&= (y_1 + x_1, \ldots, y_n + x_n) && \text{\textit{since real numbers are}} \\
& && \text{\textit{commutative under addition}} \\
&= (y_1, \ldots, y_n) + (x_1, \ldots, x_n) \\
&= \mathbf{v} + \mathbf{u}
\end{aligned}
$$

Thus the order in which vectors are added is unimportant.

Part (2) is proved in a similar manner using the associative property of real numbers. Thus addition can be performed by grouping vectors together either way.

Algebra and geometry often go hand in hand, the one assisting and clarifying the other. We shall often give geometrical "insights" to algebraic concepts in this course. We have seen how elements of $\mathbf{R}^2$ can be used to discuss locations of points in a plane and also as position vectors. A position vector in $\mathbf{R}^2$ describes a certain distance in a certain direction from the origin $(0, 0)$. Occasionally the need arises to talk about a distance in a direction from a point other than the origin. We conveniently generalize the concept of position vector for this. This discussion will lead not only to new useful mathematical tools but also to a way of visualizing the results of the previous theorem.

Consider the points $A(x_1, y_1)$ and $B(x_2, y_2)$ in $\mathbf{R}^2$ (Figure 4-12). B is at a certain distance and in a certain direction from A; the distance and direction are characterized by the direction and length of the line segment AB. Let this *directed line segment* be represented by an element of $\mathbf{R}^2$ having components $x_2 - x_1$ and $y_2 - y_1$. It is a *vector* denoted by $\overrightarrow{AB}$:

$$\overrightarrow{AB} = (x_2 - x_1, y_2 - y_1)$$

We call A the *initial point* of $\overrightarrow{AB}$ and B its *terminal point*. A position vector fits into this scheme; it is a vector having the origin as its initial point.

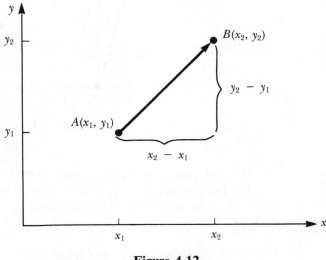

Figure 4-12

✦ *Example 1*

Consider the points $A(3, 2)$, $B(5, 4)$, and $C(-2, 7)$ in $\mathbf{R}^2$. Let us compute and illustrate the vectors $\overrightarrow{AB}$, $\overrightarrow{BC}$, and $\overrightarrow{OC}$.

We get

$$\overrightarrow{AB} = (5 - 3, 4 - 2) = (2, 2)$$
$$\overrightarrow{BC} = (-2 - 5, 7 - 4) = (-7, 3)$$

Since O is the origin, $(0, 0)$,

$$\overrightarrow{OC} = (-2 - 0, 7 - 0) = (-2, 7) \qquad (a\ position\ vector)$$

Geometrically these vectors are as shown in Figure 4-13.

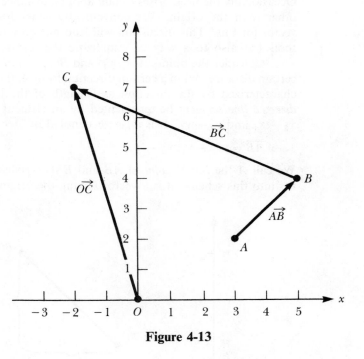

Figure 4-13   ✦

We now generalize these concepts to $\mathbf{R}^n$.

Let the directed line segment from the point $A(x_1, \ldots, x_n)$ to the point $B(y_1, \ldots, y_n)$ be denoted by $\overrightarrow{AB}$ and defined by

$$\overrightarrow{AB} = (y_1 - x_1, \ldots, y_n - x_n)$$

$\overrightarrow{AB}$ is thus an element of $\mathbf{R}^n$; it is called a *vector*. A is its initial point and B is its terminal point.

One visualizes $\overrightarrow{AB}$ in $\mathbf{R}^n$ as shown in Figure 4-14.

The following example illustrates the usefulness of vector addition for vectors of this type. It also leads to a geometrical interpretation of $\mathbf{u} + \mathbf{v} = \mathbf{v} + \mathbf{u}$.

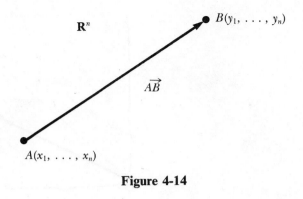

Figure 4-14

✦ *Example 2*

The vector $\overrightarrow{AB} = (1, 3)$ describes a man's route from a starting point A to a point B. At B he changes direction and travels to C according to the vector $\overrightarrow{BC} = (4, 2)$. Figure 4-15 illustrates his route. The final position C is described relative to A by the vector $\overrightarrow{AC}$.

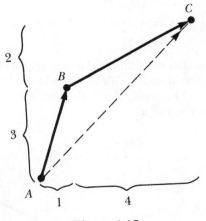

Figure 4-15

From the diagram the components of $\overrightarrow{AC}$ can be seen to be $1 + 4$ and $3 + 2$. $\overrightarrow{AC}$ is the vector $(5, 5)$. We can, in fact, get $\overrightarrow{AC}$ by adding the vectors $\overrightarrow{AB}$ and $\overrightarrow{BC}$:

$$\overrightarrow{AC} = \overrightarrow{AB} + \overrightarrow{BC}$$
$$= (1, 3) + (4, 2)$$
$$= (5, 5)$$

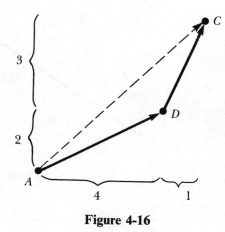

Figure 4-16

Observe that the man would have reached the same location C, described by the vector $\overrightarrow{AC} = (5, 5)$, if he had traveled along the route $(4, 2)$ first and then $(1, 3)$. We represent these vectors by AD and DC in Figure 4-16.

Here we get

$$\overrightarrow{AC} = \overrightarrow{AD} + \overrightarrow{DC}$$
$$= (4, 2) + (1, 3)$$
$$= (5, 5)$$

Let us combine the two figures into one figure and generalize, calling the vectors **u** and **v**. We get Figure 4-17.

The routes $\overrightarrow{AB} + \overrightarrow{BC}$ and $\overrightarrow{AD} + \overrightarrow{DC}$ both result in $\overrightarrow{AC}$. Thus we see that **u** + **v** = **v** + **u**.

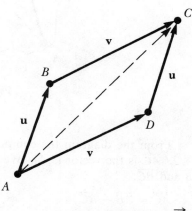

u + **v** = **v** + **u**   (both being $\overrightarrow{AC}$)

Figure 4-17

The same type of construction can be used to give a geometrical interpretation of $\mathbf{u} + (\mathbf{v} + \mathbf{w}) = (\mathbf{u} + \mathbf{v}) + \mathbf{w}$.

✦ *Example 3*

Let $\mathbf{u}$, $\mathbf{v}$, and $\mathbf{w}$ be the following vectors $\overrightarrow{AB}$, $\overrightarrow{BC}$, and $\overrightarrow{CD}$ in $\mathbf{R}^n$. We get the geometrical representations of $\mathbf{u} + (\mathbf{v} + \mathbf{w})$ and $(\mathbf{u} + \mathbf{v}) + \mathbf{w}$ shown in Figure 4-18.

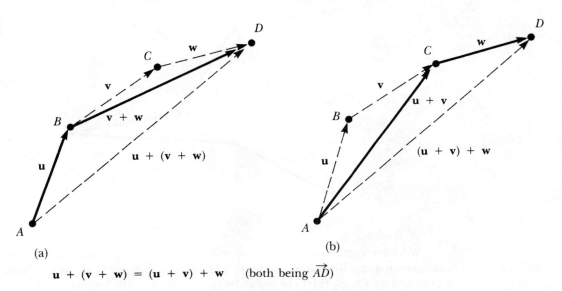

$$\mathbf{u} + (\mathbf{v} + \mathbf{w}) = (\mathbf{u} + \mathbf{v}) + \mathbf{w} \quad \text{(both being } \overrightarrow{AD})$$

Figure 4-18

In (a),

$$\mathbf{u} + (\mathbf{v} + \mathbf{w}) = \overrightarrow{AB} + (\overrightarrow{BC} + \overrightarrow{CD})$$
$$= \overrightarrow{AB} + \overrightarrow{BD}$$
$$= \overrightarrow{AD}$$

In (b),

$$(\mathbf{u} + \mathbf{v}) + \mathbf{w} = (\overrightarrow{AB} + \overrightarrow{BC}) + \overrightarrow{CD}$$
$$= \overrightarrow{AC} + \overrightarrow{CD}$$
$$= \overrightarrow{AD}$$

Thus, $\mathbf{u} + (\mathbf{v} + \mathbf{w}) = \overrightarrow{AD} = (\mathbf{u} + \mathbf{v}) + \mathbf{w}$.   ✦

Let us now look at the computational implications of the theorem. Combining the two results we see that three vectors in $\mathbf{R}^n$ can be added in any order, in any grouping. There is no need, for example, for brackets to indicate which pair is to be added first. The sum of three vectors can be obtained by adding the corresponding components of the three vectors:

$$(x_1, \ldots, x_n) + (y_1, \ldots, y_n) + (z_1, \ldots, z_n) = (x_1 + y_1 + z_1, \ldots, x_n + y_n + z_n)$$

In general, the sum of any number of vectors in **R**^{n} is independent of the order in which they are added. The sum is obtained by adding corresponding components. The sum of vectors, say **u**, **v**, **w**, and **p**, can be interpreted geometrically as shown in Figure 4-19.

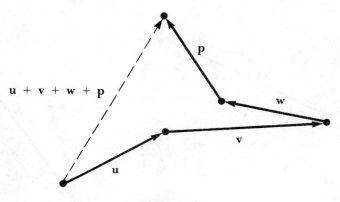

Figure 4-19

We have used row vectors in our discussion; all the results hold in an analogous manner for column vectors. The sum of a number of column vectors is obtained by adding their corresponding components. The following example illustrates the analysis of test scores using these tools.

✦ *Example 4**

A class of ten students has had five tests during the quarter. A perfect score on each of the tests is 50. The scores are listed in Table 4-1.

Table 4-1

	Test 1	Test 2	Test 3	Test 4	Test 5
Anderson	40	45	30	48	42
Boggs	20	15	30	25	10
Chittar	40	35	25	45	46
Diessner	25	40	45	40	38
Farnam	35	35	38	37	39
Gill	50	46	45	48	47
Homes	22	24	30	32	29
Johnson	35	27	20	41	30
Schomer	28	31	25	27	31
Wong	40	35	36	32	38

We can express these scores as column vectors

$$\begin{pmatrix} 40 \\ 20 \\ 40 \\ 25 \\ 35 \\ 50 \\ 22 \\ 35 \\ 28 \\ 40 \end{pmatrix}, \begin{pmatrix} 45 \\ 15 \\ 35 \\ 40 \\ 35 \\ 46 \\ 24 \\ 27 \\ 31 \\ 35 \end{pmatrix}, \begin{pmatrix} 30 \\ 30 \\ 25 \\ 45 \\ 38 \\ 45 \\ 30 \\ 20 \\ 25 \\ 36 \end{pmatrix}, \begin{pmatrix} 48 \\ 25 \\ 45 \\ 40 \\ 37 \\ 48 \\ 32 \\ 41 \\ 27 \\ 32 \end{pmatrix}, \begin{pmatrix} 42 \\ 10 \\ 46 \\ 38 \\ 39 \\ 47 \\ 29 \\ 30 \\ 31 \\ 38 \end{pmatrix}$$

where each vector is an element of $\mathbf{R}^{10}$. To obtain each person's average, we use vector addition to add the vectors, and then scalar multiplication to multiply by $\frac{1}{5}$ (dividing by the number of tests). We get

$$\frac{1}{5} \begin{pmatrix} 205 \\ 100 \\ 191 \\ 188 \\ 184 \\ 236 \\ 137 \\ 153 \\ 142 \\ 181 \end{pmatrix} = \begin{pmatrix} 41 \\ 20 \\ 38.2 \\ 37.6 \\ 36.8 \\ 47.2 \\ 27.4 \\ 30.6 \\ 28.4 \\ 36.2 \end{pmatrix} \qquad \text{\textit{column vector giving each}} \\ \text{\textit{person's average score}}$$

Row vectors are also useful; a person's complete set of scores corresponds to a row vector. For example,

(25, 40, 45, 40, 38)

is a row vector giving Diessner's scores.

To determine the average score on each test, we add all the row vectors using vector addition and then multiply this vector by $\frac{1}{10}$ using scalar multiplication. This gives

$\frac{1}{10}(335, 333, 324, 375, 350) = (33.5, 33.3, 32.4, 37.5, 35)$ *row vector giving average score on each test*

✦

The advantage of this approach to analyzing test scores is that it lends itself to implementation on the computer. A computer program can be written that will perform the desired vector additions and scalar multiplications. In fact, an important feature of microcomputers such as the Radio Shack "TRS

80," "Pet," and the "Apple" is the software (the computer programs) that can be bought for them. These programs perform data analyses like the above, very often using vector and matrix algebra.

For example, electronic spreadsheets are computer programs written to handle data in matrix form. Operations are built in to add columns or rows or to manipulate individual matrix elements. One of the most popular spreadsheets is VISICALC, put out by VISICORP of San Jose, California. This spreadsheet is able to accommodate a matrix having 63 columns and 254 rows. VISICALC is most appropriate for handling vector analysis of test scores. The following is hard-copy output of a file that was created using VISICALC to carry out the above analysis on an Apple microcomputer. Readers who are interested in a discussion of VISICALC and similar electronic spreadsheets are referred to "VISICALC and VISICLONES," *Popular Computing*, September, 1982.

NAME	TEST 1	TEST 2	TEST 3	TEST 4	TEST 5	TOTAL	AVERAGE
ANDERSON	40	45	30	48	42	205	41
BOGGS	20	15	30	25	10	100	20
CHITTAR	40	35	25	45	46	191	38.2
DIESSNER	25	40	45	40	38	188	37.6
FARNAM	35	35	38	37	39	184	36.8
GILL	50	46	45	48	47	236	47.2
HOMES	22	24	30	32	29	137	27.4
JOHNSON	35	27	20	41	30	153	30.6
SCHOMER	28	31	25	27	31	142	28.4
WONG	40	35	36	32	38	181	36.2
AVERAGE ON TEST	33.5	33.3	32.4	37.5	35		

EXERCISES

1. Determine the vector having the following initial point A and terminal point B. Sketch the vector.

   *(a) $A(1, 4)$, $B(3, 7)$ (b) $A(2, 1)$, $B(4, 5)$

   (c) $A(-1, 2)$, $B(3, 2)$ *(d) $A(2, -2)$, $B(-2, 3)$

*2. Consider the points $A(1, 1)$, $B(5, 3)$, and $C(-1, 4)$ in **R**^{2}. Determine and sketch the vectors $\overrightarrow{AB}$, $\overrightarrow{BC}$, and $\overrightarrow{OC}$.

3. Consider the points $A(1, -2)$, $B(1, 3)$, and $C(3, 5)$ in **R**^{2}. Determine and sketch the vectors $\overrightarrow{OA}$, $\overrightarrow{AB}$, and $\overrightarrow{BC}$.

*4. Sketch the vector $\overrightarrow{AB} = (3, 2)$ having initial point $A(1, 1)$. Find its terminal point.

5. Sketch the vector $\overrightarrow{AB} = (2, -1)$ having initial point $A(1, 2)$. Find its terminal point.

6. Sketch the vector $\overrightarrow{AB} = (-1, -2)$ having terminal point $B(3, 2)$. Find its initial point.

*7. Compute the vector sums $(1, 3) + (4, 3)$ and $(4, 3) + (1, 3)$. Observe that the results are the same, illustrating the commutativity of vector addition. Draw a sketch of the situation.

*8. Compute the following sums in the two distinct ways given: $(1, 2) + [(3, 1) + (-1, 6)]$ and $[(1, 2) + (3, 1)] + (-1, 6)$. Observe that the results are the same, illustrating associativity of vectors under addition.

9. Compute the vector sums $(-1, 4) + (5, 1)$ and $(5, 1) + (-1, 4)$. Observe that the results are the same.

10. Compute the sums $[(3, 1) + (0, 4)] + (5, 0)$ and $(0, 4) + [(3, 1) + (5, 0)]$. Why do you expect the final answers to be equal?

11. Prove that $(\mathbf{u} + \mathbf{v}) + (\mathbf{w} + \mathbf{z}) = (\mathbf{v} + \mathbf{z}) + (\mathbf{w} + \mathbf{u})$ for four vectors in the same vector space.

*12. A person travels from O according to the sequence of vectors $\overrightarrow{OA} = (1, 2)$, $\overrightarrow{AB} = (4, 2)$, $\overrightarrow{BC} = (-1, -3)$. Determine the person's final position vector relative to O.

13. A person travels from the point $A(2, 1)$ according to the sequence of vectors $\overrightarrow{AB} = (4, 1)$, $\overrightarrow{BC} = (-2, -2)$, $\overrightarrow{CD} = (-3, 5)$. Determine the person's final vectorial position relative to starting point A. Determine the person's final position vector (relative to the origin).

*14. A person travels from the point $A(1, 2, -1)$ according to the sequence of vectors $\overrightarrow{AB} = (1, 1, 1)$, $\overrightarrow{BC} = (2, 3, 5)$, $\overrightarrow{CD} = (-5, -4, 1)$. Determine the person's final location both relative to the starting point and relative to the origin. Suppose the person had traversed the vector $(-5, -4, 1)$ immediately after $\overrightarrow{AB}$ and then the vector $(2, 3, 5)$. Show that the person's final location relative to A and the origin would still be the same.

15. In Example 4 of this section, suppose the final grade corresponds to a comprehensive examination and is worth two tests. Using vector operations, work out the average grade of each student.

16. *(a) If $(2, -1) + (a, b) = (3, 4)$, determine the vector (a, b).

   (b) If $(1, 0, -3) + (b, q, r) + (2, 1, -3) = (1, 5, -1)$, determine the vector (b, q, r).

   (c) If $\mathbf{a}$, $\mathbf{b}$, and $\mathbf{x}$ are all elements of $\mathbf{R}^n$ and $\mathbf{a} + \mathbf{x} = \mathbf{b}$, prove that $\mathbf{x} = \mathbf{b} - \mathbf{a}$.

17. With our new understanding of vector addition—we now know how to add any finite number of vectors—we can generalize the result given in the previous section for finding the resultant of two forces. Experiments show that if any number of forces acting at a point are represented by vectors, they are equivalent to a single force, *the resultant*, which can be represented by the sum of these vectors. Using this observation, determine the resultant of the following forces (in vector form):

   *(a) forces of $(4, 2)$, $(3, 4)$, and $(7, 3)$ acting on a body

   (b) forces of $(2, 4)$, $(1, 3)$, $(5, -1)$, and $(4, 2)$ acting on a body

   (c) forces of $(1, 2, 3)$, $(-1, 4, 2)$, $(5, 1, 3)$, and $(7, -3, 2)$ acting on a body in three-space

4-3 SUBSPACES OF $\mathbf{R}^n$

In this chapter, we have introduced the concept of the vector space $\mathbf{R}^n$ and have seen three distinct ways in which we can use the elements of $\mathbf{R}^n$. $\mathbf{R}^n$ can be used to describe the locations of points in *n*-space, position vectors in *n*-space, or general directed line segments in *n*-space. In this section, we shall focus on using **R**ⁿ to describe the locations of points in *n*-space. As we proceed in this chapter to develop meaningful algebraic concepts for the vector space **R**ⁿ, we shall illustrate the importance of these concepts to each of these three areas.

Rⁿ is a set of elements called vectors on which two operations were defined, addition and scalar multiplication. Observe that on adding two elements of **R**ⁿ we get an element of **R**ⁿ. **R**ⁿ is said to be *closed under addition*. **R**ⁿ is also *closed under scalar multiplication*; the scalar multiple of an element of **R**ⁿ is again in **R**ⁿ. These two closure properties give vector spaces a certain completeness under the operations. In this section, we see that certain subsets of **R**ⁿ have these characteristics and can be interpreted as vector spaces in their own right.

To illustrate the concepts involved, let us look below at the geometrical interpretation of **R**³, three-space. The *xy* plane is embedded in this three-space. We have seen how we can interpret a plane as a vector space in itself, using **R**² to characterize its points. It is natural that the *xy* plane can be interpreted as a vector space also when considered as a subset of three-space, with certain elements of **R**³ used to describe it.

The *xy* plane is described by the elements of **R**³ of the type $(a, b, 0)$, having zero as a third component. (See Figure 4-20.)

We make the observation that if we add two elements that lie in the *xy* plane such as $(a, b, 0)$ and $(c, d, 0)$, we get element $(a + c, b + d, 0)$ that also lies in the plane. Further, if we multiply an element that lies in the plane by a scalar, we get another element in the plane; $p(a, b, 0)$ gives $(pa, pb, 0)$, which again lies in

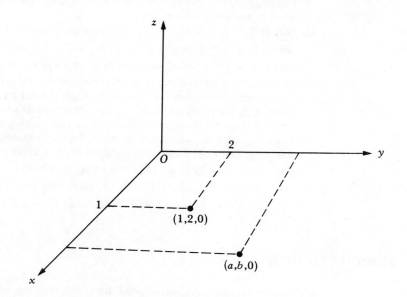

Figure 4-20

the plane. The xy plane, considered as a subset of $\mathbf{R}^3$, has operations of addition and scalar multiplication defined on it and is closed under these operations. The xy plane has all the characteristics of a vector space when considered as a subset of $\mathbf{R}^3$. We call such an embedded vector space a subspace of the larger space.

We now formally define subspace of $\mathbf{R}^n$.

Definition 4-4   A nonempty subset of the vector space $\mathbf{R}^n$ forms a *subspace* of $\mathbf{R}^n$ if it is closed under addition and under scalar multiplication.[†]

✦ *Example 1*

Consider the vector space $\mathbf{R}^3$. All vectors of the form $(a, 0, 0)$ (with zeros as second and third components) form a subset U of $\mathbf{R}^3$. Let us show that this subset is a subspace of $\mathbf{R}^3$.

Consider two arbitrary elements of U, $(a, 0, 0)$ and $(b, 0, 0)$. Their sum is $(a + b, 0, 0)$, an element of U. U is closed under addition.

Let p be an arbitrary scalar. Then $p(a, 0, 0) = (pa, 0, 0)$, an element of U. U is closed under scalar multiplication. Hence U is a subspace of $\mathbf{R}^3$.

Let us look at the geometrical interpretation of U. Let $\mathbf{R}^3$ be the set of all points in three-space. U is the x-axis.   ✦

✦ *Example 2*

Prove that the subset V of $\mathbf{R}^3$ consisting of all vectors of the form (a, a, b) is a subspace of $\mathbf{R}^3$. Interpret this result geometrically.

First of all, we see that the subset of interest is the subset of $\mathbf{R}^3$ consisting of vectors in which the first two components are the same. For example, $(1, 1, 2)$ and $(-\frac{1}{2}, -\frac{1}{2}, 3)$ would be elements of the subset; $(1, 2, 3)$ would not. We have to show that V is closed under addition and under scalar multiplication.

Let (a, a, b) and (c, c, d) be arbitrary elements of V. Then $(a, a, b) + (c, c, d) = (a + c, a + c, b + d)$. The first two components are equal; hence this is an element of V.

Let p be an arbitrary scalar. $p(a, a, b) = (pa, pa, pb)$ is an element of V. V is closed under addition and scalar multiplication; it is a subspace of $\mathbf{R}^3$.

We now interpret V geometrically. If $\mathbf{R}^3$ is three-space, then the points (a, a, b), having equal x and y components, make up the plane in Figure 4-21.
   ✦

✦ *Example 3*

Consider the subset of $\mathbf{R}^3$ consisting of vectors of the form (a, b, a^2). We show that this subset of $\mathbf{R}^3$ is not a subspace, illustrating the fact that not all subsets of $\mathbf{R}^3$ are subspaces.

The subset consists of all elements in which the third component is the square of the first. For example, $(2, 5, 4)$ is in the subset, whereas the vector $(2, 5, 3)$ is not.

[†] Note that this definition does imply that a vector space is a subspace of itself. The subspaces that are proper subsets can be distinguished by use of the term *proper* subspaces.

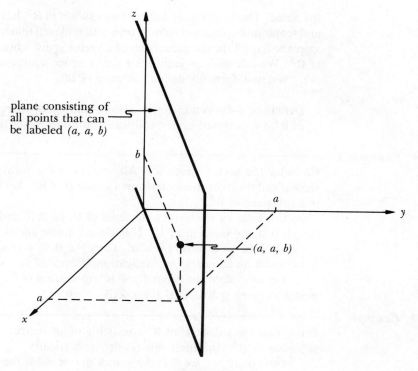

plane consisting of
all points that can
be labeled *(a, a, b)*

(a, a, b)

Figure 4-21

Let (a, b, a^2) and (c, d, c^2) be arbitrary elements of the subset:

$$(a, b, a^2) + (c, d, c^2) = (a + c, b + d, a^2 + c^2)$$

In general, the third component here is not the square of the first, $a^2 + c^2 \neq (a + c)^2$. Thus the subset is not closed under addition; it is not a subspace of **R**³. ◆

The following theorem gives us an important characteristic of all subspaces of **R**ⁿ.

Theorem 4-2   *Let V be subspace of the vector space* **R**ⁿ. *Then V contains the zero vector of* **R**ⁿ.

***Proof*:**   Let **v** be an arbitrary element of V; **v** will have n components. Let 0 be the zero scalar.

Since V is a subspace, 0**v** will be in V. But 0**v** = $(0, \ldots, 0)$, the zero vector, proving the theorem.

If we interpret **R**³ as being three-dimensional space, then this theorem tells us that all subspaces of **R**³ contain the origin (that is, they pass through the origin). We do indeed see that this is true for the subspaces we have discussed in this section.

In the following sections, we develop tools that will enable us to understand subspaces further. We shall find that all the significant subspaces of $\mathbf{R}^3$ are lines or planes through the origin, and that those of $\mathbf{R}^2$ are lines through the origin.

EXERCISES

*1. Prove that the subset of $\mathbf{R}^3$ consisting of all vectors of the form $(a, 0, b)$ forms a subspace of $\mathbf{R}^3$. Interpret this result geometrically.

 2. Prove that the subset of $\mathbf{R}^3$ consisting of all vectors of the form $(0, b, 0)$ forms a subspace of $\mathbf{R}^3$. Interpret this result geometrically.

*3. Prove that the subset of $\mathbf{R}^3$ consisting of all vectors of the form $(a, 2a, b)$ forms a subspace of $\mathbf{R}^3$. Interpret this result geometrically.

 4. Prove that the subset of $\mathbf{R}^4$ consisting of all vectors of the form $(a, 2a, b, 3b)$ forms a subspace of $\mathbf{R}^4$.

In Exercises 5–11, determine which subsets form subspaces of $\mathbf{R}^3$.

*5. The subset of vectors of the form $(a, b, a + 2)$

 6. The subset of vectors of the form $(a, 2a, 3a)$

*7. The subset of vectors of the form $(a, b, 3b)$

 8. The subset of vectors of the form $(a, b, 2)$

*9. The subset of vectors of the form (a, b, c) satisfying the condition

   (a) $a + b + c = 1$ (b) $a + b + c = 0$ (c) $ab = 0$

 10. The subset of vectors of the form $(b + c, b, c)$

*11. The subset consisting of vectors of the form $p(1, 2, 3)$ and vectors of the form $q(-1, 0, 2)$

 12. Give an example of a subset of $\mathbf{R}^2$ that is

   (a) closed under addition but not under scalar multiplication

   (b) closed under scalar multiplication but not under addition

   Such examples illustrate the independence of these two conditions.

4-4   LINEAR COMBINATIONS OF VECTORS

In Example 2 of the previous section, we discussed the subspace of $\mathbf{R}^3$ consisting of all vectors of the form (a, a, b). Observe that an arbitrary vector in this space can be written

$$(a, a, b) = a(1, 1, 0) + b(0, 0, 1)$$

All vectors in the subspace can be expressed in terms of $(1, 1, 0)$ and $(0, 0, 1)$— these vectors do, in some sense, characterize the space. In this section and the following ones, we pursue this approach to understanding vector spaces. We develop mathematical techniques for analyzing vector spaces in terms of certain vectors that lie in those spaces.

Definition 4-5   Consider m vectors $\mathbf{v}_1, \ldots, \mathbf{v}_m$ of a vector space V. We say that a vector $\mathbf{v}$ of V is a *linear combination* of these vectors if there exist m scalars, $a_1, \ldots, a_m$, such that

$$\mathbf{v} = a_1\mathbf{v}_1 + \cdots + a_m\mathbf{v}_m$$

✦ *Example 1*

In **R**^{3}, the vector $(1, 2, 9)$ is a linear combination of the three vectors $(2, 1, 0)$, $(1, 3, 0)$, and $(0, 0, 3)$, since it can be expressed

$$(1, 2, 9) = \tfrac{1}{5}(2, 1, 0) + \tfrac{3}{5}(1, 3, 0) + 3(0, 0, 3) \qquad ✦$$

The problem of determining whether or not a vector is a linear combination of given vectors becomes that of solving a system of linear equations.

✦ *Example 2*

Determine whether or not the vector $(-1, 1, 5)$ is a linear combination of the vectors $(1, 2, 3)$, $(0, 1, 4)$, and $(2, 3, 6)$.

We examine the identity

$$a_1(1, 2, 3) + a_2(0, 1, 4) + a_3(2, 3, 6) = (-1, 1, 5)$$

If scalars a_1, a_2, and a_3 can be found that satisfy this identity, $(-1, 1, 5)$ is a linear combination of the given vectors.

The identity becomes

$$(a_1, 2a_1, 3a_1) + (0, a_2, 4a_2) + (2a_3, 3a_3, 6a_3) = (-1, 1, 5)$$
$$(a_1 + 2a_3, 2a_1 + a_2 + 3a_3, 3a_1 + 4a_2 + 6a_3) = (-1, 1, 5)$$

Thus

$$
\begin{aligned}
a_1 \phantom{{}+ a_2} + 2a_3 &= -1 \\
2a_1 + a_2 + 3a_3 &= 1 \\
3a_1 + 4a_2 + 6a_3 &= 5
\end{aligned}
$$

There exists a unique solution to this system; $a_1 = 1$, $a_2 = 2$, $a_3 = -1$. Thus the vector $(-1, 1, 5)$ is a linear combination of $(1, 2, 3)$, $(0, 1, 4)$, and $(2, 3, 6)$:

$$(-1, 1, 5) = 1(1, 2, 3) + 2(0, 1, 4) - 1(2, 3, 6)$$

Since the solution to the system of equations is unique, it is possible to express $(-1, 1, 5)$ in only one way as a linear combination of the other three vectors.

✦

The following example illustrates that certain vectors can be represented in more than one way as linear combinations of others.

✦ *Example 3*

Discuss the representation of the vector $(4, 5, 5)$ as a linear combination of $(1, 2, 3)$, $(-1, 1, 4)$, and $(3, 3, 2)$.

The identity

$$a_1(1, 2, 3) + a_2(-1, 1, 4) + a_3(3, 3, 2) = (4, 5, 5)$$

becomes

$$(a_1 - a_2 + 3a_3, 2a_1 + a_2 + 3a_3, 3a_1 + 4a_2 + 2a_3) = (4, 5, 5)$$

Thus

$$\begin{aligned} a_1 - a_2 + 3a_3 &= 4 \\ 2a_1 + a_2 + 3a_3 &= 5 \\ 3a_1 + 4a_2 + 2a_3 &= 5 \end{aligned}$$

This system of equations has many solutions; $a_1 = -2r + 3$, $a_2 = r - 1$. Thus the linear combination is not unique:

$$(4, 5, 5) = (-2r + 3)(1, 2, 3) + (r - 1)(-1, 1, 4) + r(3, 3, 2)$$

for any value of r.   ✦

That it may not be possible to express a given vector as a linear combination of others is illustrated by the following example.

✦ *Example 4*

Prove that the vector $(0, 0, 2)$ is not a linear combination of the vectors $(2, 1, 0)$ and $(-1, 2, 1)$.

We examine the identity

$$a_1(2, 1, 0) + a_2(-1, 2, 1) = (0, 0, 2)$$

It becomes

$$(2a_1 - a_2, a_1 + 2a_2, a_2) = (0, 0, 2)$$

giving

$$\begin{aligned} 2a_1 - a_2 &= 0 \\ a_1 + 2a_2 &= 0 \\ a_2 &= 2 \end{aligned}$$

This system has no solution. Thus $(0, 0, 2)$ is not a linear combination of $(2, 1, 0)$ and $(-1, 2, 1)$.   ✦

Definition 4-6   The vectors $\mathbf{v}_1, \ldots, \mathbf{v}_m$ are said to *span* a vector space if every vector in the space can be expressed as a linear combination of these vectors.

Such a set of vectors in a sense defines the vector space, since every vector in the space can be obtained from this set.

✦ *Example 5* ───

Consider the subspace of $\mathbf{R}^3$ consisting of vectors of the form (a, a, b). An arbitrary vector in the subspace can be written

$$(a, a, b) = a(1, 1, 0) + b(0, 0, 1)$$

Thus the vectors $(1, 1, 0)$ and $(0, 0, 1)$ span the subspace.   ✦

✦ *Example 6* ───

Prove that the vectors $(1, 2, 0)$, $(0, 1, -1)$, and $(1, 1, 2)$ span $\mathbf{R}^3$.

Let (x, y, z) represent an arbitrary element of $\mathbf{R}^3$. We have to show that (x, y, z) can be expressed as a linear combination of the vectors. We prove that there exist scalars a_1, a_2, a_3 such that

$$(x, y, z) = a_1(1, 2, 0) + a_2(0, 1, -1) + a_3(1, 1, 2)$$

This identity may be written

$$(a_1 + a_3, 2a_1 + a_2 + a_3, -a_2 + 2a_3) = (x, y, z)$$

Thus

$$
\begin{aligned}
a_1 \quad\;\; + \; a_3 &= x \\
2a_1 + a_2 + \; a_3 &= y \\
- a_2 + 2a_3 &= z
\end{aligned}
$$

This system has solution $a_1 = 3x - y - z$, $a_2 = -4x + 2y + z$, $a_3 = -2x + y + z$. Thus the arbitrary vector (x, y, z) of $\mathbf{R}^3$ can be expressed as a linear combination of the vectors $(1, 2, 0)$, $(0, 1, -1)$, and $(1, 1, 2)$.

$$(x, y, z) = (3x - y - z)(1, 2, 0) + (-4x + 2y + z)(0, 1, -1)$$
$$+ (-2x + y + z)(1, 1, 2)$$

These three vectors span $\mathbf{R}^3$.   ✦

We have developed a way of looking at vector spaces in terms of certain vectors in the space, spanning sets of vectors. It is also useful to be able to do the converse—to use a given set of vectors to construct a vector space. Subspaces often arise in discussions of vector spaces in this manner. We now pursue this approach.

Let $\mathbf{v}_1, \ldots, \mathbf{v}_m$ be vectors in a vector space V. Let U be the set consisting of all possible linear combinations of $\mathbf{v}_1, \ldots, \mathbf{v}_m$. We say that U is *generated* by $\mathbf{v}_1, \ldots, \mathbf{v}_m$. The following theorem tells us that such a set U is a subspace of V.

Theorem 4-3 *Let* $\mathbf{v}_1, \ldots, \mathbf{v}_m$ *be a given set of vectors in a vector space* V. *The set generated by* $\mathbf{v}_1, \ldots, \mathbf{v}_m$ *is a subspace of* V. *The vectors* $\mathbf{v}_1, \ldots, \mathbf{v}_m$ *span this subspace.*

Proof: Let U be the set generated by $\mathbf{v}_1, \ldots, \mathbf{v}_m$.

Let $\mathbf{v} = a_1\mathbf{v}_1 + \cdots + a_m\mathbf{v}_m$ and $\mathbf{v}' = b_1\mathbf{v}_1 + \cdots + b_m\mathbf{v}_m$ be arbitrary elements of U and let p be an arbitrary scalar. Then

$$\mathbf{v} + \mathbf{v}' = (a_1 + b_1)\mathbf{v}_1 + \cdots + (a_m + b_m)\mathbf{v}_m$$

This is an element of U, since it is expressed as a linear combination of $\mathbf{v}_1, \ldots, \mathbf{v}_m$. Further,

$$p\mathbf{v} = pa_1\mathbf{v}_1 + \cdots + pa_m\mathbf{v}_m$$

$p\mathbf{v}$ is an element of U.

Thus U is a subspace.

The vectors $\mathbf{v}_1, \ldots, \mathbf{v}_m$ span U, since an arbitrary element of U can be expressed as a linear combination of these vectors.

✦ Example 7

Discuss the subspace of $\mathbf{R}^3$ generated by the vectors $(1, 1, 1)$ and $(2, 1, 3)$.

The subspace will consist of all possible linear combinations of these vectors. An arbitrary element of the subspace will be

$$\mathbf{v} = a_1(1, 1, 1) + a_2(2, 1, 3)$$

The vector $(5, 3, 7)$ will, for example, be in the subspace, since

$$(5, 3, 7) = (1, 1, 1) + 2(2, 1, 3)$$

The vector (x, y, z) is in the subspace if and only if there exist scalars a_1 and a_2 such that

$$(x, y, z) = a_1(1, 1, 1) + a_2(2, 1, 3)$$

This condition leads to

$$a_1 + 2a_2 = x$$
$$a_1 + a_2 = y$$
$$a_1 + 3a_2 = z$$

Solving, we get

$$\begin{pmatrix} 1 & 2 & x \\ 1 & 1 & y \\ 1 & 3 & z \end{pmatrix} \approx \begin{pmatrix} 1 & 2 & x \\ 0 & -1 & y - x \\ 0 & 1 & z - x \end{pmatrix}$$

$$\approx \begin{pmatrix} 1 & 0 & -x + 2y \\ 0 & 1 & x - y \\ 0 & 0 & -2x + y + z \end{pmatrix}$$

giving $a_1 = -x + 2y$, $a_2 = x - y$, $-2x + y + z = 0$.

The interpretation of this result is that the vector (x, y, z) is in the subspace if and only if it satisfies the condition $-2x + y + z = 0$. If it is in the subspace, it can then be represented as a linear combination of $(1, 1, 1)$ and $(2, 1, 3)$:

$$(x, y, z) = (-x + 2y)(1, 1, 1) + (x - y)(2, 1, 3)$$

All the vectors in $\mathbf{R}^3$ that satisfy $-2x + y + z = 0$ lie in a plane through the origin—this is the geometrical representation of the subspace generated by $(1, 1, 1)$ and $(2, 1, 3)$. The subspace is the plane defined by the vectors $(1, 1, 1)$ and $(2, 1, 3)$. (See Figure 4-22.)

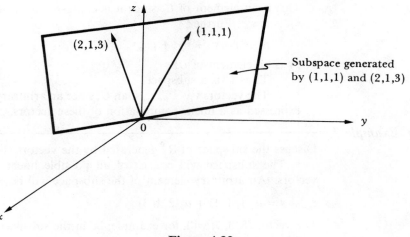

Figure 4-22

This result can be generalized.

Consider a subspace generated by two non-colinear vectors **u** and **v** (that is, they are not in the same or opposite directions). An arbitrary vector **w** in this subspace can be expressed

$$\mathbf{w} = a\mathbf{u} + b\mathbf{v}$$

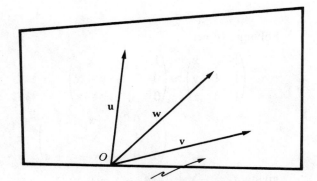

Subspace generated by **u** and **v**
is plane defined by **u** and **v**

Figure 4-23

The geometrical interpretation of this subspace is that it is the plane defined by **u** and **v**; see Figure 4-23.

EXERCISES

In Exercises 1–10, determine whether or not the first vector is a linear combination of the following vectors. Determine the combination(s) if it is.

*1. $(-1, 7)$; $(1, -1)$, $(2, 4)$

2. $(4, 0)$; $(-1, 2)$, $(3, 2)$, $(6,4)$

*3. $(6, 22)$; $(2, 3)$, $(-1, 5)$

4. $(-3, 3, 7)$; $(1, -1, 2)$, $(2, 1, 0)$, $(-1, 2, 1)$

*5. $(2, 7, 13)$; $(1, 2, 3)$, $(-1, 2, 4)$, $(1, 6, 10)$

6. $(0, 10, 8)$; $(-1, 2, 3)$, $(1, 3, 1)$, $(1, 8, 5)$

*7. $(2, 2, -2)$; $(1, 1, -1)$, $(2, 1, 3)$, $(4, 3, 1)$

8. $(4, 3, 8)$; $(-1, 0, 1)$, $(2, 1, 3)$, $(0, 1, 5)$

*9. $(-2, -4, -3)$; $(1, 1, -1)$, $(0, 1, 2)$, $(2, 3, 0)$

10. $(2, 0, 12, 3)$; $(1, 2, 3, 0)$, $(1, -1, 1, 2)$, $(3, 4, -1, 1)$

*11. Prove that the vectors $(1, 2)$ and $(3, 1)$ span $\mathbf{R}^2$.

12. Prove that the vectors $(2, 1, 0)$, $(-1, 3, 1)$, and $(4, 5, 0)$ span $\mathbf{R}^3$.

*13. Do the vectors $(3, -1)$, $(2, 3)$, and $(4, 0)$ span $\mathbf{R}^2$?

14. Do the vectors $(4, 0, 1)$, $(0, 1, 0)$, and $(0, 0, 1)$ span $\mathbf{R}^3$?

*15. Do the vectors $(1, 2, 1)$, $(-1, 3, 0)$, and $(0, 5, 1)$ span $\mathbf{R}^3$?

16. Do the vectors $(1, 1, -1)$, $(1, 3, 2)$, and $(3, 5, 0)$ span $\mathbf{R}^3$?

*17. Do the vectors $(1, -1, -1)$, $(0, 1, 2)$, and $(1, 2, 1)$ span $\mathbf{R}^3$?

*18. Give three vectors in the subspace of $\mathbf{R}^3$ generated by $(1, 2, 3)$ and $(1, 2, 0)$ other than the given vectors. Sketch the subspace.

19. Give three vectors in the subspace of $\mathbf{R}^3$ generated by $(1, 2, 1)$ and $(2, 1, 4)$. Sketch the subspace.

*20. Give three vectors in the subspace of $\mathbf{R}^3$ generated by the vector $(1, 2, 3)$. Sketch the subspace.

21. Give three vectors in the subspace of $\mathbf{R}^2$ generated by the vector $(1, 2)$. Sketch the subspace.

22. Give three vectors in the subspace of $\mathbf{R}^4$ generated by the vectors $(1, 2, -1, 3)$ and $(1, 2, 1, 1)$.

4-5 LINEAR DEPENDENCE OF VECTORS

In this section, we continue the development of vector space structure. We focus on developing concepts of dependence and independence of vectors. These will be useful tools in constructing "efficient" spanning sets of vectors—spanning sets in which there are no redundant vectors. Such a set is called a *basis* of a vector space.

Let us motivate a concept of dependence of vectors by looking at a specific example. Observe that the vector $(-2, 6, 1)$ is a linear combination of $(1, 2, 3)$, $(0, 1, 4)$, and $(-1, 2, 2)$, for it can be expressed

$$(-2, 6, 1) = (1, 2, 3) - 2(0, 1, 4) + 3(-1, 2, 2) \tag{1}$$

The above equation could be rewritten in a number of ways. Each of the vectors could in turn be represented in terms of the others:

$$(1, 2, 3) = (-2, 6, 1) + 2(0, 1, 4) - 3(-1, 2, 2)$$
$$(0, 1, 4) = \tfrac{1}{2}(1, 2, 3) - \tfrac{1}{2}(-2, 6, 1) + \tfrac{3}{2}(-1, 2, 2)$$
$$(-1, 2, 2) = \tfrac{1}{3}(-2, 6, 1) - \tfrac{1}{3}(1, 2, 3) + \tfrac{2}{3}(0, 1, 4)$$

Each of the four vectors is in fact dependent on the other three. We express this fact by writing equation (1) as

$$(-2, 6, 1) - (1, 2, 3) + 2(0, 1, 4) - 3(-1, 2, 2) = \mathbf{0}$$

This concept of dependence of vectors is made precise with the following definition.

Definition 4-7 m vectors $\mathbf{v}_1, \ldots, \mathbf{v}_m$ in a vector space V are said to be *linearly dependent* if there exist scalars $a_1, \ldots, a_m$, not all zero, such that

$$a_1\mathbf{v}_1 + \cdots + a_m\mathbf{v}_m = \mathbf{0}$$

If the vectors are not linearly dependent, they are said to be *linearly independent*.

Note that in the above definition the **0** on the right of the equation denotes the zero vector of the vector space $(0, 0, \ldots, 0)$.

✦ *Example 1* ──

Prove that the vectors $(1, 2, 3)$, $(-2, 1, 1)$, and $(8, 6, 10)$ are linearly dependent vectors in **R**^{3}.

To arrive at this conclusion, we examine the identity

$$a_1(1, 2, 3) + a_2(-2, 1, 1) + a_3(8, 6, 10) = \mathbf{0}$$

This becomes

$$(a_1 - 2a_2 + 8a_3, 2a_1 + a_2 + 6a_3, 3a_1 + a_2 + 10a_3) = \mathbf{0}$$

giving

$$a_1 - 2a_2 + 8a_3 = 0$$
$$2a_1 + a_2 + 6a_3 = 0$$
$$3a_1 + a_2 + 10a_3 = 0$$

This system has $a_1 = 4$, $a_2 = -2$, $a_3 = -1$ as one of its solutions, a set of a's not all zero. Thus the vectors are linearly dependent.   ✦

✦ **Example 2**

Determine whether the three vectors $(1, -1, 1, 2, 1)$, $(4, -1, 6, 6, 2)$, and $(-2, -1, 1, -2, -2)$ are linearly dependent or independent in $\mathbf{R}^5$.

We examine the identity

$$a_1(1, -1, 1, 2, 1) + a_2(4, -1, 6, 6, 2) + a_3(-2, -1, 1, -2, -2) = \mathbf{0}$$

This gives

$$(a_1 + 4a_2 - 2a_3, -a_1 - a_2 - a_3, a_1 + 6a_2 + a_3,$$
$$2a_1 + 6a_2 - 2a_3, a_1 + 2a_2 - 2a_3) = \mathbf{0}$$

$$a_1 + 4a_2 - 2a_3 = 0$$
$$-a_1 - a_2 - a_3 = 0$$
$$a_1 + 6a_2 + a_3 = 0$$
$$2a_1 + 6a_2 - 2a_3 = 0$$
$$a_1 + 2a_2 - 2a_3 = 0$$

This system can be shown to have the unique solution $a_1 = a_2 = a_3 = 0$. The vectors are thus linearly independent.   ✦

We now present two important general results in the analysis of vector spaces:

Theorem 4-4 *In a vector space V, any collection of vectors that contains the zero vector is linearly dependent.*

Proof: Consider the collection $\mathbf{0}, \mathbf{v}_2, \ldots, \mathbf{v}_m$. Then

$$a_1\mathbf{0} + 0\mathbf{v}_2 + \cdots + 0\mathbf{v}_m = \mathbf{0}$$

for any real number a_1. Hence this set is linearly dependent.

Theorem 4-5 *Let $\mathbf{v}_1, \ldots, \mathbf{v}_m$ be linearly dependent vectors in a vector space V. Any set of vectors that contains these vectors will be a set of linearly dependent vectors.*

Proof: The vectors $\mathbf{v}_1, \ldots, \mathbf{v}_m$ are linearly dependent; hence there exist scalars $a_1, \ldots, a_m$ that are not all zero such that

$$a_1\mathbf{v}_1 + \cdots + a_m\mathbf{v}_m = \mathbf{0}$$

Consider the set $\mathbf{v}_1, \ldots, \mathbf{v}_m, \mathbf{u}_1, \ldots, \mathbf{u}_k$, which contains the vectors $\mathbf{v}_1, \ldots, \mathbf{v}_m$. There are scalars, $a_1, \ldots, a_m, 0, \ldots, 0$ that are not all zero such that

$$a_1\mathbf{v}_1 + \cdots + a_m\mathbf{v}_m + 0\mathbf{u}_1 + \cdots + 0\mathbf{u}_k = \mathbf{0}$$

Therefore this set is linearly dependent.

EXERCISES

*1. Prove that the vectors $(1, 0)$ and $(0, 1)$ are linearly independent in $\mathbf{R}^2$.

2. Prove that the vectors $(2, 1)$ and $(1, 2)$ are linearly independent in $\mathbf{R}^2$.

*3. Prove that the vectors $(1, -2)$ and $(3, -6)$ are linearly dependent in $\mathbf{R}^2$.

4. Are the vectors $(2, 0)$, $(0, 1)$, and $(3, 4)$ linearly dependent in $\mathbf{R}^2$?

*5. Prove that the vectors $(-1, 3, 1)$, $(2, 1, 0)$, and $(1, 4, 1)$ are linearly dependent in $\mathbf{R}^3$.

6. Prove that the vectors $(2, 3, 0)$, $(-1, 4, 0)$, and $(0, 0, 2)$ are linearly independent in $\mathbf{R}^3$.

*7. Prove that the vectors $(1, 1, 1)$, $(2, 1, 0)$, $(3, 1, 4)$, and $(1, 2, -2)$ are linearly dependent in $\mathbf{R}^3$.

8. (a) Prove that the vectors $(1, 1)$ and $(0, 2)$ are linearly independent in $\mathbf{R}^2$.

   (b) Prove that the vectors $(1, 1, 2)$, $(0, -1, 3)$, and $(0, 0, -1)$ are linearly independent in $\mathbf{R}^3$.

   (c) Prove that the vectors $(-1, 1, 2, 3)$, $(0, 1, 2, 3)$, $(0, 0, -1, 2)$, and $(0, 0, 0, -3)$ are linearly independent in $\mathbf{R}^4$.

   (d) Discuss the pattern of zero components in the vectors of (a), (b), and (c). How can you use this pattern to construct any number of sets of five linearly independent vectors in $\mathbf{R}^5$? Any number of sets of n linearly independent vectors in $\mathbf{R}^n$?

   (e) Give a geometrical interpretation of the concepts involved in (a) and (b).

9. Prove that the set of vectors $(-1, 3, 1)$, $(2, 1, 0)$, $(1, 4, 1)$, $(5, 7, 9)$, and $(8, -1, 0)$ is linearly dependent. (*Hint*: Use Exercise 5 above and Theorem 4-5.)

*10. Give an expression that defines a general vector in the space spanned by the vectors $(2, 1, 0)$ and $(-1, 2, 0)$.

11. Prove that two vectors are linearly dependent if and only if it is possible to express one vector as a scalar multiple of the other.

12. Prove that the vectors $\mathbf{v}_1, \ldots, \mathbf{v}_m$ of a vector space V are linearly dependent if and only if it is possible to express one of the vectors as a linear combination of the others.

13. Let $\mathbf{v}_1$, $\mathbf{v}_2$, and $\mathbf{v}_3$ be linearly independent vectors in V and let c be a nonzero scalar. Prove that $\mathbf{v}_1$, $c\mathbf{v}_2$, and $\mathbf{v}_3$ are also linearly independent. Further prove that $\mathbf{v}_1 + c\mathbf{v}_2$, $\mathbf{v}_2$, and $\mathbf{v}_3$ are linearly independent.

14. Extend the results of Exercise 13 to m linearly independent vectors in V.

15. If $\mathbf{v}_1$, $\mathbf{v}_2$, and $\mathbf{v}_3$ are linearly dependent vectors in V and c is a nonzero scalar, prove that both the sets $\mathbf{v}_1$, $c\mathbf{v}_2$, $\mathbf{v}_3$ and $\mathbf{v}_1 + c\mathbf{v}_2$, $\mathbf{v}_2$, $\mathbf{v}_3$ are linearly dependent.

16. Extend the results of Exercise 15 to m linearly dependent vectors in V.

4-6   BASES

We now bring together the concepts of spanning set and linear independence.

Definition 4-8   A finite set of vectors $\mathbf{v}_1, \ldots, \mathbf{v}_m$ in a vector space V forms a *basis* for the space if it spans the space and is linearly independent.

In this section, we give examples of bases for various vector spaces and develop results concerning bases. We shall see that a basis is more appropriate for discussing a vector space than is a general spanning set.

✦ ***Example 1***

The set of vectors $(1, 0, \ldots, 0), (0, 1, 0, \ldots, 0), (0, \ldots, 0, 1)$ forms a basis for $\mathbf{R}^n$ called the *standard basis*.

We have to show that this set spans $\mathbf{R}^n$ and that the vectors are linearly independent.

Let $(x_1, \ldots, x_n)$ be an arbitrary element of $\mathbf{R}^n$. We can express this vector as

$$(x_1, \ldots, x_n) = x_1(1, 0, \ldots, 0) + \cdots + x_n(0, \ldots, 0, 1)$$

Hence this set of vectors spans $\mathbf{R}^n$.

Further, consider the identity

$$a_1(1, 0, \ldots, 0) + \cdots + a_n(0, \ldots, 0, 1) = \mathbf{0}$$

This implies that

$$(a_1, 0, \ldots, 0) + \cdots + (0, \ldots, 0, a_n) = \mathbf{0}$$
$$(a_1, \ldots, a_n) = \mathbf{0}$$

giving $a_1 = a_2 = \cdots = a_n = 0$.

The vectors are thus linearly independent, and the set is a basis for $\mathbf{R}^n$.

✦

For any given vector space there can be many bases. The following theorem leads up to a key result concerning all bases for a given vector space— all bases consist of the same number of vectors.

Theorem 4-6 Let $\mathbf{v}_1, \ldots, \mathbf{v}_n$ *be a basis for a given vector space. If* $\mathbf{w}_1, \ldots, \mathbf{w}_m$ *is a set of more than n vectors in that space, then this set is linearly dependent.*

***Proof*:** We examine the identity

$$a_1\mathbf{w}_1 + \cdots + a_m\mathbf{w}_m = \mathbf{0} \tag{1}$$

We shall prove that values of $a_1, \ldots, a_m$, not all zero, exist such that this identity holds, proving that the vectors are linearly dependent.

$\mathbf{v}_1, \ldots, \mathbf{v}_n$ is a basis. Thus each of $\mathbf{w}_1, \ldots, \mathbf{w}_m$ can be expressed in terms of these vectors. Let

$$\mathbf{w}_1 = b_{11}\mathbf{v}_1 + b_{12}\mathbf{v}_2 + \cdots + b_{1n}\mathbf{v}_n$$
$$\vdots$$
$$\mathbf{w}_m = b_{m1}\mathbf{v}_1 + b_{m2}\mathbf{v}_2 + \cdots + b_{mn}\mathbf{v}_n$$

Substituting for $\mathbf{w}_1, \ldots, \mathbf{w}_m$ in (1),

$$a_1(b_{11}\mathbf{v}_1 + \cdots + b_{1n}\mathbf{v}_n) + \cdots + a_m(b_{m1}\mathbf{v}_1 + \cdots + b_{mn}\mathbf{v}_n) = \mathbf{0}$$

Rearranging, we get

$$(a_1b_{11} + a_2b_{21} + \cdots + a_mb_{m1})\mathbf{v}_1 + \cdots$$
$$+ (a_1b_{1n} + a_2b_{2n} + \cdots + a_mb_{mn})\mathbf{v}_m = \mathbf{0}$$

Since $\mathbf{v}_1, \ldots, \mathbf{v}_m$ are linearly independent, this identity can only be satisfied if the coefficients are all zero. Thus

$$b_{11}a_1 + b_{21}a_2 + \cdots + b_{m1}a_m = 0$$
$$\vdots$$
$$b_{1n}a_1 + b_{2n}a_2 + \cdots + b_{mn}a_m = 0$$

Thus finding a's that satisfy (1) reduces to finding a's that are solutions to this system of n equations in m variables. Since n, the number of equations, is less than m, the number of variables, we know that many solutions exist; there must be nonzero a's that satisfy (1). Thus the vectors $\mathbf{w}_1, \ldots, \mathbf{w}_m$ are linearly dependent.

Theorem 4-7 *Any two bases for a given vector space V have the same number of vectors.*

Proof: Let $\mathbf{v}_1, \ldots, \mathbf{v}_n$ and $\mathbf{w}_1, \ldots, \mathbf{w}_m$ be two bases for V.

If we interpret $\mathbf{v}_1, \ldots, \mathbf{v}_n$ as a basis for V and $\mathbf{w}_1, \ldots, \mathbf{w}_m$ as a set of m linearly independent vectors in V, the previous theorem implies that $m \leq n$. Conversely, if we interpret $\mathbf{w}_1, \ldots, \mathbf{w}_m$ as a basis for V and $\mathbf{v}_1, \ldots, \mathbf{v}_n$ as a set of n vectors in V, the theorem tells us that $n \leq m$. Thus $n = m$.

Definition 4-9 The *dimension* of a vector space is the number of vectors that make up a basis.

Since the n vectors $(1, 0, \ldots, 0), (0, 1, 0, \ldots, 0), \ldots, (0, \ldots, 0, 1)$ form a basis for **R**^{n}, the dimension of **R**^{n} is n.

Note that we have defined a basis in terms of a *finite* set of vectors that spans the space and is linearly independent. Such a set does not exist for all vector spaces. When a basis, as defined thus, exists for a vector space, the space is said to be *finite dimensional*. If such a set of vectors does not exist, the space is said to be *infinite dimensional*. Some of the function spaces of Section 5-1 are infinite-dimensional vector spaces. We are primarily interested in finite-dimensional vector spaces in this text.

✦ *Example 2* ───────────────────────────────────

In Example 6 of Section 4-4, we saw that the vectors $(1, 2, 0)$, $(0, 1, -1)$, and $(1, 1, 2)$ span $\mathbf{R}^3$. It can be shown that these vectors are also linearly independent. They thus form a basis for $\mathbf{R}^3$.   ✦

✦ *Example 3* ───────────────────────────────────

In Example 2 of Section 4-5, we saw that the vectors $(1, -1, 1, 2, 1)$, $(4, -1, 6, 6, 2)$, and $(-2, -1, 1, -2, -2)$ are linearly independent in $\mathbf{R}^5$. They can be used to generate a subspace V of $\mathbf{R}^5$. These vectors will span that subspace. They form a basis for the subspace; its dimension will be 3.

An arbitrary vector $\mathbf{v}$ in V can be expressed

$$\mathbf{v} = a(1, -1, 1, 2, 1) + b(4, -1, 6, 6, 2) + c(-2, -1, 1, -2, -2)   ✦$$

✦ *Example 4* ───────────────────────────────────

Consider the subspace V of $\mathbf{R}^3$ consisting of vectors of the form (a, a, b). V is a plane through the origin (Example 2, Section 4-3). We can express (a, a, b) as

$$(a, a, b) = a(1, 1, 0) + b(0, 0, 1)$$

The vectors $(1, 1, 0)$ and $(0, 0, 1)$ span the space. They are also linearly independent. They form a basis for the space; its dimension is 2.   ✦

In this last example, we saw that a certain plane through the origin was a two-dimensional subspace of $\mathbf{R}^3$. The following theorem tells us more about the geometrical interpretations of subspaces of $\mathbf{R}^3$.

Theorem 4-8

(1) *The origin is a subspace of* $\mathbf{R}^3$. *This subspace is defined to be of dimension zero.*

(2) *The one-dimensional subspaces of* $\mathbf{R}^3$ *are lines through the origin.*

(3) *The two-dimensional subspaces of* $\mathbf{R}^3$ *are planes through the origin.*

Proof: For part (1), we know that the origin of $\mathbf{R}^3$ is the vector $(0, 0, 0)$, the zero vector. Consider the subset of $\mathbf{R}^3$ consisting of this single vector only. It is closed under addition and under scalar multiplication:

$$(0, 0, 0) + (0, 0, 0) = (0, 0, 0)$$

and

$$p(0, 0, 0) = (0, 0, 0)$$

for arbitrary scalar p. Thus the subset consisting of the zero vector is a subspace of $\mathbf{R}^3$. We define the dimension of this subspace to be zero.

For part (2), let us consider the one-dimensional subspaces of $\mathbf{R}^3$.

Let $\mathbf{v}$ be a basis for such a one-dimensional subspace. Every vector in the space can be expressed in the form $a\mathbf{v}$ for some scalar a. These vectors

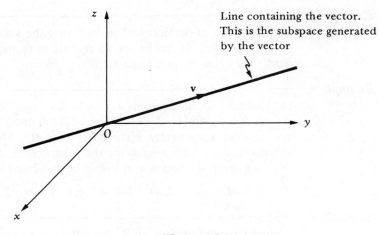

Figure 4-24

make up the line through the origin defined by the vector **v**. Thus the arbitrary one-dimensional subspace is a line through the origin. (See Figure 4-24.)

Finally, for part (3) we look at the two-dimensional subspaces of $\mathbf{R}^3$.

Let **u** and **v** be a basis for such a subspace. If **w** is an arbitrary vector in the subspace, then it can be expressed

$$\mathbf{w} = a\mathbf{u} + b\mathbf{v}$$

for scalars a and b. By the geometrical interpretation of vector addition, **w** lies in the plane of $a\mathbf{u}$ and $b\mathbf{v}$—it is, in fact, the diagonal of the parallelogram defined by $a\mathbf{u}$ and $b\mathbf{v}$. (See Figure 4-25.)

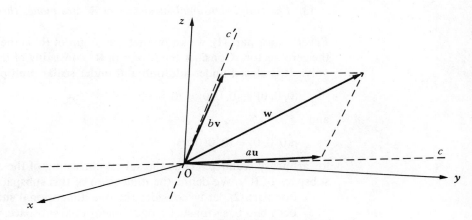

Figure 4-25

As a and b vary, the vectors $a\mathbf{u}$ and $b\mathbf{v}$ will always lie on the lines c and c'. $\mathbf{w}$ will always lie in the plane defined by these two lines. Furthermore, the plane is made up of vectors that can be expressed in the form $a\mathbf{u} + b\mathbf{v}$. Thus the arbitrary two-dimensional subspace is a plane through the origin.

✦ *Example 5*
_____

The dominant graphic display device for computers is the CRT (cathode ray tube). In this example, we discuss how shades and colors are controlled in a color monitor. Information is transmitted internally in a computer in terms of numbers. Each color and intensity has to be represented by a numerical code that the computer will recognize. We first introduce the code and then relate it to the hardware.

Experimentation in color perception has shown that any color can be generated by using a proper additive mixture of the primary colors: red (denoted r), green (denoted g), and blue (denoted b). Let $(1, 0, 0)$, $(0, 1, 0)$, and $(0, 0, 1)$ denote, respectively, a unit of illumination of red, green, and blue. r_1 units of red light would be represented by the vector $r_1(1, 0, 0)$—that is, $(r_1, 0, 0)$. g_1 units of green would be represented by $(0, g_1, 0)$, and b_1 units of blue by $(0, 0, b_1)$.

A combination of r_1, g_1, and b_1 units of red, green, and blue would be represented by the linear combination

$$r_1(1, 0, 0) + g_1(0, 1, 0) + b_1(0, 0, 1)$$

or the vector

$$(r_1, g_1, b_1)$$

Let (r_1, g_1, b_1) and (r_2, g_2, b_2) be two colors. If both are flashed onto a screen simultaneously, the resulting color will be

$$(r_1, g_1, b_1) + (r_2, g_2, b_2)$$

that is,

$$(r_1 + r_2, g_1 + g_2, b_1 + b_2)$$

Thus we have represented unit brightness of each of the primary colors by the base vectors of $\mathbf{R}^3$: $(1, 0, 0)$, $(0, 1, 0)$, and $(0, 0, 1)$. All other colors can be represented by elements of $\mathbf{R}^3$ that are linear combinations of these vectors. The color (r_1, g_1, b_1) can be varied in brightness through use of scalar multiplication. $a(r_1, g_1, b_1)$, where $a > 0$, would represent the same color with different brightness.

Thus the operation of vector addition of $\mathbf{R}^3$ corresponds to mixing colors; the operation of scalar multiplication by a positive scalar corresponds to changing the intensity. Note that this color model is a proper subset of $\mathbf{R}^3$—namely, all the nonzero vectors with nonnegative components. The other elements of $\mathbf{R}^3$ do not have physical interpretation. The color purple, for example, is represented by the vector $a(1, 0, 1)$, where $a > 0$; it is a combination of red and blue in equal quantities.

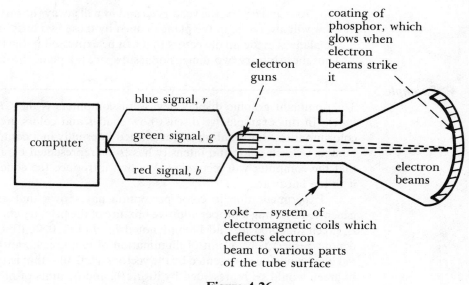

Figure 4-26

Let us now look at the hardware side; we introduce the shadow-mask color CRT, which is the one most widely used. Its construction is shown in Figure 4-26.

Three electron guns are responsible for the red, green, and blue components of color. Three beams strike the phosphor on the screen in slightly different locations; the overall effect, however, is a single colored spot. The intensity of each beam can be controlled in order to produce various colors. By scanning the screen rapidly, the electron beams generate color pictures. In a standard monitor, there are eight levels of intensity for each of the red and green primary colors and four levels of intensity for the blue. The colors available can be represented by vectors in $\mathbf{R}^3$:

$$(r, g, b)$$

where $0 \leq r \leq 8$, $0 \leq g \leq 8$, and $0 \leq b \leq 4$, all integers. There are thus 256 different intensities and colors available. [Observe that, for example, $(1, 2, 1)$ and $(2, 4, 2)$ are not different colors; since $(2, 4, 2) = 2(1, 2, 1)$, they are the same color but with different intensities.]

There is a location in the memory of the computer associated with each point on the screen. That location contains a vector (r, g, b), the numerical code for the color and intensity to be exhibited at the point. In the computer the numbers r, g, and b are represented in binary form—the computer works in binary, not decimal. As the electron beams scan the points on the screen, every 90 nanoseconds on a standard monitor, the corresponding memory locations are scanned. Their contents are read and transmitted to the electron guns, causing the desired colors to be generated on the screen. Changing the color vectors changes patterns on the screen.

Readers who are interested in an in-depth discussion of color graphics are referred to *Principles of Interactive Computer Graphics*, 2nd edition, by William M. Newman and Robert F. Sproull, McGraw-Hill Book Company, 1979, chapters 3 and 19.   ✦

We have discussed the fact that a set of vectors that spans a space characterizes the space since every vector can be expressed as a linear combination of the spanning set. In general, there may be more than one way of expressing a vector as such a linear combination. For example, if we consider the space spanned by the vectors (1, 2, 3), (−1, 1, 4), and (3, 3, 2), we see that we can express the vector (4, 5, 5), which lies in this space, in more than one way in terms of these vectors:

$$(4, 5, 5) = -(1, 2, 3) + (-1, 1, 4) + 2(3, 3, 2)$$

and

$$(4, 5, 5) = 5(1, 2, 3) - 2(-1, 1, 4) - (3, 3, 2)$$

There are many other representations of (4, 5, 5) in terms of the vectors (1, 2, 3), (−1, 1, 4), and (3, 3, 2). (See Example 3, Section 4-4.) However, if the spanning set is a basis, the following theorem tells us that each such linear combination is unique. A basis thus characterizes the space more exactly than does a general spanning set.

Theorem 4-9   *Let* $\mathbf{v}_1, \ldots, \mathbf{v}_m$ *be a basis for a vector space* V. *Then each vector in* V *can be expressed uniquely as a linear combination of these vectors.*

If $\mathbf{v}_1, \ldots, \mathbf{v}_m$ *is a spanning set that is not a basis, then each vector in* V *can be expressed in more than one way as a linear combination of these vectors.*

Proof: Assume that $\mathbf{v}_1, \ldots, \mathbf{v}_m$ is a basis for V.

By the definition of a basis we know that each vector in V can be expressed as a linear combination of $\mathbf{v}_1, \ldots, \mathbf{v}_m$. Hence we need only prove the uniqueness of the combination.

Suppose the linear combination is not unique for some vector $\mathbf{v}$; suppose there exist two distinct sets of scalars, $a_1, \ldots, a_m$ and $b_1, \ldots, b_m$, such that

$$\mathbf{v} = a_1\mathbf{v}_1 + \cdots + a_m\mathbf{v}_m$$

and

$$\mathbf{v} = b_1\mathbf{v}_1 + \cdots + b_m\mathbf{v}_m$$

Hence

$$a_1\mathbf{v}_1 + \cdots + a_m\mathbf{v}_m = b_1\mathbf{v}_1 + \cdots + b_m\mathbf{v}_m$$
$$(a_1 - b_1)\mathbf{v}_1 + \cdots + (a_m - b_m)\mathbf{v}_m = \mathbf{0}$$

Since the vectors $\mathbf{v}_1, \ldots, \mathbf{v}_m$ form a basis, they are linearly independent and each coefficient must be zero.

Thus $a_1 = b_1, \ldots, a_m = b_m$. This contradicts the assumption that the two linear combinations are distinct. Therefore, each linear combination is unique.

To prove the second part of the theorem, let $\mathbf{v}_1, \ldots, \mathbf{v}_m$ be a spanning set that is not a basis—that is, a spanning set of linearly dependent vectors. Let $\sum_{i=1}^{m} c_i \mathbf{v}_i = \mathbf{0}$, where at least one $c_i \neq 0$. Let $\mathbf{v}$ be an arbitrary vector of V and let it be expressed $\mathbf{v} = \sum_{i=1}^{m} d_i \mathbf{v}_i$ in terms of the spanning set. Then

$$\mathbf{v} = \sum_{i=1}^{m} d_i \mathbf{v}_i + \mathbf{0} = \sum_{i=1}^{m} d_i \mathbf{v}_i + \sum_{i=1}^{m} c_i \mathbf{v}_i = \sum_{i=1}^{m} (d_i + c_i) \mathbf{v}_i$$

giving another distinct expression for $\mathbf{v}$ in terms of $\mathbf{v}_1, \ldots, \mathbf{v}_m$.

We complete this discussion of bases with two further results. The following theorem gives a sufficient condition for a given set of vectors to be a basis.

Theorem 4-10 *In a vector space of dimension n, any set of n linearly independent vectors forms a basis.*

Proof: Let $\mathbf{v}_1, \ldots, \mathbf{v}_n$ be a set of n linearly independent vectors in a vector space V of dimension n. Let $\mathbf{w}$ be another vector in the space. The set $\mathbf{v}_1, \ldots, \mathbf{v}_n, \mathbf{w}$, being a set of more than n vectors in V, is linearly dependent. Thus there exist scalars $a_1, \ldots, a_n, a$, not all zero, such that

$$a_1 \mathbf{v}_1 + \cdots + a_n \mathbf{v}_n + a\mathbf{w} = \mathbf{0}$$

a cannot be zero, for if it were, $\mathbf{v}_1, \ldots, \mathbf{v}_n$ would be linearly dependent. Thus $\mathbf{w}$ can be expressed

$$\mathbf{w} = \left(-\frac{a_1}{a}\right) \mathbf{v}_1 + \cdots + \left(-\frac{a_n}{a}\right) \mathbf{v}_n$$

proving that $\mathbf{v}_1, \ldots, \mathbf{v}_n$ span V. They are given to be linearly independent; they form a basis for V.

The final theorem of this section discusses the construction of bases using given linearly independent vectors.

Theorem 4-11 *Let V be a vector space of dimension n, and let $\mathbf{v}_1, \ldots, \mathbf{v}_r$, with $r < n$, be a set of linearly independent vectors in V. Then there exist vectors $\mathbf{v}_{r+1}, \ldots, \mathbf{v}_n$ in V such that the set $\mathbf{v}_1, \ldots, \mathbf{v}_r, \mathbf{v}_{r+1}, \ldots, \mathbf{v}_n$ forms a basis for V. That is, a given set of linearly independent vectors can always be extended to form a basis for the space.*

Proof: The set $\mathbf{v}_1, \ldots, \mathbf{v}_r$ cannot span V, for if it did, the dimension would be r and not n. Hence there exists a vector $\mathbf{v}_{r+1}$ in V which is not a linear combination of $\mathbf{v}_1, \ldots, \mathbf{v}_r$. If $r + 1 = n$, the set $\mathbf{v}_1, \ldots, \mathbf{v}_{r+1}$ is a set of n linearly independent vectors in V and hence forms a basis for V. If $n > r + 1$, then a vector $\mathbf{v}_{r+2}$ can be added to the set such that the set $\mathbf{v}_1, \ldots, \mathbf{v}_{r+2}$ is linearly independent. Continuing in this way, we can add i vectors until $r + i = n$. Then the set $\mathbf{v}_1, \ldots, \mathbf{v}_r, \mathbf{v}_{r+1}, \ldots, \mathbf{v}_n$ forms a basis for V.

In later sections, we shall be interested in constructing bases, starting off from given sets of linearly independent vectors. This theorem will be useful then.

EXERCISES

*1. Prove that the vectors $(-1, 3)$ and $(2, 2)$ form a basis for $\mathbf{R}^2$.

2. Prove that the vectors $(-1, 2, 1)$, $(3, -1, 0)$, and $(2, 2, -2)$ form a basis for $\mathbf{R}^3$.

*3. Give examples of three distinct bases for $\mathbf{R}^2$.

4. Give examples of three distinct bases for $\mathbf{R}^3$.

*5. Prove that the set of vectors $(1, 2)$, $(-2, 3)$, and $(1, 1)$ spans $\mathbf{R}^2$ but is not linearly independent. Thus it is not a basis for $\mathbf{R}^2$. Prove that the vector $(3, 4)$, an element of $\mathbf{R}^2$, can be expressed in more than one way as a linear combination of these vectors.

6. Prove that the vectors $(1, 0, 0)$, $(-1, 1, 0)$, $(1, 1, 2)$, and $(1, 0, 1)$ span $\mathbf{R}^3$, are not linearly independent, and thus do not form a basis for $\mathbf{R}^3$. Illustrate the fact that the vector $(1, 1, 1)$ can be expressed in more than one way as a linear combination of these vectors. Select a subset of this set of four vectors that does form a basis. Show that $(1, 1, 1)$ is expressed uniquely as a linear combination of these basis vectors.

*7. The vector $(1, 2, 0)$ can be expressed as a linear combination of the vectors $(1, 0, 0)$ and $(0, 1, 0)$ in $\mathbf{R}^3$, for $(1, 2, 0) = (1, 0, 0) + 2(0, 1, 0)$. What is the geometrical interpretation of this equation? What is the geometrical characteristic of all vectors in $\mathbf{R}^3$ that can be expressed as linear combinations of $(1, 0, 0)$ and $(0, 1, 0)$?

8. Use Exercise 8 of the previous section to demonstrate how any number of distinct bases can be constructed for
    (a) $\mathbf{R}^2$ **(b) $\mathbf{R}^3$**   **(c) $\mathbf{R}^n$**

*9. $(1, 2, 0)$ and $(2, 3, 0)$ are vectors in $\mathbf{R}^3$. Consider the subspace of $\mathbf{R}^3$ generated by these vectors. What is the dimension of this subspace? Illustrate this subspace geometrically.

10. $(-1, 2, 1)$, $(3, 0, 1)$, and $(1, 4, 3)$ are vectors in $\mathbf{R}^3$. Prove that the set of vectors generated by these three vectors is a two-dimensional subspace of $\mathbf{R}^3$.

*11. $(1, -1, 2)$ is a vector in $\mathbf{R}^3$. Use this vector to generate a one-dimensional subspace of $\mathbf{R}^3$. Illustrate this subspace geometrically.

12. Let $\mathbf{v}_1, \ldots, \mathbf{v}_m$ be m vectors in $\mathbf{R}^n$, when $m < n$. Use these vectors to generate a subspace of $\mathbf{R}^n$. Discuss the dimension of this subspace.

*13. Prove that the vector $(1, 2, -1)$ lies in the two-dimensional subspace of $\mathbf{R}^3$ generated by the vectors $(1, 3, 1)$ and $(1, 4, 3)$.

14. Prove that the vector $(2, 1, 4)$ lies in the two-dimensional subspace of $\mathbf{R}^3$ generated by the vectors $(1, 0, 2)$ and $(1, 1, 2)$.

*15. Prove that the vector $(1, 2, -1, 3)$ lies in the three-dimensional subspace of $\mathbf{R}^4$ generated by the vectors $(1, 2, 1, 0)$, $(1, -2, 3, 1)$, and $(-2, 4, -8, 1)$.

16. Does the vector $(1, 2, -1)$ lie in the subspace of $\mathbf{R}^3$ generated by the vectors $(1, -1, 0)$ and $(3, -1, 2)$?

*17. Determine a basis for the subspace of $\mathbf{R}^3$ consisting of vectors of the form $(a, -b, 3a)$.

18. Determine a basis for the subspace of $\mathbf{R}^3$ consisting of vectors of the form $(a + b, 2a, a - b)$. Give the dimension of the subspace.

*19. Determine a basis for the subspace of $\mathbf{R}^4$ consisting of vectors of the form $(a, -a + b, 3a + 2b + c, a - c)$. Give the dimension of the subspace.

20. Consider the vector space $\mathbf{R}^2$. Prove that

(a) the zero vector forms a subspace whose dimension is defined to be zero

(b) the one-dimensional subspaces are lines through the origin

21. Prove that if $\{\mathbf{v}_1, \ldots, \mathbf{v}_n\}$ is a basis for a vector space V, it is a maximal set of linearly independent vectors in V. That is, any set of vectors in V containing this set as a proper subset is a set of linearly dependent vectors.

22. In Example 5, the primary colors are represented by linearly independent vectors— the vectors $(1, 0, 0)$, $(0, 1, 0)$, and $(0, 0, 1)$. What is the physical significance of linear independence here? Can colors other than the primary colors be used to generate all the other colors? (Filters are not allowed.)

4-7 RANK

In this section, the reader is introduced to the concept of rank. Rank enables one to relate vectors to matrices and vice versa. Rank is a unifying tool which enables us to bring together many of the concepts discussed in the course; solutions to certain systems of linear equations, singularity of a matrix, and invertibility of a matrix all come together under the umbrella of rank.

> **Definition 4-10** Let A be an $m \times n$ matrix. The rows of A may be viewed as vectors $\mathbf{r}_1, \ldots, \mathbf{r}_m$ and the columns as vectors $\mathbf{c}_1, \ldots, \mathbf{c}_n$. Each row vector will have n components and each column vector will have m components. The row vectors will span a subspace of $\mathbf{R}^n$ called the *row space* of A, and the column vectors will span a subspace of $\mathbf{R}^m$ called the *column space* of A.

✦ *Example 1* ────────────────────────────────

Consider the matrix $A = \begin{pmatrix} 1 & 2 & -1 & 2 \\ 3 & 4 & 1 & 6 \\ 5 & 4 & 1 & 0 \end{pmatrix}$. The row vectors of A are

$$\mathbf{r}_1 = (1, 2, -1, 2), \qquad \mathbf{r}_2 = (3, 4, 1, 6), \qquad \mathbf{r}_3 = (5, 4, 1, 0)$$

These vectors will span a subspace of $\mathbf{R}^4$ called the row space of A. The column vectors of A are

$$\mathbf{c}_1 = \begin{pmatrix} 1 \\ 3 \\ 5 \end{pmatrix}, \qquad \mathbf{c}_2 = \begin{pmatrix} 2 \\ 4 \\ 4 \end{pmatrix}, \qquad \mathbf{c}_3 = \begin{pmatrix} -1 \\ 1 \\ 1 \end{pmatrix}, \qquad \mathbf{c}_4 = \begin{pmatrix} 2 \\ 6 \\ 0 \end{pmatrix}$$

These vectors will span a subspace of $\mathbf{R}^3$ called the column space of A.   ✦

Theorem 4-12   *The row space and column space of a matrix A both have the same dimension.*

Proof: An outline of the proof of this theorem will be presented at the end of this section.

Definition 4-11   The dimension of the row space and the column space of a matrix A is called the *rank* of A.

✦ *Example 2* ──────────────────────────────────────

Determine the rank of the matrix $A = \begin{pmatrix} 1 & 2 & 3 \\ 0 & 1 & 2 \\ 2 & 5 & 8 \end{pmatrix}$.

We see by inspection that the third row of A can be expressed as a linear combination of the first two rows:

$$(2, 5, 8) = 2(1, 2, 3) + (0, 1, 2)$$

Hence the three rows of this matrix are linearly dependent, and its rank must be less than 3.

On examining the vectors $(1, 2, 3)$ and $(0, 1, 2)$, we find that the identity

$$a(1, 2, 3) + b(0, 1, 2) = \mathbf{0}$$

leads to

$$a = b = 0$$

Thus the vectors $(1, 2, 3)$ and $(0, 1, 2)$ are linearly independent. They form a basis for the row space of A. The rank of A is 2.   ✦

This method, implied by the definition, is not practical for determining ranks of matrices larger than the above. We shall now develop a more systematic approach for determining the rank of a matrix. The reader was introduced to the concept of the reduced echelon form of a matrix earlier and saw its importance in solving systems of linear equations. We now introduce an *echelon form*. A matrix is in echelon form if it satisfies the first three axioms of a reduced echelon form.

Definition 4-12   A matrix is in *echelon form* if

1. Any rows consisting entirely of zeros are grouped at the bottom of the matrix.

2. The first nonzero element of each other row is 1. This element is called a *leading* 1.
3. The leading 1 of each row after the first is positioned to the right of the leading 1 of the previous row.

The following matrices are all in echelon form:

$$\begin{pmatrix} 1 & 2 & 3 \\ 0 & 1 & 4 \\ 0 & 0 & 1 \end{pmatrix}, \quad \begin{pmatrix} 1 & 2 & 4 & 3 \\ 0 & 0 & 1 & 2 \\ 0 & 0 & 0 & 1 \end{pmatrix}, \quad \begin{pmatrix} 1 & -2 & 4 & 3 & 0 & 2 \\ 0 & 0 & 1 & 2 & 5 & 4 \\ 0 & 0 & 0 & 0 & 1 & 6 \end{pmatrix}$$

Note that the difference between echelon form and reduced echelon form is that the elements above leading ones need not be zeros in the former, whereas they are zeros in the latter. Computationally this means that we only have to create zeros below pivots to arrive at an echelon form. Less computation is usually involved in deriving an echelon form of a matrix than in deriving its reduced echelon form. For the application we have in mind here, we could actually use either the echelon form or the reduced echelon form, but we prefer the former because of the computation factor.

We now illustrate the reduction of a matrix to echelon form.

✦ Example 3 ─────────────────────────────────

Compute an echelon form for the matrix

$$\begin{pmatrix} 1 & 2 & 4 & -4 \\ -1 & -1 & -1 & 5 \\ 3 & 3 & 5 & -11 \end{pmatrix}$$

We get

$$\begin{pmatrix} ① & 2 & 4 & -4 \\ -1 & -1 & -1 & 5 \\ 3 & 3 & 5 & -11 \end{pmatrix} \underset{\substack{R2+R1 \\ R3+(-3)R1}}{\approx} \begin{pmatrix} 1 & 2 & 4 & -4 \\ 0 & ① & 3 & 1 \\ 0 & -3 & -7 & 1 \end{pmatrix}$$

$$\underset{R3+(3)R2}{\approx} \begin{pmatrix} 1 & 2 & 4 & -4 \\ 0 & 1 & 3 & 1 \\ 0 & 0 & 2 & 4 \end{pmatrix}$$

$$\uparrow$$

note: only created a zero below the 1

$$\underset{(\frac{1}{2})R3}{\approx} \begin{pmatrix} 1 & 2 & 4 & -4 \\ 0 & 1 & 3 & 1 \\ 0 & 0 & 1 & 2 \end{pmatrix}$$

This is an echelon form of the given matrix.   ✦

A matrix can usually be reduced to echelon form through numerous sequences of elementary matrix transformations. Unlike the situation for the reduced echelon form, different sequences of transformations can lead to different echelon forms. An echelon form of a matrix is in general not unique.

The following theorem informs us that we can tell at a glance the rank of a matrix that is in echelon form.

Theorem 4-13   *The nonzero row vectors of a matrix in echelon form form a basis for the row space. The rank of a matrix in echelon form is its number of nonzero row vectors.*

***Proof*:**   Let E be an $m \times n$ matrix in echelon form with nonzero row vectors $\mathbf{e}_1, \ldots, \mathbf{e}_r$. Consider the identity

$$a_1\mathbf{e}_1 + a_2\mathbf{e}_2 + \cdots + a_r\mathbf{e}_r = \mathbf{0}$$

where $a_1, \ldots, a_r$ are scalars.

The first nonzero component of $\mathbf{e}_1$ is 1, and it is the only vector with a nonzero component in this slot. Thus, on adding the vectors $a_1\mathbf{e}_1, \ldots, a_r\mathbf{e}_r$, we get an n-tuple having as its first component a_1. On equating to zero, $a_1 = 0$. The identity thus reduces to

$$a_2\mathbf{e}_2 + \cdots + a_r\mathbf{e}_r = \mathbf{0}$$

The first nonzero component of $\mathbf{e}_2$ is 1, and it is the only vector in this identity with a nonzero component in this slot. Thus $a_2 = 0$. Similarly, $a_3, \ldots, a_r$ are all zero. The vectors $\mathbf{e}_1, \ldots, \mathbf{e}_r$ are linearly independent vectors that span the row space of E; they thus form a basis for the row space. The rank of E is r, the number of nonzero row vectors.

The following example illustrates the proof.

✦ *Example 4*
_____

Consider the echelon matrix

$$E = \begin{pmatrix} 1 & 0 & 2 & 1 \\ 0 & 0 & 1 & 3 \\ 0 & 0 & 0 & 1 \\ 0 & 0 & 0 & 0 \end{pmatrix}$$

According to Theorem 4-13, its rank is 3. Let us prove that this is so.

The row vectors are

$$\mathbf{e}_1 = (1, 0, 2, 1)$$
$$\mathbf{e}_2 = (0, 0, 1, 3)$$
$$\mathbf{e}_3 = (0, 0, 0, 1)$$
$$\mathbf{e}_4 = (0, 0, 0, 0)$$

The identity

$$a_1(1, 0, 2, 1) + a_2(0, 0, 1, 3) + a_3(0, 0, 0, 1) = \mathbf{0}$$

gives

$$(a_1, 0, 2a_1, a_1) + (0, 0, a_2, 3a_2) + (0, 0, 0, a_3) = \mathbf{0}$$

$$(a_1, 0, 2a_1 + a_2, a_1 + 3a_2 + a_3) = \mathbf{0}$$

Since the first component must be zero, $a_1 = 0$. Since the third component is zero, $a_2 = 0$; and since the fourth component is zero, $a_3 = 0$. Thus the vectors $\mathbf{e}_1$, $\mathbf{e}_2$, and $\mathbf{e}_3$ are linearly independent. The vectors $\mathbf{e}_1$, $\mathbf{e}_2$, $\mathbf{e}_3$, and $\mathbf{e}_4$ are linearly dependent, since this set contains the zero vector. The nonzero row vectors $\mathbf{e}_1$, $\mathbf{e}_2$, and $\mathbf{e}_3$ form a basis for the row space of E. The rank of E is 3.   ◆

The following theorem enables us to determine a basis for the row space of a matrix using an echelon form of the matrix.

Theorem 4-14   *Let E be an echelon form of a matrix A. The nonzero row vectors of E form a basis for the row space of A. The rank of A is equal to the rank of E.*

Proof: Let the row vectors of A be $\mathbf{a}_1, \dots, \mathbf{a}_m$. Let $\mathbf{v}$ be an arbitrary vector in the row space of A. Thus there exist scalars $s_1, \dots, s_m$ such that

$$\mathbf{v} = s_1\mathbf{a}_1 + \cdots + s_m\mathbf{a}_m$$

Suppose we have the following sequence of elementary matrix transformations leading from A to E:

$$A \underset{R1 \leftrightarrow R2}{\approx} B \underset{(1/p)R1}{\approx} C \approx \cdots \approx E$$

where p is the pivot. Let the row vectors of B be $\mathbf{b}_1, \dots, \mathbf{b}_m$, etc. Then we have

$$
\begin{aligned}
\mathbf{v} &= s_1\mathbf{a}_1 && + s_2\mathbf{a}_2 + \cdots + s_m\mathbf{a}_m \\
&= s_2\mathbf{a}_2 && + s_1\mathbf{a}_1 + \cdots + s_m\mathbf{a}_m \\
&= s_2\mathbf{b}_1 && + s_1\mathbf{b}_2 + \cdots + s_m\mathbf{b}_m && \text{(\textbf{v} is expressed as a linear} \\
& && && \text{combination of the row} \\
& && && \text{vectors of } B) \\
&= (ps_2)\frac{1}{p}\mathbf{b}_1 && + s_1\mathbf{b}_2 + \cdots + s_m\mathbf{b}_m \\
&= ps_2\mathbf{c}_1 && + s_1\mathbf{c}_2 + \cdots + s_m\mathbf{c}_m && \text{(\textbf{v} is expressed as a linear} \\
& && && \text{combination of the row} \\
& && && \text{vectors of } C) \\
& \quad\vdots && \\
&= *\mathbf{e}_1 + \cdots + *\mathbf{e}_r && && \text{(\textbf{v} is expressed as a linear combination} \\
& && && \text{of the nonzero row vectors of} \\
& && && E, \text{ where } * \text{ are scalars)}
\end{aligned}
$$

Since $\mathbf{v}$ was an arbitrary vector in the row space of A, this implies that $\mathbf{e}_1, \dots, \mathbf{e}_r$ span the row space of A. Since they are linearly independent, they form a basis for the row space of A. The rank of A equals the rank of E. Thus to find a basis for the row space of A, we compute its echelon form E. The nonzero row vectors of E will provide such a basis.

Observe that the previous theorem also enables us to determine a basis for the column space of a matrix A. One would determine a basis for the row space of A^t using an echelon form of A^t. These vectors written in column form will be a basis for the column space of A.

The theorem also enables us to determine a basis for the subspace spanned by a given set of vectors. The vectors are written as the row vectors of a matrix, and a basis is found for the row space.

✦ *Example 5*
_____

Find a basis for the subspace of $\mathbf{R}^4$ spanned by the vectors $(1, 2, 3, 4)$, $(-1, -1, -4, -2)$, and $(3, 4, 11, 8)$.

We construct the matrix having these vectors as its row vectors:

$$\begin{pmatrix} 1 & 2 & 3 & 4 \\ -1 & -1 & -4 & -2 \\ 3 & 4 & 11 & 8 \end{pmatrix}$$

We now determine an echelon form for this matrix:

$$\begin{pmatrix} 1 & 2 & 3 & 4 \\ -1 & -1 & -4 & -2 \\ 3 & 4 & 11 & 8 \end{pmatrix} \underset{\substack{R2+R1 \\ R3+(-3)R1}}{\approx} \begin{pmatrix} 1 & 2 & 3 & 4 \\ 0 & 1 & -1 & 2 \\ 0 & -2 & 2 & -4 \end{pmatrix}$$

$$\underset{R3+(2)R2}{\approx} \begin{pmatrix} 1 & 2 & 3 & 4 \\ 0 & 1 & -1 & 2 \\ 0 & 0 & 0 & 0 \end{pmatrix}$$

The nonzero row vectors of this echelon form, $(1, 2, 3, 4)$ and $(0, 1, -1, 2)$, form a basis for the subspace of $\mathbf{R}^4$.   ✦

The concept of rank plays a role in understanding the behavior of systems of linear equations. We have seen how systems of linear equations can have a unique solution, many solutions, or no solution at all. These situations can be conveniently categorized in terms of the ranks of the augmented matrix and the matrix of coefficients.

Theorem 4-15   *Consider a system of m equations in n variables.*

(1) *If the augmented matrix and the matrix of coefficients have the same rank r and r = n, the solution is unique.*

(2) *If the augmented matrix and the matrix of coefficients have the same rank r and r < n, there are many solutions.*

(3) *If the augmented matrix and the matrix of coefficients do not have the same rank, a solution does not exist.*

***Proof*:** Let the system be

$$a_{11}x_1 + \cdots + a_{1n}x_n = y_1$$
$$\vdots$$
$$a_{m1}x_1 + \cdots + a_{mn}x_n = y_m$$

This system can be written

$$x_1 \begin{pmatrix} a_{11} \\ \vdots \\ a_{m1} \end{pmatrix} + \cdots + x_n \begin{pmatrix} a_{1n} \\ \vdots \\ a_{mn} \end{pmatrix} = \begin{pmatrix} y_1 \\ \vdots \\ y_m \end{pmatrix}$$

That is,

$$x_1 \mathbf{a}_1 + \cdots + x_n \mathbf{a}_n = \mathbf{y} \tag{1}$$

Let us now look at the three possibilities.

(1) The rank of the augmented matrix equals the rank of matrix of coefficients, r, and $r = n$. Since the ranks are the same, the last column vector $\mathbf{y}$ must be linearly dependent on $\mathbf{a}_1, \ldots, \mathbf{a}_n$. Furthermore, since $r = n$, the vectors $\mathbf{a}_1, \ldots, \mathbf{a}_n$ are linearly independent and thus form a basis for the column space of the augmented matrix. Therefore, equation (1) has a unique solution; the solution to the system is unique.

(2) The rank of the augmented matrix equals the rank of the matrix of coefficients, r, and $r < n$. Since the ranks are the same, the last column vector $\mathbf{y}$ must be linearly dependent on $\mathbf{a}_1, \ldots, \mathbf{a}_n$. However, since $r < n$, the vectors $\mathbf{a}_1, \ldots, \mathbf{a}_n$ are linearly dependent. $\mathbf{y}$ can therefore be expressed in more than one way as a linear combination of $\mathbf{a}_1, \ldots, \mathbf{a}_n$. Thus equation (1) has many solutions; the solution to the system exists but is not unique.

(3) The rank of the augmented matrix is not equal to the rank of the matrix of coefficients. This implies that the column vector $\mathbf{y}$ is linearly independent of $\mathbf{a}_1, \ldots, \mathbf{a}_n$. Thus equation (1) has no solutions; a solution to the system does not exist.

Our final theorem of this section brings together in a convenient manner numerous results that have appeared in this course.

Theorem 4-16 *Let A be an $n \times n$ matrix. The following statements are equivalent.*

(1) *A is invertible.*

(2) *A is nonsingular.*

(3) *A is of rank n.*

(4) *The system $A\mathbf{x} = \mathbf{y}$ has a unique solution for any $n \times 1$ vector $\mathbf{y}$.*

These results follow from Theorems 3-9 and 3-10 and the previous theorem.

> *Outline of the Proof of Theorem 4-12**: The row rank of a matrix is equal to its column rank.
> Let the column vectors of A be $\mathbf{a}_1, \ldots, \mathbf{a}_n$. A defines a homogeneous system of equations $A\mathbf{x} = \mathbf{0}$. From the proof of Theorem 4-15 we know that this system can be expressed in the following form:
>
> $$x_1\mathbf{a}_1 + x_2\mathbf{a}_2 + \cdots + x_n\mathbf{a}_n = \mathbf{0} \tag{1}$$
>
> Let E be an echelon form of A, having column vectors $\mathbf{e}_1, \ldots, \mathbf{e}_n$. Then E defines a system $E\mathbf{x} = \mathbf{0}$ which can be written
>
> $$x_1\mathbf{e}_1 + x_2\mathbf{e}_2 + \cdots + x_n\mathbf{e}_n = \mathbf{0} \tag{2}$$
>
> Since the systems $A\mathbf{x} = \mathbf{0}$ and $E\mathbf{x} = \mathbf{0}$ have the same solution, the x's in identities (1) and (2) are the same. This implies that the linear dependence/independence relationships between the column vectors of A and E are the same. The first r column vectors of A form a basis for the column space of A if and only if the first r column vectors of E form a basis for the column space of E. Thus column rank of A equals column rank of E.
> The final step of the proof is to observe that the row rank of E is equal to its column rank. This can be concluded by examining a specific matrix in echelon form. (See Exercise 18.) We therefore have
>
> $$\text{row rank of } A = \text{row rank of } E = \text{column rank of } E$$
> $$= \text{column rank of } A$$
>
> proving the theorem.

EXERCISES

*1. Determine the ranks of the following matrices using the definition of rank.

(a) $\begin{pmatrix} 1 & 2 \\ 3 & 4 \end{pmatrix}$ **(b)** $\begin{pmatrix} 1 & 2 & 1 \\ 2 & 4 & 2 \\ 1 & 2 & 3 \end{pmatrix}$

(c) $\begin{pmatrix} 1 & 0 & 0 \\ 0 & 1 & 0 \\ 0 & 0 & 1 \end{pmatrix}$ **(d)** $\begin{pmatrix} 2 & 1 & 3 \\ 4 & 2 & 6 \\ 2 & 1 & 3 \end{pmatrix}$

Determine the ranks of the following matrices by finding echelon forms.

2. $\begin{pmatrix} 1 & 2 & 0 \\ 0 & 1 & 1 \\ -1 & 2 & 3 \end{pmatrix}$ ***3.** $\begin{pmatrix} 1 & 2 & 3 \\ 0 & -1 & -1 \\ 3 & 4 & 7 \end{pmatrix}$

4. $\begin{pmatrix} -1 & 0 & 2 \\ 1 & 1 & 1 \\ -1 & 2 & 3 \end{pmatrix}$ ***5.** $\begin{pmatrix} 1 & 2 & 3 & 4 \\ -1 & 2 & 0 & 1 \\ 0 & 1 & 0 & 2 \end{pmatrix}$

6. $\begin{pmatrix} 1 & 3 & 4 & 1 \\ 2 & 6 & 8 & 2 \\ 0 & 1 & 2 & 1 \end{pmatrix}$ ***7.** $\begin{pmatrix} -1 & 2 & 3 & 1 \\ 0 & -2 & -6 & -2 \\ 1 & 2 & 1 & 1 \end{pmatrix}$

8. $\begin{pmatrix} -1 & 2 & 3 & 0 \\ 1 & 2 & 0 & 0 \\ -1 & 0 & 1 & 2 \\ 0 & -1 & 1 & 1 \end{pmatrix}$ ***9.** $\begin{pmatrix} 1 & 1 & 0 & -1 \\ 2 & 1 & 0 & 0 \\ 3 & 2 & 0 & -1 \\ -1 & 0 & 1 & 1 \end{pmatrix}$

10. $\begin{pmatrix} -1 & 2 & 1 & 0 \\ 1 & -2 & -1 & 0 \\ -1 & 0 & 1 & 1 \\ -2 & 0 & 2 & 2 \end{pmatrix}$

In Exercises 11–14, find bases for the subspaces spanned by the given vectors.

*11. $(2, 1, 3, 0, 4), (-1, 2, 3, 1, 0), (3, -1, 0, -1, 4)$

12. $(1, 2, 3, 4), (0, -1, 2, 3), (2, 3, 8, 11), (2, 3, 6, 8)$

*13. $(-1, 2, 0, 1, 3), (1, 2, 0, 5, 4), (3, 2, 2, 0, -1)$

14. $(0, -1, 2, 3, -1, 4, 5), (1, 4, 0, -1, 5, 2, 1), (3, 1, 4, 2, 1, 0, 0), (-2, 2, -2, 0, 3, 6, 6)$

15. Write a computer program that can be used to determine a basis for the space spanned by a given set of vectors.

16. If A and B are matrices of the same kind, prove that

$$\text{rank} (A + B) \leq \text{rank } A + \text{rank } B$$

17. Prove that a system of linear equations $A\mathbf{x} = \mathbf{y}, \mathbf{y} \neq \mathbf{0}$, has a solution if and only if $\mathbf{y}$ is in the column space of A.

18. Consider the following 4×4 matrices, all in echelon form. Show that the row rank of each matrix is equal to its column rank.

*(a) $\begin{pmatrix} 1 & 2 & 3 & 5 \\ 0 & 1 & 2 & 3 \\ 0 & 0 & 1 & 2 \\ 0 & 0 & 0 & 0 \end{pmatrix}$ **(b)** $\begin{pmatrix} 1 & 2 & 3 & 1 \\ 0 & 1 & 2 & 4 \\ 0 & 0 & 0 & 1 \\ 0 & 0 & 0 & 0 \end{pmatrix}$ **(c)** $\begin{pmatrix} 1 & 2 & 6 & 4 \\ 0 & 1 & 1 & 4 \\ 0 & 0 & 1 & 5 \\ 0 & 0 & 0 & 1 \end{pmatrix}$

4-8 DOT PRODUCTS, NORMS, ANGLES, AND DISTANCES

In this section, we add additional structure to the vector space $\mathbf{R}^n$. This additional structure leads to concepts of magnitudes of vectors and angles between vectors.

Definition 4-13   If $(x_1, \ldots, x_n)$ and $(y_1, \ldots, y_n)$ are vectors in $\mathbf{R}^n$, their *dot product* is

$$(x_1, \ldots, x_n) \cdot (y_1, \ldots, y_n) = x_1 y_1 + \cdots + x_n y_n$$

The dot product assigns to each pair of vectors a real number.

✦ *Example 1* ───

$(1, 2)$ and $(3, -1)$ are elements of $\mathbf{R}^2$. Their dot product is

$$(1, 2) \cdot (3, -1) = (1 \times 3) + (2 \times -1) = 3 - 2 = 1$$

$(0, -1, 4)$ and $(5, -2, -1)$ are elements of $\mathbf{R}^3$. Their dot product is

$$(0, -1, 4) \cdot (5, -2, -1) = (0 \times 5) + (-1 \times -2) + (4 \times -1)$$
$$= 0 + 2 - 4 = -2 \quad ✦$$

The dot product has the four properties given in the following theorem. These properties are of theoretical value in algebraic manipulations involving the dot product. The reader will, for example, use these results in discussing angles between vectors in $\mathbf{R}^n$ in this section and in handling projections in the following section. In later sections these properties become axioms for generalizing the dot product.

Theorem 4-17   *If* $\mathbf{u}$, $\mathbf{v}$, *and* $\mathbf{w}$ *are vectors in* $\mathbf{R}^n$ *and* a *is a scalar, then*

(1) $\mathbf{u} \cdot \mathbf{v} = \mathbf{v} \cdot \mathbf{u}$
(2) $(\mathbf{u} + \mathbf{v}) \cdot \mathbf{w} = \mathbf{u} \cdot \mathbf{w} + \mathbf{v} \cdot \mathbf{w}$
(3) $a\mathbf{u} \cdot \mathbf{v} = a(\mathbf{u} \cdot \mathbf{v}) = \mathbf{u} \cdot a\mathbf{v}$
(4) $\mathbf{u} \cdot \mathbf{u} \geq 0$ *and* $\mathbf{u} \cdot \mathbf{u} = 0$ *if and only if* $\mathbf{u} = \mathbf{0}$

Proof: For part (1), let $\mathbf{u} = (x_1, \ldots, x_n)$ and $\mathbf{v} = (y_1, \ldots, y_n)$. Then

$$\mathbf{u} \cdot \mathbf{v} = (x_1, \ldots, x_n) \cdot (y_1, \ldots, y_n)$$
$$= x_1 y_1 + \cdots + x_n y_n$$

Using the commutative property of real numbers under multiplication, we can write this

$$\mathbf{u} \cdot \mathbf{v} = y_1 x_1 + \cdots + y_n x_n$$
$$= \mathbf{v} \cdot \mathbf{u}$$

The proofs of the remaining parts are left as an exercise.

✦ *Example 2* ───

In working with dot products, one finds the need to simplify algebraic expressions. In such manipulations it is important to keep track of scalars and vectors.

For example, each of the following expressions will result in a scalar: $\mathbf{u} \cdot \mathbf{v}$, $(\mathbf{u} \cdot \mathbf{v})(\mathbf{u} \cdot \mathbf{w}) + 4$, $(3\mathbf{u} \cdot \mathbf{v} - 4\mathbf{w} \cdot \mathbf{v})(\mathbf{u} \cdot \mathbf{w} - \mathbf{x} \cdot \mathbf{y})$.

The following will all result in vectors: $3(\mathbf{u} \cdot \mathbf{v})\mathbf{w}$, $(\mathbf{u} \cdot \mathbf{v})\mathbf{w} + \mathbf{u}$, $(\mathbf{u} \cdot \mathbf{v} - \mathbf{v} \cdot \mathbf{w})\mathbf{w}$.

We now illustrate the use of the above theorems in simplifying algebraic expressions. Let us simplify $\left((\mathbf{u} \cdot \mathbf{u})\mathbf{v} - \dfrac{\mathbf{u} \cdot \mathbf{v}}{3}\dfrac{\mathbf{u}}{2} \right) \cdot 3\mathbf{u}$.

$$\left((\mathbf{u} \cdot \mathbf{u})\mathbf{v} - \frac{\mathbf{u} \cdot \mathbf{v}}{3}\frac{\mathbf{u}}{2} \right) \cdot 3\mathbf{u} = (\mathbf{u} \cdot \mathbf{u})\mathbf{v} \cdot 3\mathbf{u} - \frac{\mathbf{u} \cdot \mathbf{v}}{3}\frac{\mathbf{u}}{2} \cdot 3\mathbf{u}$$

$$= 3(\mathbf{u} \cdot \mathbf{u})(\mathbf{v} \cdot \mathbf{u}) - \frac{3}{3 \times 2}(\mathbf{u} \cdot \mathbf{v})(\mathbf{u} \cdot \mathbf{u})$$

$$= 3(\mathbf{u} \cdot \mathbf{u})(\mathbf{v} \cdot \mathbf{u}) - \tfrac{1}{2}(\mathbf{u} \cdot \mathbf{u})(\mathbf{v} \cdot \mathbf{u})$$

$$= \tfrac{5}{2}(\mathbf{u} \cdot \mathbf{u})(\mathbf{v} \cdot \mathbf{u}) \qquad \blacklozenge$$

Norm of a Vector

We now use the concept of dot product to define the *norm* of a vector. The norm is also called the *magnitude* or *length* of the vector.

Definition 4-14   The norm of the arbitrary vector $(x_1, \dots, x_n)$ of $\mathbf{R}^n$ is denoted by $\|(x_1, \dots, x_n)\|$ and given by

$$\|(x_1, \dots, x_n)\| = \sqrt{(x_1, \dots, x_n) \cdot (x_1, \dots, x_n)}$$
$$= \sqrt{x_1^2 + \dots + x_n^2}$$

This definition is compatible with the geometric lengths of position vectors in $\mathbf{R}^2$ and $\mathbf{R}^3$.

If $\mathbf{u}$ is the position vector (a, b) in $\mathbf{R}^2$, its length is indeed $\sqrt{a^2 + b^2}$; see Figure 4-27.

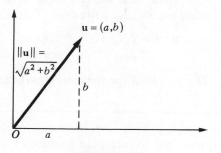

Figure 4-27

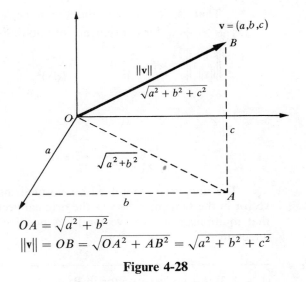

$$OA = \sqrt{a^2 + b^2}$$
$$\|\mathbf{v}\| = OB = \sqrt{OA^2 + AB^2} = \sqrt{a^2 + b^2 + c^2}$$

Figure 4-28

If $\mathbf{v}$ is the position vector (a, b, c) in $\mathbf{R}^3$, its length is $\sqrt{a^2 + b^2 + c^2}$; see Figure 4-28.

✦ *Example 3**

We have seen how forces can be represented by vectors. The magnitude of the force will be the norm of the vector. Let the vector $(4, 3)$ represent a force acting on a particle. The magnitude of this force is

$$\|(4, 3)\| = \sqrt{(4, 3) \cdot (4, 3)} = \sqrt{16 + 9} = 5 \qquad ✦$$

A vector whose norm is 1 is said to be a *unit vector*. The vector $(1, 0)$ in $\mathbf{R}^2$ is a unit vector, for

$$\|(1, 0)\| = \sqrt{(1, 0) \cdot (1, 0)} = \sqrt{(1 \times 1) + (0 \times 0)} = 1$$

Similarly, the vector $(0, 1)$ of $\mathbf{R}^2$ is a unit vector, and so are the vectors $(1, 0, 0)$, $(0, 1, 0)$, and $(0, 0, 1)$ of $\mathbf{R}^3$.

Theorem 4-18   *If $\mathbf{v}$ is any nonzero vector in $\mathbf{R}^n$, the vector $(1/\|\mathbf{v}\|)\mathbf{v}$ derived from $\mathbf{v}$ is a unit vector. This procedure of constructing a unit vector from a given nonzero vector is called normalizing the vector.*

Proof: We set out to prove that $(1/\|\mathbf{v}\|)\mathbf{v}$ is a unit vector.
Let $\mathbf{v}$ be the vector $(x_1, \ldots, x_n)$. Then

$$\|\mathbf{v}\| = \sqrt{x_1^2 + \cdots + x_n^2}$$

$(1/\|\mathbf{v}\|)\mathbf{v}$ is thus the vector $(1/\sqrt{x_1^2 + \cdots + x_n^2})(x_1, \ldots, x_n)$.

That is, it becomes the vector $(x_1/\sqrt{}, \ldots, x_n/\sqrt{})$, where $\sqrt{} = \sqrt{x_1^2 + \cdots + x_n^2}$ for convenience of notation. Hence

$$\left\|\frac{1}{\|\mathbf{v}\|}\mathbf{v}\right\| = \sqrt{\frac{x_1^2}{(\sqrt{})^2} + \cdots + \frac{x_n^2}{(\sqrt{})^2}}$$

$$= \sqrt{\frac{x_1^2 + \cdots + x_n^2}{(\sqrt{})^2}}$$

$$= \sqrt{\frac{x_1^2 + \cdots + x_n^2}{x_1^2 + \cdots + x_n^2}} = 1$$

The unit vector obtained on normalizing a given vector will be a unit vector in the same direction as the original vector; in Section 4-1 we discussed that multiplication of a vector by a positive scalar leaves it unchanged in direction.

✦ **Example 4** ───

$(1, -2, 3)$ is a nonzero vector in $\mathbf{R}^3$.

$$\|(1, -2, 3)\| = \sqrt{1^2 + 2^2 + 3^2}$$
$$= \sqrt{14}$$

Thus, according to Theorem 4-18, the vector $(1/\sqrt{14})(1, -2, 3)$ is a unit vector. Let us verify this.

$$\left\|\frac{1}{\sqrt{14}}(1, -2, 3)\right\| = \left\|\left(\frac{1}{\sqrt{14}}, \frac{-2}{\sqrt{14}}, \frac{3}{\sqrt{14}}\right)\right\|$$

$$= \sqrt{\left(\frac{1}{\sqrt{14}}\right)^2 + \left(\frac{-2}{\sqrt{14}}\right)^2 + \left(\frac{3}{\sqrt{14}}\right)^2}$$

$$= \sqrt{\frac{1}{14} + \frac{4}{14} + \frac{9}{14}}$$

$$= 1 \qquad ✦$$

Angles Between Vectors

Definition 4-15   If $\mathbf{u}$ and $\mathbf{v}$ are vectors in $\mathbf{R}^n$, the cosine of the angle between them is given by

$$\cos\theta = \frac{1}{\|\mathbf{u}\|}\mathbf{u} \cdot \frac{1}{\|\mathbf{v}\|}\mathbf{v}$$

If one interprets $\mathbf{u}$ and $\mathbf{v}$ as position vectors in $\mathbf{R}^2$ or $\mathbf{R}^3$, this definition leads to the expected angle θ, with $0 \leq \theta \leq \pi$, between the vectors. We prove this result for $\mathbf{R}^2$.

Let $\mathbf{u} = (a, b)$, $\mathbf{v} = (c, d)$ in Figure 4-29. The vector $\overrightarrow{AB} = \mathbf{v} - \mathbf{u} = (c - a, d - b)$. Thus the lengths of the sides of the triangle OAB are

$$OA = \|\mathbf{u}\| = \sqrt{a^2 + b^2}$$
$$OB = \|\mathbf{v}\| = \sqrt{c^2 + d^2}$$
$$AB = \|\mathbf{v} - \mathbf{u}\| = \sqrt{(c - a)^2 + (d - b)^2}$$

The law of cosines gives

$$(AB)^2 = (OA)^2 + (OB)^2 - 2(OA)(OB) \cos \theta$$

Thus

$$\cos \theta = \frac{(OA)^2 + (OB)^2 - (AB)^2}{2(OA)(OB)}$$

$$= \frac{a^2 + b^2 + c^2 + d^2 - (c - a)^2 - (d - b)^2}{2(OA)(OB)}$$

$$= \frac{2ac + 2db}{2(OA)(OB)} = \frac{ac + db}{(OA)(OB)}$$

$$= \frac{\mathbf{u} \cdot \mathbf{v}}{\|\mathbf{u}\| \, \|\mathbf{v}\|}$$

Using the result of Theorem 4-17, we can express this in the more meaningful form

$$\cos \theta = \frac{1}{\|\mathbf{u}\|} \mathbf{u} \cdot \frac{1}{\|\mathbf{v}\|} \mathbf{v}$$

To obtain the cosine of the angle between two nonzero vectors, one normalizes each vector and then takes their dot product.

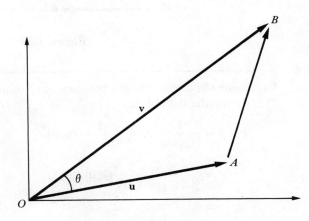

Figure 4-29

✦ *Example 5*

Determine the angle between the vectors $(1, 0)$ and $(1, 1)$ in **R**^{2}. Using the definition, we find that

$$\cos \theta = \frac{1}{\|(1, 0)\|} (1, 0) \cdot \frac{1}{\|(1, 1)\|} (1, 1)$$

$$= (1, 0) \cdot \frac{1}{\sqrt{2}} (1, 1)$$

$$= (1, 0) \cdot \left(\frac{1}{\sqrt{2}}, \frac{1}{\sqrt{2}} \right) = \frac{1}{\sqrt{2}}$$

Hence $\theta = 45°$.

If we interpret $(1, 0)$ and $(1, 1)$ as position vectors in **R**^{2}, we see that the angle between them is indeed $45°$ (Figure 4-30).

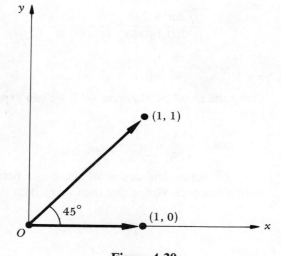

Figure 4-30   ✦

✦ *Example 6*

Determine the angle between the vectors $(2, 0)$ and $(0, 3)$ in **R**^{2}.

Using the definition,

$$\cos \theta = \frac{1}{\|(2, 0)\|} (2, 0) \cdot \frac{1}{\|(0, 3)\|} (0, 3)$$

$$= \frac{1}{\|(2, 0)\| \, \|(0, 3)\|} (2, 0) \cdot (0, 3)$$

$$= 0$$

Hence $\theta = 90°$; the vectors are at right angles.   ✦

We say that two nonzero vectors are *orthogonal* if they are at right angles.

Theorem 4-19   *Two nonzero vectors are orthogonal if and only if their dot product is zero.*

Proof: Let $\mathbf{v}_1$ and $\mathbf{v}_2$ be orthogonal vectors. Hence θ, the angle between them, is $90°$, and $\cos \theta = 0$.

$$\frac{1}{\|\mathbf{v}_1\|} \mathbf{v}_1 \cdot \frac{1}{\|\mathbf{v}_2\|} \mathbf{v}_2 = 0$$

implying that

$$\frac{1}{\|\mathbf{v}_1\| \, \|\mathbf{v}_2\|} \mathbf{v}_1 \cdot \mathbf{v}_2 = 0$$

and hence that $\mathbf{v}_1 \cdot \mathbf{v}_2 = 0$.
    Conversely, if $\mathbf{v}_1 \cdot \mathbf{v}_2 = 0$, $\cos \theta = 0$, and θ is $90°$.

✦ *Example 7* ─────────────────────────────────

The vectors $(1, 0)$ and $(0, 1)$ in $\mathbf{R}^2$ are orthogonal, since

$$(1, 0) \cdot (0, 1) = 0 + 0 = 0 \qquad ✦$$

✦ *Example 8* ─────────────────────────────────

The vectors $(1, 2, -1)$ and $(2, 4, 10)$ in $\mathbf{R}^3$ are orthogonal, since

$$(1, 2, -1) \cdot (2, 4, 10) = (1 \times 2) + (2 \times 4) + (-1 \times 10)$$
$$= 2 + 8 - 10 = 0 \qquad ✦$$

A basis consisting of unit vectors any two of which are orthogonal is called an *orthonormal basis.*

Theorem 4-20   *The standard basis for* $\mathbf{R}^n$, *namely*

$$(1, 0, \ldots, 0), \ldots, (0, \ldots, 0, 1)$$

is an orthonormal basis.

Proof: Consider the vector $(0, \ldots, 1, \ldots, 0)$. Its magnitude is

$$\sqrt{(0, \ldots, 1, \ldots, 0) \cdot (0, \ldots, 1, \ldots, 0)} = \sqrt{1} = 1$$

Thus it is a unit vector.
    Consider two distinct vectors: $\mathbf{v}_1 = (0, \ldots, 1, \ldots, 0, \ldots, 0)$ with the 1 in the ith slot and $\mathbf{v}_2 = (0, \ldots, 0, \ldots, 1, \ldots, 0)$ with the 1 in the jth

slot. Since they are both of unit magnitude,

$$\cos \theta = \mathbf{v}_1 \cdot \mathbf{v}_2$$

$$= \sqrt{(0, \ldots, 1, \ldots, 0, \ldots, 0) \cdot (0, \ldots, 0, \ldots, 1, \ldots, 0)}$$
$$= 0$$

Hence they are orthogonal.

Thus the basis **i**, **j** is an orthonormal basis for **R**^{2} (Figure 4-31), and **i**, **j**, **k** is an orthonormal basis for **R**^{3} (Figure 4-32). In **R**^{2}, $\mathbf{i} \cdot \mathbf{j} = 0$ and $\|\mathbf{i}\| = \|\mathbf{j}\| = 1$. In **R**^{3}, $\mathbf{i} \cdot \mathbf{j} = \mathbf{j} \cdot \mathbf{k} = \mathbf{i} \cdot \mathbf{k} = 0$ and $\|\mathbf{i}\| = \|\mathbf{j}\| = \|\mathbf{k}\| = 1$.

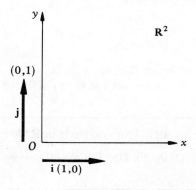

Figure 4-31

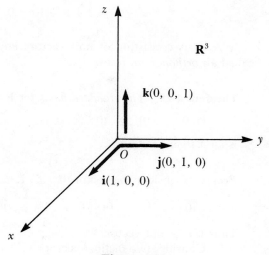

Figure 4-32

Two nonzero vectors are said to be *parallel* if the angle between them is either 0 or π—that is, if they are in the same direction or opposite directions.

✦ *Example 9*

The vectors $(1, 2)$ and $(3, 6)$ are parallel, since

$$\cos \theta = \frac{(1, 2)}{\|(1, 2)\|} \cdot \frac{(3, 6)}{\|(3, 6)\|} = \frac{15}{\sqrt{5}\sqrt{45}} = 1$$

$\theta = 0$, and the vectors are parallel; they are, in fact, in the same direction.

The vectors $(1, -2, 3)$ and $(-2, 4, -6)$ are parallel, since

$$\cos \theta = \frac{(1, -2, 3)}{\|(1, -2, 3)\|} \cdot \frac{(-2, 4, -6)}{\|(-2, 4, -6)\|}$$

$$= \frac{-2 - 8 - 18}{\sqrt{1^2 + 2^2 + 3^2}\sqrt{2^2 + 4^2 + 6^2}} = -1$$

$\theta = \pi$, and the vectors are in opposite directions.   ✦

Distance Between Points

We now use the concept of norm to define distance between points in $\mathbf{R}^n$. Once again our definition will be compatible with known results for two- and three-dimensional spaces.

Let X and Y be arbitrary points in $\mathbf{R}^n$. These points define position vectors. Let us denote the position vectors $\mathbf{x}$ and $\mathbf{y}$ in Figure 4-33. Construct the vector $-\mathbf{y}$. Observe that the sum of the vectors $\mathbf{x}$ and $-\mathbf{y}$ has magnitude equal to the distance between the points X and Y.

Denote the distance between X and Y as $d(X, Y)$, and let

$$d(X, Y) = \|\mathbf{x} - \mathbf{y}\|$$

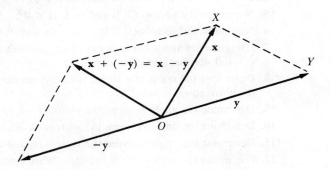

Figure 4-33

Let us express $d(X, Y)$ in terms of the components of **x** and **y**. Suppose $\mathbf{x} = (x_1, \ldots, x_n)$ and $\mathbf{y} = (y_1, \ldots, y_n)$. Then

$$\mathbf{x} - \mathbf{y} = (x_1 - y_1, \ldots, x_n - y_n)$$

and

$$d(X, Y) = \sqrt{(x_1 - y_1)^2 + \cdots + (x_n - y_n)^2}$$

Note that this reduces to the usual expression for distance when the vector space is $\mathbf{R}^2$. Then $\mathbf{x} = (x_1, x_2)$ and $\mathbf{y} = (y_1, y_2)$ and

$$d(X, Y) = \sqrt{(x_1 - y_1)^2 + (x_2 - y_2)^2}$$

✦ *Example 10*

Determine the distance between the points $(2, 4, 5)$ and $(1, 2, 3)$ of $\mathbf{R}^3$. We have that

$$\begin{aligned} d(X, Y) &= \sqrt{(x_1 - y_1)^2 + (x_2 - y_2)^2 + (x_3 - y_3)^2} \\ &= \sqrt{(2 - 1)^2 + (4 - 2)^2 + (5 - 3)^2} \\ &= \sqrt{1^2 + 2^2 + 2^2} = \sqrt{9} = 3 \quad ✦ \end{aligned}$$

✦ *Example 11*

Determine the distance between the two points $X(18, 4, 3, 3, -2)$ and $Y(14, 1, 0, 2, -3)$ of $\mathbf{R}^5$.

$$\begin{aligned} d(X, Y) &= \sqrt{(x_1 - y_1)^2 + (x_2 - y_2)^2 + (x_3 - y_3)^2 + (x_4 - y_4)^2 + (x_5 - y_5)^2} \\ &= \sqrt{(18 - 14)^2 + (4 - 1)^2 + (3 - 0)^2 + (3 - 2)^2 + (-2 + 3)^2} \\ &= \sqrt{4^2 + 3^2 + 3^2 + 1^2 + 1^2} = \sqrt{36} = 6 \quad ✦ \end{aligned}$$

EXERCISES

*1. Determine the dot product of the following elements of $\mathbf{R}^2$: $(2, 1)$ and $(3, 0)$.

2. Determine the dot product of the vectors $(-1, 3, 0)$ and $(2, 1, 5)$ in $\mathbf{R}^3$.

*3. Determine the norms of the vectors $(1, 2)$ and $(3, 4)$ in $\mathbf{R}^2$.

4. Determine the norms of the vectors $(2, 0, 0)$, $(0, 4, 0)$, and $(1, 2, 3)$ in $\mathbf{R}^3$.

*5. Normalize the vectors $(2, 3)$ and $(-1, 2)$ in $\mathbf{R}^2$.

6. Normalize the vectors $(1, 2, 3)$, $(-1, 2, 0)$, and $(0, 0, 1)$ in $\mathbf{R}^3$.

7. Using the definition of the dot product, prove that $2(1, 2, 3) \cdot (-1, 0, 4) = (1, 2, 3) \cdot 2(-1, 0, 4)$ (illustrates Theorem 4-17).

8. Prove that the norm of a vector is always greater than or equal to zero. When is the norm equal to zero?

*9. Determine the angle between the vectors $(-1, 1)$ and $(0, 1)$ in $\mathbf{R}^2$.

10. Determine the angle between the vectors $(2, 0)$ and $(1, \sqrt{3})$ in $\mathbf{R}^2$.

*11. Determine the angle between the vectors $(1, 2, 0)$ and $(0, 1, 1)$ in $\mathbf{R}^3$.

12. Prove that the vectors $(1, 2)$ and $(2, -1)$ are orthogonal in $\mathbf{R}^2$.

13. Prove that the vectors $(4, -2, 2)$ and $(2, 3, -1)$ are orthogonal in $\mathbf{R}^3$.

*14. Determine a vector orthogonal to $(2, 4)$ in $\mathbf{R}^2$.

15. Determine a vector orthogonal to $(-1, 2, 3)$ in $\mathbf{R}^3$.

16. Determine a vector orthogonal to $(\frac{1}{2}, 2, -1)$ in $\mathbf{R}^3$.

17. Prove that the set of vectors orthogonal to the vector $(1, 0, 0)$ in $\mathbf{R}^3$ forms a subspace of $\mathbf{R}^3$.

18. Prove that the set of vectors orthogonal to the vector $(1, 2, 1)$ in $\mathbf{R}^3$ forms a subspace of $\mathbf{R}^3$.

19. (a) Prove that the vectors $(1, 2, -1)$ and $(-1, -2, 1)$ are parallel.

   (b) Prove that the vectors $(2, 3, 1)$ and $(4, 6, 2)$ are parallel.

20. In each of the following, compute the distance between X and Y.

   *(a) $X(6, 5)$, $Y(2, 2)$

   (b) $X(3, 1, 1)$, $Y(1, -1, 0)$

   *(c) $X(6, 6, 2)$, $Y(-2, 4, 1)$

   (d) $X(3, 2, 3, 1)$, $Y(-2, -2, 1, -1)$

   *(e) $X(3, -2, 4)$, $Y(-1, 2, 0)$

   (f) $X(4, -1, 3, 2, 4)$, $Y(4, 5, 2, 1, 1)$

21. In Example 9, we proved that the vectors $(1, 2)$ and $(3, 6)$ are in the same direction. Note that $(3, 6) = 3(1, 2)$. Prove that, in general, two vectors $\mathbf{u}$ and $\mathbf{v}$ are in the same direction if and only if there exists a nonzero positive scalar c such that $\mathbf{u} = c\mathbf{v}$.

   Further, in Example 9 we saw that $(1, -2, 3)$ and $(-2, 4, -6)$ were in opposite directions. Note that $(-2, 4, -6) = -2(1, -2, 3)$. Prove that two vectors $\mathbf{u}$ and $\mathbf{v}$ are in opposite directions if and only if there exists a nonzero negative scalar c such that $\mathbf{u} = c\mathbf{v}$.

22. Let $\mathbf{u}, \mathbf{v}$, and $\mathbf{w}$ be vectors in $\mathbf{R}^n$, let c and d be nonzero scalars, and let α be an angle. Tell whether each of the following expressions will be a vector, scalar, or make no sense.

   *(a) $(\mathbf{u} \cdot \mathbf{v})\mathbf{w}$ **(b)** $(\mathbf{u} \cdot \mathbf{v}) \cdot \mathbf{w}$ ***(c)** $\mathbf{u} \cdot \mathbf{v} + c\mathbf{w}$

   (d) $\mathbf{u} \cdot \mathbf{w} + c$ ***(e)** $c\mathbf{u} \cdot d\mathbf{w} + \|\mathbf{w}\|\mathbf{v}$ **(f)** $\|\mathbf{u} \cdot \mathbf{v}\|$

   *(g) $c(\mathbf{u} \cdot \mathbf{v}) + d\mathbf{w}$ **(h)** $\dfrac{\mathbf{w} + \mathbf{u}}{\mathbf{w} \cdot \mathbf{u}}$ ***(i)** $\|\mathbf{u} + c\mathbf{v}\| + d$

   (j) $\|\mathbf{v}\| \cos \alpha \dfrac{\mathbf{u}}{\|\mathbf{u}\|}$ ***(k)** $\left(\mathbf{v} \cdot \dfrac{\mathbf{u}}{\|\mathbf{u}\|}\right)\dfrac{\mathbf{u}}{\|\mathbf{u}\|}$ **(l)** $\mathbf{u} - \left(\dfrac{\mathbf{u}}{\|\mathbf{u}\|} \cdot \mathbf{v}\right)\dfrac{\mathbf{v}}{\|\mathbf{v}\|}$

   [We shall see the significance of (j), (k), and (l) in the following section.]

23. Prove that if $\mathbf{u}$ is orthogonal to $\mathbf{v}$ and to $\mathbf{w}$, then $\mathbf{u}$ is orthogonal to every vector in the subspace generated by $\mathbf{v}$ and $\mathbf{w}$.

24. Let $\mathbf{u}$ be orthogonal to each of the vectors $\mathbf{v}_1, \ldots, \mathbf{v}_m$. Prove that $\mathbf{u}$ is orthogonal to each vector in the subspace generated by $\mathbf{v}_1, \ldots, \mathbf{v}_m$. We say that $\mathbf{u}$ is orthogonal to this subspace.

25. Let $\mathbf{v}$ be a vector in a space V of dimension n. Show that all the vectors in V orthogonal to $\mathbf{v}$ form a subspace of V.

26. Let U and V be two subspaces of $\mathbf{R}^n$. U is said to be orthogonal to V if and only if every vector in U is orthogonal to every vector in V. Give an example of two orthogonal subspaces of $\mathbf{R}^3$.

27. Let $\mathbf{u}, \mathbf{v}_1, \ldots, \mathbf{v}_m$ be vectors in a given vector space and $a_1, \ldots, a_m$ be scalars. Prove that

$$\mathbf{u} \cdot (a_1\mathbf{v}_1 + a_2\mathbf{v}_2 + \cdots + a_m\mathbf{v}_m) = a_1\mathbf{u} \cdot \mathbf{v}_1 + a_2\mathbf{u} \cdot \mathbf{v}_2 + \cdots + a_m\mathbf{u} \cdot \mathbf{v}_m$$

28. Let $\mathbf{v}_1, \ldots, \mathbf{v}_m$ be a set of m mutually orthogonal vectors. (Any two of these vectors are orthogonal.) Prove that $\mathbf{v}_1, \ldots, \mathbf{v}_m$ are linearly independent. [*Hint*: Use the result of Exercise 27.]

29. (a) If $\mathbf{u}$ and $\mathbf{v}$ are vectors in $\mathbf{R}^n$, then $|\mathbf{u} \cdot \mathbf{v}| \leq \|\mathbf{u}\| \, \|\mathbf{v}\|$. (The bars on the left of this equation mean absolute value.) This identity is called the *Schwarz inequality*. Prove that the Schwarz inequality holds for $\mathbf{R}^2$. [*Hint*: Let $\mathbf{u} = (a, b)$ and $\mathbf{v} = (c, d)$.]

    (b) Use the Schwarz inequality to prove that

$$-1 \leq \frac{1}{\|\mathbf{u}\|} \mathbf{u} \cdot \frac{1}{\|\mathbf{v}\|} \mathbf{v} \leq 1$$

This is a necessary condition for our definition of angle between vectors,

$$\cos \theta = \frac{1}{\|\mathbf{u}\|} \mathbf{u} \cdot \frac{1}{\|\mathbf{v}\|} \mathbf{v}$$

to be a valid one.

The reader may be interested in reading an article "Pythagoras and the Cauchy-Schwarz Inequality," by Ladnor Geissinger, in *The American Mathematical Monthly*, January, 1976. In this paper, the author uses geometric considerations to motivate the concept of dot product, leading to a definition of angle.

30. Write a program for computing the dot product of two vectors. Test your program for the vectors $(1, -1, 2, 3)$ and $(0, 1, 4, 2)$. Their dot product is 13.

31. Write a program for computing norms of vectors. Test your program on the vector $(1, 2, 3, 4)$. Its norm is 5.4772255.

32. Write a program for normalizing a vector. Test your program on the vector $(1, 0, 3, -4, 5)$. The normalized vector is

$$(0.140028, 0, 0.420084, -0.560112, 0.70014)$$

33. Write a program for determining the cosine of the angle between two given vectors. Test your program on the vectors $(1, 2, 3, 0, -1)$ and $(-1, 8, 1, 4, 1)$. The cosine of the angle is .481797173. The angle is 61.1972 degrees.

4-9   PROJECTIONS AND THE GRAM-SCHMIDT ORTHOGONALIZATION PROCESS

Let $\mathbf{u}$ and $\mathbf{v}$ be vectors with angle α between them. The *scalar projection* (or *component*) of $\mathbf{v}$ in the direction of $\mathbf{u}$ is defined to be $\|\mathbf{v}\| \cos \alpha$. The geometrical interpretation of this projection for $\mathbf{R}^2$ and $\mathbf{R}^3$ is given in Figure 4-34 as the length OA.

Note that if $90° < \alpha \leq 180°$, then $\cos \alpha$ is negative and the scalar projection will be negative (Figure 4-35).

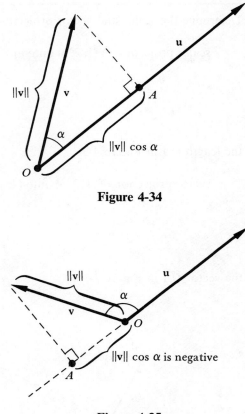

Figure 4-34

Figure 4-35

In certain contexts, we are interested in the *vector projection* of **v** onto **u**. The vector projection is defined to be

$$\|\mathbf{v}\| \cos \alpha \, \frac{\mathbf{u}}{\|\mathbf{u}\|}$$

This is the vector $\overrightarrow{OA}$ in Figures 4-34 and 4-35.

To make computation of scalar and vector projections easier, we rewrite the definitions:

$$\text{scalar projection of } \mathbf{v} \text{ onto } \mathbf{u} = \|\mathbf{v}\| \cos \alpha = \|\mathbf{v}\| \left(\frac{\mathbf{v}}{\|\mathbf{v}\|} \cdot \frac{\mathbf{u}}{\|\mathbf{u}\|} \right) = \mathbf{v} \cdot \frac{\mathbf{u}}{\|\mathbf{u}\|}$$

Note that this is the inner product of **v** with the unit vector in the direction of **u**.

$$\text{vector projection of } \mathbf{v} \text{ onto } \mathbf{u} = \|\mathbf{v}\| \cos \alpha \, \frac{\mathbf{u}}{\|\mathbf{u}\|} = \left(\mathbf{v} \cdot \frac{\mathbf{u}}{\|\mathbf{u}\|} \right) \frac{\mathbf{u}}{\|\mathbf{u}\|}$$

This is the scalar projection times the unit vector in the direction of **u**.

✦ *Example 1* ────────────────────────────────────

Determine the scalar and vector projections of $(1, 2, 3)$ onto $(4, -1, 3)$.

$$\text{scalar projection of } (1, 2, 3) \text{ onto } (4, -1, 3) = (1, 2, 3) \cdot \frac{(4, -1, 3)}{\|(4, -1, 3)\|}$$

$$= \frac{11}{\sqrt{26}}$$

$$= 2.157$$

the length OA in Figure 4-36.

$$\text{vector projection of } (1, 2, 3) \text{ onto } (4, -1, 3) = \left(\frac{11}{\sqrt{26}}\right) \frac{(4, -1, 3)}{\|(4, -1, 3)\|}$$

$$= \frac{11}{26}(4, -1, 3)$$

$$= 0.423(4, -1, 3)$$

the vector $\overrightarrow{OA}$ in Figure 4-36.

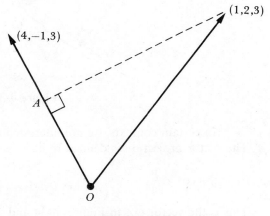

Figure 4-36   ✦

The next example paves the way for a method of constructing an orthogonal set of vectors from a given set.

✦ *Example 2* ────────────────────────────────────

Let **u** and **v** be elements of **R**^{n}. Prove that when one subtracts from **v** its vector projection onto **u**, the resulting vector is orthogonal to **u**.

By subtracting the vector projection, we get the vector

$$\mathbf{v} - \left(\mathbf{v} \cdot \frac{\mathbf{u}}{\|\mathbf{u}\|} \right) \frac{\mathbf{u}}{\|\mathbf{u}\|}$$

We are interested in proving that

$$\left[\mathbf{v} - \left(\mathbf{v} \cdot \frac{\mathbf{u}}{\|\mathbf{u}\|} \right) \frac{\mathbf{u}}{\|\mathbf{u}\|} \right] \cdot \mathbf{u} = 0$$

The left side becomes

$$\mathbf{v} \cdot \mathbf{u} - \left(\mathbf{v} \cdot \frac{\mathbf{u}}{\|\mathbf{u}\|} \right) \frac{\mathbf{u}}{\|\mathbf{u}\|} \cdot \mathbf{u} = \mathbf{v} \cdot \mathbf{u} - (\mathbf{v} \cdot \mathbf{u}) \frac{\mathbf{u}}{\|\mathbf{u}\|} \cdot \frac{\mathbf{u}}{\|\mathbf{u}\|}$$

$$= \mathbf{v} \cdot \mathbf{u} - \mathbf{v} \cdot \mathbf{u} = 0 \quad \blacklozenge$$

We now use the results of this example to construct a set of orthogonal vectors from any given set of linearly independent vectors in $\mathbf{R}^n$. The procedure is called the *Gram-Schmidt orthogonalization procedure*.

Let $\mathbf{v}_1, \ldots, \mathbf{v}_m$ be a set of linearly independent vectors in $\mathbf{R}^n$. We shall construct a set of orthogonal vectors $\mathbf{u}_1, \ldots, \mathbf{u}_m$ from this set.

Let $\mathbf{u}_1 = \mathbf{v}_1$. Next let $\mathbf{u}_2 = \mathbf{v}_2 - [\mathbf{v}_2 \cdot (\mathbf{u}_1/\|\mathbf{u}_1\|)](\mathbf{u}_1/\|\mathbf{u}_1\|)$. We constructed $\mathbf{u}_2$ by subtracting from $\mathbf{v}_2$ the vector projection of $\mathbf{v}_2$ onto $\mathbf{u}_1$. Example 2 tells us that $\mathbf{u}_2$ is orthogonal to $\mathbf{u}_1$.

We continue in the same way.

$$\mathbf{u}_3 = \mathbf{v}_3 - \left(\mathbf{v}_3 \cdot \frac{\mathbf{u}_1}{\|\mathbf{u}_1\|} \right) \frac{\mathbf{u}_1}{\|\mathbf{u}_1\|} - \left(\mathbf{v}_3 \cdot \frac{\mathbf{u}_2}{\|\mathbf{u}_2\|} \right) \frac{\mathbf{u}_2}{\|\mathbf{u}_2\|}$$

$\mathbf{u}_3$ is obtained by subtracting from $\mathbf{v}_3$ the vector projections of $\mathbf{v}_3$ onto both $\mathbf{u}_1$ and $\mathbf{u}_2$. $\mathbf{u}_3$ will be orthogonal to both $\mathbf{u}_1$ and $\mathbf{u}_2$.

The general vector $\mathbf{u}_i$ is

$$\mathbf{u}_i = \mathbf{v}_i - \left(\mathbf{v}_i \cdot \frac{\mathbf{u}_1}{\|\mathbf{u}_1\|} \right) \frac{\mathbf{u}_1}{\|\mathbf{u}_1\|} - \cdots - \left(\mathbf{v}_i \cdot \frac{\mathbf{u}_{i-1}}{\|\mathbf{u}_{i-1}\|} \right) \frac{\mathbf{u}_{i-1}}{\|\mathbf{u}_{i-1}\|}$$

The vectors $\mathbf{u}_1, \ldots, \mathbf{u}_m$ then form an orthogonal set. In order to obtain a set of unit orthogonal vectors, all one has to do is normalize each of these vectors.

This procedure can be used to construct an orthonormal basis from a given basis that defines a subspace. If $\mathbf{v}_1, \ldots, \mathbf{v}_m$ is a given basis for a subspace of $\mathbf{R}^n$, then by using this procedure one can construct an orthonormal basis for this subspace. The following example illustrates the method.

✦ *Example 3*
_____

The set $(3, 0, 0, 0)$, $(0, 1, 2, 1)$, $(0, -1, 3, 2)$ is a set of three linearly independent vectors in $\mathbf{R}^4$. The set defines a three-dimensional subspace of $\mathbf{R}^4$. Using the Gram-Schmidt orthogonalization procedure, construct an orthonormal basis for the space.

Using the notation of the above theory, let $\mathbf{v}_1 = (3, 0, 0, 0)$, $\mathbf{v}_2 = (0, 1, 2, 1)$, and $\mathbf{v}_3 = (0, -1, 3, 2)$. Then

$$\mathbf{u}_1 = \mathbf{v}_1 = (3, 0, 0, 0)$$

$$\mathbf{u}_2 = \mathbf{v}_2 - \left(\mathbf{v}_2 \cdot \frac{\mathbf{u}_1}{\|\mathbf{u}_1\|} \right) \frac{\mathbf{u}_1}{\|\mathbf{u}_1\|}$$

$$= (0, 1, 2, 1) - \left[(0, 1, 2, 1) \cdot \frac{(3, 0, 0, 0)}{3} \right] \frac{(3, 0, 0, 0)}{3}$$

$$= (0, 1, 2, 1)$$

$$\mathbf{u}_3 = \mathbf{v}_3 - \left(\mathbf{v}_3 \cdot \frac{\mathbf{u}_1}{\|\mathbf{u}_1\|} \right) \frac{\mathbf{u}_1}{\|\mathbf{u}_1\|} - \left(\mathbf{v}_3 \cdot \frac{\mathbf{u}_2}{\|\mathbf{u}_2\|} \right) \frac{\mathbf{u}_2}{\|\mathbf{u}_2\|}$$

$$= (0, -1, 3, 2) - \left[(0, -1, 3, 2) \cdot \frac{(3, 0, 0, 0)}{3} \right] \frac{(3, 0, 0, 0)}{3}$$

$$- \left[(0, -1, 3, 2) \cdot \frac{(0, 1, 2, 1)}{\sqrt{1^2 + 2^2 + 1^2}} \right] \frac{(0, 1, 2, 1)}{\sqrt{1^2 + 2^2 + 1^2}}$$

$$= (0, -1, 3, 2) - (0, 0, 0, 0) - \tfrac{7}{6}(0, 1, 2, 1)$$

$$= (0, -\tfrac{13}{6}, \tfrac{2}{3}, \tfrac{5}{6})$$

Thus the vectors $(3, 0, 0, 0)$, $(0, 1, 2, 1)$, and $(0, -\tfrac{13}{6}, \tfrac{2}{3}, \tfrac{5}{6})$ form an orthogonal set. The first two vectors are obviously in the subspace of interest; the third also lies in this subspace because it is a linear combination of the first two and $\mathbf{v}_3$. This set of vectors is therefore an orthogonal basis for the subspace. On normalizing each vector, we get the orthonormal basis $(1, 0, 0, 0)$, $(1/\sqrt{6})(0, 1, 2, 1)$, $(1/\sqrt{210})(0, -13, 4, 5)$. ✦

The following examples illustrate the use of the concept of scalar projection in the analyses of forces. The term *component* is more commonly used in this context.

✦ *Example 4**

The vector $(1, 2, 3)$ represents a force acting on a body. What is the component of the force in the direction of the vector $(4, -1, 3)$?

Its component will be the scalar projection of $(1, 2, 3)$ in the direction $(4, -1, 3)$, or $\|(1, 2, 3)\| \cos \alpha$, which has been calculated to be 2.157 (Example 1). Physically, this will be the effect of the force in this direction. ✦

✦ *Example 5**

This example of stress analysis in frameworks illustrates how a problem in engineering is approached using components of forces. The problem reduces to solving a system of linear equations.

In designing a structure such as a bridge, one must know the forces that will come into play in different parts of the structure when it is subject to various loads. In many cases the structure is composed of bars connected together to form a rigid framework. The weights of the bars are usually insignificant compared to those of the forces they carry, and they are neglected in analyzing the distribution of forces over the bars. The bars are usually pin-jointed at their ends.

Each bar is in equilibrium under the action of forces at its two ends. Hence these two forces acting along the bar are equal and opposite. The bar itself exerts equal and opposite forces at its end joints. They may both be outward or both be inward. The force in the bar is called a *stress*.

Consider the framework in Figure 4-37, which carries a load of 20 tons at its upper joint. Let us determine the stresses in the bars.

Consider the joint *A*. Intuitively, one can see that the forces exerted by the two bars at this joint must be upward in order to balance the 20 tons if the framework is going to remain in equilibrium. By the symmetry of the situation, the forces exerted by the bars are equal. Label the magnitude of these forces *p*. The magnitudes of the forces at *B* and *C* are labeled *p*, *q*, and *r*. The student should be able to see that these bars either push or pull at the joints. To determine *p*, *q*, and *r*, we analyze the situation at various joints. Each joint is in equilibrium under the forces acting on it. This is usually described mathematically by stating that the sum of the components of the forces at a joint in any two directions is 0. (The resultant vector at the joint is then 0.) We choose the directions as convenient perpendicular directions.

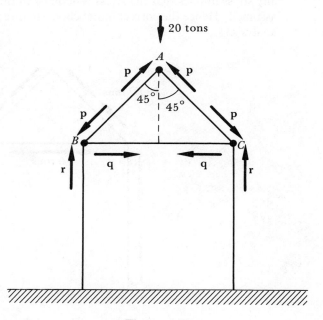

Figure 4-37

Let us examine the forces at joint A. Select directions $\uparrow$ and $\rightarrow$. We have that

$$\uparrow -20 + p \cos 45° + p \cos 45° = 0, \qquad \text{giving } p = 10\sqrt{2}$$
$$\rightarrow p \cos 45° - p \cos 45° = 0, \qquad \text{giving no information}$$

At joint B, using the same pair of directions,

$$\uparrow -p \cos 45° + r = 0, \qquad \text{giving } \frac{-p}{\sqrt{2}} + r = 0$$

$$\rightarrow -p \cos 45° + q = 0, \qquad \text{giving } \frac{-p}{\sqrt{2}} + q = 0$$

The analysis at joint C leads to the same equations as did that at joint B. Thus the problem reduces to solving the system of linear equations

$$p \qquad\qquad = 10\sqrt{2}$$
$$-\frac{p}{\sqrt{2}} + r \qquad = 0$$
$$-\frac{p}{\sqrt{2}} \qquad + q = 0$$

This system has the solution $p = 10\sqrt{2}, r = 10, q = 10$. (A negative number for any stress implies that the stress is actually in the direction opposite to the one assumed. Hence the correct initial choice of directions is not crucial to the final analysis.)

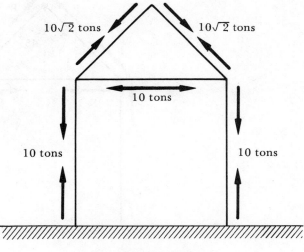

10√2 tons   10√2 tons

10 tons

10 tons   10 tons

Figure 4-38

The actual stresses in the bars themselves are necessarily equal in magnitude and in directions opposite to those of the forces exerted by the bars on the joints. The stresses in the bars are represented in Figure 4-38.

In practice, these structures can involve many bars and the mathematical analysis reduces to solving a system involving many equations in many variables. From the nature of the problem, we can expect the solution to be unique.

✦

EXERCISES

In Exercises 1–6, determine the scalar and vector projections of **u** onto **v**. Illustrate these concepts geometrically when possible.

*1. $\mathbf{u} = (1, 2)$, $\mathbf{v} = (2, 5)$

2. $\mathbf{u} = (-1, 3)$, $\mathbf{v} = (2, 4)$

*3. $\mathbf{u} = (1, 2, 3)$, $\mathbf{v} = (1, 2, 0)$

4. $\mathbf{u} = (2, 1, 4)$, $\mathbf{v} = (-1, -3, 2)$

*5. $\mathbf{u} = (1, -1, 0, 1)$, $\mathbf{v} = (2, 3, -2, 1)$

6. $\mathbf{u} = (2, -1, 3, 1)$, $\mathbf{v} = (-1, 2, 1, 3)$

7. Let $\mathbf{v} = (a_1, \dots, a_n)$ be an arbitrary element of $\mathbf{R}^n$. Prove that the components of **v** onto the vectors $(1, 0, \dots, 0), \dots, (0, \dots, 0, 1)$ are $a_1, \dots, a_n$, respectively. This illustrates the compatibility of the meaning of the term *components* introduced in Section 4-1 (the scalars that make up the vector) and our use of the term in this section.

*8. Determine a vector whose vector projection onto the vector $(1, 0)$ is $(1/\sqrt{2})(1, 0)$. Then give an arbitrary vector whose vector projection onto the vector $(1, 0)$ is $(1/\sqrt{2})(1, 0)$. Illustrate your answer geometrically.

9. Construct an orthonormal basis for the subspace of $\mathbf{R}^3$ defined by the vectors
   (a) $(1, 0, 2)$, $(-1, 0, 1)$
   (b) $(1, -1, 1)$, $(1, 2, -1)$

10. Construct the following bases for the subspaces of $\mathbf{R}^4$ defined by the vectors
   (a) $(1, 2, 3, 4)$, $(-1, 1, 0, 1)$—orthonormal basis
   (b) $(1, 0, -1, 2)$, $(1, 1, 2, 0)$, $(1, 2, -3, -1)$—orthogonal basis

*11. Construct a vector in $\mathbf{R}^4$ orthogonal to the vector $(1, 2, -1, 1)$.

12. Construct a vector in $\mathbf{R}^5$ orthogonal to the vector $(2, 0, 0, 1, 1)$.

13. If $\mathbf{v}_1$, $\mathbf{v}_2$, and $\mathbf{v}_3$ are linearly independent in $\mathbf{R}^3$, $\mathbf{u}_1 = \mathbf{v}_1$, and

$$\mathbf{u}_2 = \mathbf{v}_2 - \frac{\mathbf{v}_2 \cdot \mathbf{u}_1}{\|\mathbf{u}_1\|} \frac{\mathbf{u}_1}{\|\mathbf{u}_1\|}$$

prove that the vector $\mathbf{u}_3$ defined by

$$\mathbf{u}_3 = \mathbf{v}_3 - \frac{\mathbf{v}_3 \cdot \mathbf{u}_1}{\|\mathbf{u}_1\|} \frac{\mathbf{u}_1}{\|\mathbf{u}_1\|} - \frac{\mathbf{v}_3 \cdot \mathbf{u}_2}{\|\mathbf{u}_2\|} \frac{\mathbf{u}_2}{\|\mathbf{u}_2\|}$$

is orthogonal to both $\mathbf{u}_1$ and $\mathbf{u}_2$.

14. Discuss the significance of the fact that $\mathbf{u}_2$ is actually equal to $\mathbf{v}_2$ in Example 3.

*15. (1, 2) represents a force acting on a body. Find the component of the force
    (a) in the direction (1, 1)
    (b) in the **i** direction
    (c) in the **j** direction

  16. $(1, -1, 3)$ represents a force acting on a body. Find the component of the force
    (a) in the direction (1, 1, 1)
    (b) in the direction $(-1, 1, -1)$
    (c) in each of the directions **i, j,** and **k**

*17. Determine the stresses in the bars of the structures in Figure 4-39.

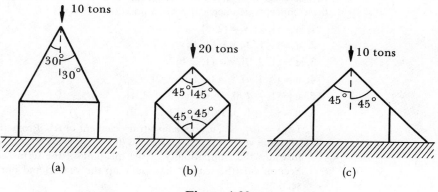

(a)   (b)   (c)

Figure 4-39

18. (a) Determine the tensions T_1 and T_2 in the strings in Figure 4-40.

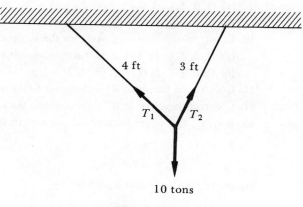

Figure 4-40

  (b) Figure 4-41 is a lamp structure attached to a wall. Determine the stresses in the bar *AB* and in the wire *BC* as functions of *w* and *θ*. Note that the tension in *BC* is greater than *w*! As *θ* decreases, this tension increases. This type of device can be used to break wires.

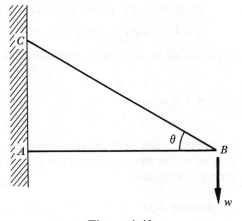

Figure 4-41

19. Write a computer program to determine scalar and vector projections. Test your program with the vectors of Example 1. The scalar and vector projections of (1, 2, 3) onto (4, −1, 3) are 2.1572776 and (1.6923076, −0.4230769, 1.2692307), respectively.

20. Write a computer program that uses the Gram-Schmidt procedure to construct an orthonormal set of vectors from two given vectors. Test your program for the vectors (1, 2, −3, 4) and (1, 2, 4, −3). An orthonormal set, constructed by normalizing the vector (1, 2, −3, 4) to get the first unit vector, is

$$(0.1825742, 0.3651484, -0.547726, 0.7302967),$$

$$(0.3853373, 0.7706746, 0.4954337, -0.1100964)$$

4-10   VECTOR PRODUCTS

In this section, we introduce various operations on vectors. These are indispensable tools in many areas of science and engineering; we shall discuss applications in some of the areas.

All vectors are considered to be in three-space. Their representations are thus elements of $\mathbf{R}^3$.

Let $\mathbf{a}$ and $\mathbf{b}$ be the vectors (a_1, a_2, a_3) and (b_1, b_2, b_3), respectively. The vector product of $\mathbf{a}$ and $\mathbf{b}$, denoted by $\mathbf{a} \times \mathbf{b}$, is the vector

$$\mathbf{a} \times \mathbf{b} = (a_2 b_3 - a_3 b_2, a_3 b_1 - a_1 b_3, a_1 b_2 - a_2 b_1)$$

✦ *Example 1* ─────────────────────────────────

Let $\mathbf{a} = (1, -2, 3)$ and $\mathbf{b} = (0, 1, 3)$. Determine $\mathbf{a} \times \mathbf{b}$.

$$\mathbf{a} \times \mathbf{b} = [(-2) \times 3 - 3 \times 1, 3 \times 0 - 1 \times 3, 1 \times 1 - (-2) \times 0]$$
$$= (-9, -3, 1) \quad ✦$$

We shall use the notation $\mathbf{i} = (1, 0, 0)$, $\mathbf{j} = (0, 1, 0)$, $\mathbf{k} = (0, 0, 1)$ in this section for the vectors of the standard basis of $\mathbf{R}^3$.

If we write the vector product $\mathbf{a} \times \mathbf{b}$ in the form

$$\mathbf{a} \times \mathbf{b} = (a_2b_3 - a_3b_2)\mathbf{i} + (a_3b_1 - a_1b_3)\mathbf{j} + (a_1b_2 - a_2b_1)\mathbf{k}$$

it can be easily verified (by expanding the determinant) that

$$\mathbf{a} \times \mathbf{b} = \begin{vmatrix} \mathbf{i} & \mathbf{j} & \mathbf{k} \\ a_1 & a_2 & a_3 \\ b_1 & b_2 & b_3 \end{vmatrix}$$

This is not an ordinary determinant, the elements of the first row being vectors rather than scalars. However, it has the algebraic properties of an ordinary determinant.

Theorem 4-21 $\mathbf{a} \times \mathbf{a} = \mathbf{0}$. *Here* **0** *is the zero vector.*

Proof: $\mathbf{a} \times \mathbf{a} = \begin{vmatrix} \mathbf{i} & \mathbf{j} & \mathbf{k} \\ a_1 & a_2 & a_3 \\ a_1 & a_2 & a_3 \end{vmatrix} \underset{R2-R3}{=} \begin{vmatrix} \mathbf{i} & \mathbf{j} & \mathbf{k} \\ 0 & 0 & 0 \\ a_1 & a_2 & a_3 \end{vmatrix}$

$$= 0\mathbf{i} + 0\mathbf{j} + 0\mathbf{k} = \mathbf{0}$$

Thus, in particular, $\mathbf{i} \times \mathbf{i} = \mathbf{j} \times \mathbf{j} = \mathbf{k} \times \mathbf{k} = \mathbf{0}$.

Theorem 4-22 $\mathbf{a} \times \mathbf{b} = -\mathbf{b} \times \mathbf{a}$. *The vector* $\mathbf{b} \times \mathbf{a}$ *is parallel to* $\mathbf{a} \times \mathbf{b}$ *and of the same magnitude as* $\mathbf{a} \times \mathbf{b}$, *but in a direction opposite to that of* $\mathbf{a} \times \mathbf{b}$.

Proof: $\mathbf{a} \times \mathbf{b} = \begin{vmatrix} \mathbf{i} & \mathbf{j} & \mathbf{k} \\ a_1 & a_2 & a_3 \\ b_1 & b_2 & b_3 \end{vmatrix} \underset{R2 \leftrightarrow R3}{=} -\begin{vmatrix} \mathbf{i} & \mathbf{j} & \mathbf{k} \\ b_1 & b_2 & b_3 \\ a_1 & a_2 & a_3 \end{vmatrix}$

$$= -\mathbf{b} \times \mathbf{a}$$

Theorem 4-23 $\mathbf{a} \cdot (\mathbf{a} \times \mathbf{b}) = \mathbf{b} \cdot (\mathbf{a} \times \mathbf{b}) = 0$. *The vector* $\mathbf{a} \times \mathbf{b}$ *is orthogonal to both* **a** *and* **b**. *Thus it is orthogonal to the vector space spanned by* **a** *and* **b**; *that is, it is orthogonal to the plane containing* **a** *and* **b** (*Figure* 4-42).

Proof: $\mathbf{a} \cdot (\mathbf{a} \times \mathbf{b}) = (a_1\mathbf{i} + a_2\mathbf{j} + a_3\mathbf{k}) \cdot \begin{vmatrix} \mathbf{i} & \mathbf{j} & \mathbf{k} \\ a_1 & a_2 & a_3 \\ b_1 & b_2 & b_3 \end{vmatrix}$

$$= \begin{vmatrix} a_1 & a_2 & a_3 \\ a_1 & a_2 & a_3 \\ b_1 & b_2 & b_3 \end{vmatrix} = 0$$

It can be proved similarly that $\mathbf{b} \cdot (\mathbf{a} \times \mathbf{b}) = 0$.

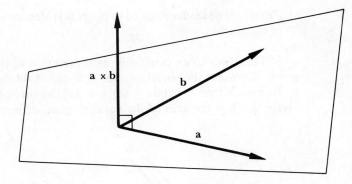

Figure 4-42

Theorem 4-24 $\mathbf{i} \times \mathbf{j} = \mathbf{k}, \mathbf{j} \times \mathbf{k} = \mathbf{i}, \mathbf{k} \times \mathbf{i} = \mathbf{j}$ (*Figure* 4-43).

***Proof*:** $\mathbf{i} \times \mathbf{j} = \begin{vmatrix} \mathbf{i} & \mathbf{j} & \mathbf{k} \\ 1 & 0 & 0 \\ 0 & 1 & 0 \end{vmatrix} = 0\mathbf{i} + 0\mathbf{j} + \mathbf{k} = \mathbf{k}$

The proofs of the other identities are similar.

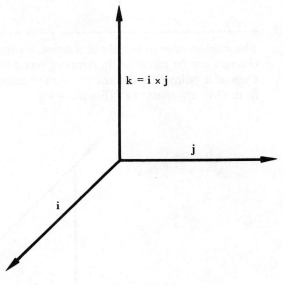

Figure 4-43

Theorem 4-25 $\|\mathbf{a} \times \mathbf{b}\| = \|\mathbf{a}\| \, \|\mathbf{b}\| \sin \theta$, *where* θ *is the angle between* **a** *and* **b**.

Proof :   The reader is asked to prove this identity in the exercises at the end of the section.

This leads to an interesting interpretation of the magnitude of the vector $\mathbf{a} \times \mathbf{b}$. Consider the parallelogram in Figure 4-44 defined by the vectors $\mathbf{a}$ and $\mathbf{b}$. Its area is base × height = $\|\mathbf{a}\| \, h = \|\mathbf{a}\| \, \|\mathbf{b}\| \sin \theta$. Thus the magnitude of the vector $\mathbf{a} \times \mathbf{b}$ is the area of the parallelogram defined by the vectors $\mathbf{a}$ and $\mathbf{b}$.

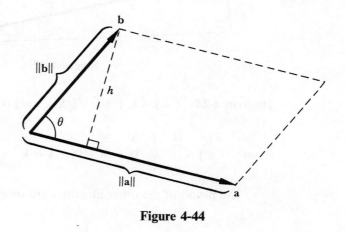

Figure 4-44

✦ ***Example 2***

The *moment* of a vector about a point, a concept that occurs frequently in mechanics, can be discussed in terms of vector products. The moment of a vector $\mathbf{f}$ about a point O is defined to be the vector $\mathbf{m} = \mathbf{r} \times \mathbf{f}$, where $\mathbf{r}$ is the vector from O to any point of $\mathbf{f}$ (Figure 4-45).

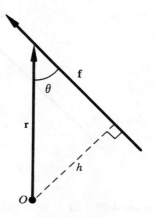

Figure 4-45

(In Exercise 7 at the end of this section the reader is asked to prove that **m** is independent of the choice of **r**.) Thus **m** is a vector in a direction perpendicular to the plane containing **r** and **f**. Its magnitude is

$$\|\mathbf{m}\| = \|\mathbf{r} \times \mathbf{f}\| = \|\mathbf{r}\| \, \|\mathbf{f}\| \sin \theta = \|\mathbf{f}\| \, h$$

where h is the perpendicular distance from O to **f**. If **f** is a force, then $\mathbf{r} \times \mathbf{f}$ is called the moment of the force about O. It represents the turning effect of the force about O. ✦

The product $\mathbf{a} \cdot (\mathbf{b} \times \mathbf{c})$ is called the *triple scalar product* of **a**, **b**, and **c**. The parentheses are not actually needed, as the vector operations can only be performed in this order.

Theorem 4-26 $\mathbf{a} \cdot \mathbf{b} \times \mathbf{c} = \mathbf{a} \times \mathbf{b} \cdot \mathbf{c}.$

Proof:

$$\mathbf{a} \cdot \mathbf{b} \times \mathbf{c} = (a_1 \mathbf{i} + a_2 \mathbf{j} + a_3 \mathbf{k}) \cdot \begin{vmatrix} \mathbf{i} & \mathbf{j} & \mathbf{k} \\ b_1 & b_2 & b_3 \\ c_1 & c_2 & c_3 \end{vmatrix} = \begin{vmatrix} a_1 & a_2 & a_3 \\ b_1 & b_2 & b_3 \\ c_1 & c_2 & c_3 \end{vmatrix}$$

and

$$\mathbf{a} \times \mathbf{b} \cdot \mathbf{c} = \begin{vmatrix} \mathbf{i} & \mathbf{j} & \mathbf{k} \\ a_1 & a_2 & a_3 \\ b_1 & b_2 & b_3 \end{vmatrix} \cdot (c_1 \mathbf{i} + c_2 \mathbf{j} + c_3 \mathbf{k})$$

$$= \begin{vmatrix} c_1 & c_2 & c_3 \\ a_1 & a_2 & a_3 \\ b_1 & b_2 & b_3 \end{vmatrix} = \begin{vmatrix} a_1 & a_2 & a_3 \\ b_1 & b_2 & b_3 \\ c_1 & c_2 & c_3 \end{vmatrix}$$

There is an interesting geometrical interpretation of the triple scalar product that leads to an interpretation of 3×3 determinants. Consider the parallelepiped in Figure 4-46 defined by **a**, **b**, and **c**. Its base area is $\|\mathbf{a} \times \mathbf{b}\|$. Its volume is $\|\mathbf{a} \times \mathbf{b}\| \, h$, where h is the height.

Consider $\left| \dfrac{\mathbf{a} \times \mathbf{b}}{\|\mathbf{a} \times \mathbf{b}\|} \cdot \mathbf{c} \right|$, the absolute value of the scalar projection of **c** onto the vector $\mathbf{a} \times \mathbf{b}$. We can see that $\left| \dfrac{\mathbf{a} \times \mathbf{b}}{\|\mathbf{a} \times \mathbf{b}\|} \cdot \mathbf{c} \right| = h.$

Thus $|\mathbf{a} \times \mathbf{b} \cdot \mathbf{c}| = \|\mathbf{a} \times \mathbf{b}\| h$; $|\mathbf{a} \times \mathbf{b} \cdot \mathbf{c}|$ is the volume of the parallelepiped defined by **a**, **b**, and **c**. Furthermore, since

$$|\mathbf{a} \times \mathbf{b} \cdot \mathbf{c}| = \left\| \begin{matrix} a_1 & a_2 & a_3 \\ b_1 & b_2 & b_3 \\ c_1 & c_2 & c_3 \end{matrix} \right\| \quad \text{(see above proof)}$$

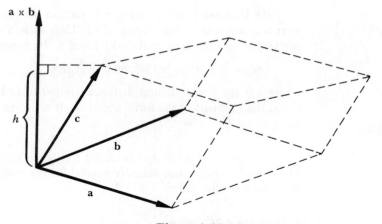

Figure 4-46

then $\begin{Vmatrix} a_1 & a_2 & a_3 \\ b_1 & b_2 & b_3 \\ c_1 & c_2 & c_3 \end{Vmatrix}$ = volume of parallelepiped with edges (a_1, a_2, a_3), (b_1, b_2, b_3), and (c_1, c_2, c_3).

EXERCISES

1. If $\mathbf{a} = (1, 2, 3)$, $\mathbf{b} = (-1, 0, 4)$, and $\mathbf{c} = (1, 2, -1)$, determine

   *(a) $\mathbf{a} \times \mathbf{b}$ (b) $\mathbf{a} \times \mathbf{c}$ *(c) $(\mathbf{a} \times \mathbf{b}) \times \mathbf{c}$

   (d) $\mathbf{a} \times (\mathbf{b} \times \mathbf{c})$ *(e) $\mathbf{a} \times \mathbf{b} \cdot \mathbf{c}$

   Comment on your results for (c) and (d). $\mathbf{a} \times \mathbf{b} \times \mathbf{c}$ is called the *triple vector product* of $\mathbf{a}$, $\mathbf{b}$, and $\mathbf{c}$.

2. If $\mathbf{a}$, $\mathbf{b}$, and $\mathbf{c}$ are arbitrary vectors in three-space, prove that

   $$\mathbf{a} \times (\mathbf{b} + \mathbf{c}) = (\mathbf{a} \times \mathbf{b}) + (\mathbf{a} \times \mathbf{c})$$

3. If $\mathbf{a}$ and $\mathbf{b}$ are arbitrary vectors in three-space and d is a scalar, prove that

   $$(d\mathbf{a}) \times \mathbf{b} = \mathbf{a} \times (d\mathbf{b}) = d(\mathbf{a} \times \mathbf{b})$$

4. If $\mathbf{a}$ and $\mathbf{b}$ are two vectors in three-space and θ is the angle between them, prove that $\|\mathbf{a} \times \mathbf{b}\| = \|\mathbf{a}\| \, \|\mathbf{b}\| \sin \theta$. [*Hint*: Let $\mathbf{a} = (a_1, a_2, a_3)$ and $\mathbf{b} = (b_1, b_2, b_3)$. Determine expressions for both sides of the identity. Use $\sin^2 \theta = 1 - \cos^2 \theta$ to find an expression for $\sin \theta$.]

*5. Determine the area of the triangle with vertices $(1, 2, 3)$, $(-1, 1, 1)$, and $(1, 0, 4)$.

*6. Find the volume of the parallelepiped defined by the vectors $(1, 2, 3)$, $(-1, 0, 4)$, and $(1, 2, 5)$.

7. In Example 2, the moment of a vector $\mathbf{f}$ about a point O was defined to be $\mathbf{m} = \mathbf{r} \times \mathbf{f}$, where $\mathbf{r}$ is a vector from O to any point of $\mathbf{f}$. If this is a valid definition, we should get the same vector $\mathbf{m}$ for all such vectors $\mathbf{r}$; that is, if $\bar{\mathbf{r}}$ is another such vector, then

we would expect $\mathbf{m} = \mathbf{r} \times \mathbf{f} = \bar{\mathbf{r}} \times \mathbf{f}$. Prove that this is so. [*Hint:* $(\mathbf{r} \times \mathbf{f}) - (\bar{\mathbf{r}} \times \mathbf{f}) = (\mathbf{r} - \bar{\mathbf{r}}) \times \mathbf{f}$ by Exercise 2 above.]

*8. Determine the moment of the vector

   (a) $(1, -1, 4)$ passing through the point $(2, 4, 4)$ about the point $(1, -2, 3)$

   (b) $(1, 2, 3)$ passing through the point $(0, 1, 2)$ about the point $(1, 2, -1)$

9. Write computer programs that can be used to determine

   (a) the vector product of two given vectors

   (b) the triple scalar product of three given vectors

   (c) the triple vector product of three given vectors

Check your answers to Exercise 1 with your programs.

4-11*   EQUATIONS OF PLANES AND LINES IN THREE-SPACE

The tools that we have developed in this chapter are useful for describing planes and lines in three-space. Let us first of all look at the equation of a plane.

Let $P_0(x_0, y_0, z_0)$ be a point in a plane, and let (a, b, c) be a *normal* to the plane. The vector (a, b, c) is perpendicular to the plane. (See Figure 4-47.)

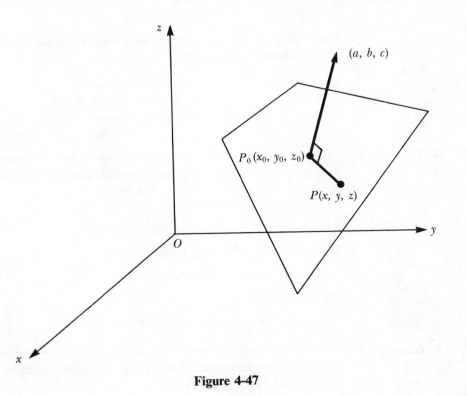

Figure 4-47

These two quantities—namely, a point and a normal vector—characterize the plane; there is only one plane through a given point having a given normal. We shall now derive the equation of the plane from this given information.

Let $P(x, y, z)$ be an arbitrary point in the plane. The vector $\overline{P_0 P}$ lies in the plane. Thus the vectors (a, b, c) and $\overline{P_0 P}$ are orthogonal. Their dot product will be zero. This observation, as we shall now see, leads to an equation for the plane.

$$\overline{P_0 P} = (x - x_0, y - y_0, z - z_0)$$
$$(a, b, c) \cdot \overline{P_0 P} = 0$$
$$(a, b, c) \cdot (x - x_0, y - y_0, z - z_0) = 0$$

$$\boxed{a(x - x_0) + b(y - y_0) + c(z - z_0) = 0}$$

This is called a *point-normal* form of the equation of the plane.

Let us now rearrange this equation. Multiplying out, we get

$$ax - ax_0 + by - by_0 + cz - cz_0 = 0$$
$$ax + by + cz - ax_0 - by_0 - cz_0 = 0$$

Combining the last three terms into a single constant d leads to a *general form* of the equation of the plane

$$\boxed{ax + by + cz + d = 0}$$

✦ **Example 1** ───────────────────────────────

Find the point-normal and general forms of the equation of the plane through the point $(1, 2, 3)$ having normal $(-1, 4, 6)$.

The point-normal form is

$$-1(x - 1) + 4(y - 2) + 6(z - 3) = 0$$

Multiplying out and simplifying, we have

$$-x + 1 + 4y - 8 + 6z - 18 = 0$$

The general form is

$$-x + 4y + 6z - 25 = 0 \qquad ✦$$

✦ **Example 2** ───────────────────────────────

Determine the equation of the plane through the three points $P_1(1, 2, 3)$, $P_2(-1, 4, 6)$, and $P_3(2, 0, 3)$.

The vectors $\overrightarrow{P_1P_2}$ and $\overrightarrow{P_1P_3}$ will lie in the plane. Thus $\overrightarrow{P_1P_2} \times \overrightarrow{P_1P_3}$ will be a normal to the plane. We get

$$\overrightarrow{P_1P_2} = (-2, 2, 3) \qquad \text{and} \qquad \overrightarrow{P_1P_3} = (1, -2, 0)$$

$$\overrightarrow{P_1P_2} \times \overrightarrow{P_1P_3} = \begin{vmatrix} \mathbf{i} & \mathbf{j} & \mathbf{k} \\ -2 & 2 & 3 \\ 1 & -2 & 0 \end{vmatrix} = (0 + 6)\mathbf{i} - (0 - 3)\mathbf{j} + (4 - 2)\mathbf{k}$$

$$= (6, 3, 2)$$

Let $(x_0, y_0, z_0) = (1, 2, 3)$ and $(a, b, c) = (6, 3, 2)$. The point-normal form of the plane is

$$6(x - 1) + 3(y - 2) + 2(z - 3) = 0$$

Simplifying this identity gives the general form

$$6x + 3y + 2z - 18 = 0$$

Observe that each of the given points $(1, 2, 3)$, $(-1, 4, 6)$, and $(2, 0, 3)$ does satisfy this equation. ◆

◆ *Example 3*

Prove that the following system of linear equations has no solution:

$$2x - y + 3z = 6$$
$$-4x + 2y - 6z = 2$$
$$x + y + 4z = 5$$

Interpret each equation as defining a plane in three-space. Comparing each equation to the general form, we see that the normals to the planes are given by the vectors $(2, -1, 3)$, $(-4, 2, -6)$, and $(1, 1, 4)$. Observe that $(-4, 2, -6) = -2(2, -1, 3)$. The normals to the first two planes are parallel; the first two planes are thus parallel. The three planes can have no points in common; the system of equations has no solution. ◆

We now turn to discuss lines. Consider a line through the point $P_0(x_0, y_0, z_0)$ in the direction defined by the vector (a, b, c). See Figure 4-48. Let $P(x, y, z)$ be any other point on the line. We get that

$$\overrightarrow{P_0P} = (x - x_0, y - y_0, z - z_0)$$

The vectors $\overrightarrow{P_0P}$ and (a, b, c) are parallel. Thus there is a scalar t such that

$$\overrightarrow{P_0P} = t(a, b, c)$$

Comparing the two expressions for $\overrightarrow{P_0P}$, we have

$$(x - x_0, y - y_0, z - z_0) = t(a, b, c)$$

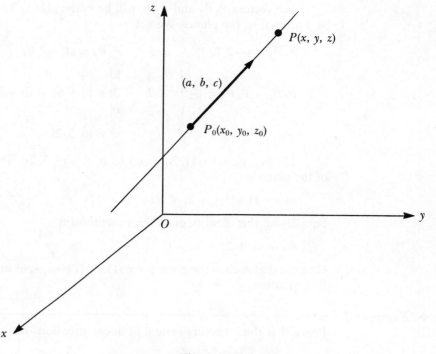

Figure 4-48

Equating the components of the vectors leads to

$$x - x_0 = ta, \qquad y - y_0 = tb, \qquad z - z_0 = tc$$

Rearranging these equations, we get the *parametric equations* of a line in three-space

$$x = x_0 + ta$$
$$y = y_0 + tb$$
$$z = z_0 + tc$$

where $-\infty < t < \infty$.

As t varies from $-\infty$ to ∞, we get all the points on the line.

✦ Example 4

Find the parametric equations of the line through the point $(1, 2, 5)$ in the direction $(4, 3, 1)$. Determine any two points on the line.

The parametric equations are

$$x = 1 + 4t$$
$$y = 2 + 3t$$
$$z = 5 + t$$

where $-\infty < t < \infty$. By letting t take on any two convenient values, we can determine two other points on the line. For example, $t = 1$ leads to the point $(5, 5, 6)$, while $t = -2$ leads to the point $(-7, -4, 3)$. ✦

✦ **Example 5**

Find the parametric equations of the line through the points $A(-1, 2, 6)$ and $B(1, 5, 4)$.

Let P_0 be the point $A(-1, 2, 6)$, and let $(a, b, c) = \overrightarrow{AB}$. Thus $(a, b, c) = (2, 3, -2)$. The parametric equations of the line are

$$x = -1 + 2t$$
$$y = \quad 2 + 3t$$
$$z = \quad 6 - 2t$$

where $-\infty < t < \infty$. ✦

The parametric equations of a line for which a, b, and c are all nonzero lead to another useful way of expressing a line, called the symmetric equations of the line. Isolating the variable t in each of the three parametric equations, we have

$$t = \frac{x - x_0}{a}, \qquad t = \frac{y - y_0}{b}, \qquad t = \frac{z - z_0}{c}$$

Equating these expressions for t, we get the *symmetric equations* of the line

$$\boxed{\frac{x - x_0}{a} = \frac{y - y_0}{b} = \frac{z - z_0}{c}}$$

In this form the line can be interpreted as the intersection of the planes

$$\frac{x - x_0}{a} = \frac{y - y_0}{b} \qquad \text{and} \qquad \frac{y - y_0}{b} = \frac{z - z_0}{c}$$

✦ **Example 6**

Determine the symmetric equations of the line through the points $A(1, 4, 3)$ and $B(-4, 2, -6)$. Let (x_0, y_0, z_0) be the point $(1, 4, 3)$ and (a, b, c) be the vector $\overrightarrow{AB}$. Thus $(a, b, c) = (-5, -2, -9)$.

The symmetric equations of the line are

$$\frac{x - 1}{-5} = \frac{y - 4}{-2} = \frac{z - 3}{-9}$$

Write the equations in the form

$$\frac{x-1}{-5}=\frac{y-4}{-2} \quad \text{and} \quad \frac{y-4}{-2}=\frac{z-3}{-9}$$

Cross multiply and simplify to get

$$-2(x-1)=-5(y-4) \quad \text{and} \quad -9(y-4)=-2(z-3)$$
$$-2x+2=-5y+20 \quad \text{and} \quad -9y+36=-2z+6$$
$$-2x+5y-18=0 \quad \text{and} \quad -9y+2z+30=0$$

The line through A and B is thus the intersection of the planes

$$-2x+5y-18=0 \quad \text{and} \quad -9y+2z+30=0 \quad \blacklozenge$$

EXERCISES

1. Determine a point-normal form and general form of the equation of the plane through each of the following points, having the given normal.
   *(a) point $(1, -2, 4)$; normal $(1, 1, 1)$
   (b) point $(-3, 5, 6)$; normal $(-2, 4, 5)$
   *(c) point $(0, 0, 0)$; normal $(1, 2, 3)$
   (d) point $(4, 5, -2)$; normal $(-1, -4, 3)$
2. Determine a general form of the equation of the plane through each of the following sets of three points.
   *(a) $P_1(1, -2, 3)$, $P_2(1, 0, 2)$, $P_3(-1, 4, 6)$
   (b) $P_1(0, 0, 0)$, $P_2(1, 2, 4)$, $P_3(-3, 5, 1)$
   *(c) $P_1(-1, -1, 2)$, $P_2(3, 5, 4)$, $P_3(1, 2, 5)$
   (d) $P_1(7, 1, 3)$, $P_2(-2, 4, -3)$, $P_3(5, 4, 1)$
*3. Prove that the planes $3x - 2y + 4z - 3 = 0$ and $-6x + 4y - 8z + 7 = 0$ are parallel.
4. Prove that the following system of linear equations has no solution:

$$x + y - 3z = 7$$
$$3x - 6y + 9z = 6$$
$$-x + 2y - 3z = 2$$

*5. Determine the equation of the plane parallel to the plane $2x - 3y + z + 4 = 0$, passing through the point $(1, 2, -3)$.
6. Determine parametric and symmetric equations of lines through the following points, in the given directions.
   *(a) point $(1, 2, 3)$; direction $(-1, 2, 4)$
   (b) point $(-3, 1, 2)$; direction $(1, 1, 1)$
   *(c) point $(0, 0, 0)$; direction $(-2, -3, 5)$
   (d) point $(-2, -4, 1)$; direction $(2, -2, 4)$

*7. Find the equation of the line through the point $(1, 2, -4)$ parallel to the line $x = 4 + 2t$, $y = -1 + 3t$, $z + 2 + t$.

8. Find the equation of a line through the point $(2, -3, 1)$ in a direction orthogonal to the line $\dfrac{x+1}{3} = \dfrac{y-1}{2} = \dfrac{z+2}{5}$.

*9. Determine the equation of the line through the point $(4, -1, 3)$ perpendicular to the plane $2x - y + 4z + 7 = 0$.

10. Give general forms for the equations of the xy plane, xz plane, and yz plane.

*11. Show that there are many planes that contain the three points $P_1(3, -5, 5)$, $P_2(-1, 1, 3)$, and $P_3(5, -8, 6)$. Interpret your conclusion geometrically.

12. Find an equation for the plane through the point $(4, -1, 5)$ in a direction perpendicular to the line $x = 1 - t$, $y = 3 + 2t$, $z = 5 - 4t$, where $-\infty < t < \infty$.

*13. Show that the line $x = 1 + t$, $y = 14 - t$, $z = 2 - t$, where $-\infty < t < \infty$, lies in the plane $2x - y + 3z + 6 = 0$.

14. Prove that the line $x = 4 + 2t$, $y + 5 + t$, $z = 7 + 2t$, where $-\infty < t < \infty$, never intersects the plane $3x + 2y - 4z + 7 = 0$.

*15. Find the equation of a line through the point $(5, -1, 2)$ in a direction perpendicular to the line $x = 5 - 2t$, $y = 2 + 3t$, $z = 2t$.

16. Prove that the lines $x = -1 - 4t$, $y = 4 + 4t$, $z = 7 + 8t$, where $-\infty < t < \infty$, and $x = 4 + 3t'$, $y = 1 - t'$, $z = 5 + 2t'$, where $-\infty < t' < \infty$, intersect at right angles. Find the point of intersection.

17. Two lines are said to be *skew* if they are not parallel and do not intersect. Show that the following two lines are skew: $x = 1 + 4t$, $y = 2 + 5t$, $z = 3 + 3t$, where $-\infty < t < \infty$, and $x = 2 + t'$, $y = 1 - 3t'$, $z = -1 - 2t'$, where $-\infty < t' < \infty$.

18. (a) Show that the set of all points on the plane $2x + 3y - 4z = 0$ forms a subspace of $\mathbf{R}^3$. Find a basis for this subspace.

 (b) Show that the set of all points on the plane $3x - 4y + 2z - 6 = 0$ does not form a subspace of $\mathbf{R}^3$.

 (c) Prove that the set of all points on the plane $ax + by + cz + d = 0$ forms a subspace of $\mathbf{R}^3$ if and only if $d = 0$.

5*

General Vector Spaces

5-1 GENERAL VECTOR SPACES

The vector space $\mathbf{R}^n$ is a set of elements called vectors, on which two operations—addition and scalar multiplication—are defined. As we have seen, the space $\mathbf{R}^n$ is closed under these operations; the sum of two vectors in $\mathbf{R}^n$ lies in $\mathbf{R}^n$, and the scalar multiple of a vector in $\mathbf{R}^n$ lies in $\mathbf{R}^n$. The vector space also has further algebraic properties. For example, we saw that vectors are commutative and associative under addition: $\mathbf{u} + \mathbf{v} = \mathbf{v} + \mathbf{u}$ and $\mathbf{u} + (\mathbf{v} + \mathbf{w}) = (\mathbf{u} + \mathbf{v}) + \mathbf{w}$. Our aim in this section will be to focus on these and additional algebraic properties of $\mathbf{R}^n$. We shall draw up a set of *axioms* based on the properties of $\mathbf{R}^n$. Any set that satisfies these axioms will have algebraic properties similar to those of the vector space $\mathbf{R}^n$. Such a set will be called a *vector space*, and its elements will be called *vectors*. The power of this approach lies in that the concepts and theorems developed for the vector space $\mathbf{R}^n$ will then apply to all vector spaces.

We now give the formal definition of a vector space. Observe that the axioms can be broken down into three convenient groups.

Definition 5-1 A set V is a *vector space* if it has defined on it operations of *addition* and *scalar multiplication* that satisfy the following conditions.

Closure Conditions

1. If $\mathbf{u}$ and $\mathbf{v}$ are elements of V, then their sum $\mathbf{u} + \mathbf{v}$ exists and is an element of V (closed under addition).
2. If c is a scalar and $\mathbf{u}$ is an element of V, then the scalar multiple $c\mathbf{u}$ exists and is an element of V (closed under scalar multiplication).

233

Addition Conditions

3. $\mathbf{u} + \mathbf{v} = \mathbf{v} + \mathbf{u}$ for all $\mathbf{u}$ and $\mathbf{v}$ in V (commutative under addition).
4. $\mathbf{u} + (\mathbf{v} + \mathbf{w}) = (\mathbf{u} + \mathbf{v}) + \mathbf{w}$ for all $\mathbf{u}$, $\mathbf{v}$, and $\mathbf{w}$ in V (associative under addition).
5. There is an element of V called the *zero vector*, denoted $\mathbf{0}$, such that $\mathbf{u} + \mathbf{0} = \mathbf{u}$ for all $\mathbf{u}$ in V.
6. For every element $\mathbf{u}$ of V there exists an element denoted $-\mathbf{u}$, called the *negative of* $\mathbf{u}$, such that $\mathbf{u} + (-\mathbf{u}) = \mathbf{0}$.

Scalar Multiplication Conditions

7. $c(\mathbf{u} + \mathbf{v}) = c\mathbf{u} + c\mathbf{v}$ for all scalars c and $\mathbf{u}$, $\mathbf{v}$ in V.
8. $(c + d)\mathbf{u} = c\mathbf{u} + d\mathbf{u}$ for all scalars c, d and $\mathbf{u}$ in V.
9. $(cd)\mathbf{u} = c(d\mathbf{u})$ for all scalars c, d and $\mathbf{u}$ in V.
10. $1\mathbf{u} = \mathbf{u}$ for all $\mathbf{u}$ in V.

The elements of a vector space are called *vectors*.

As was pointed out earlier, the set of scalars can be the set of real numbers or the set of complex numbers. We shall restrict our discussion to real numbers. The vector spaces are then called *real vector spaces*.

The above axioms of a vector space are based on the properties of $\mathbf{R}^n$. We have already seen in Sections 4-1 and 4-2 that $\mathbf{R}^n$ satisfies the first four axioms. Let us now demonstrate that $\mathbf{R}^n$ satisfies axioms 5–7. We leave the remaining axioms for the reader to check (Exercise 1).

Let $\mathbf{u} = (x_1, \ldots, x_n)$, $\mathbf{v} = (y_1, \ldots, y_n)$, and $\mathbf{w} = (z_1, \ldots, z_n)$ be elements of $\mathbf{R}^n$ and let c be a scalar.

Axiom 5: The zero vector of $\mathbf{R}^n$ is $(0, \ldots, 0)$, since

$$\mathbf{u} + (0, \ldots, 0) = (x_1, \ldots, x_n) + (0, \ldots, 0) = (x_1, \ldots, x_n) = \mathbf{u}$$

Axiom 6: Consider the vector $(-x_1, \ldots, -x_n)$. We have that

$$
\begin{aligned}
\mathbf{u} + (-x_1, \ldots, -x_n) &= (x_1, \ldots, x_n) + (-x_1, \ldots, -x_n) \\
&= (x_1 - x_1, \ldots, x_n - x_n) \\
&= (0, \ldots, 0) \\
&= \mathbf{0}
\end{aligned}
$$

Thus $-\mathbf{u} = (-x_1, \ldots, -x_n)$; the negative of $\mathbf{u}$ exists.

Axiom 7:

$$
\begin{aligned}
c(\mathbf{u} + \mathbf{v}) &= c\big[(x_1, \ldots, x_n) + (y_1, \ldots, y_n)\big] \\
&= c(x_1 + y_1, \ldots, x_n + y_n) \\
&= (cx_1 + cy_1, \ldots, cx_n + cy_n) \\
&= (cx_1, \ldots, cx_n) + (cy_1, \ldots, cy_n) \\
&= c(x_1, \ldots, x_n) + c(y_1, \ldots, y_n) \\
&= c\mathbf{u} + c\mathbf{v}
\end{aligned}
$$

We now present examples of other vector spaces—sets with appropriate operations of addition and scalar multiplication that satisfy these axioms.

✦ *Example 1*

Consider the set M_{22} of 2×2 matrices. We have defined operations of addition and scalar multiplication on this set. The set does in fact form a vector space. We shall look at axioms 1, 3, 4, 5, and 6, leaving the remainder for the reader to check in Exercise 2.

Let $\mathbf{u} = \begin{pmatrix} p & q \\ r & s \end{pmatrix}$, $\mathbf{v} = \begin{pmatrix} x & y \\ z & h \end{pmatrix}$, and $\mathbf{w} = \begin{pmatrix} a & b \\ g & t \end{pmatrix}$ represent three arbitrary elements of M_{22}.

Axiom 1:

$$\mathbf{u} + \mathbf{v} = \begin{pmatrix} p & q \\ r & s \end{pmatrix} + \begin{pmatrix} x & y \\ z & h \end{pmatrix} = \begin{pmatrix} p+x & q+y \\ r+z & s+h \end{pmatrix}$$

We see that $\mathbf{u} + \mathbf{v}$ is a 2×2 matrix, an element of M_{22}. We have closure under addition.

Axioms 3 and 4: From our previous analysis of matrix addition, we know that the elements of M_{22} are commutative and associative under addition (Theorems 2-1 and 2-2).

Axiom 5: The zero 2×2 matrix is $\begin{pmatrix} 0 & 0 \\ 0 & 0 \end{pmatrix}$, since

$$\mathbf{u} + \begin{pmatrix} 0 & 0 \\ 0 & 0 \end{pmatrix} = \begin{pmatrix} p & q \\ r & s \end{pmatrix} + \begin{pmatrix} 0 & 0 \\ 0 & 0 \end{pmatrix} = \begin{pmatrix} p & q \\ r & s \end{pmatrix} = \mathbf{u}$$

Axiom 6: The negative of $\mathbf{u}$ is $\begin{pmatrix} -p & -q \\ -r & -s \end{pmatrix}$, since

$$\mathbf{u} + \begin{pmatrix} -p & -q \\ -r & -s \end{pmatrix} = \begin{pmatrix} p & q \\ r & s \end{pmatrix} + \begin{pmatrix} -p & -q \\ -r & -s \end{pmatrix}$$

$$= \begin{pmatrix} 0 & 0 \\ 0 & 0 \end{pmatrix} = \mathbf{0}$$

Thus

$$-\mathbf{u} = \begin{pmatrix} -p & -q \\ -r & -s \end{pmatrix}$$

etc.

The set M_{22} of 2×2 matrices forms a vector space; its algebraic properties are similar to those of $\mathbf{R}^n$. It can be shown in a like manner that M_{mn}, the set of $m \times n$ matrices, forms a vector space. ✦

Let us now construct another important type of vector space, a *function vector space*.

♦ *Example 2*

Let V be the set of real-valued functions defined on the entire real line **R**. We can visualize V as shown in Figure 5-1. An element of V such as f will have the entire real line **R** as its domain and will map every element of this domain into **R**. Our aim will now be to introduce operations of addition and scalar multiplication on V, operations under which V will become a vector space.

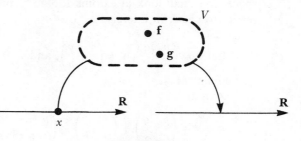

Figure 5-1

Let f and g be arbitrary elements of V. We define their sum $f + g$ to be a function such that

$$(f + g)(x) = f(x) + g(x)$$

This defines $f + g$ as a real-valued function on the entire line, since it tells us how it operates on every element of **R**. To find the value of $f + g$ at a point x, we add the value of f at x to the value of g at x. This operation is called *pointwise addition*.

We next define scalar multiplication on the elements of V. If c is an arbitrary scalar, then the scalar multiple of f, cf, is defined by

$$(cf)(x) = c[f(x)]$$

This defines cf as a real-valued function on **R**, since it tells us how it operates on every element x of **R**. To find the value of cf at a point x, multiply the value of f at x by c. This operation is called *pointwise scalar multiplication*.

To get a geometrical feel for pointwise addition and pointwise scalar multiplication, let us consider the specific functions f and g defined by $f(x) = x$ and $g(x) = x^2$ and graph $f + g$ and $3f$.

$f + g$ is defined by the equation

$$(f + g)(x) = f(x) + g(x) = x + x^2$$

and $3f$ by

$$(3f)(x) = 3[f(x)] = 3x$$

Thus

$$(f + g)(x) = x + x^2 \qquad \text{and} \qquad (3f)(x) = 3x$$

We sketch these functions in Figures 5-2 and 5-3 illustrating $(f + g)(2)$ and $3f(1)$.

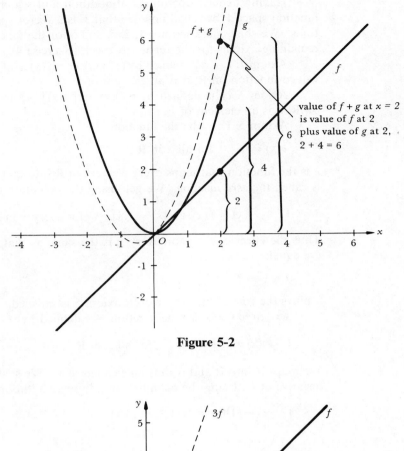

Figure 5-2

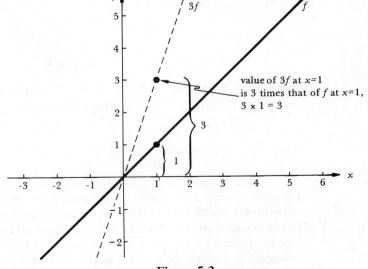

Figure 5-3

Having defined operations of addition and scalar multiplication on the function space V, we shall now see that V is a vector space under these operations. We shall check axioms 1, 2, 5, and 6 for the function space, leaving the remaining axioms for the reader to check (Exercise 4).

Axiom 1: $f + g$ is defined by $(f + g)(x) = f(x) + g(x)$. $f + g$ is thus a real-valued function on $\mathbf{R}$; it is an element of V.

Axiom 2: cf is defined by $(cf)(x) = c[f(x)]$. cf is a real-valued function on $\mathbf{R}$; it is an element of V.

Axiom 5: Let o be the function

$$o(x) = 0, \qquad \text{for all } x \text{ in } \mathbf{R}$$

o is the function that maps every element of $\mathbf{R}$ into 0; it is an element of V. o is called the *zero function*. We get, using the definition of addition,

$$(f + o)(x) = f(x) + o(x) = f(x), \qquad \text{for every } x \text{ in } \mathbf{R}$$

Thus the effect of $f + o$ on every x is the same as that of f. These functions are equal:

$$f + o = f$$

o plays the role of the zero vector; axiom 5 is satisfied.

Axiom 6: Consider the function $-f$ defined by

$$(-f)(x) = -[f(x)], \qquad \text{for all } x \text{ in } \mathbf{R}$$

$-f$ maps $\mathbf{R}$ into $\mathbf{R}$ and is thus an element of V. We shall show that $-f$ is the negative of f. Using the definition of pointwise addition,

$$
\begin{aligned}
[f + (-f)](x) &= f(x) + (-f)(x) \\
&= f(x) - [f(x)] \\
&= 0 \\
&= o(x), \qquad \text{for all } x \text{ in } \mathbf{R}
\end{aligned}
$$

The effects of the functions $[f + (-f)]$ and o are the same on all elements of the common domain. These functions are equal:

$$f + (-f) = o$$

Thus $-f$ is the negative of f; axiom 6 is satisfied.   ✦

In the previous example, we discussed a specific function vector space—the space of real-valued functions defined on the entire real line with operations of pointwise addition and pointwise scalar multiplication. There are many such function vector spaces. For example, the set of all real-valued functions on the interval $[-\pi, \pi]$ with operations of pointwise addition and pointwise scalar multiplication would be a vector space. The reader will investigate other function spaces in the exercises.

We now present a theorem that gives important useful properties of vectors. These are properties that were immediately apparent for $\mathbf{R}^n$ and were almost taken for granted in manipulating vectors in $\mathbf{R}^n$. They are not, however, so apparent for all vector spaces.

Theorem 5-1   *Let V be a vector space, $\mathbf{v}$ an arbitrary vector and c an arbitrary scalar in V, 0 the zero scalar, and $\mathbf{0}$ the zero vector of V. Then*

(1) $0\mathbf{v} = \mathbf{0}$

(2) $c\mathbf{0} = \mathbf{0}$

(3) $(-1)\mathbf{v} = -\mathbf{v}$

(4) *If $c\mathbf{v} = \mathbf{0}$, then either $c = 0$ or $\mathbf{v} = \mathbf{0}$.*

Proof: We shall prove (1) and (3), leaving the proofs of (2) and (4) for Exercise 13.

For part (1),

$$0\mathbf{v} + 0\mathbf{v} = (0 + 0)\mathbf{v} \qquad (axiom\ 8)$$
$$= 0\mathbf{v} \qquad (property\ of\ the\ scalar\ 0)$$

Let us add the negative of $0\mathbf{v}$, namely $-0\mathbf{v}$, to both sides:

$$(0\mathbf{v} + 0\mathbf{v}) + (-0\mathbf{v}) = 0\mathbf{v} + (-0\mathbf{v})$$
$$0\mathbf{v} + [0\mathbf{v} + (-0\mathbf{v})] = \mathbf{0} \qquad (axiom\ 4\ and\ axiom\ 6)$$
$$0\mathbf{v} + \mathbf{0} = \mathbf{0} \qquad (axiom\ 6)$$
$$0\mathbf{v} = \mathbf{0} \qquad (axiom\ 5)$$

For part (3),

$$(-1)\mathbf{v} + \mathbf{v} = (-1)\mathbf{v} + 1\mathbf{v} \qquad (axiom\ 10)$$
$$= [(-1) + 1]\mathbf{v} \qquad (axiom\ 8)$$
$$= 0\mathbf{v} \qquad (property\ of\ the\ scalar\ 0)$$
$$= \mathbf{0} \qquad (part\ 1\ above)$$

Thus $(-1)\mathbf{v}$ is the negative of $\mathbf{v}$ (axiom 6).

In the case of the vector space $\mathbf{R}^n$, we saw how certain subsets formed vector spaces in their own right. We called such vector spaces, embedded in the larger vector space $\mathbf{R}^n$, subspaces of $\mathbf{R}^n$. Quite naturally this same concept exists within the more general framework which we are pursuing at this time. We shall now look at subsets of general vector spaces which do in themselves form vector spaces. Subspaces of $\mathbf{R}^n$ could very often be interpreted geometrically, for example, as planes or lines. In contrast, however, many of the embedded vector spaces that we shall examine have no meaningful geometric interpretation. The tools that we now develop increasingly become solely algebraic in nature. We shall, for example, look at subsets of vector spaces of matrices and of vector spaces of functions.

Definition 5-2   Let V be a vector space. Let U be a subset of V. U is called a *subspace of* V if it forms a vector space under the operations of addition and scalar multiplication of V.

As in the case of $\mathbf{R}^n$, a subset of a general vector space forms a subspace if it is closed under the operations of addition and scalar multiplication of the larger space. Such a subset will automatically satisfy the remaining axioms of a vector space, axioms 3 through 10; these properties are "inherited" from the larger space.

✦ *Example 3*
___

Consider the set M_{22} of 2×2 matrices. We have seen that M_{22} forms a vector space under the operations of matrix addition and scalar multiplication. Let U be the subset of diagonal 2×2 matrices. We shall now demonstrate that U is a subspace of M_{22}; it is a vector space that is embedded in M_{22}.

Let $\mathbf{u} = \begin{pmatrix} p & 0 \\ 0 & q \end{pmatrix}$ and $\mathbf{v} = \begin{pmatrix} a & 0 \\ 0 & b \end{pmatrix}$ be two arbitrary elements of U (two arbitrary diagonal 2×2 matrices). Then

$$\mathbf{u} + \mathbf{v} = \begin{pmatrix} p & 0 \\ 0 & q \end{pmatrix} + \begin{pmatrix} a & 0 \\ 0 & b \end{pmatrix} = \begin{pmatrix} p+a & 0 \\ 0 & q+b \end{pmatrix}$$

Thus $\mathbf{u} + \mathbf{v}$ is a diagonal 2×2 matrix. $\mathbf{u} + \mathbf{v}$ is an element of U. U is closed under addition.

Let c be an arbitrary scalar. Then

$$c\mathbf{u} = c\begin{pmatrix} p & 0 \\ 0 & q \end{pmatrix} = \begin{pmatrix} cp & 0 \\ 0 & cq \end{pmatrix}$$

We see that $c\mathbf{u}$ is also a diagonal 2×2 matrix, an element of U. U is closed under scalar multiplication.

U is vector space embedded in the vector space M_{22}. U is a subspace of M_{22}.   ✦

The following example illustrates a subspace of a function space.

✦ *Example 4*
___

Let n be a positive integer and let P_n denote the set of all real polynomial functions of degree $\leq n$. P_n is the set of all functions that can be expressed in the form

$$f(x) = a_n x^n + a_{n-1}x^{n-1} + \cdots + a_1 x + a_0$$

where $a_0, \ldots, a_n$ are real numbers.

P_n is a subset of the vector space V of all real-valued functions defined on the entire real line. Let us prove that P_n is a subspace of V. We show that P_n is closed under addition and under scalar multiplication.

Let f and g be the polynomials

$$f(x) = a_n x^n + a_{n-1} x^{n-1} + \cdots + a_1 x + a_0$$

$$g(x) = b_n x^n + b_{n-1} x^{n-1} + \cdots + b_1 x + b_0$$

two arbitrary elements of P_n. Then

$$(f + g)(x) = f(x) + g(x) = a_n x^n + \cdots + a_1 x + a_0 + b_n x^n + \cdots + b_1 x + b_0$$

$$= (a_n + b_n)x^n + \cdots + (a_1 + b_1)x + a_0 + b_0$$

We see that $f + g$ is a polynomial of degree $\leq n$, an element of P_n. P_n is closed under addition.

Let c be an arbitrary scalar. Then cf is defined by

$$(cf)(x) = c[f(x)]$$

$$= c a_n x^n + c a_{n-1} x^{n-1} + \cdots + c a_1 x + c a_0$$

cf is a polynomial of degree $\leq n$, an element of P_n. P_n is closed under scalar multiplication.

Thus P_n is a subspace of the vector space of all real-valued functions defined on the entire real line.   ✦

We now extend the definitions of linear combination, spanning set, linear dependence/independence, basis, and dimension to general vector spaces. The results that were developed for $\mathbf{R}^n$ in terms of these concepts will then hold for all vector spaces. We illustrate these concepts for matrix and function spaces.

Consider m vectors $\mathbf{v}_1, \ldots, \mathbf{v}_m$ of a vector space V. A vector $\mathbf{v}$ in V is a *linear combination* of these vectors if there exist m scalars $a_1, \ldots, a_m$ such that

$$\mathbf{v} = a_1 \mathbf{v}_1 + \cdots + a_m \mathbf{v}_m$$

✦ **Example 5** ───

Determine whether the matrix $\begin{pmatrix} -1 & 7 \\ 8 & -1 \end{pmatrix}$ is a linear combination of the matrices $\begin{pmatrix} 1 & 0 \\ 2 & 1 \end{pmatrix}$, $\begin{pmatrix} 2 & -3 \\ 0 & 2 \end{pmatrix}$, and $\begin{pmatrix} 0 & 1 \\ 2 & 0 \end{pmatrix}$.

We are here looking at the vector space M_{22} of 2×2 matrices. The matrix $\begin{pmatrix} -1 & 7 \\ 8 & -1 \end{pmatrix}$ will be a linear combination of the other three matrices if there exist scalars a_1, a_2, and a_3 such that

$$\begin{pmatrix} -1 & 7 \\ 8 & -1 \end{pmatrix} = a_1 \begin{pmatrix} 1 & 0 \\ 2 & 1 \end{pmatrix} + a_2 \begin{pmatrix} 2 & -3 \\ 0 & 2 \end{pmatrix} + a_3 \begin{pmatrix} 0 & 1 \\ 2 & 0 \end{pmatrix}$$

We get

$$\begin{pmatrix} -1 & 7 \\ 8 & -1 \end{pmatrix} = \begin{pmatrix} a_1 & 0 \\ 2a_1 & a_1 \end{pmatrix} + \begin{pmatrix} 2a_2 & -3a_2 \\ 0 & 2a_2 \end{pmatrix} + \begin{pmatrix} 0 & a_3 \\ 2a_3 & 0 \end{pmatrix}$$

$$= \begin{pmatrix} a_1 + 2a_2 & -3a_2 + a_3 \\ 2a_1 + 2a_3 & a_1 + 2a_2 \end{pmatrix}$$

On equating corresponding coefficients, we get the following system of equations:

$$\begin{aligned}
a_1 + 2a_2 \quad\quad &= -1 \\
-3a_2 + a_3 &= 7 \\
2a_1 \quad\quad + 2a_3 &= 8 \\
a_1 + 2a_2 \quad\quad &= -1
\end{aligned}$$

This system has the unique solution $a_1 = 3$, $a_2 = -2$, and $a_3 = 1$. The given matrix is thus the following linear combination of the other three matrices:

$$\begin{pmatrix} -1 & 7 \\ 8 & -1 \end{pmatrix} = 3\begin{pmatrix} 1 & 0 \\ 2 & 1 \end{pmatrix} - 2\begin{pmatrix} 2 & -3 \\ 0 & 2 \end{pmatrix} + \begin{pmatrix} 0 & 1 \\ 2 & 0 \end{pmatrix}$$

If, on the other hand, it had turned out that the above system of linear equations had no solution, then of course the given matrix would not have been a linear combination of the remaining three matrices.   ✦

The vectors $\mathbf{v}_1, \ldots, \mathbf{v}_m$ are said to *span* a vector space if every vector in the space can be expressed as a linear combination of these vectors.

✦ **Example 6** ───────────────────────────────────

Consider the vector space M_{22} of 2×2 matrices. Let us show that the set $\begin{pmatrix} 1 & 0 \\ 0 & 0 \end{pmatrix}, \begin{pmatrix} 0 & 1 \\ 0 & 0 \end{pmatrix}, \begin{pmatrix} 0 & 0 \\ 1 & 0 \end{pmatrix}, \begin{pmatrix} 0 & 0 \\ 0 & 1 \end{pmatrix}$ spans this space.

Let $\begin{pmatrix} a & b \\ c & d \end{pmatrix}$ be an arbitrary 2×2 matrix. We can express this matrix

$$\begin{pmatrix} a & b \\ c & d \end{pmatrix} = a\begin{pmatrix} 1 & 0 \\ 0 & 0 \end{pmatrix} + b\begin{pmatrix} 0 & 1 \\ 0 & 0 \end{pmatrix} + c\begin{pmatrix} 0 & 0 \\ 1 & 0 \end{pmatrix} + d\begin{pmatrix} 0 & 0 \\ 0 & 1 \end{pmatrix}$$

proving the result.   ✦

The vectors $\mathbf{v}_1, \ldots, \mathbf{v}_m$ of a vector space V are said to be *linearly dependent* if there exist scalars $a_1, \ldots, a_m$, not all zero, such that

$$a_1\mathbf{v}_1 + \cdots + a_m\mathbf{v}_m = \mathbf{0}$$

If the vectors are not linearly dependent, they are said to be *linearly independent*.

✦ **Example 7** ───────────────────────────────────

Consider P_2, the vector space of real polynomials of degree ≤ 2. The following functions f, g, and h are elements of P_2:

$$f(x) = x^2, \qquad g(x) = x, \qquad h(x) = 1$$

We shall now show that these functions are linearly independent. Consider the following identity for scalars a_1, a_2, and a_3:

$$a_1 f(x) + a_2 g(x) + a_3 h(x) = 0$$

The functions f, g, and h will be linearly independent if we can show that this identity implies that $a_1 = a_2 = a_3 = 0$. Writing the identity in terms of the specific functions, we get

$$a_1 x^2 + a_2 x + a_3 = 0$$

This equation must hold for all x.

$x = 0$ gives $a_1 0 + a_2 0 + a_3 = 0$, implying that $a_3 = 0$

$x = 1$ gives $a_1 + a_2 = 0$

$x = -1$ gives $a_1 - a_2 = 0$

The last two equations imply that $a_1 = a_2 = 0$.

Thus

$$a_1 f(x) + a_2 g(x) + a_3 h(x) = 0$$

if and only if $a_1 = a_2 = a_3 = 0$. The functions f, g, and h are linearly independent in P_2. ✦

A finite set of vectors $\mathbf{v}_1, \ldots, \mathbf{v}_m$ in a vector space V forms a *basis* for the space if it spans the space and is linearly independent.

✦ Example 8

In the previous example, we saw that the functions $f(x) = x^2$, $g(x) = x$, and $h(x) = 1$ are linearly independent in P_2. Let us proceed one stage further in our analysis of P_2. Let us show that these three functions also span P_2 and that they therefore form a basis for P_2.

Let $p(x)$ be an arbitrary function in P_2. $p(x)$ is a polynomial of degree ≤ 2; it is of the form

$$p(x) = bx^2 + cx + d$$

Observe that p is in fact a linear combination of the functions f, g, and h:

$$p(x) = b[f(x)] + c[g(x)] + d[h(x)]$$

The functions f, g, and h thus span P_2. They form a basis for P_2. This is the *standard basis* for P_2. ✦

We are now in a position to realize the full power of the approach we have taken—namely, that of gradually generalizing the algebraic concepts of the vector space $\mathbf{R}^n$. Theorems that were developed for $\mathbf{R}^n$ hold also for general vector spaces; the algebraic concepts are identical. To illustrate the point, we focus on one important theorem and the definition that arises from that theorem.

Theorem 5-2   *Any two bases for a given vector space have the same number of vectors.*

The *dimension* of a vector space is the number of vectors that make up a basis.

We are now able to talk about the dimension of a space of matrices or the dimension of a function space in the same way as we could talk about the dimension of $\mathbf{R}^n$.

✦ ***Example 9***
_____

We have already seen that the functions $f(x) = x^2$, $g(x) = x$, and $h(x) = 1$ form a basis for P_2. We are now able to say that P_2 is a three-dimensional function space.   ✦

✦ ***Example 10***
_____

Consider the vector space M_{22} of 2×2 matrices. Let us determine a basis for this space.

We have already seen that the set $\begin{pmatrix} 1 & 0 \\ 0 & 0 \end{pmatrix}, \begin{pmatrix} 0 & 1 \\ 0 & 0 \end{pmatrix}, \begin{pmatrix} 0 & 0 \\ 1 & 0 \end{pmatrix}, \begin{pmatrix} 0 & 0 \\ 0 & 1 \end{pmatrix}$ spans M_{22}, since an arbitrary matrix can be expressed

$$\begin{pmatrix} a & b \\ c & d \end{pmatrix} = a\begin{pmatrix} 1 & 0 \\ 0 & 0 \end{pmatrix} + b\begin{pmatrix} 0 & 1 \\ 0 & 0 \end{pmatrix} + c\begin{pmatrix} 0 & 0 \\ 1 & 0 \end{pmatrix} + d\begin{pmatrix} 0 & 0 \\ 0 & 1 \end{pmatrix}$$

We now show that these matrices are also linearly independent. Consider the identity

$$p\begin{pmatrix} 1 & 0 \\ 0 & 0 \end{pmatrix} + q\begin{pmatrix} 0 & 1 \\ 0 & 0 \end{pmatrix} + r\begin{pmatrix} 0 & 0 \\ 1 & 0 \end{pmatrix} + s\begin{pmatrix} 0 & 0 \\ 0 & 1 \end{pmatrix} = \mathbf{0}$$

On simplifying,

$$\begin{pmatrix} p & 0 \\ 0 & 0 \end{pmatrix} + \begin{pmatrix} 0 & q \\ 0 & 0 \end{pmatrix} + \begin{pmatrix} 0 & 0 \\ r & 0 \end{pmatrix} + \begin{pmatrix} 0 & 0 \\ 0 & s \end{pmatrix} = \mathbf{0}$$

$$\begin{pmatrix} p & q \\ r & s \end{pmatrix} = \mathbf{0}$$

This implies that $p = q = r = s = 0$. Thus the matrices are linearly independent. The set of matrices $\begin{pmatrix} 1 & 0 \\ 0 & 0 \end{pmatrix}, \begin{pmatrix} 0 & 1 \\ 0 & 0 \end{pmatrix}, \begin{pmatrix} 0 & 0 \\ 1 & 0 \end{pmatrix}, \begin{pmatrix} 0 & 0 \\ 0 & 1 \end{pmatrix}$ forms a basis for M_{22}. This basis is called the *standard basis* for M_{22}. The vector space M_{22} is of dimension 4.   ✦

We mention at this time that vector spaces can be *infinite-dimensional*—there may not be a finite set of vectors that forms a basis. The set of all real-valued functions defined on the real line, for example, is an infinite-dimensional space. We shall not pursue the concept of infinite-dimensional vector spaces at this time. This is a topic for more advanced courses.

EXERCISES

1. We have demonstrated that $\mathbf{R}^n$ satisfies axioms 1–7 of a vector space. Verify that it also satisfies axioms 8, 9, and 10.

2. In Example 1 of this section, we proved that M_{22}, the set of 2×2 matrices, satisfies axioms 1, 3, 4, 5, and 6 of a vector space. Prove that M_{22} also satisfies the remaining axioms, thus confirming that M_{22} is indeed a vector space.

*3. (a) Let $f(x) = x + 2$ and $g(x) = x^2 - 1$. Compute the functions $f + g$, $2f$, and $3g$.
   (b) Let $f(x) = 2x$ and $g(x) = 4 - 2x$. Compute $f + g$, $3f$, and $-g$.

4. In Example 2 of this section, we introduced the concept of a function vector space. We proved that the set of real-valued functions defined over the set of reals, under the operations of pointwise addition and pointwise scalar multiplication, satisfies axioms 1, 2, 5, and 6 of a vector space. Prove that this set also satisfies the remaining axioms, thus confirming that the space is indeed a vector space.

*5. (a) Consider the set of all continuous real-valued functions over the interval $[0, 1]$. Does this set form a vector space under the operations of pointwise addition and pointwise scalar multiplication?
   (b) Consider the set of all discontinuous real-valued functions over the interval $[0, 1]$. Does this set form a vector space?

6. Determine which of the following subsets of M_{22} form subspaces:
   (a) the subset having diagonal elements zero
   (b) the subset consisting of matrices the sum of whose elements is 6 (For example, $\begin{pmatrix} 2 & 7 \\ -1 & -2 \end{pmatrix}$ would be such a matrix.)
   (c) the subset of matrices of the form $\begin{pmatrix} a & a^2 \\ b & b^2 \end{pmatrix}$
   (d) the subset of matrices of the form $\begin{pmatrix} a & a+2 \\ b & c \end{pmatrix}$
   (e) the subset of 2×2 symmetric matrices
   (f) the subset of 2×2 invertible matrices

*7. Which of the following subsets of M_{23} form subspaces?
   (a) the subset of matrices of the form $\begin{pmatrix} a & b & 0 \\ c & d & 0 \end{pmatrix}$
   (b) the subset of matrices of the form $\begin{pmatrix} a & a & a \\ b & 2b & 3b \end{pmatrix}$

8. P_3 is the vector space of polynomial functions of degree ≤ 3, and P_2 is the vector space of polynomials of degree ≤ 2. Prove that P_2 is a subspace of P_3.

9. Consider the set of polynomials of degree 2. Prove that this set is not closed under addition or under scalar multiplication. It is thus not a vector space.

*10. Let A be the set of all functions of the type $f(x) = ax^2 + bx + 3$, where a and b are real numbers. Is A a subspace of the vector space of real-valued functions defined on the set of real numbers?

11. Prove that every subspace of a vector space V has to contain the zero vector of V. This is often a quick way of deciding that some subsets cannot be subspaces. Use

this criterion to prove the following:

(a) The set of matrices of the form $\begin{pmatrix} a & 1 \\ b & c \end{pmatrix}$ is not a subspace of M_{22}.

(b) The set of functions of the form $f(x) = ax + 2$ is not a subspace of P_1.

*12. Consider the vector space of real-valued functions defined on the set of real numbers. Which of the following subsets form subspaces?

(a) subset consisting of all functions f such that $f(0) = 0$

(b) subset consisting of all functions f such that $f(0) = 3$

(c) subset consisting of all constant functions

13. Let V be a vector space, $\mathbf{v}$ a vector in V, and a, b, and c scalars. Use the axioms of a vector space to prove the following:

(a) $c\mathbf{0} = \mathbf{0}$

(b) If $c\mathbf{v} = \mathbf{0}$, then either $c = 0$ or $\mathbf{v} = \mathbf{0}$.

(c) $-(-\mathbf{v}) = \mathbf{v}$

(d) If $\mathbf{u} + \mathbf{w} = \mathbf{v} + \mathbf{w}$, then $\mathbf{u} = \mathbf{v}$.

(e) If $\mathbf{u} \neq \mathbf{0}$ and $a\mathbf{u} = b\mathbf{u}$, then $a = b$.

14. Determine whether the first matrix is a linear combination of the following matrices:

*(a) $\begin{pmatrix} 5 & 7 \\ 5 & -10 \end{pmatrix}$; $\begin{pmatrix} 1 & 2 \\ 3 & -4 \end{pmatrix}$, $\begin{pmatrix} 0 & 3 \\ 1 & 2 \end{pmatrix}$, $\begin{pmatrix} 1 & 2 \\ 0 & 0 \end{pmatrix}$

(b) $\begin{pmatrix} 7 & 6 \\ -5 & -3 \end{pmatrix}$; $\begin{pmatrix} 3 & 0 \\ 1 & 1 \end{pmatrix}$, $\begin{pmatrix} 0 & 1 \\ 3 & 4 \end{pmatrix}$, $\begin{pmatrix} 1 & 2 \\ 0 & 1 \end{pmatrix}$

*(c) $\begin{pmatrix} 4 & 1 \\ 7 & 10 \end{pmatrix}$; $\begin{pmatrix} 1 & 1 \\ 1 & 1 \end{pmatrix}$, $\begin{pmatrix} 3 & 1 \\ 0 & 0 \end{pmatrix}$, $\begin{pmatrix} -1 & -1 \\ 2 & 3 \end{pmatrix}$

15. Determine whether the first function is a linear combination of the following functions:

*(a) $f(x) = 3x^2 + 2x + 9$; $g(x) = x^2 + 1$, $h(x) = x + 3$

(b) $f(x) = 2x^2 + x - 3$; $g(x) = x^2 - x + 1$, $h(x) = x^2 + 2x - 2$

*(c) $f(x) = x^2 + 4x + 5$; $g(x) = x^2 + x - 1$, $h(x) = x^2 + 2x + 1$

16. In each of the following parts, determine whether or not the given vectors are linearly dependent in the appropriate vector spaces.

*(a) $f(x) = 2x^2 + 1$, $g(x) = x^2 + 4x$, $h(x) = x^2 - 4x + 1$ in P_2

(b) $f(x) = x^2 + 3$, $g(x) = x + 1$, $h(x) = 2x^2 - 3x + 3$ in P_2

*(c) $f(x) = x^2 + 3x - 1$, $g(x) = x + 3$, $h(x) = 2x^2 - x + 1$ in P_2

(d) $\begin{pmatrix} 1 & 0 \\ 0 & 0 \end{pmatrix}$, $\begin{pmatrix} 0 & 2 \\ 0 & 0 \end{pmatrix}$, $\begin{pmatrix} 0 & 0 \\ 3 & 0 \end{pmatrix}$, $\begin{pmatrix} 0 & 0 \\ 0 & 4 \end{pmatrix}$ in M_{22}

*(e) $\begin{pmatrix} 1 & 2 \\ 3 & 1 \end{pmatrix}$, $\begin{pmatrix} 1 & 1 \\ 1 & 1 \end{pmatrix}$, $\begin{pmatrix} 2 & 1 \\ 4 & 2 \end{pmatrix}$ in M_{22}

(f) $\begin{pmatrix} 1 & 2 \\ -1 & 0 \end{pmatrix}$, $\begin{pmatrix} 1 & 2 \\ 1 & 1 \end{pmatrix}$, $\begin{pmatrix} 1 & 2 \\ 5 & 3 \end{pmatrix}$ in M_{22}

17. Determine a basis for each of the following spaces and give the dimension of the space.
    *(a) the vector space P_3
    (b) the vector space M_{23}
    *(c) the subspace of M_{22} consisting of diagonal matrices
    (d) the subspace of M_{33} consisting of symmetric matrices

18. *(a) Is the function $f(x) = x + 5$ in the space spanned by $g(x) = x + 1$ and $h(x) = x + 3$?
    (b) Is the function $f(x) = 3x^2 + 5x + 1$ in the space spanned by $g(x) = 2x^2 + 3$ and $h(x) = x^2 + 3x - 1$?
    *(c) Give three other functions in the space spanned by $g(x) = 2x^2 + 3$ and $h(x) = x^2 + 3x - 1$.
    (d) Give a basis for the space spanned by $f(x) = 2x + 3$, $g(x) = x - 1$, and $h(x) = -x - 4$.

19. Are the following sets bases for the given vector spaces?
    *(a) $f(x) = x^2 + 2x - 1$, $g(x) = x + 3$, $h(x) = x^2 + 3x + 2$ for P_2
    (b) $f(x) = x^2 + x - 3$, $g(x) = x^2 - x + 1$, $h(x) = x^2 + x - 1$ for P_2

    *(c) $\begin{pmatrix} 1 & 2 \\ 0 & 1 \end{pmatrix}$, $\begin{pmatrix} 3 & 4 \\ 1 & 1 \end{pmatrix}$, $\begin{pmatrix} 1 & 2 \\ 1 & 1 \end{pmatrix}$, $\begin{pmatrix} 0 & 2 \\ 1 & 2 \end{pmatrix}$ for M_{22}

    (d) $\begin{pmatrix} 1 & 0 \\ 0 & 0 \end{pmatrix}$, $\begin{pmatrix} 1 & 2 \\ 0 & 0 \end{pmatrix}$, $\begin{pmatrix} 1 & 2 \\ 3 & 0 \end{pmatrix}$, $\begin{pmatrix} 1 & 2 \\ 3 & 4 \end{pmatrix}$ for M_{22}

20. Let V be the vector space of real-valued functions defined on the real line and let f, g, and h be elements of V which are twice differentiable. The function $W(x)$ defined by

$$W(x) = \begin{vmatrix} f(x) & g(x) & h(x) \\ f'(x) & g'(x) & h'(x) \\ f''(x) & g''(x) & h''(x) \end{vmatrix}$$

is called the *Wronskian* of f, g, and h. Prove that f, g, and h are linearly independent if the Wronskian is not the zero function. [If $W(x) = 0$, we cannot say anything about the linear dependence/independence of f, g, and h.] Readers who go on to study differential equations will make use of the Wronskian in that field.

21. Use the Wronskian, introduced in Exercise 20, to show that the following functions are linearly independent.
    *(a) $1, x, x^2$ (b) $1, x, e^x$ *(c) $\sin x, \cos x, x \sin x$ (d) $e^x, xe^x, x^2 e^x$

22. If U is a subspace of a vector space V, and if $\mathbf{u}$ and $\mathbf{v}$ are elements of V, but one or both not in U, can $\mathbf{u} + \mathbf{v}$ be in U? Can $c\mathbf{u}$ be in U for some nonzero scalar c if $\mathbf{u}$ is not in U?

23. Prove that a necessary and sufficient condition for a subset U of a vector space V to be a subspace is that $a\mathbf{u} + b\mathbf{v}$ be in U for all scalars a and b and all vectors $\mathbf{u}$ and $\mathbf{v}$ of U.

24. Prove:
    (a) The union of two subspaces need not be a subspace.
    (b) The intersection of two subspaces is a subspace.

5-2   INNER PRODUCT SPACES

In the previous section, we generalized the concept of the vector space $\mathbf{R}^n$. We drew up a list of axioms based on the most important properties of $\mathbf{R}^n$. Any set that satisfied these axioms was called a vector space. Such a space had algebraic properties similar to those of $\mathbf{R}^n$. We now proceed one stage further in our process of generalization; we extend the concepts of dot product, norm, angle, and distance to general vector spaces.

The dot product was a key concept on $\mathbf{R}^n$ that led to definitions of norm, angle, and distance. Our approach will be to generalize the dot product into a concept called the *inner product* for a general vector space. This in turn will be used to meaningfully define norm, angle, and distance on a general vector space. The following definition of inner product is based on the most important properties of the dot product in $\mathbf{R}^n$.

> **Definition 5-3**   An *inner product* on a vector space V is a function that associates a real number with each pair of vectors $\mathbf{u}$ and $\mathbf{v}$ of V. This real number is denoted $\langle \mathbf{u}, \mathbf{v} \rangle$. This function has to satisfy the following axioms for all vectors $\mathbf{u}$, $\mathbf{v}$, and $\mathbf{w}$ of V and all scalars a:
>
> 1. $\langle \mathbf{u}, \mathbf{v} \rangle = \langle \mathbf{v}, \mathbf{u} \rangle$ (symmetry axiom)
> 2. $\langle \mathbf{u} + \mathbf{v}, \mathbf{w} \rangle = \langle \mathbf{u}, \mathbf{w} \rangle + \langle \mathbf{v}, \mathbf{w} \rangle$ (additive axiom)
> 3. $\langle a\mathbf{u}, \mathbf{v} \rangle = a\langle \mathbf{u}, \mathbf{v} \rangle$ (homogeneity axiom)
> 4. $\langle \mathbf{u}, \mathbf{u} \rangle \geq 0$ and $\langle \mathbf{u}, \mathbf{u} \rangle = 0$ if and only if $\mathbf{u} = \mathbf{0}$
>   (positive definite axiom)

A vector space on which an inner product has been defined is called an *inner product space*.

We now demonstrate that the dot product on $\mathbf{R}^n$ does indeed have these four properties.

Let $\mathbf{u} = (x_1, \ldots, x_n)$ and $\mathbf{v} = (y_1, \ldots, y_n)$ be two arbitrary vectors in $\mathbf{R}^n$. Their dot product is

$$\mathbf{u} \cdot \mathbf{v} = (x_1, \ldots, x_n) \cdot (y_1, \ldots, y_n) = x_1 y_1 + \cdots + x_n y_n$$

Theorem 4-17 of Section 4-8 tells us that the dot product has the following properties:

$$\mathbf{u} \cdot \mathbf{v} = \mathbf{v} \cdot \mathbf{u}$$

$$(\mathbf{u} + \mathbf{v}) \cdot \mathbf{w} = \mathbf{u} \cdot \mathbf{w} + \mathbf{v} \cdot \mathbf{w}$$

$$a\mathbf{u} \cdot \mathbf{v} = a(\mathbf{u} \cdot \mathbf{v})$$

$$\mathbf{u} \cdot \mathbf{u} \geq 0 \text{ and } \mathbf{u} \cdot \mathbf{u} = 0 \text{ if and only if } \mathbf{u} = \mathbf{0}$$

If we define the inner product on $\mathbf{R}^n$ to be the dot product,

$$\langle \mathbf{u}, \mathbf{v} \rangle = \mathbf{u} \cdot \mathbf{v}$$

all the axioms are satisfied.

Any function on a vector space that satisfies the axioms of inner product does define an inner product. There can be many inner products on a given vector space. The following example illustrates an alternative inner product to the dot product on $\mathbf{R}^2$.

✦ Example 1

Let $\mathbf{u} = (x_1, x_2)$, $\mathbf{v} = (y_1, y_2)$, and $\mathbf{w} = (z_1, z_2)$ be three arbitrary vectors in $\mathbf{R}^2$. Then

$$\langle \mathbf{u}, \mathbf{v} \rangle = x_1 y_1 + 4 x_2 y_2$$

defines an inner product on $\mathbf{R}^2$. To see that this is so we check each of the four axioms.

Axiom 1:

$$\langle \mathbf{u}, \mathbf{v} \rangle = x_1 y_1 + 4 x_2 y_2 = y_1 x_1 + 4 y_2 x_2 = \langle \mathbf{v}, \mathbf{u} \rangle$$

Axiom 2:

$$\begin{aligned}
\langle \mathbf{u} + \mathbf{v}, \mathbf{w} \rangle &= \langle (x_1, x_2) + (y_1, y_2), (z_1, z_2) \rangle \\
&= \langle (x_1 + y_1, x_2 + y_2), (z_1, z_2) \rangle \\
&= (x_1 + y_1)z_1 + 4(x_2 + y_2)z_2 \\
&= x_1 z_1 + y_1 z_1 + 4 x_2 z_2 + 4 y_2 z_2 \\
&= x_1 z_1 + 4 x_2 z_2 + y_1 z_1 + 4 y_2 z_2 \\
&= \langle (x_1, x_2), (z_1, z_2) \rangle + \langle (y_1, z_1), (y_2, z_2) \rangle \\
&= \langle \mathbf{u}, \mathbf{w} \rangle + \langle \mathbf{v}, \mathbf{w} \rangle
\end{aligned}$$

Axiom 3:

$$\begin{aligned}
\langle a\mathbf{u}, \mathbf{v} \rangle &= \langle a(x_1, x_2), (y_1, y_2) \rangle \\
&= \langle (ax_1, ax_2), (y_1, y_2) \rangle \\
&= ax_1 y_1 + 4 ax_2 y_2 \\
&= a(x_1 y_1 + 4 x_2 y_2) \\
&= a \langle \mathbf{u}, \mathbf{v} \rangle
\end{aligned}$$

Axiom 4:

$$\begin{aligned}
\langle \mathbf{v}, \mathbf{v} \rangle &= \langle (y_1, y_2), (y_1, y_2) \rangle \\
&= y_1^2 + 4 y_2^2
\end{aligned}$$

≥ 0 and $= 0$ if and only if $y_1 = y_2 = 0$; that is, if $\mathbf{v} = \mathbf{0}$.

The four inner product axioms are satisfied. The above identity does indeed define an inner product on $\mathbf{R}^2$. There are many other inner products on $\mathbf{R}^2$. The dot product, however, is the most meaningful inner product on $\mathbf{R}^2$; this is the inner product, as we saw, that leads to the concepts of magnitude, angle, and distance that we usually associate with $\mathbf{R}^2$ in Euclidean geometry. We shall illustrate a non-Euclidean geometry based on the inner product of this example later in this section.   ✦

We now illustrate suitable inner products on further vector spaces.

✦ *Example 2*

Consider the vector space M_{22} of 2×2 matrices.

Let $\mathbf{u} = \begin{pmatrix} p & q \\ r & s \end{pmatrix}$, $\mathbf{v} = \begin{pmatrix} x & y \\ z & h \end{pmatrix}$, and $\mathbf{w} = \begin{pmatrix} a & b \\ g & t \end{pmatrix}$ be three arbitrary vectors in M_{22}. The following function is an inner product on M_{22}:

$$\langle \mathbf{u}, \mathbf{v} \rangle = px + qy + rz + sh$$

(One multiplies the corresponding elements of the matrices and adds.)

We verify axioms 1 and 3, leaving axioms 2 and 4 for the reader to check in Exercise 4.

Axiom 1:

$$\begin{aligned} \langle \mathbf{u}, \mathbf{v} \rangle &= px + qy + rz + sh \\ &= xp + yq + zr + hs \\ &= \langle \mathbf{v}, \mathbf{u} \rangle \end{aligned}$$

Axiom 3:

$$\begin{aligned} \langle a\mathbf{u}, \mathbf{v} \rangle &= apx + aqy + arz + ash \\ &= a[px + qy + rz + sh] \\ &= a\langle \mathbf{u}, \mathbf{v} \rangle \quad \blacklozenge \end{aligned}$$

✦ *Example 3*

Consider the vector space P_n of polynomials of degree $\leq n$. Let f, g, and h be elements of P_n. The following function defines an inner product on V.

$$\langle f, g \rangle = \int_0^1 f(x)g(x)\,dx$$

We check axioms 1 and 2, leaving axioms 3 and 4 for the reader to verify in Exercise 6.

Axiom 1:

$$\langle f, g \rangle = \int_0^1 f(x)g(x)\,dx = \int_0^1 g(x)f(x)\,dx = \langle g, f \rangle$$

Axiom 2:

$$\begin{aligned} \langle f + g, h \rangle &= \int_0^1 [f(x) + g(x)]h(x)\,dx \\ &= \int_0^1 [f(x)h(x) + g(x)h(x)]\,dx \\ &= \int_0^1 f(x)h(x)\,dx + \int_0^1 g(x)h(x)\,dx \\ &= \langle f, h \rangle + \langle g, h \rangle \quad \blacklozenge \end{aligned}$$

In the case of the vector space $\mathbf{R}^n$, we saw how the dot product led in a natural manner to the following concept of norm (or magnitude):

$$\|(x_1, \ldots, x_n)\| = \sqrt{(x_1, \ldots, x_n) \cdot (x_1, \ldots, x_n)}$$

This definition of norm fitted in with the geometric magnitudes that we expected for position vectors in $\mathbf{R}^2$ and $\mathbf{R}^3$ in Euclidean geometry. We base our definition of norm for a general vector space on this approach. The norm is defined in a manner analogous to the above, using the inner product generalization of the dot product. The norms that we thus obtain do not necessarily have geometric interpretations but do play useful roles in, for example, numerical work involving functions, where it is desirable to have some concept of magnitude of a function.

Definition 5-4   Let V be an inner product space. The *norm* (or magnitude) of a vector $\mathbf{v}$ is denoted by $\|\mathbf{v}\|$ and defined by

$$\|\mathbf{v}\| = \sqrt{\langle \mathbf{v}, \mathbf{v} \rangle}$$

✦ *Example 4*

Consider the inner product space P_n of polynomials with inner product defined by

$$\langle f, g \rangle = \int_0^1 f(x)g(x)\,dx$$

The norm (or magnitude) of f *generated* by this inner product is

$$\|f\| = \sqrt{\int_0^1 f(x)f(x)\,dx}$$

$$= \sqrt{\int_0^1 [f(x)]^2\,dx}$$

Let us, for example, find the magnitude of the vector $f(x) = 5x^2 + 1$.

$$\|f\| = \sqrt{\int_0^1 (5x^2 + 1)^2\,dx}$$

$$= \sqrt{\int_0^1 (25x^4 + 10x^2 + 1)\,dx}$$

$$= \sqrt{[5x^5 + \tfrac{10}{3}x^3 + x]_0^1}$$

$$= \sqrt{5 + \tfrac{10}{3} + 1} = \sqrt{28/3} \quad ✦$$

The dot product on $\mathbf{R}^n$ was also used to define angles between vectors in $\mathbf{R}^n$:

$$\cos \theta = \frac{\mathbf{u}}{\|\mathbf{u}\|} \cdot \frac{\mathbf{v}}{\|\mathbf{v}\|}$$

Using the inner product generalization of the dot product, we adopt this as our definition of angle between vectors in an inner product space.

Definition 5-5   Let V be an inner product space with $\mathbf{u}$ and $\mathbf{v}$ two non-zero vectors in V. The cosine of the angle between $\mathbf{u}$ and $\mathbf{v}$ is given by

$$\cos \theta = \left\langle \frac{\mathbf{u}}{\|\mathbf{u}\|}, \frac{\mathbf{v}}{\|\mathbf{v}\|} \right\rangle$$

✦ Example 5
_____

Consider the inner product space P_n of polynomials with inner product defined by

$$\langle f, g \rangle = \int_0^1 f(x)g(x)\, dx$$

The angle between two nonzero functions $\mathbf{f}$ and $\mathbf{g}$ is defined by

$$\cos \theta = \left\langle \frac{f}{\|f\|}, \frac{g}{\|g\|} \right\rangle = \int_0^1 \left(\frac{f(x)}{\|f(x)\|} \right)\left(\frac{g(x)}{\|g(x)\|} \right) dx$$

Let us determine the cosine of the angle between the functions $f(x) = 5x^2$ and $g(x) = 3x$.

$$\|5x^2\| = \sqrt{\int_0^1 (5x^2)^2\, dx} = \sqrt{\int_0^1 25x^4\, dx} = \sqrt{[5x^5]_0^1} = \sqrt{5}$$

and

$$\|3x\| = \sqrt{\int_0^1 (3x)^2\, dx} = \sqrt{\int_0^1 9x^2\, dx} = \sqrt{[3x^3]_0^1} = \sqrt{3}$$

Thus

$$\cos \theta = \int_0^1 (5x^2/\sqrt{5})(3x/\sqrt{3})\, dx$$

$$= 15/\sqrt{15} \int_0^1 x^3\, dx$$

$$= \sqrt{15}[x^4/4]_0^1$$

$$= \sqrt{15}/4 \quad ✦$$

As for the inner product space $\mathbf{R}^n$, two nonzero vectors in a general inner product space are said to be *orthogonal* if their inner product is zero.

✦ Example 6
_____

Let us show that the vectors $f(x) = 3x - 2$ and $g(x) = x$ are orthogonal in the inner product space P_n of polynomials introduced earlier.

$$\langle f, g \rangle = \int_0^1 (3x - 2)(x)\, dx$$

$$= \int_0^1 (3x^2 - 2x)\, dx = [x^3 - x^2]_0^1 = 0$$

Thus f and g are orthogonal; they are at right angles in this space.   ✦

Our final task in relation to inner product spaces is to generalize the concept of distance between elements that we had for $\mathbf{R}^n$.

Definition 5-6   Let V be an inner product space with norm defined by $\|\mathbf{v}\| = \sqrt{\langle \mathbf{v}, \mathbf{v} \rangle}$. The distance between two vectors $\mathbf{u}$ and $\mathbf{v}$ of V is denoted by $d(\mathbf{u}, \mathbf{v})$ and is defined by

$$d(\mathbf{u}, \mathbf{v}) = \|\mathbf{u} - \mathbf{v}\|$$

This definition was compatible with the Euclidean geometric concept of distance that was expected in $\mathbf{R}^2$ and $\mathbf{R}^3$. In general inner product spaces, however, this idea of distance does not have direct meaningful geometric interpretation. It is a concept that can be shown to have many of the algebraic properties that one would want "distance" to have. As for norm, it is a most useful tool in numerical and algebraic work involving functions.

✦ *Example 7*
_____

We illustrate the concept of distance for the function inner product space P_n. Let us compute the distance between the functions $f(x) = 2x^2 + 5x + 1$ and $g(x) = 2x^2 + 2x - 1$. We get

$$
\begin{aligned}
d(f, g) &= \|f(x) - g(x)\| = \|(2x^2 + 5x + 1) - (2x^2 + 2x - 1)\| \\
&= \|3x + 2\| \\
&= \sqrt{\int_0^1 (3x + 2)^2 \, dx} \\
&= \sqrt{\int_0^1 (9x^2 + 12x + 4) \, dx} \\
&= \sqrt{[3x^3 + 6x^2 + 4x]_0^1} \\
&= \sqrt{13} \quad ✦
\end{aligned}
$$

Let us now return to $\mathbf{R}^n$ and to geometry. Let us discuss the full significance of the new-found mathematical freedom we have in the many inner products that we can place on $\mathbf{R}^n$.

We now have very powerful tools at our disposal that enable us to control the mathematical geometry of $\mathbf{R}^n$. Observe that the inner product on a space controls the concepts of vector magnitude, angles, and distances; each is defined in terms of the inner product:

$$\|\mathbf{v}\| = \sqrt{\langle \mathbf{v}, \mathbf{v} \rangle}$$

$$\cos \theta = \left\langle \frac{\mathbf{u}}{\|\mathbf{u}\|}, \frac{\mathbf{v}}{\|\mathbf{v}\|} \right\rangle = \left\langle \frac{\mathbf{u}}{\sqrt{\langle \mathbf{u}, \mathbf{u} \rangle}}, \frac{\mathbf{v}}{\sqrt{\langle \mathbf{v}, \mathbf{v} \rangle}} \right\rangle$$

$$d(\mathbf{u}, \mathbf{v}) = \|\mathbf{u} - \mathbf{v}\| = \sqrt{\langle \mathbf{u} - \mathbf{v}, \mathbf{u} - \mathbf{v} \rangle}$$

We have seen how we can put many inner products on $\mathbf{R}^n$. Different inner products lead to different concepts of vector magnitude, angles, and distances—that is, to different geometries. The dot product led to Euclidean geometry. The vector magnitudes, angles, and distances derived from the dot product in $\mathbf{R}^2$ and $\mathbf{R}^3$ were those that we are accustomed to every day in two and three dimensions. We were comfortable with the geometry that came from the dot product! Let us now turn to look at some of the other geometries that we can get by putting inner products other than the dot product on $\mathbf{R}^2$. These *non-Euclidean geometries* have been studied by mathematicians and are used by scientists. In the next section, we shall see how Einstein made use of one non-Euclidean geometry to describe his theory of special relativity.

✦ *Example 8*

Let us look at a non-Euclidean geometry on $\mathbf{R}^2$. Consider $\mathbf{R}^2$ with the inner product from Example 1:

$$\langle (x_1, x_2), (y_1, y_2) \rangle = x_1 y_1 + 4 x_2 y_2$$

This inner product differs from the dot product in the appearance of a 4. Consider the vector $(0, 1)$ in this space. Its magnitude is given by

$$
\begin{aligned}
\|(0, 1)\| &= \sqrt{\langle (0, 1), (0, 1) \rangle} \\
&= \sqrt{(0 \times 0) + 4(1 \times 1)} \\
&= \sqrt{4} = 2
\end{aligned}
$$

We illustrate this vector in Figure 5-4. The magnitude of this vector in Euclidean geometry is of course 1; in our new geometry, however, it is 2.

Consider the vectors $(1, 1)$ and $(-4, 1)$. The inner product of these vectors is

$$\langle (1, 1), (-4, 1) \rangle = (1 \times -4) - 4(1 \times 1) = 0$$

Thus the vectors $(1, 1)$ and $(-4, 1)$ are orthogonal. We illustrate these vectors in Figure 5-5. Obviously the vectors do not appear to be at right angles; they are not orthogonal in the Euclidean geometry of our everyday experience. However, the angle between these vectors is a right angle in our new geometry.

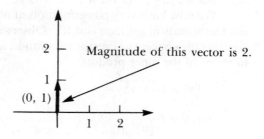

Figure 5-4

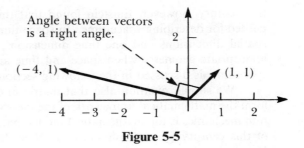

Figure 5-5

Finally, let us use the definition of distance to compute the distance between the points $(1, 0)$ and $(0, 1)$ in our space. We have that

$$
\begin{aligned}
d[(1, 0), (0, 1)] &= \|(1, 0) - (0, 1)\| \\
&= \|(1, -1)\| \\
&= \sqrt{\langle(1, -1), (1, -1)\rangle} \\
&= \sqrt{(1 \times 1) + 4(-1 \times -1)} \\
&= \sqrt{5}
\end{aligned}
$$

We illustrate this distance in Figure 5-6. The distance between these same two points in Euclidean space can be seen to be $\sqrt{2}$, by the theorem of Pythagoras. Pythagoras' theorem does not hold in our non-Euclidean geometry.

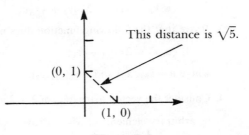

Figure 5-6   ✦

The reader should appreciate that the geometry we have developed in this example is good mathematics even though it does not describe the everyday world that we live in! Mathematics exists as an elegant field in itself without having to be useful or related to real world concepts. Even though the focus in this course has been on presenting mathematics within a background of application, the reader should be aware that much fine mathematics is not oriented toward applications. Of course, one never knows when such pure mathematics will turn out to be useful. This happened to be the case for non-Euclidean geometry. Much non-Euclidean geometry was developed by an Italian mathematician named Bernhard Riemann in the mid-nineteenth century. At the time the geometry did not describe any real physical situation. In the early part of this century, however, Einstein found that this was the type of geometry he needed for describing space-time. In space-time one uses $\mathbf{R}^4$ to represent three spatial dimensions and one time dimension. Euclidean geometry is not the appropriate geometry when space and time are "mixed." Einstein's model of space-time is discussed in the following section.

We should mention also that matrix and function inner product spaces form the mathematical framework of the theory of quantum mechanics. Quantum mechanics is the mathematical model that was developed in the early part of this century by such physicists as Niels Bohr, Max Planck, and Werner Heisenberg to describe the behavior of atoms and electrons. Thus the theory of inner product spaces forms the cornerstone of the two foremost physical theories of the twentieth century. These theories have had a great impact on all our lives; they have, for example, led to the development of both atomic energy and atomic weapons.

EXERCISES

1. Let $\mathbf{u} = (x_1, x_2)$ and $\mathbf{v} = (y_1, y_2)$ be elements of $\mathbf{R}^2$. Prove that the following function defines an inner product on $\mathbf{R}^2$:

   $$\langle \mathbf{u}, \mathbf{v} \rangle = 4x_1 y_1 + 9x_2 y_2$$

2. Let $\mathbf{u} = (x_1, x_2, x_3)$ and $\mathbf{v} = (y_1, y_2, y_3)$ be elements of $\mathbf{R}^3$. Prove that the following function defines an inner product on $\mathbf{R}^3$:

   $$\langle \mathbf{u}, \mathbf{v} \rangle = x_1 y_1 + 2x_2 y_2 + 4x_3 y_3$$

*3. Prove that the following function does not define an inner product on $\mathbf{R}^2$:

   $$\langle \mathbf{u}, \mathbf{v} \rangle = 2x_1 y_1 - x_2 y_2$$

   where $\mathbf{u} = (x_1, x_2)$ and $\mathbf{v} = (y_1, y_2)$.

4. Consider the vector space M_{22} of 2×2 matrices. Let $\mathbf{u} = \begin{pmatrix} p & q \\ r & s \end{pmatrix}$ and $\mathbf{v} = \begin{pmatrix} x & y \\ z & h \end{pmatrix}$ be arbitrary elements of M_{22}. Prove that

   $$\langle \mathbf{u}, \mathbf{v} \rangle = px + qy + rz + sh$$

   satisfies axioms 2 and 4 of the inner product. (This exercise completes Example 2 of this section.)

5. Let $\mathbf{u} = \begin{pmatrix} p & q \\ r & s \end{pmatrix}$ and $\mathbf{v} = \begin{pmatrix} x & y \\ z & h \end{pmatrix}$ be elements of the vector space M_{22}. Prove that the following is an inner product on M_{22}:

   $$\langle \mathbf{u}, \mathbf{v} \rangle = px + 2qy + 3rz + 4sh$$

*6. Consider the vector space P_n of polynomials of degree $\leq n$. Let f and g be elements of this space. Prove that the following function satisfies axioms 3 and 4 of the inner product, thus completing Example 3:

   $$\langle f, g \rangle = \int_0^1 f(x)g(x)\,dx$$

7. Let f and g be arbitrary elements of the vector space P_n. Prove that the following function defines an inner product on P for real numbers a and b if $a < b$:

$$\langle f, g \rangle = \int_a^b f(x)g(x)\, dx$$

Prove that it does not define an inner product if $a \le b$.

In Exercises 8–14, all functions are in the inner product space P_n with inner product defined by

$$\langle f, g \rangle = \int_0^1 f(x)g(x)\, dx$$

*8. Determine the magnitudes of the functions $f(x) = 7x^3$ and $g(x) = 3x^2 + 2$.

9. Prove that the functions $f(x) = x^2$ and $g(x) = 4x - 3$ are orthogonal.

*10. Compute the cosine of the angle between the functions $f(x) = 5x^2$ and $g(x) = 9x$.

11. Prove that the functions $f(x) = 1$ and $g(x) = \frac{1}{2} - x$ are orthogonal.

12. Determine a function that is orthogonal to $f(x) = 6x + 12$.

*13. Compute the distance between the functions $f(x) = x^2 + 3x + 1$ and $g(x) = x^2 + x - 3$.

14. Compute the distance between the functions $f(x) = 4x^2 + x - 3$ and $g(x) = x^2 + x - 2$.

In Exercises 15–21, all matrices are elements of the inner product space M_{22} with inner product defined by

$$\left\langle \begin{pmatrix} p & q \\ r & s \end{pmatrix}, \begin{pmatrix} a & b \\ c & d \end{pmatrix} \right\rangle = pa + qb + rc + sd$$

*15. Determine the magnitude of each of the matrices $\begin{pmatrix} 1 & 0 \\ 0 & 0 \end{pmatrix}, \begin{pmatrix} 0 & 1 \\ 0 & 0 \end{pmatrix}, \begin{pmatrix} 0 & 0 \\ 1 & 0 \end{pmatrix}, \begin{pmatrix} 0 & 0 \\ 0 & 1 \end{pmatrix}$.

16. Prove that the matrices of Exercise 15 are orthogonal when considered in pairs. Combining this result, the result of Exercise 15, and the previously derived result that these matrices form a basis for M_{22}, we see that this set forms an orthonormal basis for M_{22} (a basis of unit orthogonal matrices).

17. Prove that the matrices $\begin{pmatrix} 1 & 2 \\ -1 & 1 \end{pmatrix}$ and $\begin{pmatrix} 2 & 4 \\ 3 & -7 \end{pmatrix}$ are orthogonal.

*18. Determine a matrix orthogonal to the matrix $\begin{pmatrix} 1 & 2 \\ 3 & 4 \end{pmatrix}$.

19. Compute the magnitudes of the matrices $\begin{pmatrix} 1 & 2 \\ 3 & 4 \end{pmatrix}$ and $\begin{pmatrix} -1 & 3 \\ 0 & 2 \end{pmatrix}$.

*20. Compute the distance between the matrices $\begin{pmatrix} 4 & 3 \\ -1 & 2 \end{pmatrix}$ and $\begin{pmatrix} 2 & 1 \\ 0 & 1 \end{pmatrix}$.

21. Compute the distance between the matrices $\begin{pmatrix} 3 & 4 \\ -1 & 3 \end{pmatrix}$ and $\begin{pmatrix} 1 & 0 \\ 0 & 0 \end{pmatrix}$.

22. Consider $\mathbf{R}^2$ with the inner product of Exercise 1:

$$\langle (x_1, x_2), (y_1, y_2) \rangle = 4x_1 y_1 + 9x_2 y_2$$

*(a) Determine the magnitudes of the following elements of $\mathbf{R}^2$: (1, 0), (0, 1), (1, 1).

(b) Prove that the vectors (2, 1) and (−9, 8) are orthogonal in this space. Sketch a rough diagram of these vectors. Observe that they are not orthogonal in Euclidean space.

*(c) Determine the distance between the points (1, 0) and (0, 1) in this space. Observe that the theorem of Pythagoras does not apply.

23. Consider $\mathbf{R}^2$ with the inner product

$$\langle (x_1, x_2), (y_1, y_2) \rangle = x_1 y_1 + 16x_2 y_2$$

(a) Determine the magnitudes of the vectors (1, 0), (0, 1), and (1,1).

(b) Prove that the vectors (1, 1) and (−16, 1) are orthogonal.

(c) Compute the distance between the points (1, 0) and (0, 1).

5-3 THE MATHEMATICAL MODEL OF SPECIAL RELATIVITY

Special relativity was developed by Albert Einstein in an attempt to describe the physical world that we live in. At the time, Newtonian mechanics was the theory used to describe the motions of bodies under forces. However, experiments had led scientists to believe that the large-scale motions of bodies, such as planetary motions, were not accurately described by Newtonian mechanics. One of Einstein's main contributions to science was the development of more accurate mathematical models. First he developed special relativity, which did not incorporate gravitation; later on he incorporated gravitation in his theory of general relativity. Here we introduce the mathematical model of special relativity by describing one of the predictions of the theory and then showing how it arises out of the theory.

The nearest star to earth other than the sun is Alpha Centauri; it is about four light-years away. (A light-year is the distance light travels in one year, 5.88×10^{12} miles.) Consider a pair of twins who are separated immediately after birth. Twin 1 remains on earth, and twin 2 is flown off to Alpha Centauri in a rocket at 0.8 the speed of light. On arriving at Alpha Centauri, he immediately returns to earth at the same high speed. The mathematical theory of special relativity predicts that when he reaches earth, he will find that his twin who remained on earth is ten years old (in every sense of the word) while he is only six years old. (These times would vary according to the speed of twin 2.) We present this scenario in terms of twins since it clearly displays the phenomenon involved, the difference in the rate of aging. The same type of phenomenon is predicted by special relativity for any two people, one of whom stays on earth while the other travels to a distant star and back.

Numerous experiments have been constructed to test this hypothesis that time on earth varies from time recorded on an object moving relative to earth. We shall mention one such experiment conducted by Professor J. C. Hafele, of the Department of Physics, Washington University, St Louis, and Dr. Richard E. Keating, of the Time Service Division of the U.S. Naval Observatory, in Washington, D.C., and reported in the journal *Science*, Vol. 177, 1972. During October, 1971, four atomic clocks were flown on regularly scheduled jet flights around the world twice, once eastward and once westward. From the actual flight paths of each trip, theory (general relativity here, since gravity is involved) predicted that the flying clocks, compared with reference clocks at the U.S. Naval Observatory, would, lose 40 ± 23 nanoseconds during the eastward trip and would gain 275 ± 21 nanoseconds during the westward trip. The possible errors are due to numerical approximations made. The observed time difference during the eastward trip was 59 ± 10 nanoseconds and during the westward trip was 273 ± 7 nanoseconds. These are averages of the four clocks. The errors here allow for deficiencies in flight data. These results are considered to be in reasonable accord with the theoretical predictions, illustrating that the sort of time variation we are discussing does in fact take place. This aging variation will not be realized by humans in the near future to the extent illustrated by the trip to Alpha Centauri, since energies required to produce such high speeds in a macroscopic body are prohibitive. This is probably fortunate from the sociological viewpoint. Theoretically a man who went off on such a trip could be of an age to marry his great granddaughter when he returned! This phenomenon was experienced to a much lesser degree by astronauts who went to the moon—on arrival back on earth they were fractionally younger than they would have been if they had remained to earth.

Let us see how the prediction arises out of Einstein's model. The model is one of space-time and hence involves four coordinates: three space coordinates: x_1, x_2, and x_3 and a time coordinate x_4. We use the vector space $\mathbf{R}^4$ to represent space-time. Each element of $\mathbf{R}^4$ is called an *event*. Each event has a location in space given by x_1, x_2, and x_3 and occurs at time x_4. There are many inner products that one can put on $\mathbf{R}^4$; each would lead to a geometry for $\mathbf{R}^4$. None of these inner products, however, leads to a geometry that conforms to experimental results. If one drops the fourth inner product axiom, the positive definite requirement, one comes up with a "pseudo" inner product that leads to a geometry that fits experimental results. This geometry of special relativity is called Minkowski geometry after Hermann Minkowski, who gave this geometrical interpretation of special relativity.

Minkowski Geometry

Let $X = (x_1, x_2, x_3, x_4)$ and $Y = (y_1, y_2, y_3, y_4)$ be arbitrary elements of $\mathbf{R}^4$. Then

$$\langle X, Y \rangle = -x_1 y_1 - x_2 y_2 - x_3 y_3 + x_4 y_4$$
$$\|X\| = \sqrt{|\langle X, X \rangle|} \quad \textit{(magnitude of X)}$$

X is orthogonal to Y if $\langle X, Y \rangle = 0$. Further,

$$d(X, Y) = \|X - Y\| \quad \textit{(distance between points X and Y)}$$
$$= \sqrt{\left| -(x_1 - y_1)^2 - (x_2 - y_2)^2 - (x_3 - y_3)^2 + (x_4 - y_4)^2 \right|}$$

The minus signs prevent the $\langle X, Y \rangle$ from satisfying axiom 4 for inner products (Exercise 1). Observe the absolute value signs in the expressions for $\|X\|$ and $d(X, Y)$. These are appropriate expressions for magnitude and distance that fit the theory. Let us now describe the journey to Alpha Centauri in $\mathbf{R}^4$, using this geometry.

We draw a *space-time diagram*. For convenience, assume that Alpha Centauri lies in the direction of the x_1 axis from earth. The twin on earth advances in time, x_4, while the twin in the rocket advances in time and also moves first in the direction of increasing x_1 to Alpha Centauri and then in the direction of decreasing x_1 back to earth. The space-time diagram is shown in Figure 5-7.

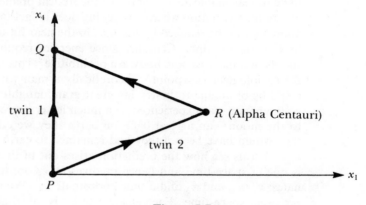

Figure 5-7

The path of twin 1 is PQ; he is advancing in time, x_4. The path of twin 2 is PR to Alpha Centauri and then RQ back to rejoin twin 1 at event Q. There is no motion in either the x_2 or x_3 direction; hence we suppress these dimensions in the diagram.

Let us first of all look at twin 1, who stays on earth. Relative to earth, the rocket travels with 0.8 the speed of light. Alpha Centauri is 4 light-years from earth. The round trip of 8 light-years, from the point of view of earth, will take (8/0.8) years, or 10 years (time = distance/velocity). The age of twin 1 at Q is 10 years.

We now examine the situation for twin 2. Let P be the origin in $\mathbf{R}^4$, (0, 0, 0, 0). (See Figure 5-8.) Since $x_4 = 10$ at Q, Q is the point (0, 0, 0, 10) and S will be (0, 0, 0, 5). SR is the spatial distance of Alpha Centauri from earth— 4 light-years. Thus R is the point (4, 0, 0, 5). Let us now use Minkowski geometry to compute the distance PR. We have that

$$d(P, R) = \|P - R\|$$
$$= \sqrt{\left| -(4 - 0)^2 - (0 - 0)^2 - (0 - 0)^2 + (5 - 0)^2 \right|}$$
$$= \sqrt{|-16 + 25|} = \sqrt{9} = 3$$

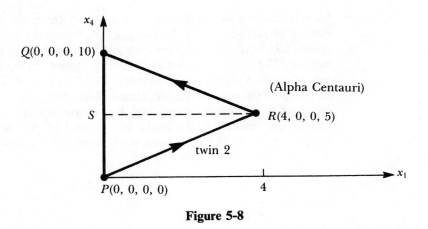

Figure 5-8

Similarly,

$$d(R, Q) = 3 \qquad (Exercise\ 2)$$

The length of $PR + RQ$ is 6.

We have completed the mathematics; now comes the physical interpretation of these results using Minkowski geometry. The theory of special relativity states that *the distance between two points on the path of an observer*, such as twin 2, *corresponds to the time recorded by the observer in traveling between the two points.*

Thus $d(P, R) = 3$ implies that twin 2 ages three years in traveling from P to R. Similarly, $d(R, Q) = 3$ implies that twin 2 ages three years in getting back to earth. The total duration of the voyage for twin 2 is thus 6 years.

Thus when the twins meet again at Q, twin 1 is ten years old, while twin 2 is six years old.

Note that this model introduces a new kind of geometry in which the straight line is not necessarily the shortest distance between two points. In Figure 5-8 the straight line distance between P and Q is 10, whereas the distance PRQ is 6, a smaller distance. In fact, it turns out that the straight line distance PQ is the longest distance between P and Q! Thus not only is the physical interpretation of this model fascinating, but it opens up a new trend in geometrical thinking.

This example illustrates well the flexibility that one has in applying mathematics. If the standard body of mathematics (inner product axioms in this case) does not fit the situation, maybe a slight modification will. Mathematicians have molded and developed mathematics to suit their needs. Mathematics is not as rigid and absolute as it is sometimes made out to be; it is a field that is continually being developed and applied in the spirit presented here.

In the general theory of relativity, gravity is taken into account and is represented by a "pseudo" inner product that involves functions. The space becomes curved in nature; curves lead to extreme distances between points instead of straight lines. Those who are interested in looking further into special relativity will find an interesting, very readable article by Alfred Schild entitled "The Clock Paradox in Relativity Theory" in the *American Mathematical Monthly*, Vol. 66, No. 1, January, 1959. *Time and the Space Traveller* by L. Marder, George, Allen and Unwin Ltd., 1971, gives a discussion of this aspect of the behavior of time and also introduces the reader to both special and general relativity. *Relativity and Cosmology*, by William J. Kaufmann, Harper and Row, 1973, discusses black holes and quasars in addition to giving the foundations of relativity.

EXERCISES

1. Prove that the "pseudo" inner product of Minkowski geometry violates axiom 4 for inner products. (*Hint*: Find element X ($\neq 0$) of $\mathbf{R}^4$ such that $\langle X, X \rangle \neq 0$.)

2. Prove that the distance between R and Q is 3 in the space-time diagram describing the voyage to Alpha Centauri.

*3. Determine the distance between the points $P(0, 0, 0, 0)$ and $M(1, 0, 0, 1)$ in Minkowski geometry. This result reveals another interesting aspect of this geometry—noncoincident points can be zero distance apart! The special theory of relativity gives a physical interpretation to such a line PM, whose points are at zero distance apart; it is the path of a photon, a light particle.

4. Prove that the vectors $(2, 0, 0, 1)$ and $(1, 0, 0, 2)$ are orthogonal in Minkowski geometry. Sketch these vectors, and observe that they are symmetrical about the "45° vector" $(1, 0, 0, 1)$. Any pair of vectors that are thus symmetrical about the vector $(1, 0, 0, 1)$ will be orthogonal in Minkowski geometry. Prove this result by demonstrating that the vectors $(a, 0, 0, b)$ and $(b, 0, 0, a)$ are orthogonal.

*5. The star Sirius is 8 light-years from earth. Sirius is the nearest star other than the sun that is visible to the naked eye from most parts of North America. It is the brightest appearing of all stars. Light reaches us from the sun in 8 minutes and from Sirius in 8 years. Suppose a rocket ship that leaves for Sirius returns to earth 20 years later. What is the duration of the voyage for a person on the space ship? What is the speed of the ship?

6. A rocket ship makes a round-trip flight to the bright star Capella, which is 45 light-years away from earth. If the time lapse on earth is 120 years, what is the length of the voyage from the traveler's viewpoint?

*7. The star cluster Pleiades in the constellation Taurus is 410 light-years from earth. A traveler to the cluster ages 40 years on a round trip from earth. By the time he returns, how many centuries will have passed on earth since he started the voyage?

8. Write a program to determine inner products of vectors using the inner product of special relativity. Use your program to determine the inner products of the following vectors.

    (a) $(1, 2, -1, 3)$ and $(2, 1, 0, 1)$
    (b) $(2, 0, 1, 4)$ and $(0, 1, 3, -2)$

9. Write a program to determine norms using the norm of special relativity. Use your program to determine the norms of the following vectors:

   (a) $(0, 1, 2, 4)$
   (b) $(-1, 2, 3, 7)$

10. Write a program to solve Exercise 7 above. Use your program to determine the number of years that have passed on earth during the following space voyages:

    (a) to the star cluster Praesepe in the constellation Cancer, 515 light-years from earth. Duration for the traveler is 50 years.

    (b) to the cluster Hyades in Taurus, 130 light-years from earth. Duration for the traveler is 20 years.

6

Mappings

In previous sections, we have discussed vector spaces and have seen applications of the theory developed. It is often of interest to relate elements of a vector space to other elements within the same space and also to elements within other spaces. In this chapter, we develop the methods for doing this. The theory developed also gives us greater insight into the properties of systems of linear equations.

6-1 MAPPINGS DEFINED BY SQUARE MATRICES

We have seen how the elements of $\mathbf{R}^2$ can be used to represent the locations of points in a plane. This is what we are actually doing when we introduce a coordinate system. In this section, we shall see, by means of examples, how a 2×2 matrix can be used to move points around in the plane.

✦ **Example 1**

Let the coordinate system in Figure 6-1 define the relative location of points in a plane. We shall use the elements of $\mathbf{R}^2$ in the form of columns rather than rows, as this permits us to perform the desired matrix multiplications.

Consider the matrix $\begin{pmatrix} 0 & -1 \\ 1 & 0 \end{pmatrix}$. We can multiply any element of $\mathbf{R}^2$ by this matrix to get another element of $\mathbf{R}^2$. For example, $\begin{pmatrix} 0 & -1 \\ 1 & 0 \end{pmatrix}\begin{pmatrix} 2 \\ 1 \end{pmatrix} = \begin{pmatrix} -1 \\ 2 \end{pmatrix}$. Interpreting $\begin{pmatrix} 2 \\ 1 \end{pmatrix}$ as a point in the plane, this matrix multiplication "takes it into" the point $\begin{pmatrix} -1 \\ 2 \end{pmatrix}$. We say that the matrix *maps* the point $\begin{pmatrix} 2 \\ 1 \end{pmatrix}$ into the point $\begin{pmatrix} -1 \\ 2 \end{pmatrix}$. In this manner the matrix can be used to map a point in the plane into another point. Let us consider the points in Figure 6-2. The point $\begin{pmatrix} 1 \\ 0 \end{pmatrix}$ is mapped into the

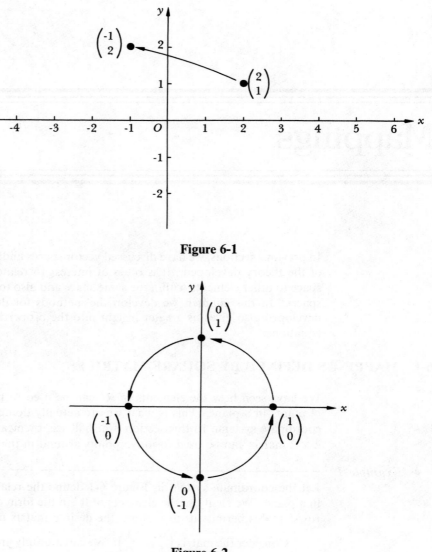

Figure 6-1

Figure 6-2

point $\begin{pmatrix} 0 \\ 1 \end{pmatrix}$, since $\begin{pmatrix} 0 & -1 \\ 1 & 0 \end{pmatrix}\begin{pmatrix} 1 \\ 0 \end{pmatrix} = \begin{pmatrix} 0 \\ 1 \end{pmatrix}$. We write $\begin{pmatrix} 1 \\ 0 \end{pmatrix} \mapsto \begin{pmatrix} 0 \\ 1 \end{pmatrix}$. Further, $\begin{pmatrix} 0 \\ 1 \end{pmatrix} \mapsto \begin{pmatrix} -1 \\ 0 \end{pmatrix}$, $\begin{pmatrix} -1 \\ 0 \end{pmatrix} \mapsto \begin{pmatrix} 0 \\ -1 \end{pmatrix}$, and $\begin{pmatrix} 0 \\ -1 \end{pmatrix} \mapsto \begin{pmatrix} 1 \\ 0 \end{pmatrix}$. (Verify these.)

At this stage, one may conjecture that the effect of the matrix multiplication is to rotate the points that make up the plane through an angle of $\pi/2$ in a counterclockwise direction. We shall prove that this is the case in the following discussion.   ◆

Rotation of a Plane

Let us determine the matrix that can be used to rotate the points that make up a plane through an angle of θ about the origin, in a counterclockwise direction.

Such a rotation about O would take the point A in Figure 6-3 into the point B. The distance OA is equal to OB for this rotation about O; let this distance be r. Let the angle AOC be α. Then we know that

$$x' = OC = r \cos(\alpha + \theta) = r \cos \alpha \cos \theta - r \sin \alpha \sin \theta$$
$$= x \cos \theta - y \sin \theta$$

$$y' = BC = r \sin(\alpha + \theta) = r \sin \alpha \cos \theta + r \cos \alpha \sin \theta$$
$$= y \cos \theta + x \sin \theta$$
$$= x \sin \theta + y \cos \theta$$

We can write these two equations in matrix form.

$$\begin{pmatrix} x' \\ y' \end{pmatrix} = \begin{pmatrix} \cos \theta & -\sin \theta \\ \sin \theta & \cos \theta \end{pmatrix} \begin{pmatrix} x \\ y \end{pmatrix}$$

Thus the matrix that maps A into B is $\begin{pmatrix} \cos \theta & -\sin \theta \\ \sin \theta & \cos \theta \end{pmatrix}$. Multiplying any point in the plane by this matrix will rotate it through an angle θ about the origin. Multiplying every point in the plane by this matrix will rotate the plane through an angle θ.

To rotate the points through $\pi/2$, we let $\theta = \pi/2$. The matrix is $\begin{pmatrix} 0 & -1 \\ 1 & 0 \end{pmatrix}$, verifying our conjecture from the previous example.

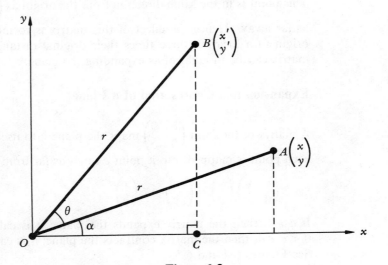

Figure 6-3

Definition 6-1   A *mapping* of a set V into a set U is a rule that associates with each element of V a single element of U.

The term *transformation* is also used for a mapping.

We have seen how the matrix $\begin{pmatrix} \cos\theta & -\sin\theta \\ \sin\theta & \cos\theta \end{pmatrix}$ defines a mapping of $\mathbf{R}^2$ into $\mathbf{R}^2$. We now continue with a discussion of the use of matrices as mappings.

✦ *Example 2* ──

Determine the geometrical effect of multiplying an element of $\mathbf{R}^2$ by the matrix $\begin{pmatrix} 3 & 0 \\ 0 & 3 \end{pmatrix}$.

Let us examine the effect of multiplying $\begin{pmatrix} 1 \\ 2 \end{pmatrix}$ by the matrix. We find that

$$\begin{pmatrix} 3 & 0 \\ 0 & 3 \end{pmatrix}\begin{pmatrix} 1 \\ 2 \end{pmatrix} = \begin{pmatrix} 3 \\ 6 \end{pmatrix}$$

The element $\begin{pmatrix} 3 \\ 6 \end{pmatrix}$ represents a point B in the plane that is in the same direction as A from the origin but at a distance three times the distance of OA from the origin: $\begin{pmatrix} 1 \\ 2 \end{pmatrix} \mapsto \begin{pmatrix} 3 \\ 6 \end{pmatrix}$. For an arbitrary point $\begin{pmatrix} x \\ y \end{pmatrix}$,

$$\begin{pmatrix} 3 & 0 \\ 0 & 3 \end{pmatrix}\begin{pmatrix} x \\ y \end{pmatrix} = \begin{pmatrix} 3x \\ 3y \end{pmatrix} = 3\begin{pmatrix} x \\ y \end{pmatrix}$$

This point is in the same direction from the origin as $\begin{pmatrix} x \\ y \end{pmatrix}$, but it is three times as far away. Hence the effect of this matrix is to move points out from the origin to a location three times their original distance from the origin. The matrix can be thought of as expanding the plane.   ✦

Expansion and Contraction of a Plane

A matrix of the form $\begin{pmatrix} c & 0 \\ 0 & c \end{pmatrix}$ maps the plane into itself in such a manner that each point is mapped into a point c times as far from the origin.

$$\begin{pmatrix} c & 0 \\ 0 & c \end{pmatrix}\begin{pmatrix} x \\ y \end{pmatrix} = \begin{pmatrix} cx \\ cy \end{pmatrix} = c\begin{pmatrix} x \\ y \end{pmatrix}$$

If $c > 1$, then the matrix expands the plane; it is called a *dilation* of $\mathbf{R}^2$. If $0 < c < 1$, then the matrix contracts the plane; it is called a *contraction* of $\mathbf{R}^2$. See Figures 6-4 and 6-5.

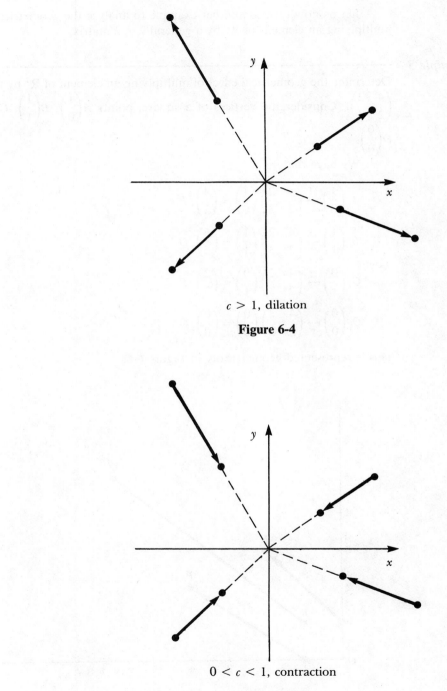

$c > 1$, dilation

Figure 6-4

$0 < c < 1$, contraction

Figure 6-5

Let us now turn to another example to analyze the geometrical effect of multiplying an element of $\mathbf{R}^2$ by a general 2×2 matrix.

✦ *Example 3*

Determine the geometrical effect of multiplying an element of $\mathbf{R}^2$ by the matrix $\begin{pmatrix} 4 & 2 \\ 2 & 3 \end{pmatrix}$. Consider the vertices of a square, points $A\begin{pmatrix} 1 \\ 0 \end{pmatrix}$, $B\begin{pmatrix} 1 \\ 1 \end{pmatrix}$, $C\begin{pmatrix} 0 \\ 1 \end{pmatrix}$, and $O\begin{pmatrix} 0 \\ 0 \end{pmatrix}$.

We have that

$$A, \begin{pmatrix} 1 \\ 0 \end{pmatrix} \mapsto \begin{pmatrix} 4 & 2 \\ 2 & 3 \end{pmatrix}\begin{pmatrix} 1 \\ 0 \end{pmatrix} = \begin{pmatrix} 4 \\ 2 \end{pmatrix}, P$$

$$B, \begin{pmatrix} 1 \\ 1 \end{pmatrix} \mapsto \begin{pmatrix} 4 & 2 \\ 2 & 3 \end{pmatrix}\begin{pmatrix} 1 \\ 1 \end{pmatrix} = \begin{pmatrix} 6 \\ 5 \end{pmatrix}, Q$$

$$C, \begin{pmatrix} 0 \\ 1 \end{pmatrix} \mapsto \begin{pmatrix} 4 & 2 \\ 2 & 3 \end{pmatrix}\begin{pmatrix} 0 \\ 1 \end{pmatrix} = \begin{pmatrix} 2 \\ 3 \end{pmatrix}, R$$

$$O, \begin{pmatrix} 0 \\ 0 \end{pmatrix} \mapsto \begin{pmatrix} 4 & 2 \\ 2 & 3 \end{pmatrix}\begin{pmatrix} 0 \\ 0 \end{pmatrix} = \begin{pmatrix} 0 \\ 0 \end{pmatrix}, O$$

This is represented geometrically in Figure 6-6.

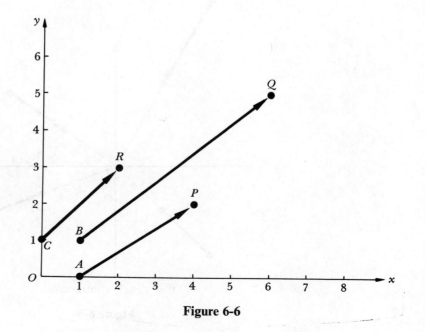

Figure 6-6

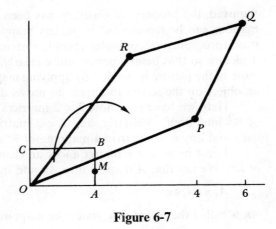

Figure 6-7

Figure 6-7 illustrates that this mapping does in fact take $OA \to OP$, $AB \to PQ$, $BC \to QR$, and $OC \to OR$, deforming the square $OABC$ into the figure $OPQR$.

We show that AB is mapped into PQ; the proofs for the other sides are similar. A general point on the line AB is M, $\begin{pmatrix} 1 \\ y \end{pmatrix}$ with $0 \le y \le 1$. Let $\begin{pmatrix} 1 \\ y \end{pmatrix} \mapsto \begin{pmatrix} x' \\ y' \end{pmatrix}$, N. Then

$$\begin{pmatrix} x' \\ y' \end{pmatrix} = \begin{pmatrix} 4 & 2 \\ 2 & 3 \end{pmatrix} \begin{pmatrix} 1 \\ y \end{pmatrix} = \begin{pmatrix} 4 + 2y \\ 2 + 3y \end{pmatrix}$$

Thus $x' = 4 + 2y$ and $y' = 2 + 3y$ with $0 \le y \le 1$. We get, on eliminating y between these identities, $3x' - 2y' = 8$ with $4 \le x' \le 6$. Thus N lies on the line $3x - 2y = 8$ with $4 \le x \le 6$. When $x = 4$, $y = 2$, giving the point P; when $x = 6$, $y = 5$, giving the point Q. Points on the line $3x - 2y = 8$ with values of x between 4 and 6 will lie on PQ between P and Q. Thus AB is mapped onto PQ.

The general geometrical effect of multiplying points in a plane by a 2×2 matrix is to map points into points such that straight lines go into straight lines or points, while the origin remains fixed (Exercise 10). Such a mapping can be illustrated by considering a geometrical figure (the square $ABCO$ in this case) and the figure into which it is transformed (called its image).   ✦

Mappings are of practical as well as theoretical value. The above square $OABC$ could represent a physical body deformed into the shape $OPQR$. Solid bodies can be represented geometrically and deformations analyzed in this manner. The fields of science that investigate such problems are *elasticity* and *plasticity*. When loads are applied to bodies, changes in shape, called *deformations*, occur. If the body returns to its original shape when the loads are removed, the property of elasticity has been displayed. In many cases, deformations can be represented by matrix mappings. Mappings are used in software (programs) of computer graphic systems. Dilations are used to magnify a picture so that detail appears more clearly, contractions to reduce it so that more of the picture is visible. By applying mappings sequentially one can cause an object on the screen to appear to move, depicting motion.

Here we have seen how 2×2 matrices can be used to define mappings of $\mathbf{R}^2$ into itself. Similarly, any 3×3 matrix defines a mapping of $\mathbf{R}^3$ into itself and any $n \times n$ matrix a mapping of $\mathbf{R}^n$ into itself.

Let A be an $n \times n$ matrix and $\mathbf{x}$ an element of $\mathbf{R}^n$. $A\mathbf{x}$ will be an element of $\mathbf{R}^n$. We say that A maps $\mathbf{x}$ into $A\mathbf{x}$ and indicate this by

$$A: \mathbf{x} \mapsto A\mathbf{x}$$

$A\mathbf{x}$ is called the *image of* $\mathbf{x}$ under the mapping.

✦ *Example 4*
_____

Interpret the matrix $A = \begin{pmatrix} 1 & 2 & 0 & -1 \\ 3 & 0 & 3 & 6 \\ 2 & 4 & 2 & 1 \\ 1 & 1 & -1 & 2 \end{pmatrix}$ as a mapping of $\mathbf{R}^4$ into

itself. Determine the images of the following two vectors $\mathbf{x}$ and $\mathbf{y}$ under this mapping.

$$\mathbf{x} = \begin{pmatrix} 1 \\ 2 \\ -1 \\ 0 \end{pmatrix}, \quad \mathbf{y} = \begin{pmatrix} 3 \\ -1 \\ 2 \\ 1 \end{pmatrix}$$

We have that

$$A: \begin{pmatrix} 1 \\ 2 \\ -1 \\ 0 \end{pmatrix} \mapsto \begin{pmatrix} 1 & 2 & 0 & -1 \\ 3 & 0 & 3 & 6 \\ 2 & 4 & 2 & 1 \\ 1 & 1 & -1 & 2 \end{pmatrix} \begin{pmatrix} 1 \\ 2 \\ -1 \\ 0 \end{pmatrix}$$

which equals $\begin{pmatrix} 5 \\ 0 \\ 8 \\ 4 \end{pmatrix}$, and

$$A: \begin{pmatrix} 3 \\ -1 \\ 2 \\ 1 \end{pmatrix} \mapsto \begin{pmatrix} 1 & 2 & 0 & -1 \\ 3 & 0 & 3 & 6 \\ 2 & 4 & 2 & 1 \\ 1 & 1 & -1 & 2 \end{pmatrix} \begin{pmatrix} 3 \\ -1 \\ 2 \\ 1 \end{pmatrix}$$

which equals $\begin{pmatrix} 0 \\ 21 \\ 7 \\ 2 \end{pmatrix}$. Thus the image of $\begin{pmatrix} 1 \\ 2 \\ -1 \\ 0 \end{pmatrix}$ is $\begin{pmatrix} 5 \\ 0 \\ 8 \\ 4 \end{pmatrix}$ and the image of

$\begin{pmatrix} 3 \\ -1 \\ 2 \\ 1 \end{pmatrix}$ is $\begin{pmatrix} 0 \\ 21 \\ 7 \\ 2 \end{pmatrix}$ under the mapping defined by A. ✦

Orthogonal Matrices

Let us look again at the matrix $\begin{pmatrix} \cos\theta & -\sin\theta \\ \sin\theta & \cos\theta \end{pmatrix}$ that defines rotation in a plane. Its columns can be viewed as unit orthogonal vectors, $\begin{pmatrix} \cos\theta \\ \sin\theta \end{pmatrix}$ and $\begin{pmatrix} -\sin\theta \\ \cos\theta \end{pmatrix}$. The matrix is an example of an orthogonal matrix, which we now formally define.

Definition 6-2 A square matrix whose columns form a set of unit mutually orthogonal vectors is called an *orthogonal matrix*. Because of their properties, orthogonal matrices play a significant role in the application of matrices. Their most significant property is given in the following theorem.

Theorem 6-1 *Let A be an n × n orthogonal matrix. A can be interpreted as defining a mapping of* **R**^{n} *into itself. This mapping preserves magnitudes and angles. Thus, if* **x** *and* **y** *are arbitrary elements of* **R**^{n} *interpreted as column vectors, then* $\|\mathbf{x}\| = \|A\mathbf{x}\|$ *(the magnitude of* **x** *is equal to the magnitude of its image, A***x***) and*

$$\frac{\mathbf{x}}{\|\mathbf{x}\|} \cdot \frac{\mathbf{y}}{\|\mathbf{y}\|} = \frac{A\mathbf{x}}{\|A\mathbf{x}\|} \cdot \frac{A\mathbf{y}}{\|A\mathbf{y}\|}$$

(the angle between **x** *and* **y** *is equal to the angle between A***x** *and A***y***).*

Proof: Let $A = (a_{ij})$, $\mathbf{x} = \begin{pmatrix} x_1 \\ \vdots \\ x_n \end{pmatrix}$, and $\mathbf{y} = \begin{pmatrix} y_1 \\ \vdots \\ y_n \end{pmatrix}$. Then

$$A\mathbf{x} = \begin{pmatrix} a_{11} & \cdots & a_{1n} \\ & \vdots & \\ a_{n1} & \cdots & a_{nn} \end{pmatrix} \begin{pmatrix} x_1 \\ \vdots \\ x_n \end{pmatrix} = \begin{pmatrix} \sum a_{1i}x_i \\ \vdots \\ \sum a_{ni}x_i \end{pmatrix}$$

$$\|\mathbf{x}\|^2 = \begin{pmatrix} x_1 \\ \vdots \\ x_n \end{pmatrix} \cdot \begin{pmatrix} x_1 \\ \vdots \\ x_n \end{pmatrix} = x_1^2 + \cdots + x_n^2$$

$$\|A\mathbf{x}\|^2 = \begin{pmatrix} \sum a_{1i}x_i \\ \vdots \\ \sum a_{ni}x_i \end{pmatrix} \cdot \begin{pmatrix} \sum a_{1j}x_j \\ \vdots \\ \sum a_{nj}x_j \end{pmatrix}$$

$$= \sum_k \left(\sum_i a_{ki}x_i \right) \left(\sum_j a_{kj}x_j \right)$$

$$= \sum_i \sum_j \left(\sum_k a_{ki}a_{kj} \right) x_i x_j$$

However, if A is orthogonal, $\sum_k a_{ki}a_{kj}$ is the inner product of the ith column and the jth column of A. Hence $\sum_k a_{ki}\,a_{kj} = \delta_{ij}$.

$$\|A\mathbf{x}\|^2 = \sum_i \sum_j \delta_{ij} x_i x_j = x_1^2 + \cdots + x_n^2 = \|\mathbf{x}\|^2$$

proving that the mapping preserves magnitudes of vectors.

The proof that the mapping preserves angles is similar. It is left as an exercise.

✦ *Example 5*

Consider the matrix $A = \begin{pmatrix} 0 & 0 & -1 \\ 1 & 0 & 0 \\ 0 & -1 & 0 \end{pmatrix}$. Its columns are the vectors $\begin{pmatrix} 0 \\ 1 \\ 0 \end{pmatrix}$, $\begin{pmatrix} 0 \\ 0 \\ -1 \end{pmatrix}$, and $\begin{pmatrix} -1 \\ 0 \\ 0 \end{pmatrix}$.

These vectors are unit mutually orthogonal vectors. Thus the matrix is an orthogonal matrix. Let us show that when this matrix is used as a mapping, it preserves magnitudes and angles. Let $\mathbf{x} = \begin{pmatrix} a \\ b \\ c \end{pmatrix}$ and $\mathbf{y} = \begin{pmatrix} p \\ q \\ r \end{pmatrix}$ be arbitrary elements of $\mathbf{R}^3$. Then $A\mathbf{x} = \begin{pmatrix} -c \\ a \\ -b \end{pmatrix}$ and $A\mathbf{y} = \begin{pmatrix} -r \\ p \\ -q \end{pmatrix}$. Since $\|\mathbf{x}\| = \sqrt{a^2 + b^2 + c^2} = \|A\mathbf{x}\|$, the mapping preserves magnitudes.

Further,

$$\frac{\mathbf{x}}{\|\mathbf{x}\|} \cdot \frac{\mathbf{y}}{\|\mathbf{y}\|} = \frac{\begin{pmatrix} a \\ b \\ c \end{pmatrix}}{\sqrt{a^2 + b^2 + c^2}} \cdot \frac{\begin{pmatrix} p \\ q \\ r \end{pmatrix}}{\sqrt{p^2 + q^2 + r^2}}$$

$$= \frac{ap + bq + cr}{\sqrt{a^2 + b^2 + c^2}\,\sqrt{p^2 + q^2 + r^2}}$$

and

$$\frac{A\mathbf{x}}{\|A\mathbf{x}\|} \cdot \frac{A\mathbf{y}}{\|A\mathbf{y}\|} = \frac{\begin{pmatrix} -c \\ a \\ -b \end{pmatrix}}{\sqrt{c^2 + a^2 + b^2}} \cdot \frac{\begin{pmatrix} -r \\ p \\ -q \end{pmatrix}}{\sqrt{r^2 + p^2 + q^2}}$$

$$= \frac{cr + ap + bq}{\sqrt{a^2 + b^2 + c^2}\,\sqrt{p^2 + q^2 + r^2}} = \frac{\mathbf{x}}{\|\mathbf{x}\|} \cdot \frac{\mathbf{y}}{\|\mathbf{y}\|}$$

Thus the mapping preserves angles.   ✦

We now complete this introduction to orthogonal matrices with two additional useful results.

Theorem 6-2   *If A is an orthogonal matrix, then $A^{-1} = A^t$.*

Thus the inverse of an orthogonal matrix always exists and can be conveniently found by taking the transpose of the matrix.

Proof: It is necessary to show that $A^t A = A A^t = I$.
Let $B = A^t$. Then

$$A^t A = BA = \left(\sum_k b_{ik}\, a_{kj}\right) = \left(\sum_k a_{ki}\, a_{kj}\right) = (\delta_{ij}) = I$$

This proves the first part of the theorem.
Let us now prove that $AA^t = I$.
Since $A^t A = I$, $|A^t A| = 1$, implying that $|A^t|\,|A| = 1$. Thus, $|A^t| \neq 0$ and $|A| \neq 0$; $(A^t)^{-1}$ and A^{-1} exist, by Theorem 3-10.
Return to the equation

$$A^t A = I$$

Multiply both sides by (A^t) to get

$$A^t A A^t = A^t$$

Premultiplying both sides by $(A^t)^{-1}$ gives

$$A A^t = I$$

proving the second part of the theorem.

Theorem 6-3   *Let A be an orthogonal matrix. Then $|A| = \pm 1$.*

Proof: $A A^t = I$ implies that $|A|\,|A^t| = 1$. But $|A| = |A^t|$. Thus $|A|^2 = 1$, implying that $|A| = \pm 1$.

✦ *Example 6* ───

Consider the orthogonal matrix $A = \begin{pmatrix} \cos \theta & -\sin \theta \\ \sin \theta & \cos \theta \end{pmatrix}$. Observe that $A^t = \begin{pmatrix} \cos \theta & \sin \theta \\ -\sin \theta & \cos \theta \end{pmatrix}$ and that

$$A^t A = \begin{pmatrix} \cos \theta & \sin \theta \\ -\sin \theta & \cos \theta \end{pmatrix} \begin{pmatrix} \cos \theta & -\sin \theta \\ \sin \theta & \cos \theta \end{pmatrix} = \begin{pmatrix} 1 & 0 \\ 0 & 1 \end{pmatrix}$$

It can be shown that $AA^t = I$, also. Thus

$$A^{-1} = A^t = \begin{pmatrix} \cos \theta & \sin \theta \\ -\sin \theta & \cos \theta \end{pmatrix}$$

Further,

$$|A| = \begin{vmatrix} \cos \theta & -\sin \theta \\ \sin \theta & \cos \theta \end{vmatrix} = \cos^2 \theta - (-\sin^2 \theta) = 1$$

illustrating Theorem 6-3.

Observe that on interchanging the columns of A one gets the orthogonal matrix $B = \begin{pmatrix} -\sin \theta & \cos \theta \\ \cos \theta & \sin \theta \end{pmatrix}$ and $|B| = -1$. In this manner, by interchanging columns, one can always get an orthogonal matrix having determinant -1 from one having determinant $+1$, and vice versa. This technique will be useful in Section 7-4 on transformations when we are interested in constructing orthogonal matrices having determinants $+1$. There orthogonal matrices will be used to describe rotations of coordinate systems.   ✦

Inverse Mappings

We have seen how certain square matrices have inverses. Let us now focus on such matrices interpreted as mappings. Let A be an $n \times n$ matrix having inverse A^{-1} (A not necessarily orthogonal). A defines a mapping of $\mathbf{R}^n$ into $\mathbf{R}^n$. Let $\mathbf{x}$ be an element of $\mathbf{R}^n$ that is mapped into the element $\mathbf{y}$,

$$A\mathbf{x} = \mathbf{y}$$

Multiplying both sides of this equation by A^{-1},

$$A^{-1}A\mathbf{x} = A^{-1}\mathbf{y}$$

gives

$$I_n\mathbf{x} = A^{-1}\mathbf{y}$$
$$\mathbf{x} = A^{-1}\mathbf{y}$$

Thus A^{-1} maps $\mathbf{y}$ into $\mathbf{x}$. A^{-1} interpreted as a mapping of $\mathbf{R}^n$ into $\mathbf{R}^n$ is the *inverse mapping* of the mapping defined by A; it brings the vector back.

A maps $\mathbf{x} \mapsto \mathbf{y}$

A^{-1} maps $\mathbf{y} \mapsto \mathbf{x}$

Given the image vector $\mathbf{y}$, one can use A^{-1} to determine the original vector $\mathbf{x}$.

✦ *Example 7*

Interpret the matrix $\begin{pmatrix} 1 & 2 \\ 3 & 4 \end{pmatrix}$ as a mapping of $\mathbf{R}^2$ into $\mathbf{R}^2$. Determine the vector that is mapped into $\begin{pmatrix} 3 \\ -1 \end{pmatrix}$ by this matrix.

We require the vector $\mathbf{x}$ such that

$$\begin{pmatrix} 1 & 2 \\ 3 & 4 \end{pmatrix} \mathbf{x} = \begin{pmatrix} 3 \\ -1 \end{pmatrix}$$

In Example 3 of Section 2-5, we found that the matrix $\begin{pmatrix} 1 & 2 \\ 3 & 4 \end{pmatrix}$ had inverse $\begin{pmatrix} -2 & 1 \\ \frac{3}{2} & -\frac{1}{2} \end{pmatrix}$. Thus

$$\mathbf{x} = \begin{pmatrix} -2 & 1 \\ \frac{3}{2} & -\frac{1}{2} \end{pmatrix} \begin{pmatrix} 3 \\ -1 \end{pmatrix} = \begin{pmatrix} -7 \\ 5 \end{pmatrix} \quad ✦$$

This discussion is another way of looking at the matrix inverse method of determining a solution to a system of linear equations (Section 2-5). Let $A\mathbf{x} = \mathbf{y}$ be a system of n equations in n variables, where A^{-1} exists. Here $\mathbf{y}$ is given and we are to find $\mathbf{x}$. We know that such a system has a unique solution given by $\mathbf{x} = A^{-1}\mathbf{y}$. Mappings enable us to discuss systems of linear equations in a very elegant and powerful manner. We shall pursue this approach for more general systems in following sections.

EXERCISES

1. Determine the matrix that can be used to rotate the points that make up a plane through each of the following angles about the origin:

   *(a) $\pi/4$ in a counterclockwise direction   (b) $\pi/2$ in a clockwise direction
   *(c) π in a counterclockwise direction

   Determine the point that $\begin{pmatrix} 2 \\ 1 \end{pmatrix}$ is mapped into in each case. This point is the *image* of $\begin{pmatrix} 2 \\ 1 \end{pmatrix}$ under the mapping.

2. Prove that the matrix $\begin{pmatrix} \cos\theta & \sin\theta \\ \sin\theta & -\cos\theta \end{pmatrix}$ is an orthogonal matrix.

*3. Determine the matrix that can be used to map each point of a plane into its mirror image in the x axis (the matrix that maps $\begin{pmatrix} 1 \\ 1 \end{pmatrix} \mapsto \begin{pmatrix} 1 \\ -1 \end{pmatrix}, \begin{pmatrix} 1 \\ 0 \end{pmatrix} \mapsto \begin{pmatrix} 1 \\ 0 \end{pmatrix}, \begin{pmatrix} -2 \\ -2 \end{pmatrix} \mapsto \begin{pmatrix} -2 \\ 2 \end{pmatrix}$, etc.).

4. (a) Determine the matrix that can be used to map each point of the plane into its mirror image in the y axis.

(b) Determine the matrix that can be used to map each point $\begin{pmatrix} x \\ y \end{pmatrix}$ into the point $\begin{pmatrix} y \\ x \end{pmatrix}$. Give the geometric interpretation of this mapping.

5. Determine the effect of the matrix $\begin{pmatrix} c & 0 \\ 0 & c \end{pmatrix}$, where $c < 0$, as a mapping of $\mathbf{R}^2$ into itself.

6. Determine the matrix that rotates the points in the plane through an angle of $\pi/2$ counterclockwise about the origin, and at the same time expands the points to twice the distance from the origin. (*Hint*: Determine the two matrices and multiply.) Is this equivalent to a rotation followed by an expansion? Is this equivalent to an expansion followed by a rotation? Discuss this from the geometrical and algebraic viewpoints.

*7. Determine the general matrix that could be used to define a rotation and expansion about the origin in a plane. Show geometrically and algebraically that this is equivalent both to a rotation followed by an expansion and to an expansion followed by a rotation.

8. Discuss the following matrices as mappings by considering the deformations of a square with vertices $\begin{pmatrix} 0 \\ 0 \end{pmatrix}, \begin{pmatrix} 1 \\ 0 \end{pmatrix}, \begin{pmatrix} 1 \\ 1 \end{pmatrix}$, and $\begin{pmatrix} 0 \\ 1 \end{pmatrix}$:

*(a) $\begin{pmatrix} 0 & -1 \\ 1 & 0 \end{pmatrix}$ **(b)** $\begin{pmatrix} 2 & 0 \\ 0 & 2 \end{pmatrix}$ ***(c)** $\begin{pmatrix} 3 & 0 \\ 1 & 4 \end{pmatrix}$

(d) $\begin{pmatrix} 4 & -1 \\ 1 & 5 \end{pmatrix}$ ***(e)** $\begin{pmatrix} -2 & -3 \\ 0 & 4 \end{pmatrix}$ **(f)** $\begin{pmatrix} -2 & -4 \\ -4 & -1 \end{pmatrix}$

9. Discuss the matrix $\begin{pmatrix} 1 & k \\ 0 & 1 \end{pmatrix}$ as a mapping of $\mathbf{R}^2$ into itself by considering its effect on the rectangle having vertices $\begin{pmatrix} 0 \\ 0 \end{pmatrix}, \begin{pmatrix} a \\ 0 \end{pmatrix}, \begin{pmatrix} a \\ b \end{pmatrix}$, and $\begin{pmatrix} 0 \\ b \end{pmatrix}$, where $a > 0$ and $b > 0$. Such a mapping is a *shear in the x-direction with factor k*. A shear in the y-direction with factor k is described by the matrix $\begin{pmatrix} 1 & 0 \\ k & 1 \end{pmatrix}$.

10. Prove that a 2×2 matrix always maps straight lines into straight lines or points. Prove that the origin is mapped into the origin.

*11. Determine the matrix that can be used to rotate the points that make up a three-dimensional space through an angle of $\pi/2$ about the z axis. (You may consider either direction.)

*12. Determine the matrix that expands three-dimensional space outward from the origin so that each point moves to a point three times as far away.

13. Prove that each of the following matrices is an orthogonal matrix. Determine the inverse of each matrix.

(a) $\begin{pmatrix} 1 & 0 \\ 0 & 1 \end{pmatrix}$ **(b)** $\begin{pmatrix} 0 & -1 \\ 1 & 0 \end{pmatrix}$ **(c)** $\begin{pmatrix} \dfrac{1}{\sqrt{2}} & \dfrac{1}{\sqrt{2}} \\ -\dfrac{1}{\sqrt{2}} & \dfrac{1}{\sqrt{2}} \end{pmatrix}$

(d) $\begin{pmatrix} 1 & 0 & 0 \\ 0 & 0 & -1 \\ 0 & 1 & 0 \end{pmatrix}$ **(e)** $\begin{pmatrix} 0 & \dfrac{1}{\sqrt{2}} & -\dfrac{1}{\sqrt{2}} \\ -\dfrac{2}{\sqrt{6}} & \dfrac{1}{\sqrt{6}} & \dfrac{1}{\sqrt{6}} \\ \dfrac{1}{\sqrt{3}} & \dfrac{1}{\sqrt{3}} & \dfrac{1}{\sqrt{3}} \end{pmatrix}$

14. Prove that when an orthogonal matrix is used as a mapping, it preserves angles.

15. Let A and B be orthogonal matrices of the same kind. Prove that AB is also orthogonal. Interpret this result geometrically from the point of view of preserving magnitudes and angles.

*16. The matrix $\begin{pmatrix} 2 & 1 \\ 4 & 3 \end{pmatrix}$ has inverse $\begin{pmatrix} \frac{3}{2} & -\frac{1}{2} \\ -2 & 1 \end{pmatrix}$. Interpret $\begin{pmatrix} 2 & 1 \\ 4 & 3 \end{pmatrix}$ as a mapping of $\mathbf{R}^2$ into $\mathbf{R}^2$. Determine the vectors that are mapped into $\begin{pmatrix} 2 \\ -4 \end{pmatrix}$ and $\begin{pmatrix} 0 \\ 2 \end{pmatrix}$.

17. The matrix $A = \begin{pmatrix} 0 & 3 & 3 \\ 1 & 2 & 3 \\ 1 & 4 & 6 \end{pmatrix}$ has inverse $\begin{pmatrix} 0 & 2 & -1 \\ 1 & 1 & -1 \\ -\frac{2}{3} & -1 & 1 \end{pmatrix}$. Interpret A as a mapping of $\mathbf{R}^3$ into $\mathbf{R}^3$. Determine the vectors that are mapped into $\begin{pmatrix} 1 \\ 0 \\ 0 \end{pmatrix}$ and $\begin{pmatrix} 3 \\ 6 \\ -3 \end{pmatrix}$.

Interpret your results in terms of solutions to systems of linear equations.

18. Write a program that can be used to perform rotations in a plane. The data should include the relevant angle and point. The output should be the image point. Check your answers to Exercise 1 using your program.

6-2 LINEAR MAPPINGS

In the previous section, we saw how any $n \times n$ matrix can be used to define a mapping of $\mathbf{R}^n$ into itself. In this section, we shall generalize this concept to mappings between vector spaces.

Consider the vector spaces $\mathbf{R}^n$ and $\mathbf{R}^m$ and an $m \times n$ matrix A. (We do not exclude the possibility that $n = m$.) A can be used to define a mapping of $\mathbf{R}^n$ into $\mathbf{R}^m$. Let $\mathbf{x}$ be an arbitrary element of $\mathbf{R}^n$ interpreted as a column vector. Then $A\mathbf{x}$ exists and is a column vector, an element of $\mathbf{R}^m$. We say that A maps the element $\mathbf{x}$ of $\mathbf{R}^n$ into the element $A\mathbf{x}$ of $\mathbf{R}^m$. We indicate this, as in the case of mappings of $\mathbf{R}^n$ into itself,

$$A: \mathbf{x} \longmapsto A\mathbf{x}$$

$A\mathbf{x}$ is the *image* of $\mathbf{x}$ under the mapping.

✦ **Example 1**

Consider the 2×3 matrix $\begin{pmatrix} 1 & -1 & 0 \\ 2 & 1 & 3 \end{pmatrix}$. This matrix can be used to define a mapping of $\mathbf{R}^3$ into $\mathbf{R}^2$ (Figure 6-8).

Let $\begin{pmatrix} x \\ y \\ z \end{pmatrix}$ be an arbitrary element of $\mathbf{R}^3$. We have that

$$\begin{pmatrix} 1 & -1 & 0 \\ 2 & 1 & 3 \end{pmatrix} : \begin{pmatrix} x \\ y \\ z \end{pmatrix} \longmapsto \begin{pmatrix} 1 & -1 & 0 \\ 2 & 1 & 3 \end{pmatrix} \begin{pmatrix} x \\ y \\ z \end{pmatrix}$$

$$= \begin{pmatrix} x - y \\ 2x + y + 3z \end{pmatrix}$$

The image of the element $\begin{pmatrix} x \\ y \\ z \end{pmatrix}$ of $\mathbf{R}^3$ under this mapping is the element $\begin{pmatrix} x - y \\ 2x + y + 3z \end{pmatrix}$ of $\mathbf{R}^2$. The image of a specific element of $\mathbf{R}^3$, such as $\begin{pmatrix} 1 \\ -2 \\ 3 \end{pmatrix}$, would be $\begin{pmatrix} 1 + 2 \\ 2 - 2 + 9 \end{pmatrix}$, or $\begin{pmatrix} 3 \\ 9 \end{pmatrix}$.

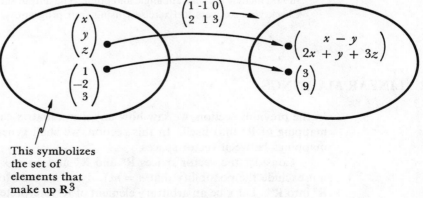

This symbolizes the set of elements that make up $\mathbf{R}^3$

Figure 6-8

✦

Let A be an $m \times n$ matrix interpreted as a mapping of $\mathbf{R}^n$ into $\mathbf{R}^m$. If $\mathbf{x}$ and $\mathbf{y}$ are elements of $\mathbf{R}^n$ and c is a scalar, then we know from the properties of matrices that

$$A(\mathbf{x} + \mathbf{y}) = A\mathbf{x} + A\mathbf{y}$$

and

$$A(c\mathbf{x}) = cA\mathbf{x}$$

(Exercise 8). These results are very significant. The first equation implies that A maps the vector $\mathbf{x} + \mathbf{y}$, an element of $\mathbf{R}^n$ that is the sum of $\mathbf{x}$ and $\mathbf{y}$, into the vector $A\mathbf{x} + A\mathbf{y}$, an element of $\mathbf{R}^m$ that is the sum of $A\mathbf{x}$ and $A\mathbf{y}$. Thus the operation of addition is preserved on these spaces.

The second equation implies that A maps the vector $c\mathbf{x}$, an element of $\mathbf{R}^n$ that is the scalar multiple of $\mathbf{x}$ by c, into the element $cA\mathbf{x}$, the scalar multiple of $A\mathbf{x}$ by c. The operation of scalar multiplication is thus preserved. Figure 6-9 illustrates these properties.

Mappings between vector spaces that preserve addition and scalar multiplication are of great importance in the analyses and applications of vector spaces. These mappings preserve the mathematical structures of the vector spaces. They are called linear mappings. The formal definition of a linear mapping follows.

Definition 6-3   Let V and U be vector spaces. f is a *linear mapping* of V into U if and only if

$$f(\mathbf{v}_1 + \mathbf{v}_2) = f(\mathbf{v}_1) + f(\mathbf{v}_2)$$

and

$$f(c\mathbf{v}_1) = cf(\mathbf{v}_1)$$

for arbitrary vectors $\mathbf{v}_1$ and $\mathbf{v}_2$ of V and arbitrary scalar c. V is called the *domain* of f.

Thus a matrix interpreted as a mapping is a linear mapping.

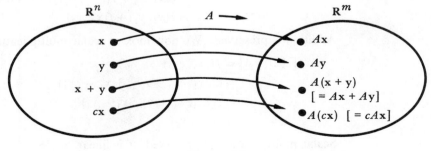

Figure 6-9

✦ *Example 2*

Consider the mapping f of $\mathbf{R}^3$ into $\mathbf{R}^2$ defined by

$$f(x, y, z) = (x - y, 3z)$$

This mapping takes the vector (x, y, z) of $\mathbf{R}^3$ into the vector $(x - y, 3z)$ of $\mathbf{R}^2$. For example, the image of the vector $(1, 2, 3)$ is $(1 - 2, 3(3))$, or $(-1, 9)$. Observe that we have expressed the elements of $\mathbf{R}^3$ and $\mathbf{R}^2$ as row vectors here. This is customary when discussing a mapping f, one reason being that row vectors take up less text space than column vectors.

Let us show that this mapping is linear. Let (x_1, y_1, z_1) and (x_2, y_2, z_2) be arbitrary elements of $\mathbf{R}^3$, and let c be a scalar. Then

$$
\begin{aligned}
f[(x_1, y_1, z_1) + (x_2, y_2, z_2)] &= f(x_1 + x_2, y_1 + y_2, z_1 + z_2) \text{ by vector addition} \\
&= (x_1 + x_2 - y_1 - y_2, 3z_1 + 3z_2) \text{ by definition of } f \\
&= (x_1 - y_1, 3z_1) + (x_2 - y_2, 3z_2) \\
&= f(x_1, y_1, z_1) + f(x_2, y_2, z_2)
\end{aligned}
$$

Thus f preserves addition. Let us now look at scalar multiplication.

$$
\begin{aligned}
f[c(x_1, y_1, z_1)] &= f(cx_1, cy_1, cz_1) \text{ by scalar multiplication} \\
&= (cx_1 - cy_1, 3cz_1) \text{ by definition of } f \\
&= c(x_1 - y_1, 3z_1) \\
&= cf(x_1, y_1, z_1)
\end{aligned}
$$

f preserves scalar multiplication. f is linear.   ✦

✦ *Example 3*

Let us show that the following mapping of $\mathbf{R}^2$ into $\mathbf{R}^3$ is linear:

$$f(x, y) = (2x, x + y, 3x - y)$$

Let (x_1, y_1) and (x_2, y_2) be elements of $\mathbf{R}^2$ and let c be a scalar. Then

$$
\begin{aligned}
f[(x_1, y_1) + (x_2, y_2)] &= f(x_1 + x_2, y_1 + y_2) \\
&= (2x_1 + 2x_2, x_1 + x_2 + y_1 + y_2, 3x_1 + 3x_2 - y_1 - y_2) \\
&= (2x_1, x_1 + y_1, 3x_1 - y_1) + (2x_2, x_2 + y_2, 3x_2 - y_2) \\
&= f(x_1, y_1) + f(x_2, y_2)
\end{aligned}
$$

Addition is preserved. We now look at scalar multiplication.

$$
\begin{aligned}
f[c(x_1, y_1)] &= f(cx_1, cy_1) \\
&= (2cx_1, cx_1 + cy_1, 3cx_1 - cy_1) \\
&= c(2x_1, x_1 + y_1, 3x_1 - y_1) \\
&= cf(x_1, y_1)
\end{aligned}
$$

Scalar multiplication is preserved. f is linear.   ✦

The following example illustrates a nonlinear mapping.

✦ *Example 4*

Examine the mapping $f(x, y, z) = (xy, z)$ of $\mathbf{R}^3$ into $\mathbf{R}^2$ for linearity.

Let (x_1, y_1, z_1) and (x_2, y_2, z_2) be two arbitrary elements of $\mathbf{R}^3$. We have that

$$
\begin{aligned}
f[(x_1, y_1, z_1) + (x_2, y_2, z_2)] &= f(x_1 + x_2, y_1 + y_2, z_1 + z_2) \\
&= ((x_1 + x_2)(y_1 + y_2), z_1 + z_2) \\
&\qquad \text{by definition of } f \\
&= (x_1 y_1 + x_2 y_2 + x_1 y_2 + x_2 y_1, z_1 + z_2)
\end{aligned}
$$

and

$$
\begin{aligned}
f(x_1, y_1, z_1) + f(x_2, y_2, z_2) &= (x_1 y_1, z_1) + (x_2 y_2, z_2) \\
&= (x_1 y_1 + x_2 y_2, z_1 + z_2)
\end{aligned}
$$

We see that, in general,

$$
f[(x_1, y_1, z_1) + (x_2, y_2, z_2)] \neq f(x_1, y_1, z_1) + f(x_2, y_2, z_2)
$$

Since the first linearity condition is not satisfied, vector addition is not preserved; f is not linear. In practice, one need not check the second condition. We do so here to illustrate the manner in which this condition is violated for this mapping also. Let c be an arbitrary scalar. Then

$$
\begin{aligned}
f[c(x_1, y_1, z_1)] &= f[(cx_1, cy_1, cz_1)] \\
&= (cx_1 cy_1, cz_1) \text{ by definition of } f \\
&= (c^2 x_1 y_1, cz_1)
\end{aligned}
$$

However,

$$
cf(x_1, y_1, z_1) = c(x_1 y_1, z_1) = (cx_1 y_1, cz_1)
$$

Thus, in general, $f[c(x_1, y_1, z_1)] \neq cf(x_1, y_1, z_1)$; the second condition is also violated.   ✦

Section 5-1 on general vector spaces is a prerequisite for Examples 5 and 6.

✦ *Example 5*

Consider the following mapping f of P_2 into P_1. Let us show that f is a linear mapping.

$$
f(a_2 x^2 + a_1 x + a_0) = (a_2 + a_1)x + a_0
$$

Let $a_2x^2 + a_1x + a_0$ and $b_2x^2 + b_1x + b_0$ be elements of P_2, and let c be a scalar. Then

$$f[(a_2x^2 + a_1x + a_0) + (b_2x^2 + b_1x + b_0)]$$
$$= f[(a_2 + b_2)x^2 + (a_1 + b_1)x + a_0 + b_0]$$
$$= (a_2 + b_2 + a_1 + b_1)x + a_0 + b_0$$
$$= (a_2 + a_1)x + a_0 + (b_2 + b_1)x + b_0$$
$$= f(a_2x^2 + a_1x + a_0) + f(b_2x^2 + b_1x + b_0)$$

f preserves addition.

$$f[c(a_2x^2 + a_1x + a_0)] = f(ca_2x^2 + ca_1x + ca_0)$$
$$= (ca_2 + ca_1)x + ca_0$$
$$= c[(a_2 + a_1)x + a_0]$$
$$= cf(a_2x^2 + a_1x + a_0)$$

f preserves scalar multiplication. f is linear.   ✦

✦ **Example 6** ─────────────────────────────────

Let V be the vector space of functions that possess derivatives of all orders on some interval (a, b). Let D be the operation of taking the derivative. D can be interpreted as a mapping of V into itself. For example,

$$D(4x^3 - 3x^2 + 2x + 1) = 12x^2 - 6x + 2$$

D maps the element $4x^3 - 3x^2 + 2x + 1$ of V into the element $12x^2 - 6x + 2$ of V. The following properties of differentiation imply that D is a linear mapping:

$$D(f + g) = Df + Dg$$

and

$$D(cf) = cDf$$

for elements f and g of V and a scalar c.   ✦

Linear Mappings in Computer Graphics*

Computer graphics is the field that studies the creation and manipulation of pictures with the aid of computers. The impact of computer graphics is felt in many homes through video games; its uses in research, industry, and business are vast and are ever expanding. Architects use computer graphics to explore designs, molecular biologists display pictures of molecules to gain insight into their structure, pilots are trained using flight simulators, and transportation engineers use computer-generated maps in their planning work—to mention but a few applications.

The manipulation of pictures in computer graphics is carried out using sequences of mappings. (The term *transformation* is also commonly used for a mapping in computer graphics.) There are four classes of mappings that are of prime importance: *rotation, scaling, translation,* and *perspective*. We have already looked at rotations; let us now focus on scaling and translation.

A *scaling* is a mapping of $\mathbf{R}^2$ into $\mathbf{R}^2$ defined by a matrix $\begin{pmatrix} c & 0 \\ 0 & d \end{pmatrix}$ where $c > 0$ and $d > 0$. Observe that dilation and contraction are special cases of such a mapping, when $c = d$. In general, a scaling distorts a figure, since x and y do not change in the same manner.

As an example of scaling, consider the effect of the matrix $\begin{pmatrix} 3 & 0 \\ 0 & \frac{1}{2} \end{pmatrix}$ on the triangle having vertices $\begin{pmatrix} 1 \\ 2 \end{pmatrix}$, $\begin{pmatrix} 2 \\ 8 \end{pmatrix}$, and $\begin{pmatrix} 3 \\ 2 \end{pmatrix}$. We get

$$\begin{pmatrix} 1 \\ 2 \end{pmatrix} \mapsto \begin{pmatrix} 3 \\ 1 \end{pmatrix}, \qquad \begin{pmatrix} 2 \\ 8 \end{pmatrix} \mapsto \begin{pmatrix} 6 \\ 4 \end{pmatrix}, \qquad \begin{pmatrix} 3 \\ 2 \end{pmatrix} \mapsto \begin{pmatrix} 9 \\ 1 \end{pmatrix}$$

The triangle *ABC* is distorted into triangle *PQR* in Figure 6-10.

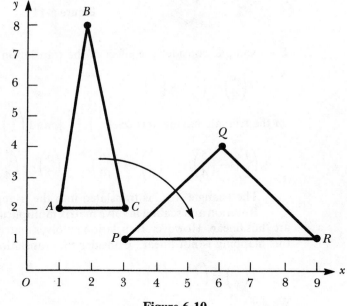

Figure 6-10

Let us now *translate* the point *A* to the point *B* in Figure 6-11. It can be seen that

$$\begin{pmatrix} x' \\ y' \end{pmatrix} = \begin{pmatrix} x \\ y \end{pmatrix} + \begin{pmatrix} h \\ k \end{pmatrix}$$

Let us represent the mapping by the letter *T* and write

$$T\begin{pmatrix} x \\ y \end{pmatrix} = \begin{pmatrix} x \\ y \end{pmatrix} + \begin{pmatrix} h \\ k \end{pmatrix}$$

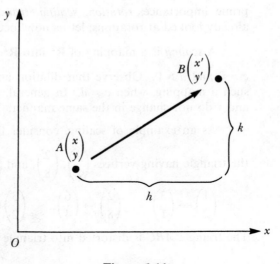

Figure 6-11

For example, consider the effect of the translation

$$T\begin{pmatrix} x \\ y \end{pmatrix} = \begin{pmatrix} x \\ y \end{pmatrix} + \begin{pmatrix} 4 \\ 2 \end{pmatrix}$$

on the triangle having vertices $\begin{pmatrix} 1 \\ 2 \end{pmatrix}$, $\begin{pmatrix} 2 \\ 8 \end{pmatrix}$, and $\begin{pmatrix} 3 \\ 2 \end{pmatrix}$. We get

$$\begin{pmatrix} 1 \\ 2 \end{pmatrix} \mapsto \begin{pmatrix} 5 \\ 4 \end{pmatrix}, \qquad \begin{pmatrix} 2 \\ 8 \end{pmatrix} \mapsto \begin{pmatrix} 6 \\ 10 \end{pmatrix}, \qquad \begin{pmatrix} 3 \\ 2 \end{pmatrix} \mapsto \begin{pmatrix} 7 \\ 4 \end{pmatrix}$$

The triangle ABC is translated into the triangle LMN in Figure 6-12.

Rotation and scaling involve matrix multiplication; both of these mappings are thus linear. However, translation involves matrix addition. Let us show that this mapping is not linear. Consider the translation

$$T\begin{pmatrix} x \\ y \end{pmatrix} = \begin{pmatrix} x \\ y \end{pmatrix} + \begin{pmatrix} h \\ k \end{pmatrix}$$

We get

$$T\left[\begin{pmatrix} x_1 \\ y_1 \end{pmatrix} + \begin{pmatrix} x_2 \\ y_2 \end{pmatrix} \right] = T\begin{pmatrix} x_1 + x_2 \\ y_1 + y_2 \end{pmatrix} = \begin{pmatrix} x_1 + x_2 \\ y_1 + y_2 \end{pmatrix} + \begin{pmatrix} h \\ k \end{pmatrix}$$

while

$$T\begin{pmatrix} x_1 \\ y_1 \end{pmatrix} + T\begin{pmatrix} x_2 \\ y_2 \end{pmatrix} = \begin{pmatrix} x_1 \\ y_1 \end{pmatrix} + \begin{pmatrix} h \\ k \end{pmatrix} + \begin{pmatrix} x_2 \\ y_2 \end{pmatrix} + \begin{pmatrix} h \\ k \end{pmatrix} = \begin{pmatrix} x_1 + x_2 \\ y_1 + y_2 \end{pmatrix} + 2\begin{pmatrix} h \\ k \end{pmatrix}$$

Thus, T does not preserve vector addition; it is not linear. It can be shown that T does not preserve scalar multiplication either.

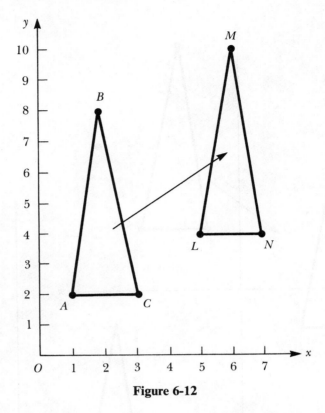

Figure 6-12

Mathematicians have, however, discovered a way of representing translation by matrix multiplication, thus making it into a linear mapping. This is convenient in that it then enables one to accomplish a sequence of rotations, scalings, and translations by means of matrix multiplications. To accomplish this we have to slightly modify the coordinates of points and the rotation and scaling matrices. We use *homogeneous coordinates* to represent points; a third coordinate 1 is added to each point. A point X, rotation matrix R, scaling matrix S, and translation matrix T are then defined as follows:

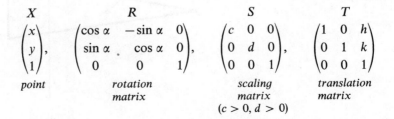

$$
\begin{array}{cccc}
X & R & S & T \\
\begin{pmatrix} x \\ y \\ 1 \end{pmatrix}, & \begin{pmatrix} \cos\alpha & -\sin\alpha & 0 \\ \sin\alpha & \cos\alpha & 0 \\ 0 & 0 & 1 \end{pmatrix}, & \begin{pmatrix} c & 0 & 0 \\ 0 & d & 0 \\ 0 & 0 & 1 \end{pmatrix}, & \begin{pmatrix} 1 & 0 & h \\ 0 & 1 & k \\ 0 & 0 & 1 \end{pmatrix} \\
\text{point} & \begin{array}{c}\text{rotation}\\\text{matrix}\end{array} & \begin{array}{c}\text{scaling}\\\text{matrix}\\(c>0, d>0)\end{array} & \begin{array}{c}\text{translation}\\\text{matrix}\end{array}
\end{array}
$$

Thus, for example, the image of X under first a translation and then a rotation would be the point found on performing the matrix multiplication RTX.

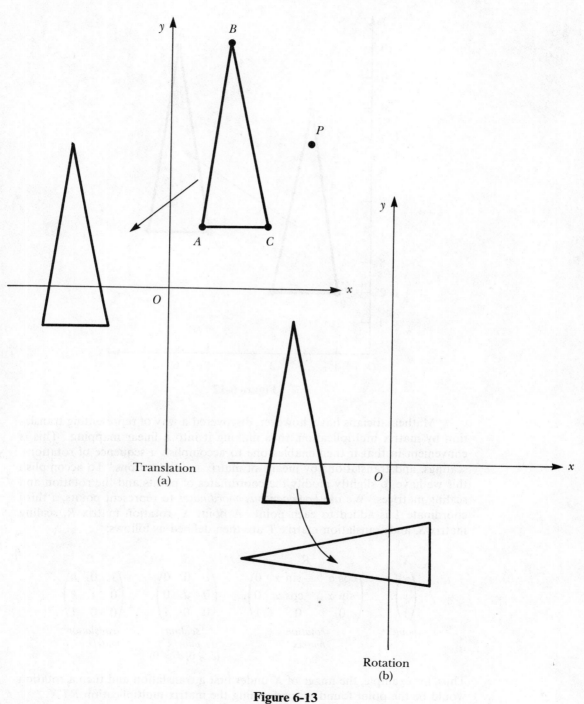

Translation
(a)

Rotation
(b)

Figure 6-13

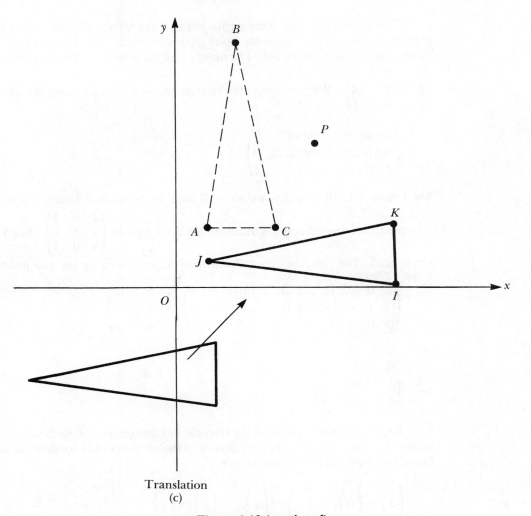

Translation
(c)

Figure 6-13 (*continued*)

Note that these matrices are not commutative under multiplication. It does matter in which sequence the transformations are performed.

We complete this section by illustrating how these mappings are used to rotate a geometrical figure about a point other than the origin. Let us rotate the triangle having vertices $A\begin{pmatrix}1\\2\end{pmatrix}$, $B\begin{pmatrix}2\\8\end{pmatrix}$, and $C\begin{pmatrix}3\\2\end{pmatrix}$ in a counterclockwise direction about the point $P\begin{pmatrix}5\\4\end{pmatrix}$. The original situation is illustrated in Figure 6-13(a).

We accomplish this desired rotation in three steps. We first of all perform a translation that takes the point P to the origin. The triangle is translated using the same mapping. The matrix that accomplishes this translation is $\begin{pmatrix} 1 & 0 & -5 \\ 0 & 1 & -4 \\ 0 & 0 & 1 \end{pmatrix}$. We now rotate the triangle about the origin using the matrix

$$\begin{pmatrix} \cos(\pi/2) & -\sin(\pi/2) & 0 \\ \sin(\pi/2) & \cos(\pi/2) & 0 \\ 0 & 0 & 1 \end{pmatrix}$$

See Figure 6-13(b). Finally, we move P back to its original location and the triangle to its desired rotated location using the matrix $\begin{pmatrix} 1 & 0 & 5 \\ 0 & 1 & 4 \\ 0 & 0 & 1 \end{pmatrix}$. See Figure 6-13(c). The rotation about P can thus be performed by the one product

$$\begin{pmatrix} 1 & 0 & 5 \\ 0 & 1 & 4 \\ 0 & 0 & 1 \end{pmatrix} \begin{pmatrix} \cos(\pi/2) & -\sin(\pi/2) & 0 \\ \sin(\pi/2) & \cos(\pi/2) & 0 \\ 0 & 0 & 1 \end{pmatrix} \begin{pmatrix} 1 & 0 & -5 \\ 0 & 1 & -4 \\ 0 & 0 & 1 \end{pmatrix}$$

or

$$\begin{pmatrix} 0 & -1 & 9 \\ 1 & 0 & -1 \\ 0 & 0 & 1 \end{pmatrix}$$

Expressing the vertices of the triangle in homogeneous coordinate form, we see that the homogeneous coordinates of the vertices of the rotated triangle, found by using the above matrix, are

$$\begin{pmatrix} 1 \\ 2 \\ 1 \end{pmatrix} \mapsto \begin{pmatrix} 7 \\ 0 \\ 1 \end{pmatrix}, \quad \begin{pmatrix} 2 \\ 8 \\ 1 \end{pmatrix} \mapsto \begin{pmatrix} 1 \\ 1 \\ 1 \end{pmatrix}, \quad \begin{pmatrix} 3 \\ 2 \\ 1 \end{pmatrix} \mapsto \begin{pmatrix} 7 \\ 2 \\ 1 \end{pmatrix}$$

In Cartesian coordinates, the vertices of the rotated triangle are $H\begin{pmatrix} 7 \\ 0 \end{pmatrix}$, $I\begin{pmatrix} 1 \\ 1 \end{pmatrix}$, and $J\begin{pmatrix} 7 \\ 2 \end{pmatrix}$. See Figure 6-13(c).

We can generalize this result. The single matrix that can be used to perform rotation through an angle α in a counterclockwise direction about the point (h, k) is the following product:

$$\begin{pmatrix} 1 & 0 & h \\ 0 & 1 & k \\ 0 & 0 & 1 \end{pmatrix} \begin{pmatrix} \cos\alpha & -\sin\alpha & 0 \\ \sin\alpha & \cos\alpha & 0 \\ 0 & 0 & 1 \end{pmatrix} \begin{pmatrix} 1 & 0 & -h \\ 0 & 1 & -k \\ 0 & 0 & 1 \end{pmatrix}$$

On multiplying out, we have the general rotation matrix

$$\begin{pmatrix} \cos\alpha & -\sin\alpha & -h\cos\alpha + k\sin\alpha + h \\ \sin\alpha & \cos\alpha & -h\sin\alpha - k\cos\alpha + k \\ 0 & 0 & 1 \end{pmatrix}$$

The fourth linear mapping used in computer graphics—*perspective*—enables one to see an object in a manner that allows for depth. Readers who are interested in reading more about computer graphics are referred to *Principles of Interactive Computer Graphics*, 2nd edition, by William M. Newman and Robert F. Sproull, McGraw-Hill, 1979, and *Interactive Computer Graphics*, by Wolfgang K. Giloi, Prentice-Hall, Inc., 1978.

EXERCISES

1. Interpret the matrix $\begin{pmatrix} 1 & 2 \\ -1 & 3 \\ 1 & 2 \end{pmatrix}$ as a mapping of $\mathbf{R}^2$ into $\mathbf{R}^3$. What are the images of the following elements under this mapping?

   *(a) $\begin{pmatrix} -1 \\ 1 \end{pmatrix}$ (b) $\begin{pmatrix} 2 \\ 3 \end{pmatrix}$ *(c) $\begin{pmatrix} 1 \\ 4 \end{pmatrix}$ (d) $\begin{pmatrix} -3 \\ 3 \end{pmatrix}$

   Observe that $\begin{pmatrix} 1 \\ 4 \end{pmatrix} = \begin{pmatrix} -1 \\ 1 \end{pmatrix} + \begin{pmatrix} 2 \\ 3 \end{pmatrix}$. Comment on a similar relation between the images of these three vectors. Observe that $\begin{pmatrix} -3 \\ 3 \end{pmatrix} = 3\begin{pmatrix} -1 \\ 1 \end{pmatrix}$. Comment on a similar relation between the images of $\begin{pmatrix} -3 \\ 3 \end{pmatrix}$ and $\begin{pmatrix} -1 \\ 1 \end{pmatrix}$.

2. Let A be an $m \times n$ matrix and $\mathbf{c}$ a column vector having n components. Prove that the mapping f of $\mathbf{R}^n$ into $\mathbf{R}^m$ defined by $f : \mathbf{x} \mapsto A\mathbf{x} + \mathbf{c}$ is not linear, if $\mathbf{c} \neq \mathbf{0}$.

3. Prove that the mapping f of $\mathbf{R}^3$ into $\mathbf{R}^3$ defined by $f(x, y, z) = (0, y, 0)$ is linear. Illustrate this mapping geometrically.

*4. Prove that the mapping f of $\mathbf{R}^3$ into $\mathbf{R}^2$ defined by $f(x, y, z) = (2x, y + z)$ is linear. Determine the images of the elements $(1, 2, 3)$ and $(-1, 4, 3)$ under f.

5. Prove that the mapping of $\mathbf{R}^2$ into $\mathbf{R}^3$ defined by $f(x, y) = (3x + y, 2y, x - y)$ is linear. Determine the images of the elements $(1, 2)$ and $(2, -1)$ under f.

6. Prove that the mapping f of $\mathbf{R}^2$ into $\mathbf{R}$ defined by $f(x, y) = x + a$, where a is constant, is not linear.

7. Prove that the following mappings of $\mathbf{R}^3$ into $\mathbf{R}^2$ are not linear:
   (a) $f(x, y, z) = (x^2, y^2)$ (b) $f(x, y, z) = (x + 2, y)$

8. A is an $m \times n$ matrix, $\mathbf{x}$ and $\mathbf{y}$ are elements of $\mathbf{R}^n$ in column form, and c is a scalar. Prove that
   (a) $A(\mathbf{x} + \mathbf{y}) = A\mathbf{x} + A\mathbf{y}$ (b) $A(c\mathbf{x}) = cA\mathbf{x}$
   Thus A interpreted as a mapping of $\mathbf{R}^n \to \mathbf{R}^m$ preserves addition and scalar multiplication.

In Exercises 9–14, determine whether or not the given mappings are linear.

*9. $f(x, y, z) = (2x, y)$ of $\mathbf{R}^2$ into $\mathbf{R}^2$

10. $f(x, y) = (x - y)$ of $\mathbf{R}^2$ into $\mathbf{R}$

*11. $f(x, y) = (x^2, y + 3, 0)$ of $\mathbf{R}^2$ into $\mathbf{R}^3$

12. $f(x) = (x, 2x, 3x)$ of $\mathbf{R}$ into $\mathbf{R}^3$

*13. $f(x, y) = (x, y^2)$ of $\mathbf{R}^2$ into $\mathbf{R}^2$

14. $f(x, y, z) = (x + 2y, x + y + z, 3z)$ of $\mathbf{R}^3$ into $\mathbf{R}^3$

*15. Determine whether the following mapping f of P_2 into P_2 is linear:

$$f(a_2x^2 + a_1x + a_0) = (a_2 + a_1)x^2 + a_1x + 2a_0$$

16. Determine whether the following mapping of P_2 into P_1 is linear:

$$f(a_2x^2 + a_1x + a_0) = a_2x + a_1$$

*17. Determine whether the following mapping of P_1 into P_1 is linear:

$$f(a_1x + a_0) = a_1x + 4$$

18. Determine whether the following mapping of P_3 into P_2 is linear:

$$f(a_3x^3 + a_2x^2 + a_1x + a_0) = 4x^2 + a_0$$

19. Let V be the vector space of functions that possess derivatives of all orders on an interval (a, b). Let D^2 be the operation of taking the second derivative. Prove that D^2 is a linear mapping of V into V.

*20. Find the image of the triangle having vertices $\begin{pmatrix} 1 \\ 1 \end{pmatrix}$, $\begin{pmatrix} 3 \\ 2 \end{pmatrix}$, and $\begin{pmatrix} 7 \\ 4 \end{pmatrix}$ under the mapping defined by the matrix $\begin{pmatrix} 2 & 0 \\ 0 & 4 \end{pmatrix}$.

21. Find the image of the triangle having vertices $\begin{pmatrix} 1 \\ 2 \end{pmatrix}$, $\begin{pmatrix} 3 \\ 4 \end{pmatrix}$, and $\begin{pmatrix} 4 \\ 6 \end{pmatrix}$ under the translation that takes the vertex $\begin{pmatrix} 1 \\ 2 \end{pmatrix}$ to $\begin{pmatrix} 2 \\ -3 \end{pmatrix}$.

*22. Determine the matrix that can be used to rotate a figure through $\pi/2$ about the point $\begin{pmatrix} 5 \\ 1 \end{pmatrix}$. Find the image of the square having vertices $\begin{pmatrix} 0 \\ 0 \end{pmatrix}, \begin{pmatrix} 1 \\ 0 \end{pmatrix}, \begin{pmatrix} 1 \\ 1 \end{pmatrix}$, and $\begin{pmatrix} 0 \\ 1 \end{pmatrix}$ under this rotation. (Rotation is counterclockwise.)

23. Show by means of an example that scaling and rotation are not commutative. Are translation and rotation commutative?

*24. Consider the translation defined by the matrix $\begin{pmatrix} 1 & 0 & h \\ 0 & 1 & k \\ 0 & 0 & 1 \end{pmatrix}$. Does this mapping have an inverse? If so, determine it.

25. Consider the scaling defined by $\begin{pmatrix} c & 0 & 0 \\ 0 & d & 0 \\ 0 & 0 & 1 \end{pmatrix}$. Does this mapping have an inverse? If so, determine it.

26. Find the image of the triangle having the vertices $\begin{pmatrix} 1 \\ 6 \\ 1 \end{pmatrix}$, $\begin{pmatrix} 3 \\ 0 \\ 1 \end{pmatrix}$, and $\begin{pmatrix} 4 \\ 6 \\ 1 \end{pmatrix}$ in homogeneous coordinates, under the sequence of mappings T, R, and S (in that order) defined below.

$$T = \begin{pmatrix} 1 & 0 & 4 \\ 0 & 1 & -3 \\ 0 & 0 & 1 \end{pmatrix}, \qquad R = \begin{pmatrix} 0 & 1 & 0 \\ -1 & 0 & 0 \\ 0 & 0 & 1 \end{pmatrix}, \qquad S = \begin{pmatrix} 3 & 0 & 0 \\ 0 & 5 & 0 \\ 0 & 0 & 1 \end{pmatrix}$$

6-3 KERNEL AND RANGE

There are two further vector spaces that are important in the discussion of linear mappings. These are called the kernel and range of a linear mapping. In this section, we introduce and discuss the significance of these spaces.

The following theorem presents an important property of all linear mappings. This result leads up to the concept of a kernel.

Theorem 6-4   *Let f be a linear mapping of V into U. The image of the zero vector of V is the zero vector of U.*

Proof: Let $\mathbf{0}_V$ and $\mathbf{0}_U$ be the zero vectors of V and U, respectively, and let 0 be the zero scalar. Let $\mathbf{v}$ be any vector in V with image $\mathbf{u}$ in U. Then we know that $0\mathbf{v} = \mathbf{0}_V$ and that $0\mathbf{u} = \mathbf{0}_U$ (Theorem 5-1). By the linearity of f, we have

$$f(\mathbf{0}_V) = f(0\mathbf{v}) = 0f(\mathbf{v}) = 0\mathbf{u} = \mathbf{0}_U$$

The image of the zero vector of V is the zero vector of U. We shall henceforth for convenience denote both zero vectors by $\mathbf{0}$.

We now introduce kernel and range.

Definition 6-4   Let f be a linear mapping of a vector space V into a vector space U.

The set of vectors in V that are mapped onto the zero vector is called the *kernel* of f. The kernel is denoted $\ker(f)$.

The set of all vectors in U that are the images of vectors in V is called the *range* of f.

By Theorem 6-4 we know that the kernel will always contain the zero vector of V. There could be other vectors in the kernel also. The following theorem tells us that both the kernel and the range are subspaces.

Theorem 6-5   *Let f be a linear mapping of a vector space V into a vector space U. Then*

(1) *The kernel of f is a subspace of V.*
(2) *The range of f is a subspace of U.*

These concepts are illustrated in Figures 6-14 and 6-15.

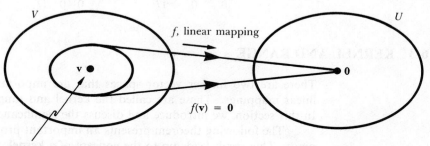

f, linear mapping

$f(\mathbf{v}) = \mathbf{0}$

Kernel of f. (Kernel is a subspace.)

Figure 6-14

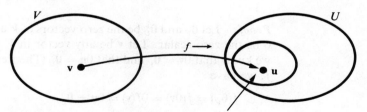

Range of f. Every element of this set is
the image of a vector in V. The range
may be the whole of U or just part of U.
(Range is a subspace.)

Figure 6-15

***Proof*:**   We prove (1) and leave the proof of (2) for Exercise 2.
Let $\mathbf{v}_1$ and $\mathbf{v}_2$ be elements of the kernel of f and let c be a scalar. Thus $f(\mathbf{v}_1) = \mathbf{0}$ and $f(\mathbf{v}_2) = \mathbf{0}$. On adding these equations, we get

$$f(\mathbf{v}_1) + f(\mathbf{v}_2) = \mathbf{0}$$
$$f(\mathbf{v}_1 + \mathbf{v}_2) = \mathbf{0} \quad \text{by the linearity of } f$$

Thus $\mathbf{v}_1 + \mathbf{v}_2$ is in the kernel of f.
Further, since $f(\mathbf{v}_1) = \mathbf{0}$,

$$cf(\mathbf{v}_1) = \mathbf{0}$$
$$f(c\mathbf{v}_1) = \mathbf{0} \quad \text{by the linearity of } f$$

Thus $c\mathbf{v}_1$ is in the kernel of f. The kernel is closed under addition and under scalar multiplication. It is a subspace.

The following example illustrates these concepts for a specific mapping.

✦ *Example 1*
_____

Consider the mapping f of $\mathbf{R}^3$ into $\mathbf{R}^3$ defined by $f\colon (x, y, z) \mapsto (x, y, 0)$. We shall prove that this mapping is linear and interpret it geometrically.

Let (x_1, y_1, z_1) and (x_2, y_2, z_2) be arbitrary elements of $\mathbf{R}^3$ and let c be an arbitrary scalar. Then

$$\begin{aligned}
f[(x_1, y_1, z_1) + (x_2, y_2, z_2)] &= f[(x_1 + x_2, y_1 + y_2, z_1 + z_2)] \\
&= (x_1 + x_2, y_1 + y_2, 0) \\
&= (x_1, y_1, 0) + (x_2, y_2, 0) \\
&= f[(x_1, y_1\, z_1)] + f[(x_2, y_2, z_2)]
\end{aligned}$$

The first requirement of a linear mapping is satisfied—it preserves addition. Further,

$$\begin{aligned}
f[c(x_1, y_1, z_1)] &= f[(cx_1, cy_1, cz_1)] \\
&= (cx_1, cy_1, 0) \\
&= c(x_1, y_1, 0) \\
&= cf[(x_1, y_1, z_1)]
\end{aligned}$$

Thus f preserves the operations of scalar multiplication; it is linear.

Geometrically we have Figure 6-16. f maps the vector (x, y, z) into the

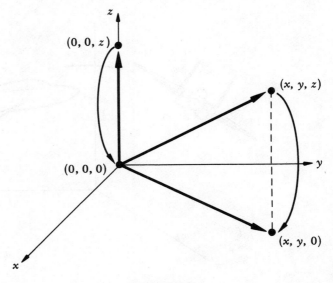

Figure 6-16

vector $(x, y, 0)$. This vector lies in the xy plane. f projects $\mathbf{R}^3$ onto the xy plane; it is called a *projection mapping*.

The range of f consists of all vectors of the type $(x, y, 0)$. Thus the range is the xy plane. It is a subspace of $\mathbf{R}^3$.

The kernel of f is the subset of $\mathbf{R}^3$ that is mapped onto the vector $(0, 0, 0)$. Since $f:(x, y, z) \mapsto (x, y, 0)$, (x, y, z) will be in the kernel if and only if $x = y = 0$. Thus the kernel consists of all vectors of the form $(0, 0, z)$. Geometrically this is the z axis, a one-dimensional subspace of $\mathbf{R}^3$.   ✦

We complete this discussion with a brief look at an application of projection mappings. The world in which we live has three spacial dimensions. When we observe an object, however, we get a two-dimensional impression of that object, the view changing from location to location. Projection mappings can be used to illustrate what three-dimensional objects look like from various locations. Such mappings are used in architecture, the auto industry, and the aerospace industry, for example. The outline of the object of interest, relative to a suitable coordinate system, is fed into a computer. The computer program contains an appropriate projection mapping that maps the object onto a plane. The output gives a two-dimensional view of the object, the outline being graphed by the computer. In this manner various mappings can be used to lead to various perspectives of an object. The General Electric Plant in Daytona Beach, Florida, uses such a computer graphics system for simulating aircraft. The Grumman Aerospace Corporation in Bethpage, New York, uses such a graphics system in designing aircraft. We illustrate these concepts in Figure 6-17 for an aircraft. The projection mapping used is onto the xz plane.

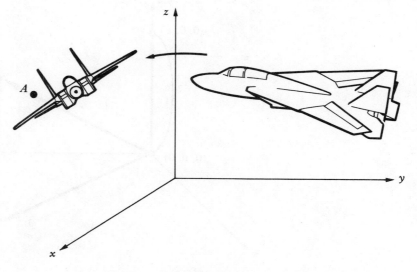

Figure 6-17

The image represents the view an observer at A has of the aircraft; it would be graphed out by the computer.

The following theorem gives us information about the range of a linear mapping of $\mathbf{R}^n$ into $\mathbf{R}^m$ defined by an $m \times n$ matrix.

Theorem 6-6 *Let an $m \times n$ matrix A define a linear mapping of $\mathbf{R}^n$ into $\mathbf{R}^m$. The range of the mapping is the column space of A.*

Proof: We remind the reader that the column space of A is the subspace of $\mathbf{R}^m$ spanned by the column vectors of A (Section 4-7).

Let $\mathbf{y}$ be an arbitrary element of the range. There exists an element $\mathbf{x}$ of $\mathbf{R}^n$ such that

$$A\mathbf{x} = \mathbf{y}$$

Write this equation

$$
\begin{aligned}
a_{11}x_1 + \cdots + a_{1n}x_n &= y_1 \\
&\vdots \\
a_{m1}x_1 + \cdots + a_{mn}x_n &= y_m
\end{aligned}
$$

This system can be written

$$
x_1 \begin{pmatrix} a_{11} \\ \vdots \\ a_{m1} \end{pmatrix} + \cdots + x_n \begin{pmatrix} a_{1n} \\ \vdots \\ a_{mn} \end{pmatrix} = \begin{pmatrix} y_1 \\ \vdots \\ y_m \end{pmatrix}
$$

Thus $\mathbf{y}$, an arbitrary element in the range, can be expressed as a linear combination of the column vectors of A. These column vectors span the range. The range of the mapping is the column space of A.

Note that since the dimension of the column space of A is the rank of A, we now have the following result.

The dimension of the range of a matrix mapping A is the rank of A.

In Example 1 we arrived at the kernel and range of the projection mapping by observation. The following example illustrates how these spaces can be determined, in general, for a mapping defined by a matrix. The above theorem is used in computing the range.

✦ Example 2

Interpret the following matrix A as a linear mapping of $\mathbf{R}^3$ into $\mathbf{R}^3$. Determine its kernel and range.

$$
A = \begin{pmatrix} 1 & 2 & 3 \\ 0 & -1 & 1 \\ 1 & 1 & 4 \end{pmatrix}
$$

The kernel will consist of all elements of $\mathbf{R}^3$ such that

$$\begin{pmatrix} 1 & 2 & 3 \\ 0 & -1 & 1 \\ 1 & 1 & 4 \end{pmatrix} \begin{pmatrix} x_1 \\ x_2 \\ x_3 \end{pmatrix} = \begin{pmatrix} 0 \\ 0 \\ 0 \end{pmatrix}$$

that is, elements of $\mathbf{R}^3$ such that x_1, x_2 and x_3 satisfy the following system of homogeneous equations:

$$x_1 + 2x_2 + 3x_3 = 0$$
$$- x_2 + x_3 = 0$$
$$x_1 + x_2 + 4x_3 = 0$$

On solving this system, we get many solutions, $\begin{pmatrix} -5r \\ r \\ r \end{pmatrix}$, where r is a param-

eter. The kernel is the one-dimensional subspace of $\mathbf{R}^3$ having basis $\begin{pmatrix} -5 \\ 1 \\ 1 \end{pmatrix}$.

Let us now determine the range. Theorem 6-6 tells us that the range is spanned by the column vectors of A. Let us find a basis for the range. We compute an echelon form of A^t. The nonzero vectors of this echelon form will be a basis for the row space of A^t. These vectors lead to a basis for the column space of A and thus a basis for the range.

$$A^t = \begin{pmatrix} 1 & 0 & 1 \\ 2 & -1 & 1 \\ 3 & 1 & 4 \end{pmatrix} \underset{\substack{\text{R2}+(-2)\text{R1} \\ \text{R3}+(-3)\text{R1}}}{\approx} \begin{pmatrix} 1 & 0 & 1 \\ 0 & -1 & -1 \\ 0 & 1 & 1 \end{pmatrix}$$

$$\underset{(-1)\text{R2}}{\approx} \begin{pmatrix} 1 & 0 & 1 \\ 0 & 1 & 1 \\ 0 & 1 & 1 \end{pmatrix} \underset{\text{R3}+(-1)\text{R2}}{\approx} \begin{pmatrix} 1 & 0 & 1 \\ 0 & 1 & 1 \\ 0 & 0 & 0 \end{pmatrix}$$

The row vectors $(1, 0, 1)$ and $(0, 1, 1)$ form a basis for the row space of A^t. Thus the column vectors $\begin{pmatrix} 1 \\ 0 \\ 1 \end{pmatrix}, \begin{pmatrix} 0 \\ 1 \\ 1 \end{pmatrix}$ form a basis for the column space of A.

The range is a two-dimensional subspace of $\mathbf{R}^3$ having basis $\begin{pmatrix} 1 \\ 0 \\ 1 \end{pmatrix}, \begin{pmatrix} 0 \\ 1 \\ 1 \end{pmatrix}$.

It will be the set of all vectors of the form $a\begin{pmatrix} 1 \\ 0 \\ 1 \end{pmatrix} + b\begin{pmatrix} 0 \\ 1 \\ 1 \end{pmatrix}$; that is, of the form

$$\begin{pmatrix} a \\ b \\ a + b \end{pmatrix}. \quad \blacklozenge$$

The following theorem gives an important relationship between the ,izes" of the kernel and of the range of a linear mapping.

Theorem 6-7   *Let f be a linear mapping of a vector space V into a vector space U. Then*

$$\textit{dimension of kernel of } f + \textit{dimension of range of } f$$
$$= \textit{dimension of } V, \textit{the domain of } f$$

Proof: Let us assume that the kernel consists of more than the zero vector, and that it is not the whole of V (the reader is asked to consider these special cases in Exercise 7).

Let $\mathbf{v}_1, \ldots, \mathbf{v}_m$ be a basis for the kernel. Complete this set to a basis $\mathbf{v}_1, \ldots, \mathbf{v}_n$ for V by adding the vectors $\mathbf{v}_{m+1}, \ldots, \mathbf{v}_n$. We shall show that the vectors $f(\mathbf{v}_{m+1}), \ldots, f(\mathbf{v}_n)$ form a basis for the range, thus verifying the theorem.

Let $\mathbf{v}$ be an arbitrary vector in V. Then $\mathbf{v}$ can be expressed

$$\mathbf{v} = a_1 \mathbf{v}_1 + \cdots + a_m \mathbf{v}_m + a_{m+1} \mathbf{v}_{m+1} + \cdots + a_n \mathbf{v}_n$$

and

$$f(\mathbf{v}) = f(a_1 \mathbf{v}_1 + \cdots + a_m \mathbf{v}_m + a_{m+1} \mathbf{v}_{m+1} + \cdots + a_n \mathbf{v}_n)$$

The linearity of f gives (Exercise 5)

$$f(\mathbf{v}) = a_1 f(\mathbf{v}_1) + \cdots + a_m f(\mathbf{v}_m) + a_{m+1} f(\mathbf{v}_{m+1}) + \cdots + a_n f(\mathbf{v}_n)$$

Since $\mathbf{v}_1, \ldots, \mathbf{v}_m$ are in the kernel of f, this reduces to

$$f(\mathbf{v}) = a_{m+1} f(\mathbf{v}_{m+1}) + \cdots + a_n f(\mathbf{v}_n)$$

$f(\mathbf{v})$ is an arbitrary vector in the range. Thus $f(\mathbf{v}_{m+1}), \ldots, f(\mathbf{v}_n)$ span the range.

It remains to prove that these vectors are linearly independent. Consider the identity

$$b_{m+1} f(\mathbf{v}_{m+1}) + \cdots + b_n f(\mathbf{v}_n) = \mathbf{0} \tag{1}$$

where $b_{m+1}, \ldots, b_n$ are arbitrary scalars, labeled thus for convenience. The linearity of f implies that

$$f(b_{m+1} \mathbf{v}_{m+1} + \cdots + b_n \mathbf{v}_n) = \mathbf{0}$$

that is, the vector $b_{m+1} \mathbf{v}_{m+1} + \cdots + b_n \mathbf{v}_n$ is an element of the kernel. Thus it can be expressed as a linear combination of $\mathbf{v}_1, \ldots, \mathbf{v}_m$. Let

$$b_{m+1} \mathbf{v}_{m+1} + \cdots + b_n \mathbf{v}_n = c_1 \mathbf{v}_1 + \cdots + c_m \mathbf{v}_m$$

That is,

$$c_1 \mathbf{v}_1 + \cdots + c_m \mathbf{v}_m - b_{m+1} \mathbf{v}_{m+1} - \cdots - b_n \mathbf{v}_n = \mathbf{0}$$

Since the vectors $\mathbf{v}_1, \ldots, \mathbf{v}_m, \mathbf{v}_{m+1}, \ldots, \mathbf{v}_n$ are linearly independent, this can only be satisfied if the coefficients are all zero. In particular, $b_{m+1} = \cdots = b_n = 0$. Thus, in (1), $f(\mathbf{v}_{m+1}), \ldots, f(\mathbf{v}_n)$ are linearly independent. They are a basis for the range, and the theorem has been proven.

Thus the "bigger" the kernel of f, the "smaller" the range, and vice versa.

✦ Example 3 ───

Let us look further at the linear mappings discussed in Examples 1 and 2 of this section.

Consider the linear mapping of $\mathbf{R}^3$ into $\mathbf{R}^3$ defined by $f:(x, y, z) \to (x, y, 0)$. The kernel of this mapping was found to be the z axis, a one-dimensional subspace of $\mathbf{R}^3$. The range was found to be the xy plane, a two-dimensional subspace of $\mathbf{R}^3$. We see that the dimension of the kernel plus the dimension of the range is 3, the dimension of the domain of f.

Consider the linear mapping of $\mathbf{R}^3$ into $\mathbf{R}^3$ defined by the matrix $\begin{pmatrix} 1 & 2 & 3 \\ 0 & -1 & 1 \\ 1 & 1 & 4 \end{pmatrix}$. We found that the kernel was the one-dimensional space with basis $\begin{pmatrix} -5 \\ 1 \\ 1 \end{pmatrix}$. The range was a two-dimensional space with basis $\begin{pmatrix} 1 \\ 0 \\ 1 \end{pmatrix}, \begin{pmatrix} 0 \\ 1 \\ 1 \end{pmatrix}$. The dimension of the domain, $\mathbf{R}^3$, is 3. The dimension of the kernel plus the dimension of the range is equal to the dimension of the domain.   ✦

Let f be a linear mapping of a vector space V into a vector space U. Let us now consider the set of vectors that are mapped onto a *nonzero* vector $\mathbf{u}$ which is in the range of f. These vectors will be the solution set of the equation $f(\mathbf{v}) = \mathbf{u}$. Unlike the set of solutions to $f(\mathbf{v}) = \mathbf{0}$ (the kernel of f), we find that this set does not form a subspace.

Let $\mathbf{v}_1$ and $\mathbf{v}_2$ be solutions. Then $f(\mathbf{v}_1) = \mathbf{u}$ and $f(\mathbf{v}_2) = \mathbf{u}$. Adding these equations,

$$f(\mathbf{v}_1) + f(\mathbf{v}_2) = 2\mathbf{u}$$
$$f(\mathbf{v}_1 + \mathbf{v}_2) = 2\mathbf{u} \qquad \text{since } f \text{ is linear}$$

Thus $\mathbf{v}_1 + \mathbf{v}_2$ is not a solution to $f(\mathbf{v}) = \mathbf{u}$. The set of solutions to $f(\mathbf{v}) = \mathbf{u}$ does not form a subspace (Figure 6-18).

It is rather unfortunate that the solutions to $f(\mathbf{v}) = \mathbf{u}$, where $\mathbf{u} \neq \mathbf{0}$, do not form a subspace, since we have so many tools at our disposal for discussing subspaces! However, we can relate the solutions of $f(\mathbf{v}) = \mathbf{u}$ to the solutions of $f(\mathbf{v}) = \mathbf{0}$ and thus to some extent use vector space concepts to understand the behavior of the set of solutions. For this discussion we need the following concept of a coset.

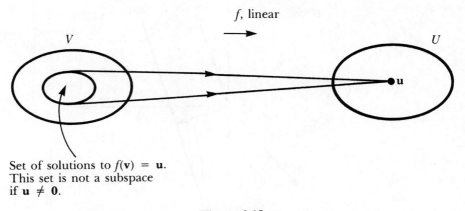

Set of solutions to $f(\mathbf{v}) = \mathbf{u}$.
This set is not a subspace
if $\mathbf{u} \neq \mathbf{0}$.

Figure 6-18

Definition 6-5   Let W be a subspace of a vector space V and $\mathbf{v}_1$ a specific
vector in V. The set of vectors of the form $\mathbf{v}_1 + \mathbf{w}$, where $\mathbf{w}$ is in W, is called a
coset and is written $\mathbf{v}_1 + W$.

One can interpret the coset $\mathbf{v}_1 + W$ as being the set obtained by adding
the vector $\mathbf{v}_1$ to every vector in W in turn. The net effect is to add $\mathbf{v}_1$ to the
subspace W. Geometrically, $\mathbf{v}_1 + W$ is the set obtained by sliding the subspace
W by an amount defined by the vector $\mathbf{v}_1$. (See Figure 6-19.) We say that the
sets W and $\mathbf{v}_1 + W$ are *parallel sets*.

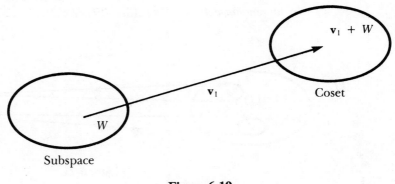

Figure 6-19

✦ *Example 4*

Consider the vector space $\mathbf{R}^3$ and the subspace W of vectors of the form
$a(1, 2, 3)$, where a is a scalar. Let $\mathbf{v}_1$ be the vector $(1, 1, 0)$. Then $\mathbf{v}_1 + W$ is
the set of vectors of the form $(1, 1, 0) + a(1, 2, 3)$. $\mathbf{v}_1 + W$ is the set of vectors
$(a + 1, 2a + 1, 3a)$. Geometrically, we get the picture in Figure 6-20.

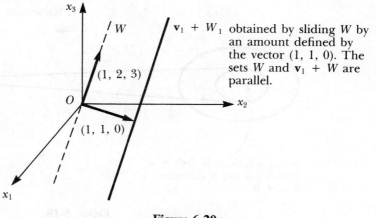

$\mathbf{v}_1 + W_1$ obtained by sliding W by an amount defined by the vector $(1, 1, 0)$. The sets W and $\mathbf{v}_1 + W$ are parallel.

Figure 6-20

We now use the concept of coset to arrive at the following important way of interpreting the set of solutions to $f(\mathbf{v}) = \mathbf{u}$.

Theorem 6-8   *Let f be a linear mapping of a vector space V into a vector space U and $\mathbf{u}$ be a given vector in U, which is in the range of f. The set of solutions S to $f(\mathbf{v}) = \mathbf{u}$ can be expressed*

$$S = \mathbf{v}_1 + \ker(f)$$

where $\mathbf{v}_1$ is any solution to $f(\mathbf{v}) = \mathbf{u}$.

Thus the set of solutions is a coset. The set is the kernel of f displaced by an amount defined by $\mathbf{v}_1$. (See Figure 6-21.)

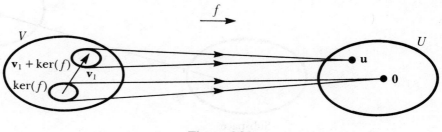

Figure 6-21

Proof: We prove first that $\mathbf{v}_1 + \ker(f)$ is a subset of S and then that S is a subset of $\mathbf{v}_1 + \ker(f)$, thus proving the equality of these two sets.

Let $\mathbf{v}$ be a vector in the coset $\mathbf{v}_1 + \ker(f)$. Thus there is a vector $\tilde{\mathbf{v}}$ in $\ker(f)$ such that $\mathbf{v} = \mathbf{v}_1 + \tilde{\mathbf{v}}$. Since f is linear,

$$f(\mathbf{v}) = f(\mathbf{v}_1 + \tilde{\mathbf{v}}) = f(\mathbf{v}_1) + f(\tilde{\mathbf{v}}) = \mathbf{u} + \mathbf{0} = \mathbf{u}$$

Thus $\mathbf{v}$ is in S.

Conversely, let **v** be in S. Thus $f(\mathbf{v}) = \mathbf{u}$. Since $\mathbf{v}_1$ is in S, we also have $f(\mathbf{v}_1) = \mathbf{u}$. Subtracting the second equation from the first, $f(\mathbf{v}) - f(\mathbf{v}_1) = \mathbf{0}$. Since f is linear, this implies that $f(\mathbf{v} - \mathbf{v}_1) = \mathbf{0}$. Thus $\mathbf{v} - \mathbf{v}_1$ is in the kernel of f. Call this vector $\tilde{\mathbf{v}}$; $\mathbf{v} - \mathbf{v}_1 = \tilde{\mathbf{v}}$. This gives

$$\mathbf{v} = \mathbf{v}_1 + \tilde{\mathbf{v}}$$

where $\tilde{\mathbf{v}}$ is in the kernel of f. Thus **v** is in the coset $\mathbf{v}_1 + \ker(f)$. implying that S is a subset of $\mathbf{v}_1 + \ker(f)$.

Thus $S = \mathbf{v}_1 + \ker(f)$.

EXERCISES

1. Interpret the following matrices as mappings between appropriate vector spaces. Determine the kernel and range of each mapping.

   *(a) $\begin{pmatrix} 1 & 2 \\ 3 & 0 \end{pmatrix}$ (b) $\begin{pmatrix} 2 & 0 \\ 3 & 0 \end{pmatrix}$ *(c) $\begin{pmatrix} 2 & 4 \\ 4 & 8 \end{pmatrix}$

   (d) $\begin{pmatrix} 1 & 2 \\ -1 & 3 \end{pmatrix}$ *(e) $\begin{pmatrix} 1 & 2 & 3 \\ 0 & 1 & 2 \end{pmatrix}$ (f) $\begin{pmatrix} 1 & 0 & 0 \\ 0 & 2 & 0 \\ 0 & 0 & 3 \end{pmatrix}$

   *(g) $\begin{pmatrix} 0 & 1 & 0 \\ 0 & 2 & 0 \\ 0 & 0 & 4 \end{pmatrix}$ (h) $\begin{pmatrix} 1 & 2 & 1 \\ -1 & -2 & 0 \\ 2 & 4 & 1 \end{pmatrix}$ (i) $\begin{pmatrix} 1 & 1 & 5 \\ 0 & 1 & 3 \\ 2 & 1 & 7 \end{pmatrix}$

2. Prove that the range of a linear mapping is a subspace.
3. Determine the kernel and range of each of the following linear mappings:
   *(a) f mapping $\mathbf{R}^3$ into $\mathbf{R}^3$ defined by $f(x, y, z) = (x, 0, 0)$
   (b) f mapping $\mathbf{R}^3$ into $\mathbf{R}^2$ defined by $f(x, y, z) = (x + y, z)$
   *(c) f mapping $\mathbf{R}^3$ into $\mathbf{R}$ defined by $f(x, y, z) = x + y + z$
4. Let A and B be $m \times n$ and $p \times m$ matrices, respectively. B may be interpreted as a linear mapping of the range of A into $\mathbf{R}^p$. BA thus defines a mapping of $\mathbf{R}^n$ into $\mathbf{R}^p$. Prove that this *composite mapping* is linear.
5. Let f be a linear mapping of a vector space V into a vector space U, let $\mathbf{v}_1, \ldots, \mathbf{v}_m$ be m arbitrary vectors in V, and let $a_1, \ldots, a_m$ be m arbitrary scalars. Prove that $f(a_1 \mathbf{v}_1 + \cdots + a_m \mathbf{v}_m) = a_1 f(\mathbf{v}_1) + \cdots + a_m f(\mathbf{v}_m)$.
6. Prove that a mapping f between vector spaces is linear if and only if

   $$f(a\mathbf{u} + b\mathbf{v}) = af(\mathbf{u}) + bf(\mathbf{v})$$

   for arbitrary vectors **u**, **v** and arbitrary scalars a, b. This is an alternative definition for a linear mapping.
7. Let f be a linear mapping of a vector space V into a vector space U. Prove that

   dimension of kernel of f + dimension of range of f = dimension of V

   when
   (a) the kernel of f consists of only the zero vector
   (b) the kernel of f is the whole of V

8. Let A be an $m \times n$ matrix. Interpret A as a mapping of $\mathbf{R}^n$ into $\mathbf{R}^m$ and A^t as a mapping of $\mathbf{R}^m$ into $\mathbf{R}^n$. Show that the dimension of the range of A is equal to the dimension of the range of A^t.

*9. Consider the subspace of $\mathbf{R}^2$ of vectors of the form $a(2, 1)$. Sketch the coset $(-1, 2) + a(2, 1)$.

10. Consider the subspace of $\mathbf{R}^2$ of vectors of the form $a(-2, 3)$. Sketch the coset $(-1, -1) + a(-2, 3)$.

*11. Consider the subspace of $\mathbf{R}^3$ of vectors of the form $a(1, -3, 2)$. Sketch the coset $(1, 3, 1) + a(1, -3, 2)$.

12. Consider the subspace of $\mathbf{R}^3$ of vectors of the form $a(1, -1, 3)$. Sketch the coset $(1, -2, 1) + a(1, -1, 3)$.

*13. Let f be the projection mapping of $\mathbf{R}^3$ into $\mathbf{R}^3$ defined by $f(x, y, z) = (x, y, 0)$. Find the set of vectors that are mapped by f onto the vector $(1, 2, 0)$. Sketch this set.

14. Let f be the linear mapping of $\mathbf{R}^2$ into $\mathbf{R}^2$ defined by $f(x, y) = (x - y, 2y - 2x)$. Find the set of vectors that are mapped by f onto the vector $(2, -4)$. Sketch this set.

*15. Let f be the linear mapping of $\mathbf{R}^2$ into $\mathbf{R}^2$ defined by $f(x, y) = (2x, 3x)$. Find and sketch the set of vectors that are mapped by f onto the vector $(4, 6)$.

16. Let f be the linear mapping of $\mathbf{R}^3$ into $\mathbf{R}^2$ defined by $f(x, y, z) = (x - y, x + z)$. Find and sketch the set of vectors that are mapped by f onto the vector $(1, 4)$.

6-4 MAPPINGS AND SYSTEMS OF EQUATIONS

Linear mappings and the concepts of kernel and range play an important role in the analyses of systems of linear equations. They also enable one to "see" geometrically what is happening in a given system. Let f be a linear mapping of a vector space V into a vector space U. We have seen that the kernel of f is a subspace. We now use this information to arrive at the following important result.

Theorem 6-9   *The set of solutions to a homogeneous system of m equations in n variables, $A\mathbf{x} = \mathbf{0}$, is a subspace of $\mathbf{R}^n$.*

Proof: Interpret A as a linear mapping of $\mathbf{R}^n$ into $\mathbf{R}^m$. The set of solutions is the set of vectors in $\mathbf{R}^n$ that are mapped onto the zero vector. The set of solutions is the kernel of the mapping and is thus a subspace of $\mathbf{R}^n$.

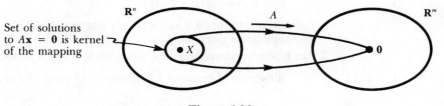

Set of solutions to $A\mathbf{x} = \mathbf{0}$ is kernel of the mapping

Figure 6-22

We are now able to visualize the set of solutions to a system of m homogeneous equations in n variables as shown in Figure 6-22.

✦ *Example 1*
_____

Consider the system of linear homogeneous equations

$$x_1 + 2x_2 + 3x_3 = 0$$
$$- x_2 + x_3 = 0$$
$$x_1 + x_2 + 4x_3 = 0$$

Let us solve this system and interpret the set of solutions in a geometrical manner. We get, using the Gauss-Jordan method,

$$\begin{pmatrix} 1 & 2 & 3 & 0 \\ 0 & -1 & 1 & 0 \\ 1 & 1 & 4 & 0 \end{pmatrix} \approx \begin{pmatrix} 1 & 2 & 3 & 0 \\ 0 & -1 & 1 & 0 \\ 0 & -1 & 1 & 0 \end{pmatrix}$$

$$\approx \begin{pmatrix} 1 & 2 & 3 & 0 \\ 0 & 1 & -1 & 0 \\ 0 & -1 & 1 & 0 \end{pmatrix} \approx \begin{pmatrix} 1 & 0 & 5 & 0 \\ 0 & 1 & -1 & 0 \\ 0 & 0 & 0 & 0 \end{pmatrix}$$

Thus

$$x_1 \quad\quad + 5x_3 = 0$$
$$x_2 - x_3 = 0$$

giving

$$x_1 = -5x_3$$
$$x_2 = x_3$$

Let us assign the value r to x_3. An arbitrary solution is thus

$$x_1 = -5r, \quad\quad x_2 = r, \quad\quad x_3 = r$$

Let us express an arbitrary solution as an element of $\mathbf{R}^3$, $\begin{pmatrix} -5r \\ r \\ r \end{pmatrix}$. The set defined by this arbitrary vector is the kernel of the linear mapping defined by the matrix

$$\begin{pmatrix} 1 & 2 & 3 \\ 0 & -1 & 1 \\ 1 & 1 & 4 \end{pmatrix}$$

It is customary to express this solution as an element of $\mathbf{R}^3$ in row form,

$$(-5r, r, r)$$

when actually discussing solutions. Observe that this arbitrary solution can be written

$$r(-5, 1, 1)$$

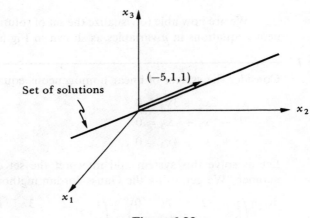

Figure 6-23

Thus the set of solutions is the one-dimensional subspace of $\mathbf{R}^3$ spanned by the vector $(-5, 1, 1)$. It is the line defined by the vector $(-5, 1, 1)$ in Figure 6-23.

In Example 2 of the previous section we discussed the linear mapping defined by the matrix of coefficients of this system. Observe that the set of solutions is indeed the kernel of this mapping.   ✦

Let us now look at nonhomogeneous systems through the eyes of linear mappings. Let $A\mathbf{x} = \mathbf{y}$ be a nonhomogeneous system of m equations in n variables. If f is a linear mapping, we know that the set of solutions to $f(\mathbf{v}) = \mathbf{u}$, where $\mathbf{u} \neq \mathbf{0}$, does not form a subspace. If we interpret A as a linear mapping of $\mathbf{R}^n$ into $\mathbf{R}^m$, the set of solutions to $A\mathbf{x} = \mathbf{y}$ will be the vectors in $\mathbf{R}^n$ that are mapped by A onto $\mathbf{y}$. This subset of $\mathbf{R}^n$ will thus not be a subspace. The picture of the set of solutions is shown in Figure 6-24.

This way of looking at systems of nonhomogeneous equations leads to the following result.

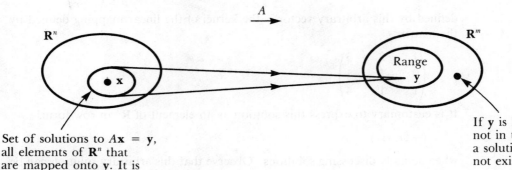

Figure 6-24

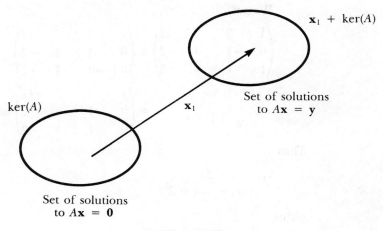

Figure 6-25

Theorem 6-10   *The set S of solutions to a nonhomogeneous system of linear equations $A\mathbf{x} = \mathbf{y}$ can be expressed*

$$S = \mathbf{x}_1 + \text{ker}(A)$$

where $\mathbf{x}_1$ is any particular solution to $A\mathbf{x} = \mathbf{y}$.

Proof: We know that if f is a linear mapping, then the set S of solutions to $f(\mathbf{v}) = \mathbf{u}$ can be expressed

$$S = \mathbf{v}_1 + \text{ker}(f)$$

where $\mathbf{v}_1$ is any solution to $f(\mathbf{v}) = \mathbf{u}$.

If we interpret A as a linear mapping, this result implies that

$$S = \mathbf{x}_1 + \text{ker}(A)$$

Since $\text{ker}(A)$ is the set of solutions to $A\mathbf{x} = \mathbf{0}$, this theorem enables us to relate the set of solutions of $A\mathbf{x} = \mathbf{y}$ to the set of solutions of its *associated homogeneous system* $A\mathbf{x} = \mathbf{0}$. Thus geometrically the set of solutions to $A\mathbf{x} = \mathbf{y}$ is obtained by sliding the set of solutions to $A\mathbf{y} = \mathbf{0}$ by an amount defined by the vector $\mathbf{x}_1$. The set of solutions of $A\mathbf{x} = \mathbf{y}$ is parallel to the set of solutions of $A\mathbf{x} = \mathbf{0}$.

Geometrically, we have Figure 6-25. Let us now put this theory into practice.

✦ *Example 2* ───

Let us solve and analyze the solutions to the following system.

$$\begin{aligned}
x_1 + 2x_2 + 3x_3 &= 11 \\
- x_2 + x_3 &= -2 \\
x_1 + x_2 + 4x_3 &= 9
\end{aligned}$$

We get

$$\begin{pmatrix} 1 & 2 & 3 & 11 \\ 0 & -1 & 1 & -2 \\ 1 & 1 & 4 & 9 \end{pmatrix} \approx \begin{pmatrix} 1 & 2 & 3 & 11 \\ 0 & -1 & 1 & -2 \\ 0 & -1 & 1 & -2 \end{pmatrix}$$

$$\approx \begin{pmatrix} 1 & 2 & 3 & 11 \\ 0 & 1 & -1 & 2 \\ 0 & -1 & 1 & -2 \end{pmatrix} \approx \begin{pmatrix} 1 & 0 & 5 & 7 \\ 0 & 1 & -1 & 2 \\ 0 & 0 & 0 & 0 \end{pmatrix}$$

Thus

$$\begin{aligned} x_1 \qquad + 5x_3 &= 7 \\ x_2 - \quad x_3 &= 2 \end{aligned}$$

giving

$$\begin{aligned} x_1 &= -5x_3 + 7 \\ x_2 &= \quad x_3 + 2 \end{aligned}$$

The arbitrary solution is $x_1 = -5r + 7$, $x_2 = r + 2$, $x_3 = r$.

We can express this solution in the following manner:

$$(-5r + 7, r + 2, r)$$

Let us "pull this solution apart." We separate the constant part from the part involving the parameter r. The constant part will be the particular solution $\mathbf{x}_1$, and the part involving r will be the arbitrary solution to the associated homogeneous equation.

$$\underbrace{(-5r + 7, r + 2, r)}_{\substack{arbitrary\ solution \\ to\ the\ given \\ nonhomogeneous \\ system}} = \underbrace{(7, 2, 0)}_{\substack{a\ particular \\ solution\ to \\ the\ given \\ nonhomogeneous \\ system}} + \underbrace{r(-5, 1, 1)}_{\substack{arbitrary\ solution \\ to\ the\ associated \\ homogeneous \\ system}}$$

By substituting, one can verify that $(7, 2, 0)$ is indeed a solution to the given system. Observe that we solved the associated homogeneous system in Example 1 and found that $r(-5, 1, 1)$ was the arbitrary solution.

The set of solutions to the nonhomogeneous system can be represented geometrically by sliding the vector space with basis $(-5, 1, 1)$ in the direction and distance defined by the vector $(7, 2, 0)$. See Figure 6-26.   ✦

If we have a number of nonhomogeneous systems, $A\mathbf{x} = \mathbf{y}_1$, $A\mathbf{x} = \mathbf{y}_2$, $A\mathbf{x} = \mathbf{y}_3$, etc., all having the same matrix of coefficients A, we now have a convenient way of visualizing how their respective sets of solutions are related. Let $\mathbf{x}_1$, $\mathbf{x}_2$, $\mathbf{x}_3$, etc., be particular solutions to the systems. Then the sets of

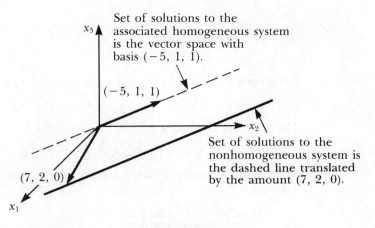

Set of solutions to the
associated homogeneous system
is the vector space with
basis $(-5, 1, 1)$.

$(-5, 1, 1)$

x_3

x_2

Set of solutions to the
nonhomogeneous system is
the dashed line translated
by the amount $(7, 2, 0)$.

$(7, 2, 0)$

x_1

Figure 6-26

solutions are $\mathbf{x}_1 + \ker(A)$, $\mathbf{x}_2 + \ker(A)$, $\mathbf{x}_3 + \ker(A)$, etc. These sets will be "parallel" sets, each being the kernel of A translated to various locations.

Thus, for example, each set of solutions to a system of the type

$$x_1 + 2x_2 + 3x_3 = a_1$$
$$- x_2 + x_3 = a_2$$
$$x_1 + x_2 + 4x_3 = a_3$$

will be a straight line parallel to the one defined by the vector $(-5, 1, 1)$.

✦ **Example 3**

Discuss and analyze the set of solutions to the system

$$x_1 - 2x_2 + 3x_3 + x_4 = 1$$
$$2x_1 - 3x_2 + 2x_3 - x_4 = 4$$
$$3x_1 - 5x_2 + 5x_3 \qquad = 5$$
$$x_1 - x_2 - x_3 - 2x_4 = 3$$

We get

$$
\begin{pmatrix}
1 & -2 & 3 & 1 & 1 \\
2 & -3 & 2 & -1 & 4 \\
3 & -5 & 5 & 0 & 5 \\
1 & -1 & -1 & -2 & 3
\end{pmatrix}
\approx
\begin{pmatrix}
1 & -2 & 3 & 1 & 1 \\
0 & 1 & -4 & -3 & 2 \\
0 & 1 & -4 & -3 & 2 \\
0 & 1 & -4 & -3 & 2
\end{pmatrix}
$$

$$
\approx
\begin{pmatrix}
1 & 0 & -5 & -5 & 5 \\
0 & 1 & -4 & -3 & 2 \\
0 & 0 & 0 & 0 & 0 \\
0 & 0 & 0 & 0 & 0
\end{pmatrix}
$$

giving

$$x_1 \quad - 5x_3 - 5x_4 = 5$$
$$x_2 - 4x_3 - 3x_4 = 2$$

Expressing the variables x_1 and x_2 in terms of x_3 and x_4,

$$x_1 = 5x_3 + 5x_4 + 5, \qquad x_2 = 4x_3 + 3x_4 + 2$$

The arbitrary solution is

$$(5r + 5s + 5, 4r + 3s + 2, r, s)$$

Separating the constant vector and the vectors involving the parameters r and s, we get

$$\underbrace{(5r + 5s + 5, 4r + 3s + 2, r, s)}_{\substack{\text{arbitrary solution to} \\ \text{the given nonhomogeneous} \\ \text{system}}} = \underbrace{(5, 2, 0, 0)}_{\substack{\text{particular} \\ \text{solution to} \\ \text{given nonhomo-} \\ \text{geneous system}}} + \underbrace{r(5, 4, 1, 0) + s(5, 3, 0, 1)}_{\substack{\text{arbitrary solution to the} \\ \text{associated homogeneous} \\ \text{system}}}$$

Observe that the set of solutions to the associated homogeneous system forms a two-dimensional subspace of $\mathbf{R}^4$, a plane with basis (5, 4, 1, 0) and (5, 3, 0, 1). The set of solutions to the given system is the plane, translated in a manner described by the vector (5, 2, 0, 0). ✦

EXERCISES

Consider the following systems of equations. The solutions are given in each case. Analyze the solutions by representing each set as the sum of a particular solution and an arbitrary solution of the corresponding system of homogeneous equations.

*1. $x_1 - x_2 + x_3 = 1$ Solutions: $x_1 = x_2 = r$
    $2x_1 - 2x_2 + 3x_3 = 3$ $x_3 = 1$
    $x_1 - x_2 - x_3 = -1$

2. $x_1 + x_2 + x_3 = 3$ Solutions: $x_1 = 0$
    $2x_1 + 3x_2 + x_3 = 5$ $x_2 = 1$
    $x_1 - x_2 - 2x_3 = -5$ $x_3 = 2$

*3. $x_1 - 2x_2 + 3x_3 = 1$ Solutions: $x_1 = 1 + r$
    $3x_1 - 4x_2 + 5x_3 = 3$ $x_2 = 2r$
    $2x_1 - 3x_2 + 4x_3 = 2$ $x_3 = r$

4. $x_1 - x_2 + x_3 = 3$ Solutions: $x_1 = 3 + r - s$
    $-2x_1 + 2x_2 - 2x_3 = -6$ $x_2 = r$
                                               $x_3 = s$

*5. $x_1 - x_2 - x_3 + 2x_4 = 4$ Solutions: $x_1 = -2 - 4r + 3s$
    $2x_1 - x_2 + 3x_3 - x_4 = 2$ $x_2 = -6 - 5r + 5s$
                                               $x_3 = r$
                                               $x_4 = s$

6. $x_1 + 2x_2 - x_3 + x_4 + 2x_5 = 0$ Solutions: $x_1 = -19 - 32r - 23s$
$\quad\quad x_2 + 2x_3 - 3x_4 - 3x_5 = 2$ $\quad\quad\quad\quad x_2 = 8 + 13r + 9s$
$\quad\quad\quad\quad\quad x_3 + 5x_4 + 3x_5 = -3$ $\quad\quad\quad x_3 = -3 - 5r - 3s$
$\quad\quad\quad\quad\quad\quad\quad\quad\quad\quad\quad\quad\quad\quad\quad x_4 = r$
$\quad\quad\quad\quad\quad\quad\quad\quad\quad\quad\quad\quad\quad\quad\quad x_5 = s$

7. If we interpret the matrix $A = \begin{pmatrix} 1 & 2 & 3 \\ 0 & -1 & 1 \\ 1 & 1 & 4 \end{pmatrix}$ as a mapping of $\mathbf{R}^3$ into $\mathbf{R}^3$, its range is

the set of vectors of the form $\begin{pmatrix} x \\ y \\ x + y \end{pmatrix}$, that is, the plane $z = x + y$. Use this given

information to determine whether solutions exist to the following systems of equations $A\mathbf{x} = \mathbf{y}$, $\mathbf{y}$ taking on the various values. (You need not determine the solutions if they exist.)

*(a) $\mathbf{y} = \begin{pmatrix} 1 \\ 1 \\ 2 \end{pmatrix}$ **(b)** $\mathbf{y} = \begin{pmatrix} -1 \\ 2 \\ 3 \end{pmatrix}$ ***(c)** $\mathbf{y} = \begin{pmatrix} 3 \\ 2 \\ 5 \end{pmatrix}$ **(d)** $\mathbf{y} = \begin{pmatrix} 2 \\ 4 \\ 5 \end{pmatrix}$

8. In this section we solved nonhomogeneous systems, arriving at an arbitrary solution. This solution was decomposed into a particular solution $\mathbf{x}_1$ and an arbitrary solution of the associated homogeneous system. Is the particular solution $\mathbf{x}_1$ unique? Discuss. If $\mathbf{x}_1$ is not unique, determine a vector other than $(7, 2, 0)$ that could be used in Example 2 of this section.

*9. Determine a vector other than $(5, 2, 0, 0)$ that could be used as a particular solution in the analysis of the set of solutions to Example 3 of this section.

10. The relationship between the set of solutions S to $A\mathbf{x} = \mathbf{y}$ and the set of solutions $\ker(A)$ to $A\mathbf{x} = \mathbf{0}$ is $S = \mathbf{x}_1 + \ker(A)$. Here $\mathbf{x}_1$ is any particular solution to $A\mathbf{x} = \mathbf{y}$. Discuss how this equation can be used to define a mapping of $\ker(A)$ onto S. There will be such a mapping associated with every such $\mathbf{x}_1$. Is this mapping linear? Consider the mapping associated with the vector $(7, 2, 0)$ in Example 2 of this section. What is the image of the vector $(-10, 2, 2)$ under this mapping?

*11. Consider the mapping associated with the vector $(5, 2, 0, 0)$ in Example 3 of this section. (See the discussion of mapping in Exercise 10.) What is the image of the vector $(5, 5, 2, -1)$ under this mapping?

12. Construct a nonhomogeneous system of equations that has matrix of coefficients $\begin{pmatrix} 1 & 2 & 1 \\ 0 & 1 & 2 \\ 1 & 1 & -1 \end{pmatrix}$ and particular solution $(1, -1, 4)$.

*13. Construct a nonhomogeneous system of equations that has matrix of coefficients $\begin{pmatrix} 2 & 1 & 0 \\ 3 & 3 & 1 \\ 0 & 1 & 1 \end{pmatrix}$ and particular solution $(2, 0, 3)$.

14. Prove that the solution to a nonhomogeneous system of linear equations is unique if and only if the kernel of the matrix of coefficients is the zero vector, i.e., if and only if the corresponding homogeneous system of equations has as its unique solution the zero vector.

6-5* LINEAR ALGEBRA AND DIFFERENTIAL EQUATIONS

This section is intended for those readers who studied the sections on general vector spaces and are interested in the insights the concepts of linear algebra can give to the behavior of certain types of differential equations.

In practice, the initial mathematical descriptions of many real situations are in the form of equations that involve derivatives of a function of interest. Such equations are called *differential equations*. One must solve the differential equation to find the function. We shall discuss a certain class of differential equations called *linear differential equations*. These equations arise frequently in applications. We mention a few such applications at this time.

1. *Motion of a vibrating spring* (from *Applied Differential Equations*, 3rd ed., by Murray R. Spiegel, Prentice-Hall, 1981, page 228). Consider a weight W suspended from a spring. The weight is pulled down and then released. Let $x(t)$ denote the position of the weight at time t, relative to its initial equilibrium position. (See Figure 6-27.) The motion of the weight is described by the differential equation

$$\frac{d^2x}{dt^2} + \alpha \frac{dx}{dt} + \beta x = 0$$

Here α and β are constants associated with the spring and damping effects. One wants to solve this differential equation to determine $x(t)$.

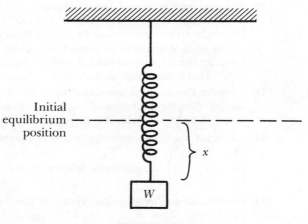

Initial equilibrium position

Figure 6-27

2. *Currents in electric circuits* (from *Differential Equations*, by John A. Tierney, Allyn & Bacon, Inc., 1979, page 162). Consider a current I flowing in the electrical circuit. E. is a source of E volts (possibly a generator), C is a capacitor (this element stores electrical energy in the circuit), L is

an inductor (produces a voltage drop), and R is a resistance (possibly a lightbulb or electric heater). I and E will be functions of time. (See Figure 6-28.) Their behavior is governed by the differential equation

$$L \frac{d^2 I}{dt^2} + R \frac{dI}{dt} + \frac{I}{C} = \frac{dE}{dt}$$

Here L, R, and C are constants, in appropriate units, that represent the "sizes" of the inductor, resistance, and capacitor. $E(t)$ is known, so dE/dt is known. One wants to solve the differential equation to find $I(t)$, the function that describes the behavior of the current.

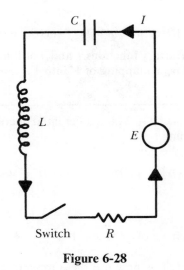

Figure 6-28

3. *An application in economics* (from *Applied Differential Equations*, 3rd ed., by Murray R. Spiegel, Prentice-Hall, 1981, page 250). Prices change in time as a result of inflation. The following differential equation describes how price $P(t)$ at time t is related to an inflation factor $F(t)$ in a model used in economics:

$$\frac{d^2 P}{dt^2} + kP = \frac{dF}{dt} + C$$

Here k and C are appropriate constants. $F(t)$ is known, and thus dF/dt is known. One wants to solve the differential equation to find $P(t)$.

We now commence our algebraic discussion.

Let V be the vector space of functions that possess derivatives of all orders on an interval (a, b). Let D be the operation of taking the derivative. We know from previous work that D is a linear mapping of V into itself.

Let D^k denote the kth derivative. We shall be interested in linear combinations of derivatives where the coefficients are themselves elements of V. An expression of the form $D^2 + a_1 D + a_0$, where a_1 and a_0 are elements of V, is called a *differential operator of order two*—the highest derivative involved is two. For the sake of clarity, we shall limit our discussion here to differential operators of order two; however, the concepts can be extended to higher order operators.

The differential operator $D^2 + a_1 D + a_0$ can be used as a mapping of V into V. By the properties of derivatives, we get

$$(D^2 + a_1 D + a_0)(f + g) = (D^2 + a_1 D + a_0)f + (D^2 + a_1 D + a_0)g$$

and

$$(D^2 + a_1 D + a_0)(cf) = c(D^2 + a_1 D + a_0)f$$

for arbitrary functions f and g of V and arbitrary scalar c. Thus $D^2 + a_1 D + a_0$ is a linear mapping of V into V.

✦ Example 1
_____

$D^2 + 4x^2 D + 2x$ is a differential operator. Let us find the image of x^2 under this mapping.

We get

$$
\begin{aligned}
(D^2 + 4x^2 D + 2x)x^2 &= D^2 x^2 + 4x^2 Dx^2 + 2xx^2 \\
&= 2 + 4x^2(2x) + 2x^3 \\
&= 2 + 10x^3
\end{aligned}
$$

The image of x^2 is $2 + 10x^3$.   ✦

Let us now use these tools to study differential equations. Consider the following differential equation:

$$\frac{d^2 y}{dx^2} + a_1 \frac{dy}{dx} + a_0 y = 0$$

where a_1 and a_0 are elements of V. Let us express this equation in terms of differential operators, $D^2 y + a_1 Dy + a_0 y = 0$. This equation may be written

$$(D^2 + a_1 D + a_0)y = 0$$

This is the form of the differential equation that we shall be interested in. In the previous section, the reader saw how the concepts of mappings and kernels, with certain general theorems, led to insights into the behavior of solutions to systems of linear equations. We shall now apply those same theorems in an analogous manner to differential equations. This illustrates the importance of developing results within a general framework—the same results can then be applied in many seemingly unrelated areas. The main results only have to be derived once, without duplication in the various areas. Also, observations arrived at in one area can lead to insights in another area. The pattern we follow for the differential equation $(D^2 + a_1 D + a_0)y = 0$ is analogous to that for the linear system of equations $A\mathbf{x} = \mathbf{0}$. We now present the first result for differential equations.

Theorem 6-11   *The set of solutions to the differential equation*

$$(D^2 + a_1 D + a_0)y = 0$$

is a subspace of V.

Proof: Interpret $D^2 + a_1 D + a_0$ as a linear mapping of V into V. The set of solutions is the subset of V that is mapped onto the zero function 0, the zero vector of V. This subset is the kernel of the linear mapping and is thus a subspace of V. See Figure 6-29.

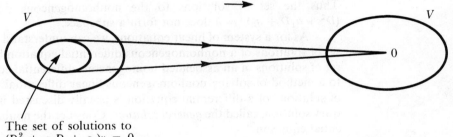

$D^2 + a_1 D + a_0$, a linear mapping

The set of solutions to
$(D^2 + a_1 D + a_0)y = 0$
is the kernel of the mapping.
It is a subspace.

Figure 6-29

The subspace will in fact be of dimension 2. (We do not prove this.)

Continuing the analogy between the discussion of the differential equation $(D^2 + a_1 D + a_0)y = 0$ and the linear system $A\mathbf{x} = \mathbf{0}$, we call the differential equation

$$(D^2 + a_1 D + a_0)y = 0$$

a *homogeneous linear differential equation.*

Our discussion of homogeneous equations led to a discussion of nonhomogeneous equations, equations of the form $A\mathbf{x} = \mathbf{y}$. Let us now consider the *nonhomogeneous linear differential equation*

$$(D^2 + a_1 D + a_0)y = h$$

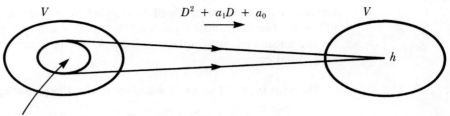

Set of solutions to
$(D^2 + a_1D + a_0)y = h$. It is
not a subspace if $h \neq 0$.

Figure 6-30

where h is an element of V. (See Figure 6-30.) The set of solutions—that is, the set of y's that satisfy this differential equation—is those elements that are mapped onto h by the linear mapping $D^2 + a_1D + a_0$. We know that if f is a linear mapping, the set of solutions to $f(\mathbf{v}) = \mathbf{u}$ does not form a subspace. Thus the set of solutions to the nonhomogeneous differential equation $(D^2 + a_1D + a_0)y = h$ does not form a subspace.

As for a system of linear equations, we can understand the behavior of the set of solutions of a nonhomogeneous differential equation by relating it to the set of solutions of an associated homogeneous differential equation. This leads to a method of solving nonhomogeneous linear differential equations. The set of solutions of a differential equation is usually discussed in terms of an arbitrary solution, called the *general solution*. Consider the nonhomogeneous differential equation

$$(D^2 + a_1D + a_0)y = h$$

The associated homogeneous differential equation is

$$(D^2 + a_1D + a_0)y = 0$$

This equation is called the *complementary equation*, and its general solution is the *complementary solution*.

The general solution to a nonhomogeneous differential equation will be discussed in terms of a *particular solution* to the differential equation and the complementary solution. Denote the particular solution y_p and the complementary solution y_c.

Theorem 6-12   *Consider the nonhomogeneous differential equation*

$$(D^2 + a_1D + a_0)y = h$$

The general solution may be obtained by finding a particular solution y_p of this equation and adding it to the complementary solution y_c:

$$y = y_p + y_c$$

Proof: We know that if f is a linear mapping, then the set S of solutions to $f(\mathbf{v}) = \mathbf{u}$ can be expressed

$$S = \mathbf{v}_1 + \ker(f)$$

where $\mathbf{v}_1$ is any solution to $f(\mathbf{v}) = \mathbf{u}$.

If we interpret $D^2 + a_1 D + a_0$ as a linear mapping, this result implies that

$$S = y_p + \ker(D^2 + a_1 D + a_0)$$

Representing S by an arbitrary element and $\ker(D^2 + a_1 D + a_0)$ by an arbitrary element, we get the conclusion of the theorem,

$$y = y_p + y_c$$

This result implies that the set of solutions to $(D^2 + a_1 D + a_0)y = h$ can be obtained by sliding the set U of solutions to $(D^2 + a_1 D + a_0)y = 0$ by an amount y_p. We can now visualize the set of solutions to a nonhomogeneous differential equation as shown in Figure 6-31.

For a number of nonhomogeneous differential equations each having the same differential operator, $(D^2 + a_1 D + a_0)y = h_1$, $(D^2 + a_1 D + a_0)y = h_2$, $(D^2 + a_1 D + a_0)y = h_3$, etc., the sets of solutions will all be "parallel." Each will be U translated to various locations. The sets of solutions will be $y_{p_1} + U$, $y_{p_2} + U$, etc., where the y_p's are particular solutions.

To illustrate the use of these tools, we shall find general solutions to *constant coefficient* linear differential equations. These are differential equations of the type already considered, with the coefficients being constants.

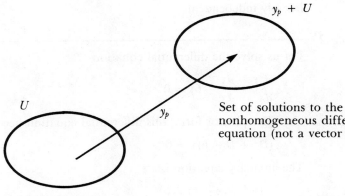

$y_p + U$

U

y_p

Set of solutions to the nonhomogeneous differential equation (not a vector space)

Set of solutions to the associated homogeneous differential equation (a vector space)

Figure 6-31

We shall present results, our aim being to illustrate the role of linear algebra in differential equations; the technical details involving calculus are left to be presented in a differential equations course. Let us first look at a homogeneous differential equation

$$(D^2 + a_1D + a_0)y = 0$$

Associated with this equation is the *auxiliary equation*, obtained by replacing the operator D by a variable m:

$$(m^2 + a_1m + a_0) = 0$$

The following theorem tells us how to determine the general solution once the auxiliary equation has been solved.

Theorem 6-13   *Let r_1 and r_2 be the roots of the auxiliary equation. The general solution will be of one of three types, depending on whether the roots are real distinct, real repeated, or complex:*

(1) $c_1e^{r_1x} + c_2e^{r_2x}$ *if r_1 and r_2 are real and distinct.*
(2) $c_1e^{r_1x} + c_2xe^{r_1x}$ *if r_1 and r_2 are real and equal.*
(3) $c_1e^{ax}\cos bx + c_2e^{ax}\sin bx$ *if r_1 and r_2 are complex with*
     $r_1 = a + ib, r_2 = a - ib.$

Observe that in all three cases the subspace of solutions is a two-dimensional space. In (1) the functions e^{r_1x} and e^{r_2x} are linearly independent and thus form a basis for the space of solutions. In (2) the functions e^{r_1x} and xe^{r_1x} are linearly independent. In (3) the functions $e^{ax}\cos bx$ and $e^{ax}\sin bx$ are linearly independent.

✦ *Example 2*

Let us solve the differential equation

$$\frac{d^2y}{dx^2} + \frac{dy}{dx} - 6y = 0$$

In operator form, this differential equation becomes

$$(D^2 + D - 6)y = 0$$

The auxiliary equation is

$$(m^2 + m - 6) = 0$$

$$(m - 2)(m + 3) = 0$$

The roots of the auxiliary equation are $r_1 = 2, r_2 = -3$, real and distinct. We have category (1) of Theorem 6-13. The general solution is

$$y = c_1e^{2x} + c_2e^{-3x}$$

For example, if we let $c_1 = 1$ and $c_2 = -1$, then $y = e^{2x} - e^{-3x}$ would be a solution. The set of solutions is a two-dimensional vector space spanned by e^{2x} and e^{-3x}.   ✦

We now turn to a discussion of nonhomogeneous differential equations. The general solution is found according to Theorem 6-13 by adding a particular solution to the general solution of the associated homogeneous differential equation (the complementary solution). We have already discussed the method of finding the complementary solution. Numerous techniques have been developed for finding a particular solution, techniques often associated with various classes of nonhomogeneous differential equations. It does not lie within our domain to go into the various analytical methods. The particular solution for each differential equation that we shall consider will be arrived at by observation. In each case the reader should verify that it is a solution by substituting into the differential equation.

✦ Example 3

Solve the nonhomogeneous differential equation

$$\frac{d^2 y}{dx^2} + 9y = 2x$$

Let us first of all look for a particular solution. Observe that if we take y to be a linear function in x then $d^2 y/dx^2 = 0$. For such a function, the equation reduces to $9y = 2x$, giving $y = 2x/9$. We are led to the particular solution $y_p = 2x/9$. On checking, we see that this function does indeed satisfy the differential equation.

We now determine the complementary solution. The equation may be written in operator form

$$(D^2 + 9)y = 2x$$

The auxiliary equation is $m^2 + 9 = 0$, having roots $\pm 3i$. The roots are complex; the complementary function is

$$y_c = c_1 \cos 3x + c_2 \sin 3x$$

Thus the general solution is

$$y = \underbrace{2x/9}_{\substack{particular \\ solution}} + \underbrace{c_1 \cos 3x + c_2 \sin 3x}_{\substack{complementary \\ solution}}   ✦$$

In this section, we have focused on second-order linear differential equations. The general results that we have developed, however, extend to linear differential equations of all orders. The set of solutions to a homogeneous differential equation of order n forms a vector space of dimension n. The solutions of a nonhomogeneous equation can be expressed in terms of a particular solution and the solutions to the associated homogeneous equation.

EXERCISES

*1. Interpret D^2 as a linear mapping of P, the vector space of all polynomials, into itself. Find the kernel of D^2.

2. Interpret D^3 as a linear mapping of P into itself. Find the kernel of D^3. Determine the kernel of D^n when D^n is interpreted as a mapping of P onto itself.

*3. Consider D^2 as a linear mapping of P_4, the vector space of polynomials of degree ≤ 4, into itself. Find the kernel and the range of D^2 and illustrate that the dimension of the domain equals the dimension of the kernel plus the dimension of the range.

4. Let $D^2 + D$ be a linear mapping of V, the vector space of functions that possess derivatives of all orders on an interval, into itself. Prove that the functions 1 and e^{-x} lie in the kernel. Since the kernel is of dimension 2, these functions will form a basis.

*5. Let $D^2 + D - 2$ be a linear mapping of V into itself. Prove that the functions e^x and e^{-2x} lie in the kernel. Since the kernel is of dimension 2, these functions form a basis. (V is defined in Exercise 4.)

6. (a) Solve the homogeneous differential equation

$$\frac{d^2y}{dx^2} + 4\frac{dy}{dx} + 3 = 0$$

(b) Use your answer to (a) to find a basis for the kernel of the differential operator $D^2 + 4D + 3$.

*7. (a) Solve the homogeneous differential equation

$$\frac{d^2y}{dx^2} + 2\frac{dy}{dx} - 8 = 0$$

(b) Use your answer to (a) to get a basis for the kernel of the differential operator $D^2 + 2D - 8$.

In Exercises 8–15, solve the given nonhomogeneous differential equations.

8. $\dfrac{d^2y}{dx^2} + 5\dfrac{dy}{dx} + 6y = 8$ ***9.** $\dfrac{d^2y}{dx^2} + 4\dfrac{dy}{dx} + 4y = 3$

10. $\dfrac{d^2y}{dx^2} + 25y = 7$ ***11.** $\dfrac{d^2y}{dx^2} - 3\dfrac{dy}{dx} = 8$

12. $\dfrac{d^2y}{dx^2} - 7\dfrac{dy}{dx} + 12y = 24$ ***13.** $\dfrac{d^2y}{dx^2} - 2\dfrac{dy}{dx} + 6y = 6$

14. $\dfrac{d^2y}{dx^2} + 8\dfrac{dy}{dx} = 3e^{2x}$ ***15.** $\dfrac{d^2y}{dx^2} + 4y = \sin 3x$

16. (a) Prove that if y_{p_1} is a particular solution to $(D^2 + a_1 D + a_0)y = h_1$ and y_{p_2} is a particular solution to $(D^2 + a_1 D + a_0)y = h_2$, then $y_{p_1} + y_{p_2}$ is a particular solution to the differential equation $(D^2 + a_1 D + a_0)y = h$, where $h = h_1 + h_2$.

This technique of splitting h into parts can be useful for finding a particular solution. Use this method to find a particular and then the general solution to $(D^2 - 9)y = x + 3e^x$.

(b) Prove that if y_p is a particular solution to $(D^2 + a_1 D + a_0)y = h$, then cy_p is a particular solution to $(D^2 + a_1 D + a_0)y = ch$ for all scalars c. Demonstrate that this result holds for the differential equations $(D^2 - 4)y = x$ and $(D^2 - 4)y = 12x$.

Once again in (a) and (b) we see operations of addition and scalar multiplication of vectors being in a certain sense preserved.

*17. On the vector space of functions that possess derivatives of all orders, construct a differential operator that has the vector space with basis e^{4x} as kernel. Observe that this leads to a homogeneous differential equation with general solution ce^{4x}.

*18. Construct a differential operator that has the vector space spanned by the vectors 1 and e^x as kernel. Observe that this leads to a homogeneous differential equation with general solution $c_1 + c_2 e^x$.

19. (a) Consider the third-order homogeneous linear differential equation

$$(D^3 + a_2 D^2 + a_1 D + a_0)y = 0$$

Prove that the set of solutions forms a subspace of the vector space of functions that possess derivatives of all order. It will, in fact, be a subspace of dimension 3. In general, such a differential equation of order n will have an n-dimensional vector space of solutions.

(b) Prove that the set of solutions to a nonhomogeneous linear differential equation of order three does not form a vector space. This is true for nonhomogeneous linear differential equations of all orders.

20. Interpret D and D^2 as linear mappings of P_5, the vector space of polynomials of degree ≤ 5, onto itself. Prove that the range of D^2 is a proper subset of the range of D. Show that in each case the dimension of the domain equals the dimension of the kernel plus the dimension of the range. Thus, quite naturally, as the dimension of the kernel increases, that of the range must decrease.

21. Prove that if x is in the kernel of a constant coefficient linear differential operator of degree 2, then 1 will also be. What is the implication of this result for differential equations?

22. Prove that if xe^x is in the kernel of a constant coefficient linear differential operator of degree 2, then e^x will also be. What is the implication of this result for differential equations?

23. In Exercise 14 of the previous section, the reader proved that the solution to a nonhomogeneous system of linear equations is unique if and only if the corresponding homogeneous equation has as its unique solution the zero vector. The analogous result for differential equations should be that the solution to a nonhomogeneous linear differential equation is unique if and only if the complementary function consists of only the zero function. Is this result true for differential equations? Is it relevant?

24. Project: Discuss the method of undetermined coefficients for finding particular solutions using the concepts of linear algebra.

6-6* ONE-TO-ONE MAPPINGS, ONTO MAPPINGS, AND ISOMORPHISMS

There are various useful classes of mappings. We now introduce three of these classes—namely, *one-to-one* mappings, *onto* mappings, and *isomorphisms*. One-to-one and onto mappings exist between sets that are not necessarily vector spaces. We shall define these mappings in their most general setting. However, in linear algebra we are primarily interested in sets that are vector spaces and in linear mappings between vector spaces. We shall thus be interested in one-to-one linear mappings and onto linear mappings between vector spaces. Isomorphisms are special linear mappings.

> **Definition 6-6** Let f be a mapping of a set V into a set W. If two distinct elements of V always have distinct images in W, then f is said to be *one-to-one*.

Thus if $\mathbf{v}_1$ and $\mathbf{v}_2$ are elements of V such that $\mathbf{v}_1 \neq \mathbf{v}_2$, f is one-to-one if $f(\mathbf{v}_1) \neq f(\mathbf{v}_2)$. We illustrate this definition in Figure 6-32. The following equivalent way of expressing the definition is most useful for checking whether or not a given mapping is one-to-one.

f is one-to-one if and only if $f(v_1) = f(v_2)$ implies that $v_1 = v_2$.

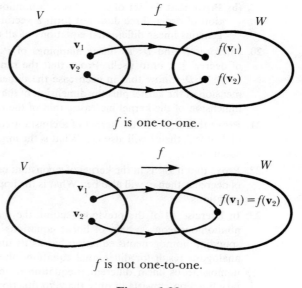

f is one-to-one.

f is not one-to-one.

Figure 6-32

✦ *Example 1* ──

Consider the linear mapping $f(x, y) = (-y, 2x)$ of $\mathbf{R}^2$ into $\mathbf{R}^2$. Let us prove that this mapping is one-to-one.

Let (x_1, y_1) and (x_2, y_2) be elements of $\mathbf{R}^2$ such that $f(x_1, y_1) = f(x_2, y_2)$. Then

$$(-y_1, 2x_1) = (-y_2, 2x_2)$$

giving

$$-y_1 = -y_2 \quad \text{and} \quad 2x_1 = 2x_2$$

Thus $y_1 = y_2$, $x_1 = x_2$, and $(x_1, y_1) = (x_2, y_2)$. The mapping is one-to-one. ◆

The following example illustrates a mapping that is not one-to-one.

◆ *Example 2*

Consider the projection mapping $f(x, y, z) = (x, y, 0)$ of $\mathbf{R}^3$ into $\mathbf{R}^3$.

Let (x_1, y_1, z_1) and (x_2, y_2, z_2) be elements of $\mathbf{R}^3$ such that $f(x_1, y_1, z_1) = f(x_2, y_2, z_2)$. This implies that

$$(x_1, y_1\ 0) = (x_2, y_2, 0)$$

giving

$$x_1 = x_2 \quad \text{and} \quad y_1 = y_2$$

z_1 and z_2 can be distinct. Thus $f(x_1, y_1, z_1) = f(x_2, y_2, z_2)$ does not imply that $(x_1, y_1, z_1) = (x_2, y_2, z_2)$. In fact, we see, for example, that

$$f(1, 2, 3) = f(1, 2, 4) = (1, 2, 0)$$

If the mapping does not always take distinct vectors into distinct images, the mapping is not one-to-one. ◆

The one-to-one property of a linear mapping is related to the kernel of the mapping. The following theorem gives the relationship.

Theorem 6-14   *A linear mapping f is one-to-one if and only if the kernel is the zero vector.*

Proof: Let us first of all assume that the mapping is one-to-one and show that the kernel consists of only the zero vector. Since the mapping is one-to-one, only one vector can be mapped onto the zero vector in the range; the kernel must consist of a single vector. The zero vector is in the kernel. Thus the kernel consists of only the zero vector.

We now prove the converse—namely, that if the kernel is the zero vector, then the mapping is one-to-one. Let $f(\mathbf{u}_1) = \mathbf{v}$ and $f(\mathbf{u}_2) = \mathbf{v}$. Then $f(\mathbf{u}_1) = f(\mathbf{u}_2)$, $f(\mathbf{u}_1) - f(\mathbf{u}_2) = \mathbf{0}$, $f(\mathbf{u}_1 - \mathbf{u}_2) = \mathbf{0}$. Thus $\mathbf{u}_1 - \mathbf{u}_2$ is in the kernel, $\mathbf{u}_1 - \mathbf{u}_2 = \mathbf{0}$, $\mathbf{u}_1 = \mathbf{u}_2$. The mapping is thus one-to-one.

Let us now look at the second class of mappings—onto mappings.

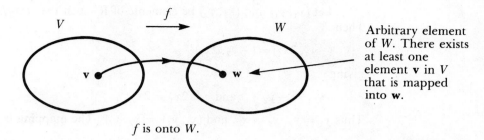

f is onto W.

Figure 6-33

Definition 6-7   Let f be a mapping of a set V into a set W. f is said to be *onto* W if every element in W is the image of at least one element of V. (See Figure 6-33.)

The mapping f will in fact be onto W if W is the range of f.

✦ *Example 3*
_____

Consider the linear mapping $f(x, y, z) = (x + y, 2z)$ of $\mathbf{R}^3$ into $\mathbf{R}^2$. Since $x + y$ can be any real number and so can $2z$, the range of f is $\mathbf{R}^2$. Thus f is a linear mapping of $\mathbf{R}^3$ onto $\mathbf{R}^2$.   ✦

✦ *Example 4*
_____

Let us return to the projection mapping $f(x, y, z) = (x, y, 0)$ of $\mathbf{R}^3$ into $\mathbf{R}^3$. We shall see that this is a mapping that is not onto.

The range of f is the set of vectors of the type $(x, y, 0)$, having 0 as the last component; that is, the range is the xy plane. Thus every vector of $\mathbf{R}^3$ is not the image of a vector in $\mathbf{R}^3$; for example, $(1, 2, 3)$ is not the image of any vector.

The mapping f is not onto $\mathbf{R}^3$.   ✦

We now bring the concepts of one-to-one and onto together in the following definition.

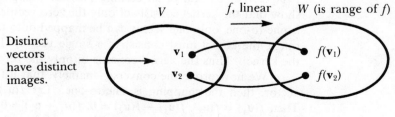

f is an isomorphism.

Figure 6-34

Definition 6-8   Let f be a linear mapping of a vector space V into a vector space W that is both one-to-one and onto. f is said to be an *isomorphism*. (See Figure 6-34.)

✦ *Example 5*

Let V be the subspace of $\mathbf{R}^3$ consisting of all vectors of the form (a, a, b). Consider the linear mapping f of V into $\mathbf{R}^2$ defined by

$$f(a, a, b) = (a, b)$$

We first of all demonstrate that f is one-to-one. Let (a_1, a_1, b_1) and (a_2, a_2, b_2) be elements of V such that $f(a_1, a_1, b_1) = f(a_2, a_2, b_2)$. This implies that

$$(a_1, b_1) = (a_2, b_2)$$

giving

$$a_1 = a_2 \quad \text{and} \quad b_1 = b_2$$

Therefore,

$$(a_1, a_1, b_1) = (a_2, a_2, b_2)$$

The mapping is one-to-one.

Furthermore, the range of f consists of all vectors of the form (a, b). The range of f is $\mathbf{R}^2$.

Thus f is both one-to-one and onto; it is an isomorphism.

Having been through the formal procedure of verifying that f is an isomorphism, let us now get a geometrical feel for the implication of the result. If $\mathbf{R}^3$ is interpreted as 3-space, then the points (a, a, b), having equal x and y components, make up the plane in Figure 6-35(a). $\mathbf{R}^2$ of course describes points in a plane—Figure 6-35(b). The plane embedded in $\mathbf{R}^3$ is physically identical to the plane standing alone. The mathematical descriptions of the two planes differ in that one describes a plane in the way it is embedded in 3-space relative to a certain coordinate system while the other gives a stand-alone description of a plane in 2-space. The isomorphism f relates the two mathematical descriptions of the planes and in so doing relates them physically also. The point P, for example, in the embedded plane corresponds to the point Q in the stand-alone plane.   ✦

Two vector spaces are said to be *isomorphic* if there exists an isomorphism of the one onto the other. When such spaces have geometric interpretation, their "shapes" will be the same, as was seen in the above example.

We now present important theorems from vector space theory. These theorems will make the above observations and techniques concerning isomorphic vector spaces more precise.

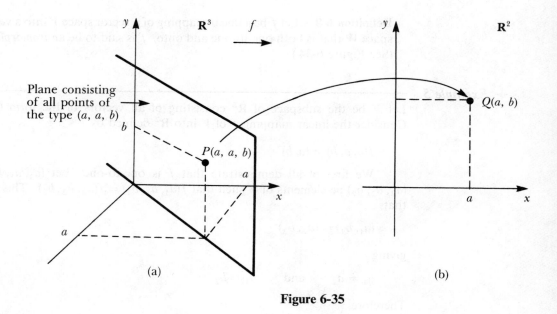

Figure 6-35

Theorem 6-15   *The kernel of an isomorphism consists of only the zero vector. Thus the kernel of an isomorphism is of dimension 0.*

Proof: The kernel of a one-to-one mapping has been seen to be the zero vector. Thus the kernel of an isomorphism is the zero vector.

Theorem 6-16

(1) *Let V and W be vector spaces of the same dimension having bases $v_1, \ldots, v_n$ and $w_1, \ldots, w_n$. The following mapping f defines an isomorphism of V onto W:*

$$f(a_1 v_1 + \cdots + a_n v_n) = a_1 w_1 + \cdots + a_n w_n$$

(2) *Two finite-dimensional vector spaces are isomorphic if and only if they are of the same dimension.*

Proof: In part (1), $a_1 v_1 + \cdots + a_n v_n$ represents an arbitrary vector of V. Thus f defines a mapping of V into W. It can be shown (Exercise 33) that f is linear, one-to-one, and onto. It is thus an isomorphism.

We see from part (1) that two vector spaces of the same dimension are isomorphic. For part (2) it remains to prove that the converse also holds; that is, given an isomorphism f of V onto W, then V and W have the same dimension. We know that

dimension of $V =$

dimension of kernel of f + dimension of range of f

Since f is an isomorphism, the dimension of the kernel is 0 and its range is W. Thus

$$\text{dimension of } V = \text{dimension of } W$$

✦ Example 6

Consider the vector spaces P_3 and M_{22}. These are both of dimension 4; they are thus isomorphic. Let us construct an isomorphism of P_3 onto M_{22}. Take convenient bases for P_3 and M_{22}, $x^3, x^2, x, 1$ and $\begin{pmatrix} 1 & 0 \\ 0 & 0 \end{pmatrix}, \begin{pmatrix} 0 & 1 \\ 0 & 0 \end{pmatrix}, \begin{pmatrix} 0 & 0 \\ 1 & 0 \end{pmatrix}, \begin{pmatrix} 0 & 0 \\ 0 & 1 \end{pmatrix}$. Theorem 6-16 tells us that the following mapping f is an isomorphism of P_3 onto M_{22}:

$$f(a_3 x^3 + a_2 x^2 + a_1 x + a_0) = a_3 \begin{pmatrix} 1 & 0 \\ 0 & 0 \end{pmatrix} + a_2 \begin{pmatrix} 0 & 1 \\ 0 & 0 \end{pmatrix} + a_1 \begin{pmatrix} 0 & 0 \\ 1 & 0 \end{pmatrix}$$

$$+ a_0 \begin{pmatrix} 0 & 0 \\ 0 & 1 \end{pmatrix}$$

$$= \begin{pmatrix} a_3 & a_2 \\ a_1 & a_0 \end{pmatrix}$$

The image of the element $2x^3 - 3x^2 + x - 4$ of P_3, for example, under this isomorphism is $\begin{pmatrix} 2 & -3 \\ 1 & -4 \end{pmatrix}$. ✦

An implication of the last theorem is that isomorphic vector spaces are identical from the algebraic point of view. An isomorphism is a linear mapping; thus such spaces have similar operations of addition and scalar multiplication. Furthermore, since they have the same dimension, they are of the same "size." This was seen to be the case for the isomorphic vector spaces of Example 5, both two-dimensional. Finally, Theorem 6-16 implies that every real finite-dimensional vector space is isomorphic to $\mathbf{R}^n$ for some n. Thus every real finite-dimensional vector space is identical from the algebraic viewpoint to some $\mathbf{R}^n$. In developing the algebraic structures for $\mathbf{R}^n$, we have thus developed the structures for all real finite-dimensional vector spaces.

The following final result of this section tells us how we can identify a given vector space V with an appropriate $\mathbf{R}^n$ through an isomorphism. This association enables us to discuss the vector space by discussing $\mathbf{R}^n$; it in effect defines a coordinate system on the space.

Theorem 6-17 *Let V be a vector space of dimension n with basis $\mathbf{v}_1, \ldots, \mathbf{v}_n$. Let $\mathbf{v} = a_1 \mathbf{v}_1 + \cdots + a_n \mathbf{v}_n$ be an arbitrary vector in V. Then the mapping*

$$f(a_1 \mathbf{v}_1 + \cdots + a_n \mathbf{v}_n) = (a_1, \ldots, a_n)$$

is an isomorphism of V onto $\mathbf{R}^n$. *This isomorphism is called a coordinate mapping of V onto* $\mathbf{R}^n$. $(a_1, \ldots, a_n)$ *is called the coordinate vector of* $\mathbf{v}$ *relative to the basis* $\mathbf{v}_1, \ldots, \mathbf{v}_n$.

We summarize this important result in Figure 6-36.

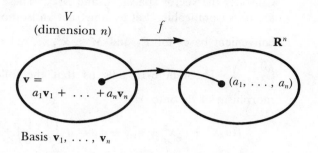

Coordinate mapping

Figure 6-36

Proof: Take the standard basis $(1, \ldots, 0), \ldots, (0, \ldots, 1)$ for $\mathbf{R}^n$. From Theorem 6-16 we know that the following mapping of V into $\mathbf{R}^n$ is an isomorphism:

$$f(a_1\mathbf{v}_1 + \cdots + a_n\mathbf{v}_n) = a_1(1, \ldots, 0) + \cdots + a_n(0, \ldots, 1)$$
$$= (a_1, \ldots, a_n)$$

This is the coordinate mapping.

✦ **Example 7** _____

Consider the vector space P_2. This space is of dimension 3. Let us construct the coordinate mapping associated with the standard basis $x^2, x, 1$. Let $ax^2 + bx + c$ be an arbitrary vector in this space. Then the coordinate mapping is defined by

$$f(ax^2 + bx + c) = (a, b, c)$$

The image of the vector $3x^2 + 2x - 4$ of P_2, for example, is $(3, 2, -4)$.

We can use $\mathbf{R}^3$, together with this correspondence, to discuss P_2. We shall use this approach to further our understanding of linear mappings on P_2 in the following section.   ✦

Let $\mathbf{u}_1, \ldots, \mathbf{u}_n$ and $\mathbf{w}_1, \ldots, \mathbf{w}_n$ be bases for a vector space V. Each basis will define a coordinate mapping of V onto $\mathbf{R}^n$. Thus each vector $\mathbf{v}$ in V will have a coordinate vector relative to each basis. Let us now discuss how such coordinate vectors are related.

Since $\mathbf{w}_1, \ldots, \mathbf{w}_n$ is a basis for V, each of the vectors $\mathbf{u}_1, \ldots, \mathbf{u}_n$ can be expressed as a linear combination of these vectors. Let

$$\mathbf{u}_i = \sum c_{ij}\mathbf{w}_j$$

The coefficients make up a matrix C. The relationship between the two bases is described by the matrix C.

Let $\mathbf{v}$ be a vector in V. $\mathbf{v}$ can be expressed as a linear combination of each set of base vectors. Let

$$\mathbf{v} = \sum a_i\mathbf{u}_i \quad \text{and} \quad \mathbf{v} = \sum b_i\mathbf{v}_i$$

The coordinate vector of $\mathbf{v}$ relative to the first basis is $(a_1, \ldots, a_n)$, and relative to the second basis it is $(b_1, \ldots, b_n)$. If we express these vectors in column form as $\mathbf{a}$ and $\mathbf{b}$, the relationship between $\mathbf{a}$ and $\mathbf{b}$ is

$$\boxed{\mathbf{b} = C^t\mathbf{a}}$$

The matrix C^t defines the *coordinate transformation* associated with the change of basis. We do not derive this result.

◆ **Example 8**

Consider the vector space P_2. Let x^2, x, 1 and x^2, $(x + 1)$, 2 be two bases for this space. Let us express the first basis in terms of the second. We get

$$x^2 = 1x^2 + 0(x + 1) + 0(2)$$
$$x = 0x^2 + 1(x + 1) - \tfrac{1}{2}(2)$$
$$1 = 0x^2 + 0(x + 1) + \tfrac{1}{2}(2)$$

[An expression such as $x = 0x^2 + 1(x + 1) - \tfrac{1}{2}(2)$ can be found either by inspection or by solving $x = px^2 + q(x + 1) - r(2)$ for p, q, and r by comparing coefficients.]

The coefficients of x^2, $x + 1$, and 2 give the following matrix C:

$$C = \begin{pmatrix} 1 & 0 & 0 \\ 0 & 1 & -\tfrac{1}{2} \\ 0 & 0 & \tfrac{1}{2} \end{pmatrix}$$

The relationship between coordinate vectors relative to the two bases is defined by C^t.

$$\begin{pmatrix} b_1 \\ b_2 \\ b_3 \end{pmatrix} = \begin{pmatrix} 1 & 0 & 0 \\ 0 & 1 & 0 \\ 0 & -\tfrac{1}{2} & \tfrac{1}{2} \end{pmatrix} \begin{pmatrix} a_1 \\ a_2 \\ a_3 \end{pmatrix} \tag{1}$$

For example, consider the vector $3x^2 + 4x + 12$ in P_2. Its coordinate vector relative to the standard basis x^2, x, 1 is $(3, 4, 12)$. Writing this vector in column form, for the coordinate vector relative to the basis x^2, $(x + 1)$, 2 we have

$$\begin{pmatrix} 1 & 0 & 0 \\ 0 & 1 & 0 \\ 0 & -\frac{1}{2} & \frac{1}{2} \end{pmatrix} \begin{pmatrix} 3 \\ 4 \\ 12 \end{pmatrix} = \begin{pmatrix} 3 \\ 4 \\ 4 \end{pmatrix}$$

Thus we have that $3x^2 + 4x + 12$ can be expressed in terms of the basis x^2, $(x + 1)$, 2 as

$$3x^2 - 4(x + 1) + 4(2)$$

Note that the elements of P_2 are usually expressed relative to the standard basis x^2, x, 1. Using (1), we can compute the coordinate vectors of any number of vectors rapidly, relative to the basis x^2, $(x + 1)$, 2. This is often the way in which such coordinate transformations are used.   ✦

EXERCISES

In Exercises 1–26, determine whether the given mappings are (a) one-to-one, (b) onto, (c) isomorphisms.

*1. The matrix $\begin{pmatrix} 0 & -1 \\ 1 & 0 \end{pmatrix}$ mapping of $\mathbf{R}^2$ into $\mathbf{R}^2$

2. The matrix $\begin{pmatrix} 2 & 0 \\ 0 & 2 \end{pmatrix}$ mapping of $\mathbf{R}^2$ into $\mathbf{R}^2$

*3. The matrix $\begin{pmatrix} 2 & 0 & 0 \\ 0 & 2 & 0 \\ 0 & 0 & 0 \end{pmatrix}$ mapping of $\mathbf{R}^3$ into $\mathbf{R}^3$

4. The matrix $\begin{pmatrix} 1 & 2 \\ 3 & 0 \end{pmatrix}$ mapping of $\mathbf{R}^2$ into $\mathbf{R}^2$

*5. The matrix $\begin{pmatrix} 1 & 2 \\ 3 & 4 \end{pmatrix}$ mapping of $\mathbf{R}^2$ into $\mathbf{R}^2$

6. The matrix $\begin{pmatrix} 1 & 2 \\ 2 & 4 \end{pmatrix}$ mapping of $\mathbf{R}^2$ into $\mathbf{R}^2$

*7. The matrix $\begin{pmatrix} 6 & 2 \\ 3 & 1 \end{pmatrix}$ mapping of $\mathbf{R}^2$ into $\mathbf{R}^2$

8. $f(x, y) = x + y$ of $\mathbf{R}^2$ into $\mathbf{R}$
*9. $f(x, y) = (2x, 2y)$ of $\mathbf{R}^2$ into $\mathbf{R}^2$
10. $f(x, y) = (x, 3x)$ of $\mathbf{R}^2$ into $\mathbf{R}^2$
*11. $f(x, y) = (x^2, y)$ of $\mathbf{R}^2$ into $\mathbf{R}^2$
12. $f(x, y, z) = (x + y, x + z)$ of $\mathbf{R}^3$ into $\mathbf{R}^2$
*13. $f(x, y, z) = (2x, x + y, y + z)$ of $\mathbf{R}^3$ into $\mathbf{R}^3$

14. The matrix $\begin{pmatrix} 1 & 1 & 3 \\ 0 & 1 & 4 \end{pmatrix}$ mapping of $\mathbf{R}^3$ into $\mathbf{R}^2$

*15. The matrix $\begin{pmatrix} 1 & 2 \\ 3 & 0 \\ 1 & 1 \end{pmatrix}$ mapping of $\mathbf{R}^2$ into $\mathbf{R}^3$

16. The mapping $f(a, b, c, d) = \begin{pmatrix} a & b \\ c & d \end{pmatrix}$ of $\mathbf{R}^4$ into M_{22}

*17. $f(a, b, c) = \begin{pmatrix} a & b \\ c & 0 \end{pmatrix}$ of $\mathbf{R}^3$ into M_{22}

18. $f\begin{pmatrix} a & b \\ c & d \end{pmatrix} = \begin{pmatrix} a & 0 \\ 0 & d \end{pmatrix}$ of M_{22} into M_{22}

*19. $f\begin{pmatrix} a & b \\ c & d \end{pmatrix} = (a, b, c, d)$ of M_{22} into $\mathbf{R}^4$

20. $f\begin{pmatrix} a & b \\ c & d \end{pmatrix} = a + b + c + d$ of M_{22} into $\mathbf{R}$

*21. $f\begin{pmatrix} a & b & c \\ d & e & f \end{pmatrix} = (a, b, c, d, e, f)$ of M_{23} into $\mathbf{R}^6$

22. $f\begin{pmatrix} a & b \\ c & d \end{pmatrix} = \begin{pmatrix} a & 0 \\ 0 & a \end{pmatrix}$ of M_{22} into M_{22}

*23. $f\begin{pmatrix} a & b \\ c & d \end{pmatrix} = (a + 2b, c + 3d)$ of M_{22} into $\mathbf{R}^2$

24. $f(ax^2 + bx + c) = (a, b, 0)$ of P_2, the vector space of polynomials of degree ≤ 2, into $\mathbf{R}^3$

*25. $f(ax^2 + bx + c) = (a, b)$ of P_2 into $\mathbf{R}^2$

26. $f(ax^3 + bx^2 + cx + d) = \begin{pmatrix} a & c \\ b & d \end{pmatrix}$ of P_3 into M_{22}

In Exercises 27–32, construct coordinate mappings of the given vector spaces onto the appropriate $\mathbf{R}^n$, relative to the stated bases.

*27. P_3 onto $\mathbf{R}^4$, relative to the basis $x^3, x^2, x, 1$ of P_3. Find the coordinate vector of $3x^3 + 2x^2 + x + 2$.

28. P_1 onto $\mathbf{R}^2$, relative to the basis $x, 1$ of P_1. Find the coordinate vector of $2x + 5$.

*29. P_2 onto $\mathbf{R}^3$, relative to the basis $x^2, (x + 1), 1$ of P_2. Find the coordinate vector of $x^2 + 2x + 6$.

30. P_1 onto $\mathbf{R}^2$, relative to the basis $(x + 4), 3$ of P_1. Find the coordinate vector of $4x + 6$.

*31. M_{22} onto $\mathbf{R}^4$, relative to the basis $\begin{pmatrix} 1 & 0 \\ 0 & 0 \end{pmatrix}, \begin{pmatrix} 0 & 1 \\ 0 & 0 \end{pmatrix}, \begin{pmatrix} 0 & 0 \\ 1 & 0 \end{pmatrix}, \begin{pmatrix} 0 & 0 \\ 0 & 1 \end{pmatrix}$ of M_{22}. Find the coordinate vector of $\begin{pmatrix} 1 & 2 \\ 3 & 4 \end{pmatrix}$.

32. M_{22} onto $\mathbf{R}^4$, relative to the basis $\begin{pmatrix} 2 & 0 \\ 0 & 0 \end{pmatrix}, \begin{pmatrix} 0 & 1 \\ 1 & 0 \end{pmatrix}, \begin{pmatrix} 0 & 0 \\ 1 & 0 \end{pmatrix}, \begin{pmatrix} 0 & 0 \\ 0 & 1 \end{pmatrix}$ of M_{22}. Find the coordinate vector of $\begin{pmatrix} 4 & 2 \\ 3 & 5 \end{pmatrix}$.

33. Let V and W be vector spaces with bases $\mathbf{v}_1, \ldots, \mathbf{v}_n$ and $\mathbf{w}_1, \ldots, \mathbf{w}_n$, respectively. Prove that the following mapping f of V into W is an isomorphism:

$$f(a_1\mathbf{v}_1 + a_2\mathbf{v}_2 + \cdots + a_n\mathbf{v}_n) = (a_1\mathbf{w}_1 + a_2\mathbf{w}_2 + \cdots + a_n\mathbf{w}_n)$$

(This result was stated in Theorem 6-16.)

34. The result of Exercise 33 can be used to construct isomorphisms between vector spaces of the same dimension (see Example 6). Use this result to construct isomorphisms between the following vector spaces:

*(a) the subspace of $\mathbf{R}^3$ consisting of vectors of the form $(a, a + b, a - b)$ and $\mathbf{R}^2$

(b) the subspace of $\mathbf{R}^3$ consisting of vectors of the form $(a - b, a + b, a + 2b)$ and $\mathbf{R}^2$

*(c) the subspace of $\mathbf{R}^3$ consisting of vectors of the form $(a + b, a, b)$ and the subspace of $\mathbf{R}^4$ consisting of vectors of the form $(a, a + b, a + 2b, a + 3b)$

(d) the subspace of $\mathbf{R}^3$ consisting of vectors of the form $(a, 2a, -a)$ and the subspace of $\mathbf{R}^2$ consisting of vectors of the form $(a, 3a)$

*(e) the set of diagonal 2×2 matrices and $\mathbf{R}^2$

(f) the set of symmetric 2×2 matrices and $\mathbf{R}^3$

35. Let the matrix A define a linear mapping of $\mathbf{R}^n$ into $\mathbf{R}^n$. Prove that A is an isomorphism if and only if $|A| \neq 0$. This is a convenient test; apply it to the matrices of Exercises 1–7 to check whether they define isomorphisms.

36. Let A be a nonsingular matrix that defines an isomorphism of $\mathbf{R}^n$ onto $\mathbf{R}^n$ (Exercise 35). Prove that A^{-1} also defines an isomorphism of $\mathbf{R}^n$ onto itself.

*37. Consider the vector space P_2. $x^2, x, 1$ and $3x^2, (x - 1), 4$ are bases for P_2. Find the matrix equation that describes the coordinate transformation from the standard basis $x^2, x, 1$ to the basis $3x^2, (x - 1), 4$. Use this transformation to find the coordinate vectors of $3x^2 + 4x + 8$, $6x^2 + 4$, $8x + 12$, and $3x^2 + 4x + 4$, relative to the basis $3x^2, (x - 1), 4$.

38. Consider the vector space P_1. $x, 1$ and $(x + 2), 3$ are bases for P_1. Find the matrix equation that describes the coordinate transformation from the standard basis $x, 1$ to $(x + 2), 3$. Use this transformation to find the coordinate vectors of $3x + 3$, $6x$, $6x + 9$, and $12x - 3$, relative to the basis $(x + 2), 3$.

*39. Consider the vector space M_{22}. $\begin{pmatrix} 1 & 0 \\ 0 & 0 \end{pmatrix}, \begin{pmatrix} 0 & 1 \\ 0 & 0 \end{pmatrix}, \begin{pmatrix} 0 & 0 \\ 1 & 0 \end{pmatrix}, \begin{pmatrix} 0 & 0 \\ 0 & 1 \end{pmatrix}$ and $\begin{pmatrix} 2 & 0 \\ 0 & 0 \end{pmatrix}, \begin{pmatrix} 0 & -1 \\ 1 & 0 \end{pmatrix}, \begin{pmatrix} 0 & 2 \\ 4 & 0 \end{pmatrix}, \begin{pmatrix} 0 & 0 \\ 0 & 3 \end{pmatrix}$ are bases for M_{22}. Find the matrix equation that describes the coordinate transformation from the standard basis to the second basis. Use this transformation to find the coordinate vectors of $\begin{pmatrix} 2 & 3 \\ 6 & 3 \end{pmatrix}$, $\begin{pmatrix} -2 & 9 \\ 6 & 6 \end{pmatrix}$, and $\begin{pmatrix} 0 & 12 \\ 9 & 0 \end{pmatrix}$, relative to the second basis.

40. Consider the vector space R^2. (1, 0), (0, 1) and (2, 3), (1, 2) are bases for $\mathbf{R}^2$. Find the matrix equation that describes the coordinate transformation from the first basis to the second. Use this transformation to find the coordinate vectors of (1, 2), (3, −1), (2, 4), and (5, 6), relative to the second basis.

*41. Consider the vector space $\mathbf{R}^3$. (1, 0, 0), (0, 1, 0), (0, 0, 1) and (1, 2, 0), (2, 1, 0), (0, 0, 4) are bases for $\mathbf{R}^3$. Find the matrix equation that describes the coordinate transformation from the first basis to the second. Use this transformation to find the coordinate vectors of (3, 6, 4), (0, 6, 8), (6, 0, 0), and (12, 3, −4), relative to the second basis.

42. The reader will no doubt have observed that when one of the bases is a standard basis, it would be easier to express the second basis in terms of the standard basis rather than vice versa as we seem to have been doing. To get the desired coordinate transformation, we have to select the basis transformation the "awkward way around." If one expresses the second basis in terms of the standard basis, one gets C^{-1} as the matrix of coefficients. One can then compute C, knowing that $(C^{-1})^{-1} = C$. Use this method to compute C for the situation of Example 8.

6-7*   MATRIX REPRESENTATIONS OF LINEAR MAPPINGS

In previous sections, we discussed mappings that were defined in terms of matrices. This led to a general definition of a linear mapping. In this section, we discuss the fact that any linear mapping between two finite-dimensional vector spaces has a *matrix representation*; that is, there is a matrix that can be used to perform the mapping.

The following theorem, giving important insights into the behavior of linear mappings, leads up to this concept.

> **Theorem 6-18**   *Let U and V be vector spaces, let* $\mathbf{u}_1, \ldots, \mathbf{u}_n$ *be a basis for U, and let f be a linear mapping of U into V. f is completely determined if its value on each base vector is known. The range of f is spanned by* $f(\mathbf{u}_1), \ldots, f(\mathbf{u}_n)$.

Thus, defining a linear mapping on a set of base vectors automatically defines it on the whole space.

Proof: Let $\mathbf{u}$ be an element of U. Since $\mathbf{u}_1, \ldots, \mathbf{u}_n$ is a basis of U, there exist scalars $a_1, \ldots, a_n$ such that

$$\mathbf{u} = a_1\mathbf{u}_1 + \cdots + a_n\mathbf{u}_n$$

Thus

$$f(\mathbf{u}) = f(a_1\mathbf{u}_1 + \cdots + a_n\mathbf{u}_n)$$

Since f is linear,

$$f(\mathbf{u}) = a_1 f(\mathbf{u}_1) + \cdots + a_n f(\mathbf{u}_n)$$

Hence $f(\mathbf{u})$ is known if $f(\mathbf{u}_1), \ldots, f(\mathbf{u}_n)$ are known.

Further, $f(\mathbf{u})$ may be taken to be an arbitrary element in the range. It can be expressed as a linear combination of $f(\mathbf{u}_1), \ldots, f(\mathbf{u}_n)$. Thus these vectors span the range of f.

We shall now discuss matrix representations of mappings.

Theorem 6-19   *Let f be a linear mapping of a vector space U into a vector space V, and let $\mathbf{u}_1, \ldots, \mathbf{u}_n$ and $\mathbf{v}_1, \ldots, \mathbf{v}_m$ be bases for U and V, respectively. Let $f(\mathbf{u}_i) = \sum_{j=1}^{m} a_{ij}\mathbf{v}_j$ for $i = 1$ to n. Then the matrix A^t, the transpose of the matrix of coefficients defined by the right sides of these equations, is the matrix representation of f.*

Here we have examined the effect of f on each base vector of U, expressing the result in each case as a linear combination of the base vectors of V. The matrix of coefficients must characterize f completely, since it indicates the effect of f on each base vector. The theorem states that the transpose of this matrix of coefficients is the matrix that can be used to perform the role of f. We now prove this and in so doing indicate how the matrix is used to define the mapping.

Proof: Let $\mathbf{u}$ be an arbitrary element of U. It can thus be represented as a linear combination of the vectors $\mathbf{u}_1, \ldots, \mathbf{u}_n$,

$$\mathbf{u} = \sum_{i=1}^{n} x_i \mathbf{u}_i$$

We have that

$$f(\mathbf{u}) = f\left(\sum_{i=1}^{n} x_i \mathbf{u}_i \right)$$

Since f is linear, this can be written

$$f(\mathbf{u}) = \sum_{i=1}^{n} x_i f(\mathbf{u}_i) = \sum_{i=1}^{n} x_i \left(\sum_{j=1}^{m} a_{ij}\mathbf{v}_j \right)$$

The sums being finite, this equation may be rearranged to read

$$f(\mathbf{u}) = \sum_{j=1}^{m} \left(\sum_{i=1}^{n} a_{ij}x_i \right)\mathbf{v}_j$$

$$= \sum_{j=1}^{m} y_j \mathbf{v}_j$$

where $y_j = \sum_{i=1}^{n} a_{ij}x_i.$ \hfill (1)

Here $(x_1, \ldots, x_n)$ is the coordinate vector of $\mathbf{u}$ in $\mathbf{R}^n$ and $(y_1, \ldots, y_m)$ is the coordinate vector of $f(\mathbf{u})$ in $\mathbf{R}^m$. Writing these vectors in column form as $\mathbf{x}$ and $\mathbf{y}$, we see that (1) is equivalent to

$$\mathbf{y} = A^t\mathbf{x}$$

A^t can thus be used to map the coordinate vector **x** of **u** onto the coordinate vector **y** of $f(\mathbf{u})$. In this manner the linear mapping f of U into V can be discussed through the mapping A^t of $\mathbf{R}^n$ into $\mathbf{R}^m$.

✦ Example 1

Consider the linear mapping f of $\mathbf{R}^3$ into $\mathbf{R}^2$ defined by $f(x, y, z) = (2x + y, y - z)$. Let us determine a matrix that "does the same job" as f.

In this context, we interpret the elements of $\mathbf{R}^3$ and $\mathbf{R}^2$ as column vectors. Let us use the standard bases for $\mathbf{R}^3$ and $\mathbf{R}^2$, $\begin{pmatrix}1\\0\\0\end{pmatrix}, \begin{pmatrix}0\\1\\0\end{pmatrix}, \begin{pmatrix}0\\0\\1\end{pmatrix}$ and $\begin{pmatrix}1\\0\end{pmatrix}, \begin{pmatrix}0\\1\end{pmatrix}$.

We determine the effect of f on each of these base vectors:

$$f\left[\begin{pmatrix}1\\0\\0\end{pmatrix}\right] = \begin{pmatrix}2\\0\end{pmatrix} = 2\begin{pmatrix}1\\0\end{pmatrix} + 0\begin{pmatrix}0\\1\end{pmatrix}$$

$$f\left[\begin{pmatrix}0\\1\\0\end{pmatrix}\right] = \begin{pmatrix}1\\1\end{pmatrix} = 1\begin{pmatrix}1\\0\end{pmatrix} + 1\begin{pmatrix}0\\1\end{pmatrix}$$

$$f\left[\begin{pmatrix}0\\0\\1\end{pmatrix}\right] = \begin{pmatrix}0\\-1\end{pmatrix} = 0\begin{pmatrix}1\\0\end{pmatrix} - 1\begin{pmatrix}0\\1\end{pmatrix}$$

The matrix of coefficients is $\begin{pmatrix}2 & 0\\1 & 1\\0 & -1\end{pmatrix}$. The matrix representation of f is the transpose of this matrix, $\begin{pmatrix}2 & 1 & 0\\0 & 1 & -1\end{pmatrix}$.

Let us show that this matrix does in fact accomplish the same thing as f:

$$\begin{pmatrix}2 & 1 & 0\\0 & 1 & -1\end{pmatrix}\begin{pmatrix}x\\y\\z\end{pmatrix} = \begin{pmatrix}2x + y\\y - z\end{pmatrix} \quad ✦$$

✦ Example 2

f is a linear mapping of a vector space U with basis $\mathbf{u}_1, \mathbf{u}_2$ into a vector space V with basis $\mathbf{v}_1, \mathbf{v}_2, \mathbf{v}_3$. f is defined by $f(\mathbf{u}_1) = 2\mathbf{v}_1 - \mathbf{v}_2 + 3\mathbf{v}_3$, $f(\mathbf{u}_2) = 3\mathbf{v}_1 + 2\mathbf{v}_2 - 4\mathbf{v}_3$. Determine the matrix representation of f relative to these bases for U and V. Use this matrix to determine the image of the element $3\mathbf{u}_1 + 2\mathbf{u}_2$ under f. Verify your answer by finding the image of $3\mathbf{u}_1 + 2\mathbf{u}_2$ directly, using the linear properties of f.

The effect of f on the base vectors of U is given by

$$f(\mathbf{u}_1) = 2\mathbf{v}_1 - \mathbf{v}_2 + 3\mathbf{v}_3$$
$$f(\mathbf{u}_2) = 3\mathbf{v}_1 + 2\mathbf{v}_2 - 4\mathbf{v}_3$$

The matrix of coefficients of the right side is $\begin{pmatrix} 2 & -1 & 3 \\ 3 & 2 & -4 \end{pmatrix}$.

The matrix representation of f relative to these bases will be the transpose of this matrix, $\begin{pmatrix} 2 & 3 \\ -1 & 2 \\ 3 & -4 \end{pmatrix}$.

We now find the image of $3\mathbf{u}_1 + 2\mathbf{u}_2$ using this matrix. The coordinate vector is $\begin{pmatrix} 3 \\ 2 \end{pmatrix}$. We get that

$$\begin{pmatrix} 2 & 3 \\ -1 & 2 \\ 3 & -4 \end{pmatrix} \begin{pmatrix} 3 \\ 2 \end{pmatrix} = \begin{pmatrix} 12 \\ 1 \\ 1 \end{pmatrix}$$

The image of the vector $3\mathbf{u}_1 + 2\mathbf{u}_2$ has coordinate vector $\begin{pmatrix} 12 \\ 1 \\ 1 \end{pmatrix}$ in terms of the basis $\mathbf{v}_1, \mathbf{v}_2, \mathbf{v}_3$ of V. It is the vector $12\mathbf{v}_1 + \mathbf{v}_2 + \mathbf{v}_3$.

We now check this result by examining the effect of f on $3\mathbf{u}_1 + 2\mathbf{u}_2$ directly. Using the linearity of f, we get

$$\begin{aligned} f(3\mathbf{u}_1 + 2\mathbf{u}_2) &= f(3\mathbf{u}_1) + f(2\mathbf{u}_2) \\ &= 3f(\mathbf{u}_1) + 2f(\mathbf{u}_2) \\ &= 3(2\mathbf{v}_1 - \mathbf{v}_2 + 3\mathbf{v}_3) + 2(3\mathbf{v}_1 + 2\mathbf{v}_2 - 4\mathbf{v}_3) \\ &= 12\mathbf{v}_1 + \mathbf{v}_2 + \mathbf{v}_3 \quad \blacklozenge \end{aligned}$$

✦ *Example 3**

Let $D = d/dx$ be the operation of taking the derivative. D is a linear mapping of P_2 into itself. Let us determine the matrix representation of D relative to the basis $x^2, x, 1$ of P_2.

We examine the effect of D on the base vectors:

$$\begin{aligned} D(x^2) &= 2x = 0x^2 + 2x + 0(1) \\ D(x) &= 1 \ = 0x^2 + 0x + 1(1) \\ D(1) &= 0 \ = 0x^2 + 0x + 0(1) \end{aligned}$$

The matrix of coefficients is $\begin{pmatrix} 0 & 2 & 0 \\ 0 & 0 & 1 \\ 0 & 0 & 0 \end{pmatrix}$.

The matrix representation of D is the transpose of this matrix, $\begin{pmatrix} 0 & 0 & 0 \\ 2 & 0 & 0 \\ 0 & 1 & 0 \end{pmatrix}$.

We now illustrate the use of this matrix in place of D. Let $a_2x^2 + a_1x + a_0$ be an arbitrary element of P_2. This element has coordinate vector (a_2, a_1, a_0), relative to the basis x^2, x, 1. We find that

$$\begin{pmatrix} 0 & 0 & 0 \\ 2 & 0 & 0 \\ 0 & 1 & 0 \end{pmatrix} \begin{pmatrix} a_2 \\ a_1 \\ a_0 \end{pmatrix} = \begin{pmatrix} 0 \\ 2a_2 \\ a_1 \end{pmatrix}$$

Thus the image of $a_2 x^2 + a_1 x + a_0$ under D has coordinate vector $\begin{pmatrix} 0 \\ 2a_2 \\ a_1 \end{pmatrix}$, relative

to the basis x^2, x, 1. It is the element $2a_2 x + a_0$.

The matrix can be used in this way to perform the mapping defined by D on P_2. As a check, we verify that $D(a_2 x^2 + a_1 x + a_0) = 2a_2 x + a_1$.   ✦

✦ *Example 4** ──

Consider the following linear mapping of P_2 into P_1:

$$f(a_2 x^2 + a_1 x + a_0) = (a_2 + a_1)x - a_0$$

Let us find the matrix representation of f relative to the bases x^2, x, 1 of P_2 and x, 1 of P_1. Consider the effect of f on each basis vector of P_2:

$$f(x^2) = 1x + 0(1)$$
$$f(x) = 1x + 0(1)$$
$$f(1) = 0x + (-1)(1)$$

The matrix of coefficients is $\begin{pmatrix} 1 & 0 \\ 1 & 0 \\ 0 & -1 \end{pmatrix}$. Thus the matrix representation of f

relative to these bases of P_2 and P_1 is $\begin{pmatrix} 1 & 1 & 0 \\ 0 & 0 & -1 \end{pmatrix}$.

Let us check the result. Let $a_2 x^2 + a_1 x + a_0$ be an arbitrary element of P_2. Represent this element by its coordinate vector, in the form of a column

vector in $\mathbf{R}^3$, $\begin{pmatrix} a_2 \\ a_1 \\ a_0 \end{pmatrix}$. Matrix multiplication gives

$$\begin{pmatrix} 1 & 1 & 0 \\ 0 & 0 & -1 \end{pmatrix} \begin{pmatrix} a_2 \\ a_1 \\ a_0 \end{pmatrix} = \begin{pmatrix} a_2 + a_1 \\ -a_0 \end{pmatrix}$$

The image vector has coordinate vector $\begin{pmatrix} a_2 + a_1 \\ -a_0 \end{pmatrix}$, relative to the basis x, 1

of P_1. The image vector is thus $(a_2 + a_1)x - a_0$, confirming the matrix representation.

We can conveniently picture the situation as shown in Figure 6-37. Part (a) describes the linear mapping f of P_2 into P_1 by showing that the image of an arbitrary element $a_2 x^2 + a_1 x + a_0$ of P_2 is the element $(a_2 + a_1)x - a_0$

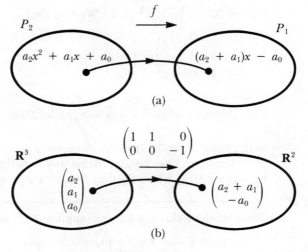

Figure 6-37

of P_1. Part (b) is completely analogous to part (a). $a_2x^2 + a_1x + a_0$ is represented by an element of $\mathbf{R}^3$, $\begin{pmatrix} a_2 \\ a_1 \\ a_0 \end{pmatrix}$; f is represented by the matrix $\begin{pmatrix} 1 & 1 & 0 \\ 0 & 0 & -1 \end{pmatrix}$. The image of $\begin{pmatrix} a_2 \\ a_1 \\ a_0 \end{pmatrix}$ under the matrix mapping is $\begin{pmatrix} a_2 + a_1 \\ -a_0 \end{pmatrix}$, which corresponds to the element $(a_2 + a_1)x - a_0$ of P_1. Part (b) is in effect a *coordinate representation* of part (a). ✦

One important characteristic of such coordinate representations is that they lend themselves to implementation on a computer. Since elements of $\mathbf{R}^n$ and matrices can be handled on a computer, it is often convenient to use such coordinate representations in computations involving linear mappings.

The following example emphasizes the fact that the matrix representation that one gets for a linear mapping depends on the bases selected for the spaces.

✦ Example 5*

Let us reconsider the linear mapping that was discussed in the previous example, this time relative to the bases x^2, $(x + 1)$, 2 of P_2 and x, 1 of P_1:

$$f(a_2x^2 + a_1x + a_0) = (a_2 + a_1)x - a_0$$

Consider the effect of f on each base vector of P_2:

$$f(x^2) = 1x + 0(1)$$
$$f(x + 1) = 1x - 1(1)$$
$$f(2) = 0x - 2(1)$$

The matrix of coefficients is $\begin{pmatrix} 1 & 0 \\ 1 & -1 \\ 0 & -2 \end{pmatrix}$. The matrix representation of f relative to these bases is $\begin{pmatrix} 1 & 1 & 0 \\ 0 & -1 & -2 \end{pmatrix}$.

This representation differs from the one derived in the previous example. Let us check the result. It is necessary to express the arbitrary element $a_2 x^2 + a_1 x + a_0$ of P_2 relative to the basis $x^2, (x + 1), 2$. We get

$$a_2 x^2 + a_1 x + a_0 = a_2 x^2 + a_1(x + 1) + \tfrac{1}{2}(a_0 - a_1)(2)$$

[This expression can be found using the techniques of the previous section or by solving $a_2 x^2 + a_1 x + a_0 = p x^2 + q(x + 1) + r(2)$ for p, q, and r.]

The coordinate vector of an arbitrary vector in P_2, relative to the basis $x^2, (x + 1), 2$, is $\begin{pmatrix} a_2 \\ a_1 \\ \tfrac{1}{2}(a_0 - a_1) \end{pmatrix}$. Matrix multiplication gives

$$\begin{pmatrix} 1 & 1 & 0 \\ 0 & -1 & -2 \end{pmatrix} \begin{pmatrix} a_2 \\ a_1 \\ \tfrac{1}{2}(a_0 - a_1) \end{pmatrix} = \begin{pmatrix} a_2 + a_1 \\ -a_0 \end{pmatrix}$$

leading to the appropriate element $(a_2 + a_1)x - a_0$ of P_1.   ✦

When linear mappings arise in applications, a goal is often to determine a simple matrix representation in order to simplify the problem.

We now discuss how matrix representations relative to different bases are related. Let f be a linear mapping of a vector space U into a vector space V. Let $\mathbf{u}_1, \ldots, \mathbf{u}_n$ and $\mathbf{v}_1, \ldots, \mathbf{v}_m$ be bases for U and V, repsectively. Let A be the matrix representation of f relative to these bases. Let $\bar{\mathbf{u}}_1, \ldots, \bar{\mathbf{u}}_n$ and $\bar{\mathbf{v}}_1, \ldots, \bar{\mathbf{v}}_m$ be another pair of bases for U and V. Let $\bar{A}$ be the matrix representation of f relative to these bases. We shall see how A and $\bar{A}$ are related.

Since $\bar{\mathbf{u}}_1, \ldots, \bar{\mathbf{u}}_n$ form a basis for U, each of $\mathbf{u}_1, \ldots, \mathbf{u}_n$ can be expressed as a linear combination of these vectors. Let

$$\mathbf{u}_i = \sum c_{ij} \bar{\mathbf{u}}_j$$

Similarly, each of $\mathbf{v}_1, \ldots, \mathbf{v}_m$ can be expressed in terms of $\bar{\mathbf{v}}_1, \ldots, \bar{\mathbf{v}}_m$. Let

$$\mathbf{v}_i = \sum d_{ij} \bar{\mathbf{v}}_j$$

The coefficients in these two expressions form matrices C and D. The two matrix representations of f—namely, A and $\bar{A}$—are related as follows through C and D:

$$\boxed{\bar{A} = (D^t)A(C^t)^{-1}}$$

We do not derive this result.

✦ *Example 6** ────────────────────────────────

Let us illustrate this formula using the results of the two previous examples. Consider the vector spaces P_2 and P_1 with f a linear mapping of P_2 into P_1 defined by

$$f(a_2 x^2 + a_1 x + a_0) = (a_2 + a_1)x - a_0$$

In Example 4 we found that the matrix representation of f relative to the standard bases x^2, x, 1 of P_2 and x, 1 of P_1 was $\begin{pmatrix} 1 & 1 & 0 \\ 0 & 0 & -1 \end{pmatrix}$.

Let us start from this matrix representation and use the above formula to compute the matrix representation relative to the bases x^2, $(x + 1)$, 2 of P_2 and x, 1 of P_1. We should arrive at the matrix $\begin{pmatrix} 1 & 1 & 0 \\ 0 & -1 & -2 \end{pmatrix}$ found in Example 5.

The bases for P_2 are x^2, x, 1 and x^2, $(x + 1)$, 2. We get

$$x^2 = 1x^2 + 0(x + 1) + 0(2)$$
$$x = 0x^2 + 1(x + 1) - \tfrac{1}{2}(2)$$
$$1 = 0x^2 + 0(x + 1) + \tfrac{1}{2}(2)$$

Thus

$$C = \begin{pmatrix} 1 & 0 & 0 \\ 0 & 0 & -\tfrac{1}{2} \\ 0 & 0 & \tfrac{1}{2} \end{pmatrix}$$

The basis is unchanged for P_1. For this situation,

$$x = 1x + 0(1)$$
$$1 = 0x + 1(1)$$

Thus D, as we might expect, is the identity matrix; $D = \begin{pmatrix} 1 & 0 \\ 0 & 1 \end{pmatrix}$.

The formula for change of matrix representation gives

$$\bar{A} = \begin{pmatrix} 1 & 0 \\ 0 & 1 \end{pmatrix}^t \begin{pmatrix} 1 & 1 & 0 \\ 0 & 0 & -1 \end{pmatrix} \left[\begin{pmatrix} 1 & 0 & 0 \\ 0 & 1 & -\tfrac{1}{2} \\ 0 & 0 & \tfrac{1}{2} \end{pmatrix}^t \right]^{-1}$$

$$= \begin{pmatrix} 1 & 1 & 0 \\ 0 & 0 & -1 \end{pmatrix} \begin{pmatrix} 1 & 0 & 0 \\ 0 & 1 & 0 \\ 0 & -\tfrac{1}{2} & \tfrac{1}{2} \end{pmatrix}^{-1}$$

$$= \begin{pmatrix} 1 & 1 & 0 \\ 0 & 0 & -1 \end{pmatrix} \begin{pmatrix} 1 & 0 & 0 \\ 0 & 1 & 0 \\ 0 & 1 & 2 \end{pmatrix} = \begin{pmatrix} 1 & 1 & 0 \\ 0 & -1 & -2 \end{pmatrix}$$

The matrix representation of f agrees with that found in Example 5.   ✦

Consider the special case of a linear mapping f of a vector space U into itself. Let $\mathbf{u}_1, \ldots, \mathbf{u}_n$ be a basis for U. f will have a matrix representation A relative to this basis, using the same vectors as bases for both the domain and image space. Let $\bar{\mathbf{u}}_1, \ldots, \bar{\mathbf{u}}_n$ be another basis for U. Let f have matrix representation $\bar{A}$ relative to this basis. Observe that now $C = D$ and the rule relating matrix representations becomes

$$\bar{A} = (D^t)A(D^t)^{-1}$$

This rule, which transforms A into $\bar{A}$, is called a *similarity transformation*. We shall study such transformations in the following chapter. There we shall find that if A is symmetric, we can find a matrix D such that $\bar{A}$ will be a diagonal matrix. Thus if a linear mapping has a symmetric matrix representation, it is always possible to find another basis for which the mapping has a diagonal matrix representation.

EXERCISES

*1. Find the matrix representations of the following linear mappings of $\mathbf{R}^3$ into $\mathbf{R}^2$ relative to the standard bases for these spaces:

  (a) $f(x, y, z) = (x, z)$ **(b)** $f(x, y, z) = (x + y, 2x - y)$

2. Determine the matrix representations of the following linear mappings of $\mathbf{R}^3$ into itself relative to the standard basis of $\mathbf{R}^3$.

  (a) $f(x, y, z) = (x, y, z)$ **(b)** $f(x, y, z) = (x, 0, 0)$

The mapping (b) is also called a *projection mapping*. It projects $\mathbf{R}^3$ onto the one-dimensional subspace, the x axis.

*3. Let f be a linear mapping of $\mathbf{R}^3$ into itself defined by its value on the base vectors thus:

$$f\left[\begin{pmatrix}1\\0\\0\end{pmatrix}\right] = \begin{pmatrix}0\\2\\0\end{pmatrix}, f\left[\begin{pmatrix}0\\1\\0\end{pmatrix}\right] = \begin{pmatrix}3\\0\\0\end{pmatrix}, f\left[\begin{pmatrix}0\\0\\1\end{pmatrix}\right] = \begin{pmatrix}0\\0\\-1\end{pmatrix}$$

By Theorem 6-18 of this section, f is defined completely. Determine the matrix representation of f relative to the standard basis of $\mathbf{R}^3$ and use it to find the image of the element $\begin{pmatrix}1\\2\\3\end{pmatrix}$ under f.

4. Let f be a linear mapping of $\mathbf{R}^3$ into $\mathbf{R}^3$ defined by

$$f\left[\begin{pmatrix}1\\0\\0\end{pmatrix}\right] = \begin{pmatrix}1\\1\\0\end{pmatrix}, f\left[\begin{pmatrix}0\\1\\0\end{pmatrix}\right] = \begin{pmatrix}0\\0\\1\end{pmatrix}, f\left[\begin{pmatrix}0\\0\\1\end{pmatrix}\right] = \begin{pmatrix}2\\1\\0\end{pmatrix}$$

Find the matrix representation of f relative to the standard basis of $\mathbf{R}^3$. Use this matrix representation to determine the image of the element $\begin{pmatrix}-1\\1\\2\end{pmatrix}$ of $\mathbf{R}^3$ under f.

*5. Let f be a linear mapping from a vector space U with basis $\mathbf{u}_1, \mathbf{u}_2, \mathbf{u}_3, \mathbf{u}_4$ into a vector space V with basis $\mathbf{v}_1, \mathbf{v}_2, \mathbf{v}_3$, defined by $f(\mathbf{u}_1) = \mathbf{v}_1 + \mathbf{v}_2, f(\mathbf{u}_2) = 3\mathbf{v}_1 - 2\mathbf{v}_2 + \mathbf{v}_3, f(\mathbf{u}_3) = \mathbf{v}_1 + 2\mathbf{v}_2 - \mathbf{v}_3, f(\mathbf{u}_4) = 2\mathbf{v}_1 + \mathbf{v}_2 - 2\mathbf{v}_3$. Find the matrix representation of f relative to these bases. Use the matrix representation to determine the image of the element $\mathbf{u}_1 + 2\mathbf{u}_2 - \mathbf{u}_3 + 3\mathbf{u}_4$ under f. Verify your answers by finding the image directly using the linear properties of f.

6. Let f be the linear mapping from a vector space V with basis $\mathbf{v}_1, \mathbf{v}_2, \mathbf{v}_3$ into a vector space W with basis $\mathbf{w}_1, \mathbf{w}_2$, defined by $f(\mathbf{v}_1) = \mathbf{w}_1, f(\mathbf{v}_2) = \mathbf{w}_2, f(\mathbf{v}_3) = \mathbf{0}$. Find the matrix representation of f relative to these bases. Use this matrix representation to determine the image of the element $2\mathbf{v}_1 - 3\mathbf{v}_2 + 4\mathbf{v}_3$ under f. Verify your answer by finding the image directly using the linear properties of f.

*7. As in Example 3, the functions 1, $2x$, and $3x^2$ can be interpreted as base vectors for the three-dimensional vector space P_2. Determine the matrix representation of D, the differential operator relative to this basis. Note that this vector space is the same vector space as the one in Example 3. This exercise and Example 3 illustrate how a linear mapping D has two distinct matrix representations relative to two distinct bases. Observe that the order in which the vectors are listed affects the matrix representation.

8. Let V be the vector space generated by the functions $\sin t$ and $\cos t$ defined on the interval $0 \leq t \leq \pi$. If D is the operation of taking the derivative and D^2 is that of taking the second derivative, find the matrix representation of the linear mapping $D^2 + 2D + 1$ of V into itself relative to the basis $\sin t, \cos t$ of V. What is the image of the element $3 \sin t + \cos t$ under this mapping?

9. (a) Let V and W be vector spaces and U be a subspace of V. Is it always possible to construct a linear mapping of V into W that has U as kernel?

(b) Construct a linear mapping of $\mathbf{R}^2$ into $\mathbf{R}^3$ that has the subspace consisting of vectors of the type $r(1, 3)$ as kernel.

10. Find the matrix representations of the following linear mappings. The bases to be taken are x, 1 for P_1 and x^2, x, 1 for P_2.

*(a) $f(a_2x^2 + a_1x + a_0) = (a_1 + a_0)x^2 + (a_1 - a_0)x$ of P_2 into P_2

(b) $f(a_1x + a_0) = a_0x^2 + a_1x + a_0$ of P_1 into P_2

*(c) $f(a_2x^2 + a_1x + a_0) = a_0x^2 + a_1x + a_2$ of P_2 into P_2

*11. Find the matrix representation of the following linear mapping of P_2 into P_1 relative to the basis $(x^2 + x)$, x, 1 of P_2 and x, 1 of P_1:

$$f(a_2x^2 + a_1x + a_0) = a_2x + a_0$$

12. Find the matrix representation of the following linear mapping of P_1 into P_1 relative to the bases $(x + 1)$, 2 of P_1:

$$f(a_1x + a_0) = a_0x - a_1$$

13. Consider the linear mapping f of P_2 into P_1 defined by

$$f(a_2x^2 + a_1x + a_0) = a_2x + (a_1 + a_0)$$

Find the matrix representation of f relative to the standard bases x^2, x, 1 of P_2 and x, 1 of P_1. Use a transformation to find the matrix representation relative to the bases x^2, x, 1 of P_2 and $2x$, $(x + 1)$ of P_1.

*14. Consider the linear mapping f of P_1 into itself defined by $f(a_1 x + a_0) = (a_1 + a_0)x - a_1$. Find the matrix representation of f relative to the standard basis $x, 1$ of P_1. Then use a transformation to find the matrix representation relative to the basis $(x + 1)$, $(x - 1)$ of P_1.

15. Consider the linear mapping $f(x, y) = (2x, x + y)$ of $\mathbf{R}^2$ into $\mathbf{R}^2$. Find the matrix representation of this mapping relative to the standard basis for $\mathbf{R}^2$. Use a transformation to find the matrix representation relative to the basis $(1, 1)$, $(2, 1)$ of $\mathbf{R}^2$.

*16. Consider the linear mapping $f(x, y) = (x - y, x + y)$ of $\mathbf{R}^2$ into itself. Find the matrix representation of this mapping relative to the standard basis of $\mathbf{R}^2$. Then use a transformation to find the matrix representation relative to the basis $(-2, 1)$, $(1, 2)$ of $\mathbf{R}^2$.

7
Eigenvalues and Eigenvectors

In this chapter, the reader will be introduced to the theory of eigenvalues and eigenvectors. These are important tools for the mathematician and the physical or social scientist. Examples of applications will be given from such diverse areas as demography, weather prediction, trade routes in medieval Russia, and normal modes of oscillating systems.

7-1 INTRODUCTION TO EIGENVALUES AND EIGENVECTORS

In this section, we introduce the basic concepts involved in eigenvalues and eigenvectors.

Definition 7-1   A nonzero vector $\mathbf{x}$ in $\mathbf{R}^n$ is said to be an *eigenvector* of an $n \times n$ matrix A if there exists a scalar λ such that $A\mathbf{x} = \lambda\mathbf{x}$. λ is called the *eigenvalue* corresponding to $\mathbf{x}$.

Let us look at the geometrical significance of an eigenvector. Let A be a 2×2 matrix with eigenvector $\mathbf{x}$ and corresponding eigenvalue λ; therefore, $A\mathbf{x} = \lambda\mathbf{x}$. We can interpret A as a mapping in the plane and $\mathbf{x}$ as the location of a point in the plane. $\lambda\mathbf{x}$ will correspond to a point in the plane that is in the same direction from the origin as $\mathbf{x}$ or the opposite direction (depending on the sign of λ), but that is at a distance $|\lambda|$ times as far away. Figure 7-1 illustrates this concept when $\lambda > 1$.

The eigenvectors of A may be interpreted geometrically as the points in the plane whose directions from O are unchanged or reversed after mapping. If $\lambda = 1$, the point remains fixed under the mapping; the vector is then said to be *invariant* under the matrix operation.

If there exists such a λ for a given $\mathbf{x}$, then it is uniquely determined: Suppose there exist two such λ's, λ_1 and λ_2. Then $\lambda_1\mathbf{x} = A\mathbf{x} = \lambda_2\mathbf{x}$, implying that $\lambda_1 = \lambda_2$. The set of all eigenvalues is called the *spectrum* of A.

345

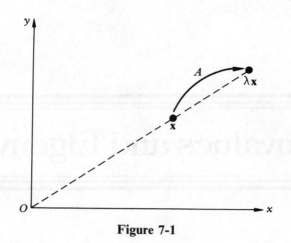

Figure 7-1

In this section we develop a method for determining the eigenvalues and eigenvectors of a matrix, and in following sections we shall see their uses.

The equation $A\mathbf{x} = \lambda\mathbf{x}$ may be rewritten

$$A\mathbf{x} - \lambda\mathbf{x} = \mathbf{0}$$

giving

$$(A - \lambda I_n)\mathbf{x} = \mathbf{0}$$

This represents the set of homogeneous linear equations

$$\begin{pmatrix} a_{11} - \lambda & \cdots & a_{1n} \\ & \vdots & \\ a_{n1} & \cdots & a_{nn} - \lambda \end{pmatrix} \begin{pmatrix} x_1 \\ \vdots \\ x_n \end{pmatrix} = \mathbf{0}$$

$\mathbf{x} = \mathbf{0}$ is a solution. Since we are looking for eigenvectors, which have been defined to be nonzero vectors, we are not interested in this solution. Further solutions can exist only if the matrix of coefficients is singular (Theorem 3-9). Hence, solving the equation $|A - \lambda I_n| = 0$ for λ gives any eigenvalues of A. The polynomial $|A - \lambda I_n|$ is called the *characteristic polynomial* of A, and the equation $|A - \lambda I_n| = 0$ is called the *characteristic equation* of A. The eigenvalues are substituted back into the equation $(A - \lambda I_n)\mathbf{x} = \mathbf{0}$ to give the corresponding eigenvectors.

✦ *Example 1* ─────────────────────────────────

Find the eigenvalues and eigenvectors of the matrix $\begin{pmatrix} 1 & 6 \\ 5 & 2 \end{pmatrix}$.

The characteristic equation $|A - \lambda I_2| = 0$ is

$$\left| \begin{pmatrix} 1 & 6 \\ 5 & 2 \end{pmatrix} - \lambda \begin{pmatrix} 1 & 0 \\ 0 & 1 \end{pmatrix} \right| = 0$$

Thus

$$\begin{vmatrix} 1 - \lambda & 6 \\ 5 & 2 - \lambda \end{vmatrix} = 0$$

Expanding,

$$\lambda^2 - 3\lambda - 28 = 0$$
$$(\lambda - 7)(\lambda + 4) = 0$$

The roots of this equation are 7 and -4; thus the eigenvalues of A are 7 and -4.

The corresponding eigenvectors are now found. For $\lambda = 7$, $(A - \lambda I_2)\mathbf{x} = \mathbf{0}$ becomes

$$\left[\begin{pmatrix} 1 & 6 \\ 5 & 2 \end{pmatrix} - 7 \begin{pmatrix} 1 & 0 \\ 0 & 1 \end{pmatrix} \right] \begin{pmatrix} x_1 \\ x_2 \end{pmatrix} = \mathbf{0}$$

which reduces to

$$\begin{pmatrix} -6 & 6 \\ 5 & -5 \end{pmatrix} \begin{pmatrix} x_1 \\ x_2 \end{pmatrix} = \mathbf{0}$$

This is the matrix form of the system of equations

$$-6x_1 + 6x_2 = 0$$
$$5x_1 - 5x_2 = 0$$

The solutions are $x_1 = x_2$. Hence any vector of the type $r \begin{pmatrix} 1 \\ 1 \end{pmatrix}$, where r is a nonzero scalar, is an eigenvector corresponding to the eigenvalue 7.

For $\lambda = -4$, $(A - \lambda I_2)\mathbf{x} = \mathbf{0}$ becomes

$$\left[\begin{pmatrix} 1 & 6 \\ 5 & 2 \end{pmatrix} + 4 \begin{pmatrix} 1 & 0 \\ 0 & 1 \end{pmatrix} \right] \begin{pmatrix} x_1 \\ x_2 \end{pmatrix} = \mathbf{0}$$

Simplified,

$$\begin{pmatrix} 5 & 6 \\ 5 & 6 \end{pmatrix} \begin{pmatrix} x_1 \\ x_2 \end{pmatrix} = \mathbf{0}$$

implying that $x_1 = -\frac{6}{5} x_2$. Hence any vector of the type $s \begin{pmatrix} -\frac{6}{5} \\ 1 \end{pmatrix}$, where s is a nonzero scalar, is an eigenvector for $\lambda = -4$.

Here vectors of the forms $\begin{pmatrix} r \\ r \end{pmatrix}$ and $\begin{pmatrix} -\frac{6}{5}s \\ s \end{pmatrix}$, where $r \neq 0$ and $s \neq 0$, are eigenvectors of $\begin{pmatrix} 1 & 6 \\ 5 & 2 \end{pmatrix}$ with eigenvalues 7 and -4, respectively. These vectors can be interpreted geometrically as points on the lines OA and OB, respectively. The effect of matrix multiplication is to map these points into other points on the lines, as shown in Figure 7-2. This only happens for points on these lines. A point

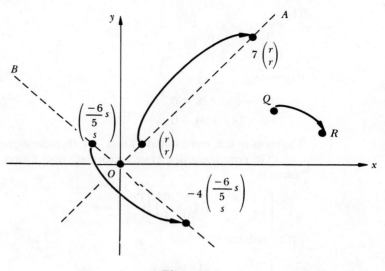

Figure 7-2

such as Q, which does not lie on one of these lines, would be mapped into a point such as R, which does not lie on OQ.

Note that the eigenvectors in this example can only be determined up to magnitude; if $A\mathbf{x} = \lambda\mathbf{x}$, then $cA\mathbf{x} = c\lambda\mathbf{x}$, implying that $A(c\mathbf{x}) = \lambda(c\mathbf{x})$ for any nonzero scalar c. Hence, if $\mathbf{x}$ is an eigenvector for eigenvalue λ, $c\mathbf{x}$ is an eigenvector of this same eigenvalue.   ✦

From this discussion we might conjecture the following result.

Theorem 7-1   *Let A be an $n \times n$ matrix and λ an eigenvalue of A. The subset consisting of all eigenvectors corresponding to λ together with the zero vector is a subspace of $\mathbf{R}^n$ called the eigenspace of the eigenvalue λ.*

Proof: In order to prove that the eigenspace is a subspace, we have to show that it is closed under addition and scalar multiplication.

Let $\mathbf{x}_1$ and $\mathbf{x}_2$ be two elements of the eigenspace corresponding to λ. Then $A\mathbf{x}_1 = \lambda\mathbf{x}_1$ and $A\mathbf{x}_2 = \lambda\mathbf{x}_2$. Hence

$$A\mathbf{x}_1 + A\mathbf{x}_2 = \lambda\mathbf{x}_1 + \lambda\mathbf{x}_2$$

or

$$A(\mathbf{x}_1 + \mathbf{x}_2) = \lambda(\mathbf{x}_1 + \mathbf{x}_2)$$

Thus $\mathbf{x}_1 + \mathbf{x}_2$ is an element of the eigenspace of λ. The eigenspace is closed under addition.

Further, since $A\mathbf{x}_1 = \lambda\mathbf{x}_1$, $cA\mathbf{x}_1 = c\lambda\mathbf{x}_1$, implying that $A(c\mathbf{x}_1) = \lambda(c\mathbf{x}_1)$. $c\mathbf{x}_1$ is an element of the eigenspace.

Hence the eigenspace is closed under scalar multiplication. Thus it is a subspace.

We now illustrate the results of this theorem with an example.

✦ *Example 2*

Determine the eigenvalues and eigenvectors of the matrix

$$\begin{pmatrix} 5 & 4 & 2 \\ 4 & 5 & 2 \\ 2 & 2 & 2 \end{pmatrix}$$

The characteristic equation is

$$\left| \begin{pmatrix} 5 & 4 & 2 \\ 4 & 5 & 2 \\ 2 & 2 & 2 \end{pmatrix} - \lambda \begin{pmatrix} 1 & 0 & 0 \\ 0 & 1 & 0 \\ 0 & 0 & 1 \end{pmatrix} \right| = 0$$

This reduces to $(\lambda - 10)(\lambda - 1)^2 = 0$, giving the eigenvalues to be 10 and 1.

For $\lambda = 10$, the eigenvectors are given by

$$\begin{pmatrix} -5 & 4 & 2 \\ 4 & -5 & 2 \\ 2 & 2 & -8 \end{pmatrix} \begin{pmatrix} x_1 \\ x_2 \\ x_3 \end{pmatrix} = \mathbf{0}$$

This system of linear equations has solutions $x_1 = 2r$, $x_2 = 2r$, $x_3 = r$. The eigenspace is the one-dimensional space $r \begin{pmatrix} 2 \\ 2 \\ 1 \end{pmatrix}$.

For $\lambda = 1$, the eigenvectors are given by

$$\begin{pmatrix} 4 & 4 & 2 \\ 4 & 4 & 2 \\ 2 & 2 & 1 \end{pmatrix} \begin{pmatrix} x_1 \\ x_2 \\ x_3 \end{pmatrix} = \mathbf{0}$$

This system has solutions $x_1 = -s - \frac{1}{2}t$, $x_2 = s$, $x_3 = t$. Hence the eigenvectors corresponding to the eigenvalue 1 are nonzero vectors of the form $\begin{pmatrix} s - \frac{1}{2}t \\ s \\ t \end{pmatrix}$. They, together with the zero vector, form a two-dimensional subspace of $\mathbf{R}^3$.

We can get two base vectors for this space by letting $s = 1$ and $t = 0$, and then letting $s = 0$ and $t = 1$. The base vectors would be $\begin{pmatrix} -1 \\ 1 \\ 0 \end{pmatrix}$ and $\begin{pmatrix} -\frac{1}{2} \\ 0 \\ 1 \end{pmatrix}$.

✦

Eigenvectors and eigenvalues are important tools in linear algebra. Their use is illustrated in the following sections.

EXERCISES

Determine the eigenvalues and corresponding eigenspaces of each of the following matrices. Use the Gram-Schmidt orthogonalization process to determine an orthonormal basis for each eigenspace of the matrices in Exercises 4, 5, and 9.

*1. $\begin{pmatrix} 5 & 4 \\ 1 & 2 \end{pmatrix}$

2. $\begin{pmatrix} 2 & 1 \\ -1 & 4 \end{pmatrix}$

*3. $\begin{pmatrix} 3 & 2 & -2 \\ -3 & -1 & 3 \\ 1 & 2 & 0 \end{pmatrix}$

4. $\begin{pmatrix} 1 & -2 & 2 \\ -2 & 1 & 2 \\ -2 & 0 & 3 \end{pmatrix}$

*5. $\begin{pmatrix} 1 & 0 & 0 \\ -2 & 1 & 2 \\ -2 & 0 & 3 \end{pmatrix}$

6. $\begin{pmatrix} 1 & 0 & 0 \\ -2 & 5 & -2 \\ -2 & 4 & -1 \end{pmatrix}$

*7. $\begin{pmatrix} 15 & 7 & -7 \\ -1 & 1 & 1 \\ 13 & 7 & -5 \end{pmatrix}$

8. $\begin{pmatrix} 5 & -2 & 2 \\ 4 & -3 & 4 \\ 4 & -6 & 7 \end{pmatrix}$

*9. $\begin{pmatrix} 4 & 2 & -2 & 2 \\ 1 & 3 & 1 & -1 \\ 0 & 0 & 2 & 0 \\ 1 & 1 & -3 & 5 \end{pmatrix}$

10. $\begin{pmatrix} 3 & 5 & -5 & 5 \\ 3 & 1 & 3 & -3 \\ -2 & 2 & 0 & 2 \\ 0 & 4 & -6 & 8 \end{pmatrix}$

*11. $\begin{pmatrix} 0.5 & 1.5 & -1.5 & 1.5 \\ 3.5 & 3.5 & 3.5 & -3.5 \\ -1 & 1 & 0 & 1 \\ -3 & -1 & -4 & 5 \end{pmatrix}$

Determine the eigenvalues and corresponding eigenspaces of the following matrices. Interpret your results geometrically. (*Hint*: Use Example 2, Section 6-1.)

*12. $\begin{pmatrix} 3 & 0 \\ 0 & 3 \end{pmatrix}$

13. $\begin{pmatrix} -5 & 0 & 0 \\ 0 & -5 & 0 \\ 0 & 0 & -5 \end{pmatrix}$

*14. Show that the matrix $\begin{pmatrix} 1 & 1 \\ -2 & -1 \end{pmatrix}$ has no real eigenvalues. Readers who have a knowledge of complex numbers will observe that it does have complex eigenvalues. In fact, all matrices have eigenvalues in the complex number system.

15. Prove that the matrix $\begin{pmatrix} 0 & -1 \\ 1 & 0 \end{pmatrix}$ has no real eigenvalues and thus no eigenvectors. Interpret your results geometrically. (*Hint*: Use Example 1, Section 6-1.)

16. Draw a diagram to illustrate the geometrical significance of
    (a) a negative eigenvalue   (b) a fractional eigenvalue
17. Show that if A is an upper triangular matrix

$$\begin{pmatrix} a_{11} & a_{12} & \cdots & a_{1n} \\ 0 & a_{22} & \cdots & a_{2n} \\ & & \vdots & \\ 0 & \cdots & 0 & a_{nn} \end{pmatrix}$$

its eigenvalues are $a_{11}, \ldots, a_{nn}$.
18. Prove that if A is a square matrix, A^t has the same eigenvalues as A.
19. Prove that if $\lambda_1, \ldots, \lambda_n$ are the eigenvalues of A, then $c\lambda_1, \ldots, c\lambda_n$ are the eigenvalues of cA; $c \neq 0$.
20. If A has eigenvalues $\lambda_1, \ldots, \lambda_n$, prove that A^2 has eigenvalues $\lambda_1^2, \ldots, \lambda_n^2$ and that A^m has eigenvalues $\lambda_1^m, \ldots, \lambda_n^m$.
21. Prove that a matrix A can have zero as an eigenvalue if and only if it is singular.

7-2 APPLICATIONS IN DEMOGRAPHY AND IN WEATHER PREDICTION

We illustrate the role of eigenvalues and eigenvectors in demography and in weather prediction using the population distribution model of Section 2-7 and a weather model.

Assume that the population distribution at a certain time is given by the row vector $\mathbf{x}_0$ and that the stochastic matrix P gives the transition probabilities for one year. We found that the population distribution after one year was $\mathbf{x}_1$, where $\mathbf{x}_1 = \mathbf{x}_0 P$, and that after n years it was $\mathbf{x}_n$, where $\mathbf{x}_n = \mathbf{x}_0 P^n$.

We would like to know if the relative populations become *stable* after a finite number of years—that is, if the number of people in both city and suburbia remains unchanged. Mathematically this condition of stability would be expressed by the existence of a nonnegative integer n, where $\mathbf{x}_n = \mathbf{x}_{n+1}$. If this happens at the initial year ($\mathbf{x}_0 = \mathbf{x}_1$), then $\mathbf{x}_0 = \mathbf{x}_0 P$. If this happens after n years, since $\mathbf{x}_{n+1} = \mathbf{x}_0 P^{n+1}$, then $\mathbf{x}_n = \mathbf{x}_n P$.

Remembering that $\mathbf{x}_n$ is a vector and P is a matrix, we see that this equation is similar to the type of equation that defines eigenvectors and eigenvalues. The only differences are that we have row vectors in place of column vectors and that we multiply the matrix on the left rather than on the right. $\mathbf{x}_n$ in this equation is a *left eigenvector* with eigenvalue 1. The eigenvectors previously considered are actually *right eigenvectors*. The theory of left eigenvectors is similar to that of right eigenvectors. For any matrix, the eigenvalues are the same set of numbers for both left and right eigenvectors.

Let us return to the equation $\mathbf{x}_n = \mathbf{x}_n P$. If such an $\mathbf{x}_n$ is going to exist, then P must have eigenvalue 1. This leads us to investigate the possibility of a stochastic matrix having an eigenvalue 1. We arrive at the following remarkable result.

Theorem 7-2   1 *is an eigenvalue for every stochastic matrix.*

Proof: Let $P = (p_{ij})$ be an arbitrary $n \times n$ stochastic matrix. We know that P will have 1 as an eigenvalue if and only if $|P - 1I_n| = 0$. We get

$$|P - 1I_n| = \begin{vmatrix} p_{11} - 1 & p_{12} & \cdots & p_{1n} \\ p_{21} & p_{22} - 1 & \cdots & p_{2n} \\ & & \vdots & \\ p_{n1} & p_{n2} & \cdots & p_{nn} - 1 \end{vmatrix}$$

If we add all the remaining columns to the first column, we find that the determinant is equal to

$$\begin{vmatrix} \sum p_{1i} - 1 & p_{12} & \cdots & p_{1n} \\ \sum p_{2i} - 1 & p_{22} - 1 & \cdots & p_{2n} \\ & & \vdots & \\ \sum p_{ni} - 1 & p_{n2} & \cdots & p_{nn} - 1 \end{vmatrix}$$

Since the sum of the elements in each row of a stochastic matrix is 1, each element in the first column of this determinant is zero. Hence the determinant is zero, proving the result.

We are thus assured that the matrix of transition probabilities P has the necessary unit eigenvalue.

Let us look at the condition $\mathbf{x}_n = \mathbf{x}_n P$ for finite stability. Is there a necessary condition on the matrix P for stability to occur? We find that there is.

Theorem 7-3   *A necessary condition for finite stability sometime after the first year is that P be singular.*

Proof: We investigate the case of stability after the first year. The condition of immediate stability is that $\mathbf{x}_0$ be an eigenvector with eigenvalue 1.

Assume that $\mathbf{x}_n = \mathbf{x}_n P$ for some $n > 0$ with a given $\mathbf{x}_0$. This may be written $\mathbf{x}_0 P^n = \mathbf{x}_0 P^{n+1}$. Interchanging sides, we have

$$\mathbf{x}_0 P^{n+1} = \mathbf{x}_0 P^n$$
$$\mathbf{x}_0(P - I)P^n = \mathbf{0}$$
$$\mathbf{x}_0(P - I)P^n = 0\mathbf{x}_0(P - I)$$

Thus a necessary condition is that P^n have eigenvalue zero [for the left eigenvector $\mathbf{x}_0(P - I)$].

P^n will have eigenvalue zero if and only if $|P^n| = 0$; that is, if and only if $|P| \cdots |P|$ (n times) $= 0$. Thus a necessary condition for stability in a finite number of steps is that P be singular.

Let us apply the result of this theorem to Example 2 of Section 2-7:

$$P = \begin{pmatrix} 0.96 & 0.04 \\ 0.01 & 0.99 \end{pmatrix}$$

P is nonsingular; hence, if the present trend continues, the situation will not stabilize in a finite number of years.

There are various classes of transition matrices leading to various classes of Markov chains. We now introduce the reader to one of these classes, illustrating the role of eigenvectors.

A transition matrix P of a Markov chain is said to be *regular* if for some power of P all the components are positive. The chain is then called a *regular Markov chain*. The transition matrix of the population movement model, $\begin{pmatrix} 0.96 & 0.04 \\ 0.01 & 0.99 \end{pmatrix}$, is a regular matrix.

The following theorem, which we present without proof, gives the important role of eigenvectors in the theory of regular Markov chains.

Theorem 7-4   *Consider a regular Markov chain having transition matrix P and initial state $\mathbf{x}_0$.*

(1) *The sequence of distributions of states $\mathbf{x}_0, \mathbf{x}_1 = \mathbf{x}_0 P, \mathbf{x}_2 = \mathbf{x}_0 P^2, \ldots$ approaches a vector $\mathbf{x}$ that satisfies $\mathbf{x}P = \mathbf{x}$. This limit vector is thus a left eigenvector of P corresponding to the eigenvalue 1.*

(2) *The sequence of matrices $P, P^2, P^3, \ldots$ approaches a stochastic matrix T. The rows of T are all identical, a row being a left eigenvector of P corresponding to the eigenvalue 1.*

The implication of (1) is that no matter what the initial distribution $\mathbf{x}_0$, a regular Markov chain will tend to stabilize as it approaches $\mathbf{x}$. The long-term behavior becomes predictable. Part (2) tells us that, no matter what the initial state, the long-term probability of a state S_i occurring is the same.

Let us now illustrate these results for our population movement model.

The transition matrix is the regular matrix $P = \begin{pmatrix} 0.96 & 0.04 \\ 0.01 & 0.99 \end{pmatrix}$ and $\mathbf{x}_0 = (57,633,\ 71,549)$. We find the left eigenvector corresponding to the eigenvalue 1.

$\mathbf{x}P = \mathbf{x}$ gives $\mathbf{x}(P - I) = \mathbf{0}$. Let $\mathbf{x} = (c, d)$. Thus

$$(c, d)\begin{pmatrix} 0.96 - 1 & 0.04 \\ 0.01 & 0.99 - 1 \end{pmatrix} = \mathbf{0}$$

implying that $-0.04c + 0.01d = 0$; $4c = d$. Left eigenvectors are thus of the form $(c, 4c)$.

In our specific problem, let us assume no population increase. Therefore, the sum of the components of $(c, 4c)$ is $57,633 + 71,549 = 129,182$. We have that $5c = 129{,}182$, giving $c = 25{,}836.4$. The limit vector is

$$\text{city} \qquad \text{suburbia}$$
$$\mathbf{x} = (25{,}836.4, \quad 103{,}345.6)$$

The populations of city and suburbia will get closer and closer to this distribution yearly if the conditions remain unchanged. (The units are thousands.)

Further, we know that P^n approaches a matrix T. Each row of T is identical, being a vector of the type $(c, 4c)$. Since T is stochastic, the sum of the elements in each row is 1. Thus $c + 4c = 1$, giving $c = 0.2$.

$$T = \begin{pmatrix} 0.2 & 0.8 \\ 0.2 & 0.8 \end{pmatrix}$$

We get the matrix sequence

$$P \qquad\qquad\qquad P^2$$
$$\begin{pmatrix} 0.96 & 0.04 \\ 0.01 & 0.99 \end{pmatrix}, \begin{pmatrix} 0.922 & 0.078 \\ 0.0195 & 0.9805 \end{pmatrix},$$

$$P^3 \qquad\qquad\qquad T$$
$$\begin{pmatrix} 0.8859 & 0.1141 \\ 0.028525 & 0.97147 \end{pmatrix}, \cdots \to \begin{pmatrix} 0.2 & 0.8 \\ 0.2 & 0.8 \end{pmatrix}$$

Let us interpret this result. We focus on the $(1, 2)$ element in each matrix—a similar interpretation will apply to the other elements. We get the sequence

$$0.04, 0.078, 0.1141, \ldots \to 0.8$$

Remembering that the $(1, 2)$ element of P^n gives the probability of going from state 1 (city) to state 2 (suburbia) in n steps, we see that the probability of moving from city to suburbia increases annually from 0.04 to 0.078 to 0.1141, etc., getting closer and closer to 0.8. The elements of T give the long-term probabilities. The long-term probability of living in suburbia is 0.8. This is independent of initial location, since $t_{12} = t_{22}$. The long-term probability of living in a city is 0.2. Here, also, this is independent of initial location, a characteristic of regular Markov chains.

Readers who are interested in the proof of Theorem 7-4 are referred to *Finite Mathematical Structures*, by J. G. Kemeny, H. Mirkil, J. L. Snell, and G. L. Thompson, Prentice-Hall, Chapter 6.

We now discuss an interesting application of Markov chains in a model that describes rainfall in Tel Aviv.

✦ *Example 1*

K. R. Gabriel and J. Neumann have developed "A Markov Chain Model for Daily Rainfall Occurrence at Tel Aviv," *Quart. J. R. Met. Soc.*, **88** (1962), 90–95. The probabilities used were based on data of daily rainfall in Tel Aviv (Nahami Street) for the 27 years 1923/24–1949/50. Days were classified as wet or dry according to whether or not there had been recorded at least 0.1 mm of precipitation in the 24-hour period from 8 A.M. to 8 A.M. the following day. A Markov chain was constructed for each of the months November through April, these months constituting the rainy season. We discuss the chain developed for November. The model assumes that the probability of rainfall on any day depends only on whether the previous day was wet or dry. The statistics accumulated over the years for November were

Preceding Day	Actual Day Wet
Wet	117 out of 195
Dry	80 out of 615

Thus the probability of a wet day following a wet day is $117/195 = 0.6$. The probability of a wet day following a dry day is $80/615 = 0.13$. The matrix of transition probabilities, P, for the weather pattern in November is thus

$$\begin{array}{cc} & \begin{array}{cc} Actual\ Day \end{array} \\ & \begin{array}{cc} Wet & Dry \end{array} \\ Preceding\ Day \begin{array}{c} Wet \\ Dry \end{array} & \begin{pmatrix} 0.6 & 0.4 \\ 0.13 & 0.87 \end{pmatrix} \end{array}$$

On any given day, one can use P to predict the weather on a future day. For example, if today, a Wednesday in November, is dry, we can determine the probability that Saturday will be wet. Saturday is three days away, so the various probabilities will be given by P^3. It can be shown that

$$P^3 = \begin{pmatrix} 0.32 & 0.68 \\ 0.22 & 0.78 \end{pmatrix}$$

(We perform all computations to six significant figures in this model but round off to two for ease of reading.) The probability of Saturday being wet is $p_{21}^{(3)}$; that is, 0.22. Observe that the matrix of transition probabilities, P, is regular. Let us determine a left eigenvector corresponding to the eigenvalue 1.

$$(a, b)\begin{pmatrix} 0.6 & 0.4 \\ 0.13 & 0.87 \end{pmatrix} = (a, b)$$

$$(a, b)\begin{pmatrix} -0.4 & 0.4 \\ 0.13 & -0.13 \end{pmatrix} = 0$$

$$-0.4a + 0.13b = 0$$

$$a = 0.325b$$

Thus left eigenvectors corresponding to the eigenvalue 1 are of the form $b(0.325, 1)$.

To construct a left eigenvector, the sum of whose components is 1, we let $b = 1/1.325$. This gives $(0.25, 0.75)$. Thus the sequence P, P^2, P^3, P^4, ... approaches the matrix $T = \begin{pmatrix} 0.25 & 0.75 \\ 0.25 & 0.75 \end{pmatrix}$. The convergence is quite rapid, as the following sample of powers of P illustrates:

$$P^2 \qquad\qquad P^4 \qquad\qquad P^6 \qquad\qquad P^8$$

$$\begin{pmatrix} 0.41 & 0.59 \\ 0.19 & 0.81 \end{pmatrix}, \begin{pmatrix} 0.28 & 0.72 \\ 0.23 & 0.77 \end{pmatrix}, \begin{pmatrix} 0.25 & 0.75 \\ 0.24 & 0.76 \end{pmatrix}, \begin{pmatrix} 0.25 & 0.75 \\ 0.25 & 0.75 \end{pmatrix}$$

We can interpret the limit matrix thus:

A Day in the Distant Future
Wet   Dry

$$A \ Given \ Day \quad \begin{matrix} \text{Wet} \\ \text{Dry} \end{matrix} \begin{pmatrix} 0.25 & 0.75 \\ 0.25 & 0.75 \end{pmatrix}$$

The long-term prediction for a future day in November is that the probability of its being wet is 0.25 and of its being dry is 0.75. These are independent of whether the current day is wet or dry.

The model can also be used to discuss wet and dry spells. A wet spell of length k is defined as a sequence of k wet days preceded and followed by dry days. Dry spells are defined analogously. Let us look at a wet spell in terms of the diagram in Figure 7-3. The diagram is to be interpreted from left to right. Let the current day denoted by the point A be wet, W. The previous day B was dry, D. At A, one is interested in going to C, not C'; the following day is to be wet. One wants to progress to E, $k - 1$ days later, and finally to

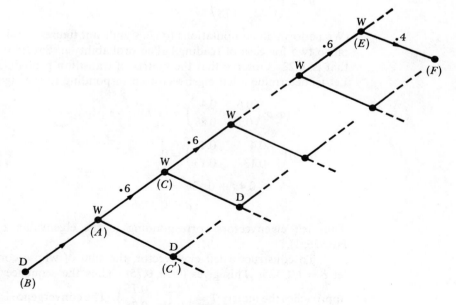

Figure 7-3

F. The probabilities are marked. It can be seen that the probability of reaching *F* is

$$\underbrace{0.6 \times 0.6 \times \cdots \times 0.6}_{k-1 \text{ times}} \times 0.4$$

Consider now an arbitrary matrix of transition probabilities *P*. If today is wet and yesterday was dry, the probability that today is the first day of a wet spell of length *k* is $(p_{11})^{k-1}p_{12}$. Similarly, if today is dry and yesterday was wet, the probability that today is the first day of a dry spell of length *k* is $(p_{22})^{k-1}p_{21}$.

Using this model, we find that the probability that a given wet day (with the preceding day dry) is the first day of a wet spell of length 4, for example, is $(0.6)^3(0.4) = 0.086$.   ✦

EXERCISES

1. Prove that if **x** is a left eigenvector of a matrix *A*, $\mathbf{x}^t$ is a right eigenvector of A^t. Prove further that the set of eigenvalues corresponding to left eigenvectors is identical to the set of eigenvalues for right eigenvectors.

*2. We return to the model of population flow between metropolitan areas and nonmetropolitan areas of Exercise 6, Section 2-7. The populations of metropolitan and nonmetropolitan areas in 1971 were 129, 182 and 69,723 (in thousands of persons one year old or over). The matrix of transition probabilities was

$$\begin{array}{cc}
 & \text{\textit{Final}} \\
 & \begin{array}{cc} \text{Metro} & \text{Nonmetro} \end{array} \\
\textit{Initial} \quad \begin{array}{c} \text{Metro} \\ \text{Nonmetro} \end{array} & \begin{pmatrix} 0.99 & 0.01 \\ 0.02 & 0.98 \end{pmatrix}
\end{array}$$

Determine the figures that the population distributions will approach if conditions remain unchanged.

3. The model of Exericse 7, Section 2-7, was a refinement of the above model in that the metro population was broken down into city and suburbia. In that model, the populations of city, suburbia, and nonmetro in 1971 were 57,633, 71,549, and 69,723 (in thousands of persons one year old or over), respectively. The stochastic matrix giving the probabilities of moves was

$$\begin{array}{cc}
 & \text{\textit{Final}} \\
 & \begin{array}{ccc} \text{City} & \text{Suburb} & \text{Nonmetro} \end{array} \\
\textit{Initial} \quad \begin{array}{c} \text{City} \\ \text{Suburb} \\ \text{Nonmetro} \end{array} & \begin{pmatrix} 0.96 & 0.03 & 0.01 \\ 0.01 & 0.98 & 0.01 \\ 0.015 & 0.005 & 0.98 \end{pmatrix}
\end{array}$$

Determine the figures that the population distributions will approach if conditions remain unchanged.

4. We return to the genetic model of Exercise 12, Section 2-7. In that model, offspring of guinea pigs were crossed with hybrids only. The transition matrix *P* for that model

was

$$
\begin{array}{c}
 & \begin{array}{ccc} AA & Aa & aa \end{array} \\
\begin{array}{c} AA \\ Aa \\ aa \end{array} &
\begin{pmatrix}
\frac{1}{2} & \frac{1}{2} & 0 \\
\frac{1}{4} & \frac{1}{2} & \frac{1}{4} \\
0 & \frac{1}{2} & \frac{1}{2}
\end{pmatrix}
\end{array}
$$

Prove that P is regular. Let $P^n \to T$. Determine T. What information can you get from T?

*5. This exercise is based on the weather model of Example 1, this section. The statistics for the month of December in Tel Aviv for the years 1923/24–1949/50 are

Preceding Day	Actual Day Wet
Wet	213 out of 326
Dry	117 out of 511

Use these statistics to construct a model for the December weather pattern.

(a) If a Thursday in December in Tel Aviv is wet, what is the probability that the following Saturday will also be wet?

(b) Determine the matrix that describes the long-term prediction of the weather in December.

(c) If today is a dry December day in Tel Aviv and yesterday was wet, what is the probability that today is the first day of a dry spell of length 5?

(d) If today is a dry December day and yesterday was wet, what is the probability that today is the first day of a dry spell of length m to be followed by a wet spell of length n?

6. (a) Let A be a 2×2 stochastic matrix that is not the identity matrix. Prove that the vectors $c \begin{pmatrix} 1 \\ 1 \end{pmatrix}$ are the only right eigenvectors for the eigenvalue 1, where c is any scalar.

(b) Let A be an $n \times n$ stochastic matrix. Prove that the vectors $c \begin{pmatrix} 1 \\ \vdots \\ 1 \end{pmatrix}$ are eigenvectors for the eigenvalue 1. Can you interpret the result?

*7. A psychologist conducts an experiment in which 20 rats are placed at random in a compartment that has been divided into rooms labeled 1, 2, and 3 as shown in Figure 7-4. Observe that there are four doors in the arrangement. There are 3 possible states for the rats: they can be in room 1, 2, or 3. Let us assume that the rats move from room to room. A rat in room 1 has the probabilities $p_{11} = 0$, $p_{12} = \frac{2}{3}$, and $p_{13} = \frac{1}{3}$ of moving to the various rooms, based on the distribution of doors. This

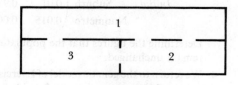

Figure 7-4

approach leads to the following matrix that defines a Markov chain describing the movement of the rats:

$$P = \begin{pmatrix} 0 & \frac{2}{3} & \frac{1}{3} \\ \frac{2}{3} & 0 & \frac{1}{3} \\ \frac{1}{2} & \frac{1}{2} & 0 \end{pmatrix}$$

Predict the distribution of the rats at the end of the experiment. At the end of the experiment, what is the probability that a given marked rat will be in room 2?

8. 40 rats are placed at random in a compartment having four rooms, as shown in Figure 7-5. Construct a Markov chain model to describe the movements of the rats between the rooms. Predict the equilibrium distribution of the rats. What is the probability that a given marked rat will be in room 4 at the end of the experiment?

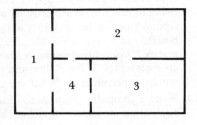

Figure 7-5

*9. Two car rental companies A and B are competing for customers at certain airports. A study has been made of customer satisfaction with the various companies. These results are expressed in the following matrix Q:

$$Q = \begin{matrix} & \begin{matrix} A & \quad B \end{matrix} \\ \begin{matrix} A \\ B \end{matrix} & \begin{pmatrix} 75\% & 25\% \\ 20\% & 80\% \end{pmatrix} \end{matrix}$$

The elements of Q are to be interpreted as follows. The first row implies that 75% of those currently using rental company A are satisfied with A and intend to use A next time, while 25% are dissatisfied with A and intend to use B next time. The second row implies that 20% of those currently using B are dissatisfied with B and intend to switch to A, while 80% will remain with B.

Modify the matrix A to obtain a transition matrix P that defines a Markov chain model of the rental patterns. If the current rental trends continue, how will the eventual rental distribution settle? Express the distribution in percentages that use A and B.

10. A market research group has been studying the buying patterns for three competing products I, II, and III. The results of the analysis are contained in the following matrix A:

$$A = \begin{matrix} & \begin{matrix} \text{I} & \quad \text{II} & \quad \text{III} \end{matrix} \\ \begin{matrix} \text{I} \\ \text{II} \\ \text{III} \end{matrix} & \begin{pmatrix} 80\% & 5\% & 15\% \\ 20\% & 75\% & 5\% \\ 5\% & 5\% & 90\% \end{pmatrix} \end{matrix}$$

Thus, for example, row I implies that of those people currently using product I, 80% will remain with product I while 5% will switch to II and 15% will switch to III, etc. Construct a stochastic matrix from A that will define a Markov chain model of these trends. If the current patterns continue, determine the likely eventual distribution of sales, in terms of percentages.

*11. A company developing a certain area of land in Arizona offers two types of houses, ranch style and split level. Residents are polled as to their satisfaction. The results of the poll are contained in the following matrix A:

$$A = \begin{array}{c} \\ R \\ S \end{array} \begin{array}{cc} R & S \\ \begin{pmatrix} 90\% & 10\% \\ 15\% & 85\% \end{pmatrix} \end{array}$$

R indicates ranch style, S split level. The first row of A implies that 90% of those living in ranch style are satisfied while 10% wish they had bought split level. The second row implies that 15% of those currently living in split level wish they had bought ranch style while 85% are satisfied with split level. Construct a stochastic matrix P from A that defines a Markov chain model of these preferences. Compute an eigenvector X for P, the sum of whose components is 100; the elements then represent percentages. What use can the developers make of X?

12. Write a computer program to illustrate the fact that the sequence of states in the population movement model of this section does indeed approach

$$(25836.4, \quad 103345.6)$$

and that the sequence $P, P^2, P^3, \ldots$ approaches $\begin{pmatrix} 0.2 & 0.8 \\ 0.2 & 0.8 \end{pmatrix}$.

13. Write a computer program for exactly determining the left eigenvector and limit matrix T for the sequence arising from a given 2×2 stochastic matrix. Test your program on the matrix $\begin{pmatrix} 0.96 & 0.04 \\ 0.01 & 0.99 \end{pmatrix}$ of the population movement model, where the vector is (25836.4, 103345.6) and the matrix is $\begin{pmatrix} 0.2 & 0.8 \\ 0.2 & 0.8 \end{pmatrix}$. [*Hint:* Let the given stochastic matrix be $\begin{pmatrix} a & 1-a \\ b & 1-b \end{pmatrix}$, and let the left eigenvector be (c, d). Then $(c, d) \begin{pmatrix} a-1 & 1-a \\ b & -b \end{pmatrix} = \mathbf{0}$ leads to algebraic expressions for c and d.]

14. Write a computer program that can be used to give results from the weather model of this section. Let the program give the probability that any given future day in November will be wet or dry, and the probability of a wet or dry spell of any given length in November. Using your model from Exercise 5, extend your program to include December, also.

7-3 SIMILARITY TRANSFORMATIONS— DIAGONALIZATION OF SYMMETRIC MATRICES

In this section and the following one, we introduce and see applications of an important class of transformations, similarity transformations. These transformations often enter into a mathematical analysis when a coordinate transformation is involved. Eigenvalues and eigenvectors play a central role in the discussion.

Definition 7-2   If A is a square matrix and C is a nonsingular matrix of the same kind as A, then the transformation of A into $C^{-1} AC$ is called a *similarity transformation*. Two square matrices of the same kind, A and B, are said to be *similar* if there exists a nonsingular matrix C such that $B = C^{-1} AC$.

✦ *Example 1*
_____

Let $A = \begin{pmatrix} 1 & 2 \\ 3 & 4 \end{pmatrix}$ and $C = \begin{pmatrix} 2 & 5 \\ 1 & 3 \end{pmatrix}$. C is a nonsingular matrix. Its inverse can be shown to be $\begin{pmatrix} 3 & -5 \\ -1 & 2 \end{pmatrix}$. Let us transform A into $C^{-1}AC$:

$$C^{-1} AC = \begin{pmatrix} 3 & -5 \\ -1 & 2 \end{pmatrix} \begin{pmatrix} 1 & 2 \\ 3 & 4 \end{pmatrix} \begin{pmatrix} 2 & 5 \\ 1 & 3 \end{pmatrix}$$

$$= \begin{pmatrix} -12 & -14 \\ 5 & 6 \end{pmatrix} \begin{pmatrix} 2 & 5 \\ 1 & 3 \end{pmatrix} = \begin{pmatrix} -38 & -102 \\ 16 & 43 \end{pmatrix}$$

The matrix $\begin{pmatrix} 1 & 2 \\ 3 & 4 \end{pmatrix}$ has been transformed into the matrix $\begin{pmatrix} -38 & -102 \\ 16 & 43 \end{pmatrix}$ using a similarity transformation involving the matrix $\begin{pmatrix} 2 & 5 \\ 1 & 3 \end{pmatrix}$.   ✦

Theorem 7-5   *Similar matrices have the same eigenvalues.*

Proof: Let A and B be similar matrices. Hence there exists a matrix C such that $B = C^{-1}AC$.
   The characteristic polynomial of B is $|B - \lambda I|$.
   Substituting for B and using the properties of determinants, we get

$$|B - \lambda I| = |C^{-1}AC - \lambda I| = |C^{-1}(A - \lambda I) C|$$
$$= |C^{-1}| \, |A - \lambda I| \, |C| = |A - \lambda I|$$

Hence the characteristic polynomials of A and B are identical; their eigenvalues will also be identical.

As the definition implies, any nonsingular matrix C can be used to define a similarity transformation. In practice, one is often interested in transforming a given matrix into a diagonal matrix, if possible. The following theorem tells us when this is possible, and the form that the transforming matrix C takes in such a transformation.

Theorem 7-6   *An $n \times n$ matrix A can be transformed into a diagonal matrix B if and only if it has n linearly independent eigenvectors. If A is such a* matrix, the matrix C whose columns consist of n linearly independent eigenvectors of A can be used to transform A into a diagonal matrix $B = C^{-1}AC$. The diagonal matrix B will then have eigenvalues of A as diagonal elements.

Proof: First, assume that A has eigenvalues $\lambda_1, \ldots, \lambda_n$ (which need not be distinct) with corresponding linearly independent eigenvectors $\mathbf{v}_1, \ldots, \mathbf{v}_n$.

Let C be the matrix with $\mathbf{v}_1, \ldots, \mathbf{v}_n$ as column vectors. Write $C = (\mathbf{v}_1, \ldots, \mathbf{v}_n)$. Then, since $A\mathbf{v}_1 = \lambda\mathbf{v}_1, \ldots, A\mathbf{v}_n = \lambda_n\mathbf{v}_n$, matrix multiplication gives

$$A(\mathbf{v}_1 \cdots \mathbf{v}_n) = (\lambda_1\mathbf{v}_1 \cdots \lambda_n\mathbf{v}_n)$$

$$= (\mathbf{v}_1 \cdots \mathbf{v}_n) \begin{pmatrix} \lambda_1 & & 0 \\ & \ddots & \\ 0 & & \lambda_n \end{pmatrix}$$

Hence

$$(\mathbf{v}_1 \cdots \mathbf{v}_n)^{-1} A(\mathbf{v}_1 \cdots \mathbf{v}_n) = \begin{pmatrix} \lambda_1 & & 0 \\ & \ddots & \\ 0 & & \lambda_n \end{pmatrix}$$

A can be transformed into a diagonal matrix through use of the matrix C.

Let us now prove the converse. Let C be a nonsingular matrix that can be used to transform A into a diagonal form $C^{-1}AC$. Let

$$C^{-1}AC = \begin{pmatrix} \gamma_1 & & 0 \\ & \ddots & \\ 0 & & \gamma_n \end{pmatrix}$$

Thus

$$AC = C \begin{pmatrix} \gamma_1 & & 0 \\ & \ddots & \\ 0 & & \gamma_n \end{pmatrix}$$

Let C have columns $\mathbf{u}_1, \ldots, \mathbf{u}_n$. Then

$$A(\mathbf{u}_1 \cdots \mathbf{u}_n) = (\mathbf{u}_1 \cdots \mathbf{u}_n) \begin{pmatrix} \gamma_1 & & 0 \\ & \ddots & \\ 0 & & \gamma_n \end{pmatrix}$$

This implies that $A\mathbf{u}_1 = \gamma_1\mathbf{u}_1, \ldots, A_n\mathbf{u}_n = \gamma_n\mathbf{u}_n$. $\mathbf{u}_1, \ldots, \mathbf{u}_n$ are eigenvectors of A with corresponding eigenvalues $\gamma_1, \ldots, \gamma_n$. Since C is nonsingular, it is of rank n; thus the n eigenvectors $\mathbf{u}_1, \ldots, \mathbf{u}_n$ are linearly independent. This proves that if an $n \times n$ matrix can be transformed into a diagonal matrix through a similarity transformation, it has n linearly independent eigenvectors.

✦ *Example 2*

Let us transform the matrix $A = \begin{pmatrix} 1 & 0 & 0 \\ -2 & 5 & -2 \\ -2 & 4 & -1 \end{pmatrix}$ into a diagonal matrix using a similarity transformation.

We determine the eigenvalues of A:

$$|A - \lambda I| = 0 \Rightarrow \begin{vmatrix} 1-\lambda & 0 & 0 \\ -2 & 5-\lambda & -2 \\ -2 & 4 & -1-\lambda \end{vmatrix} = 0$$

$$\Rightarrow (1-\lambda) \begin{vmatrix} 5-\lambda & -2 \\ 4 & -1-\lambda \end{vmatrix} = 0$$

$$\Rightarrow (1-\lambda)[(5-\lambda)(-1-\lambda)+8] = 0$$

$$\Rightarrow (1-\lambda)^2(\lambda - 3) = 0$$

The eigenvalues are 1, 1, and 3.

We now determine the eigenspaces. For $\lambda = 1$,

$$\begin{pmatrix} 0 & 0 & 0 \\ -2 & 4 & -2 \\ -2 & 4 & -2 \end{pmatrix} \begin{pmatrix} x_1 \\ x_2 \\ x_3 \end{pmatrix} = 0 \Rightarrow -2x_1 + 4x_2 - 2x_3 = 0$$

giving $x_1 = 2x_2 - x_3$. Eigenvectors are $\begin{pmatrix} 2r-s \\ r \\ s \end{pmatrix}$. Two linearly independent eigenvectors in this space are $\begin{pmatrix} 2 \\ 1 \\ 0 \end{pmatrix}$ and $\begin{pmatrix} -1 \\ 0 \\ 1 \end{pmatrix}$.

For $\lambda = 3$,

$$\begin{pmatrix} -2 & 0 & 0 \\ -2 & 2 & -2 \\ -2 & 4 & -4 \end{pmatrix} \begin{pmatrix} x_1 \\ x_2 \\ x_3 \end{pmatrix} = 0 \Rightarrow \begin{cases} -2x_1 & = 0 \\ -2x_1 + 2x_2 - 2x_3 = 0 \\ -2x_1 + 4x_2 - 4x_3 = 0 \end{cases}$$

Solving, we get $x_1 = 0$, $x_2 = x_3$. Eigenvectors are $t\begin{pmatrix} 0 \\ 1 \\ 1 \end{pmatrix}$. Thus A has three linearly independent eigenvectors $\begin{pmatrix} 2 \\ 1 \\ 0 \end{pmatrix}$, $\begin{pmatrix} -1 \\ 0 \\ 1 \end{pmatrix}$, and $\begin{pmatrix} 0 \\ 1 \\ 1 \end{pmatrix}$.

Let $C = \begin{pmatrix} 2 & -1 & 0 \\ 1 & 0 & 1 \\ 0 & 1 & 1 \end{pmatrix}$. Then C^{-1} can be found to be $\begin{pmatrix} 1 & -1 & 1 \\ 1 & -2 & 2 \\ -1 & 2 & -1 \end{pmatrix}$.

Performing a similarity transformation on A, we have

$$C^{-1}AC = \begin{pmatrix} 1 & -1 & 1 \\ 1 & -2 & 2 \\ -1 & 2 & -1 \end{pmatrix} \begin{pmatrix} 1 & 0 & 0 \\ -2 & 5 & -2 \\ -2 & 4 & -1 \end{pmatrix} \begin{pmatrix} 2 & -1 & 0 \\ 1 & 0 & 1 \\ 0 & 1 & 1 \end{pmatrix}$$

$$= \begin{pmatrix} 1 & -1 & 1 \\ 1 & -2 & 2 \\ -3 & 6 & -3 \end{pmatrix} \begin{pmatrix} 2 & -1 & 0 \\ 1 & 0 & 1 \\ 0 & 1 & 1 \end{pmatrix} = \begin{pmatrix} 1 & 0 & 0 \\ 0 & 1 & 0 \\ 0 & 0 & 3 \end{pmatrix}$$

This matrix has the eigenvalues of A as its diagonal elements.   ✦

Not every matrix has sufficient eigenvectors to be diagonalized, as the following example shows.

✦ *Example 3* _____

Determine the eigenvalues and eigenvectors of the matrix

$$\begin{pmatrix} 2 & 1 & 0 \\ 0 & 2 & 0 \\ 0 & 0 & 3 \end{pmatrix}$$

The eigenvalues are given by

$$|A - \lambda I| = 0 \Rightarrow \begin{vmatrix} 2 - \lambda & 1 & 0 \\ 0 & 2 - \lambda & 0 \\ 0 & 0 & 3 - \lambda \end{vmatrix} = 0 \Rightarrow (2 - \lambda)^2 (3 - \lambda) = 0$$

The eigenvalues are 2, 2, and 3.

Let us determine the corresponding eigenvectors. For $\lambda = 2$,

$$\begin{pmatrix} 0 & 1 & 0 \\ 0 & 0 & 0 \\ 0 & 0 & 1 \end{pmatrix} \begin{pmatrix} x_1 \\ x_2 \\ x_3 \end{pmatrix} = 0 \Rightarrow \begin{cases} x_2 = 0 \\ x_3 = 0 \end{cases}$$

Eigenvectors are $r \begin{pmatrix} 1 \\ 0 \\ 0 \end{pmatrix}$.

For $\lambda = 3$,

$$\begin{pmatrix} -1 & 1 & 0 \\ 0 & -1 & 0 \\ 0 & 0 & 0 \end{pmatrix} \begin{pmatrix} x_1 \\ x_2 \\ x_3 \end{pmatrix} = 0 \Rightarrow \begin{cases} -x_1 = 0 \\ -x_2 = 0 \end{cases}$$

Eigenvectors are $s \begin{pmatrix} 0 \\ 0 \\ 1 \end{pmatrix}$. Thus the 3×3 matrix has two one-dimensional eigenspaces. The 3×3 matrix does not have three linearly independent eigenvectors. It cannot be diagonalized through a similarity transformation.

✦

Let us now look at certain properties of symmetric matrices. We find that all symmetric matrices have sufficient eigenvectors to be diagonalized through a similarity transformation.

The following theorem sums up the properties of eigenspaces of symmetric matrices.

Theorem 7-7

(1) *The eigenvalues of a symmetric matrix are all real numbers. (Eigenvalues of a matrix can, in general, be complex numbers.)*

(2) *n linearly independent eigenvectors exist for every $n \times n$ symmetric matrix.*

(3) *The eigenspaces of a symmetric matrix corresponding to distinct eigenvalues are all orthogonal. (Two subspaces are said to be orthogonal if an arbitrary vector in the one subspace is orthogonal to an arbitrary vector in the other subspace.)*

(4) *The dimension of an eigenspace of a symmetric matrix is the multiplicity of the eigenvalue as a root of the characteristic equation.*

Proof: The reader is asked to prove part (3) in the exercises. The proof of the remainder of the theorem is beyond the reader's scope here.

Thus if A is a symmetric matrix, it can be transformed into a diagonal matrix B through a similarity transformation $C^{-1} AC$, where C has linearly independent eigenvectors of A as column vectors. Furthermore, by using eigenvectors that form an orthonormal basis for each eigenspace, one can select the transformation matrix C to be an orthogonal matrix. It is often desirable to use an orthogonal matrix as the transformation matrix—norms and angles are then preserved. Note that since the inverse of an orthogonal matrix is equal to its transpose, the similarity transformation then assumes the simpler form $C^t AC$.

✦ *Example 4*
_____

Transform the symmetric matrix $A = \begin{pmatrix} \frac{2}{3} & -\frac{1}{2} & 0 \\ -\frac{1}{2} & \frac{3}{2} & 0 \\ 0 & 0 & 3 \end{pmatrix}$ into a diagonal form using an orthogonal similarity transformation.

The eigenvalues of A are 1, 2, and 3, and corresponding eigenvectors are $r(1, 1, 0)$, $s(-1, 1, 0)$, and $t(0, 0, 1)$, respectively, for arbitrary constants r, s, and t. These vectors are already orthogonal, since they are in distinct eigenspaces. Normalizing, we get $(1/\sqrt{2}, 1/\sqrt{2}, 0)$, $(-1/\sqrt{2}, 1/\sqrt{2}, 0)$, and $(0, 0, 1)$. Hence an orthogonal transforming matrix C is

$$C = \begin{pmatrix} \dfrac{1}{\sqrt{2}} & -\dfrac{1}{\sqrt{2}} & 0 \\ \dfrac{1}{\sqrt{2}} & \dfrac{1}{\sqrt{2}} & 0 \\ 0 & 0 & 1 \end{pmatrix}$$

The similarity transformation that will give the diagonal matrix B is

$$B = C^{-1}AC = C^{t}AC$$

Thus

$$B = \begin{pmatrix} \dfrac{1}{\sqrt{2}} & -\dfrac{1}{\sqrt{2}} & 0 \\ \dfrac{1}{\sqrt{2}} & \dfrac{1}{\sqrt{2}} & 0 \\ 0 & 0 & 1 \end{pmatrix} \begin{pmatrix} \dfrac{3}{2} & -\dfrac{1}{2} & 0 \\ -\dfrac{1}{2} & \dfrac{3}{2} & 0 \\ 0 & 0 & 3 \end{pmatrix} \begin{pmatrix} \dfrac{1}{\sqrt{2}} & -\dfrac{1}{\sqrt{2}} & 0 \\ \dfrac{1}{\sqrt{2}} & \dfrac{1}{\sqrt{2}} & 0 \\ 0 & 0 & 1 \end{pmatrix}$$

$$= \begin{pmatrix} \dfrac{1}{\sqrt{2}} & \dfrac{1}{\sqrt{2}} & 0 \\ -\dfrac{1}{\sqrt{2}} & \dfrac{1}{\sqrt{2}} & 0 \\ 0 & 0 & 1 \end{pmatrix} \begin{pmatrix} \dfrac{3}{2} & -\dfrac{1}{2} & 0 \\ -\dfrac{1}{2} & \dfrac{3}{2} & 0 \\ 0 & 0 & 3 \end{pmatrix} \begin{pmatrix} \dfrac{1}{\sqrt{2}} & -\dfrac{1}{\sqrt{2}} & 0 \\ \dfrac{1}{\sqrt{2}} & \dfrac{1}{\sqrt{2}} & 0 \\ 0 & 0 & 1 \end{pmatrix}$$

$$= \begin{pmatrix} 1 & 0 & 0 \\ 0 & 2 & 0 \\ 0 & 0 & 3 \end{pmatrix}$$

Hence A is similar to the diagonal matrix $\begin{pmatrix} 1 & 0 & 0 \\ 0 & 2 & 0 \\ 0 & 0 & 3 \end{pmatrix}$. The diagonal elements of this matrix are indeed the eigenvalues of A. If one is only interested in determining a diagonal matrix, it can be determined immediately from the eigenvalues of A. However, a knowledge of the actual transformation is important in many applications of this theory. Such transformations arise, for example, when coordinate transformations are involved in a mathematical model, as in the next section. ✦

✦ *Example 5* ───────────────────────────────────

Determine a diagonal matrix similar to the symmetric matrix $\begin{pmatrix} 5 & 4 & 2 \\ 4 & 5 & 2 \\ 2 & 2 & 2 \end{pmatrix}$.

We have already determined the eigenvalues of this matrix in Example 2 of Section 7-1. They are 1, 1, and 10. Hence a diagonal matrix similar to this matrix is

$$\begin{pmatrix} 1 & 0 & 0 \\ 0 & 1 & 0 \\ 0 & 0 & 10 \end{pmatrix}$$

Note that there is a two-dimensional eigenspace corresponding to the repeated eigenvalue 1. The Gram-Schmidt orthogonalization process would have to be used to determine an orthogonal transformation matrix. The reader is asked to determine such a matrix in Exercise 4.   ✦

✦ *Example 6**

This example illustrates the importance of transformations to matrix representations.

We have discussed that the matrix representations A and $\bar{A}$ of a linear mapping f of a vector space into itself relative to two distinct bases are related as follows:

$$\bar{A} = (D^t)A(D^t)^{-1}$$

D is the matrix that relates the bases. If D is an orthogonal matrix, then $D^t = D^{-1}$ and this equation becomes

$$\bar{A} = (D^t)A(D)$$

Furthermore, if A is symmetric, D can be selected to make $\bar{A}$ diagonal.

Let us consider a specific example. Consider the linear mapping of P_2 into itself, defined as follows relative to the basis x^2, x, 1 of P_2:

$$f(x^2) = \tfrac{3}{2}x^2 - \tfrac{1}{2}x + 0(1)$$
$$f(x) = -\tfrac{1}{2}x^2 + \tfrac{3}{2}x + 0(1)$$
$$f(1) = 0x^2 + 0x + 3(1)$$

The matrix representation of f relative to this basis is

$$\begin{pmatrix} \tfrac{3}{2} & -\tfrac{1}{2} & 0 \\ -\tfrac{1}{2} & \tfrac{3}{2} & 0 \\ 0 & 0 & 3 \end{pmatrix}$$

This is the symmetric matrix of Example 4. If we select

$$D = \begin{pmatrix} \dfrac{1}{\sqrt{2}} & -\dfrac{1}{\sqrt{2}} & 0 \\ \dfrac{1}{\sqrt{2}} & \dfrac{1}{\sqrt{2}} & 0 \\ 0 & 0 & 1 \end{pmatrix}$$

we get that the diagonal matrix representation $\bar{A}$ of f is

$$\begin{pmatrix} 1 & 0 & 0 \\ 0 & 2 & 0 \\ 0 & 0 & 3 \end{pmatrix} \quad \blacklozenge$$

EXERCISES

1. In each of the following, transform A into $C^{-1}AC$ using the given nonsingular matrix C.

*(a) $A = \begin{pmatrix} 1 & 2 \\ -1 & 3 \end{pmatrix}$, $C = \begin{pmatrix} 2 & 5 \\ 1 & 3 \end{pmatrix}$

(b) $A = \begin{pmatrix} 0 & 4 \\ 3 & 2 \end{pmatrix}$, $C = \begin{pmatrix} 2 & 1 \\ 7 & 4 \end{pmatrix}$

*(c) $A = \begin{pmatrix} -1 & 2 \\ 3 & 1 \end{pmatrix}$, $C = \begin{pmatrix} -4 & 6 \\ -2 & 2 \end{pmatrix}$

(d) $A = \begin{pmatrix} 3 & 2 \\ -1 & 4 \end{pmatrix}$, $C = \begin{pmatrix} 2 & -7 \\ -1 & 4 \end{pmatrix}$

*(e) $A = \begin{pmatrix} 1 & 0 & 2 \\ -1 & 3 & 4 \\ 0 & 1 & 3 \end{pmatrix}$, $C = \begin{pmatrix} 0 & 3 & 3 \\ 1 & 2 & 3 \\ 1 & 4 & 6 \end{pmatrix}$

(f) $A = \begin{pmatrix} 2 & 1 & 0 \\ 3 & 0 & 2 \\ 4 & 1 & -1 \end{pmatrix}$, $C = \begin{pmatrix} 1 & 2 & 3 \\ 0 & 1 & 2 \\ 4 & 5 & 3 \end{pmatrix}$

*(g) $A = \begin{pmatrix} -1 & 2 & 1 \\ 0 & 4 & 1 \\ 2 & 0 & 3 \end{pmatrix}$, $C = \begin{pmatrix} 1 & 2 & 0 \\ 2 & 1 & -1 \\ 3 & 1 & 1 \end{pmatrix}$

2. Transform (if possible) each of the following matrices into diagonal form using a similarity transformation involving eigenvectors. Give the transformations.

*(a) $\begin{pmatrix} 5 & 4 \\ 1 & 2 \end{pmatrix}$ (b) $\begin{pmatrix} 2 & 1 \\ 2 & 3 \end{pmatrix}$ *(c) $\begin{pmatrix} 4 & -1 \\ 2 & 1 \end{pmatrix}$

(d) $\begin{pmatrix} 1 & 1 \\ 0 & 1 \end{pmatrix}$ *(e) $\begin{pmatrix} 15 & 7 & -7 \\ -1 & 1 & 1 \\ 13 & 7 & -5 \end{pmatrix}$ (f) $\begin{pmatrix} 5 & -2 & 2 \\ 4 & -3 & 4 \\ 4 & -6 & 7 \end{pmatrix}$

*(g) $\begin{pmatrix} 1 & 0 & 0 \\ -2 & 1 & 2 \\ -2 & 0 & 3 \end{pmatrix}$ (h) $\begin{pmatrix} 3 & 0 & 0 \\ 1 & 2 & 0 \\ 0 & 0 & -4 \end{pmatrix}$ *(i) $\begin{pmatrix} 1 & -2 & 2 \\ 4 & 5 & -4 \\ 0 & -2 & 3 \end{pmatrix}$

3. Transform each of the following symmetric matrices into diagonal form using an orthogonal similarity transformation. Give the transformation in each case.

*(a) $\begin{pmatrix} 1 & 2 \\ 2 & 1 \end{pmatrix}$ (b) $\begin{pmatrix} 11 & 2 \\ 2 & 14 \end{pmatrix}$ *(c) $\begin{pmatrix} 3 & 1 \\ 1 & 3 \end{pmatrix}$

(d) $\begin{pmatrix} -1 & -8 \\ -8 & 11 \end{pmatrix}$ *(e) $\begin{pmatrix} \frac{1}{2} & -\frac{3}{2} & 0 \\ -\frac{3}{2} & \frac{1}{2} & 0 \\ 0 & 0 & -2 \end{pmatrix}$ (f) $\begin{pmatrix} \frac{3}{2} & -\frac{1}{2} & 0 \\ -\frac{1}{2} & \frac{3}{2} & 0 \\ 0 & 0 & 1 \end{pmatrix}$

*(g) $\begin{pmatrix} 0 & 2 & 0 \\ 2 & 0 & 0 \\ 0 & 0 & 1 \end{pmatrix}$ (h) $\begin{pmatrix} 9 & -3 & 3 \\ -3 & 6 & -6 \\ 3 & -6 & 6 \end{pmatrix}$ (i) $\begin{pmatrix} 1 & 2 & -2 \\ 2 & 4 & -4 \\ -2 & -4 & 4 \end{pmatrix}$

4. Determine an orthogonal transformation matrix for Example 5.

5. Prove that if A and B are related through a similarity transformation, then $|A| = |B|$; that is, the determinant of a matrix is invariant under a similarity transformation.

6. If A is a symmetric matrix, we know that it is similar to a diagonal matrix. Is such a diagonal matrix unique? (*Hint*: Does the order of the column vectors in the transforming matrix matter?)

7. Let A be a $n \times n$ matrix and let C be an invertible $n \times n$ matrix. Show that
   (a) $C^{-1}A^2C = (C^{-1}AC)^2$
   (b) $C^{-1}A^nC = (C^{-1}AC)^n$ for n a positive integer

8. Two $n \times n$ matrices A and B are said to be *orthogonally similar* if there exists an orthogonal matrix C such that $B = C^{-1}AC$. Show that if A is symmetric and if A and B are orthogonally similar, then B is symmetric.

9. Show that if A and B are orthogonally similar and B and C are orthogonally similar, then A and C are orthogonally similar.

10. Prove that the eigenspaces of a symmetric matrix corresponding to distinct eigenvalues are orthogonal.

11. Let A be an $n \times n$ matrix with eigenvalues $\lambda_1, \ldots, \lambda_n$ and corresponding orthogonal eigenvectors $\mathbf{v}_1, \ldots, \mathbf{v}_n$. Let C be the matrix having $\mathbf{v}_1, \ldots, \mathbf{v}_n$ as column vectors. Prove that

$$C^t AC = \begin{pmatrix} \lambda_1\|\mathbf{v}_1\|^2 & & 0 \\ & \ddots & \\ 0 & & \lambda_n\|\mathbf{v}_n\|^2 \end{pmatrix}$$

Illustrate this method for the matrix of Example 3.

12. Write a computer program to perform a similarity transformation $C^{-1}AC$, where the matrices A and C are the input. Test your program on Example 1 and use it to check your answers to Exercise 1.

7-4 COORDINATE TRANSFORMATIONS

In geometry, a change in coordinate systems is often desirable when it is found that a geometrical figure can be examined more easily in an alternative coordinate system. After formulating a geometric representation of a physical system, one often discovers that there is a more suitable coordinate system, perhaps one that displays certain properties of the situation. In many cases, the change in the coordinates of points from one system to the other (called a *coordinate transformation*) is carried out using matrices, as the following discussions illustrate.

Rotation of Coordinates

Consider the two rectangular coordinate systems xy and $x'y'$ in the plane (Figure 7-6). They have a common origin O, the one system being obtained from the other by a rotation through an angle θ. Let us determine how coordinates of points in the two systems are related.

Consider the point A in the plane. The location of A can be given relative to each coordinate system. In the xy coordinate system, the x value of A is given by the length OB, and the y value by AB. Thus, in the xy coordinate system, A is the point $\begin{pmatrix} x \\ y \end{pmatrix}$ with $x = OB$ and $y = AB$. In the $x'y'$ coordinate system, the x' value of A is the length of OC, and the y' value the length of AC. In the $x'y'$ coordinate system, A is the point $\begin{pmatrix} x' \\ y' \end{pmatrix}$ with $x' = OC$ and $y' = AC$.

We need to know how these two representations of A, $\begin{pmatrix} x \\ y \end{pmatrix}$ and $\begin{pmatrix} x' \\ y' \end{pmatrix}$, are related.

Let $AOC = \alpha$. Then we have

$$x = OB = OA \cos(\alpha + \theta) = OA \cos \alpha \cos \theta - OA \sin \alpha \sin \theta$$
$$= OC \cos \theta - AC \sin \theta = x' \cos \theta - y' \sin \theta$$

$$y = AB = OA \sin(\alpha + \theta) = OA \sin \alpha \cos \theta + OA \cos \alpha \sin \theta$$
$$= AC \cos \theta + OC \sin \theta = y' \cos \theta + x' \sin \theta$$
$$= x' \sin \theta + y' \cos \theta$$

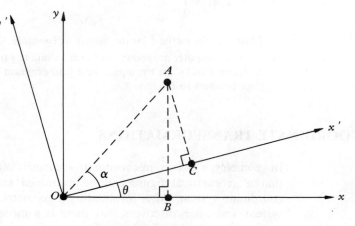

Figure 7-6

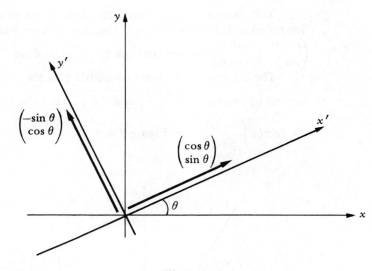

Figure 7-7

We can write these equations in matrix form:

$$\begin{pmatrix} x \\ y \end{pmatrix} = \begin{pmatrix} \cos\theta & -\sin\theta \\ \sin\theta & \cos\theta \end{pmatrix}\begin{pmatrix} x' \\ y' \end{pmatrix}$$

This equation defines the coordinate transformation. (Note the similarity between this coordinate transformation and the mapping that defines a rotation of the plane through an angle θ. This arises from the fact that one can interpret a transformation either as moving points in a fixed coordinate system or as changing the coordinate system around fixed points.)

Observe in Figure 7-7 that the column vectors of the matrix that defines the rotation give the directions of the new axes. The rotation matrix is an orthogonal matrix having determinant 1.

In general, if C is an orthogonal 2×2 matrix having determinant 1, it can be used to define a coordinate transformation in $\mathbf{R}^2$:[†]

$$\begin{pmatrix} x \\ y \end{pmatrix} = C\begin{pmatrix} x' \\ y' \end{pmatrix}$$

This will be a rotation about the origin with the columns of C giving the directions of the new axes. The first column of C will define the direction of the x' axis; the second column will give the direction of the y' axis.

✦ Example 1

Discuss the coordinate transformation

$$\begin{pmatrix} x \\ y \end{pmatrix} = \begin{pmatrix} \sqrt{3}/2 & -\tfrac{1}{2} \\ \tfrac{1}{2} & \sqrt{3}/2 \end{pmatrix}\begin{pmatrix} x' \\ y' \end{pmatrix}$$

[†] Orthogonal matrices having determinant -1 lead to a reorientation of the axes—that is, to new coordinate systems of the type . They are not in general of interest to us.

The matrix of the transformation is an orthogonal matrix having determinant 1; it thus defines a rotation of coordinates. Comparing it with $\begin{pmatrix} \cos\theta & -\sin\theta \\ \sin\theta & \cos\theta \end{pmatrix}$, we see that $\cos\theta = \sqrt{3}/2$ and $\sin\theta = \frac{1}{2}$. θ is thus 30°.

The columns of the rotation matrix give the x' axis to be in the direction defined by the vector $\begin{pmatrix} \sqrt{3}/2 \\ \frac{1}{2} \end{pmatrix}$ and the y' axis to be in the direction defined by the vector $\begin{pmatrix} -\frac{1}{2} \\ \sqrt{3}/2 \end{pmatrix}$. See Figure 7-8.

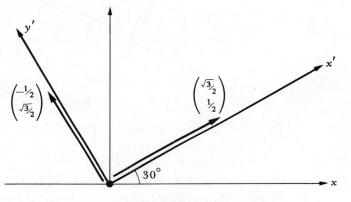

Figure 7-8

Translation of Coordinates

Consider two rectangular coordinate systems xy and $x'y'$ in the plane, where one system is obtained from the other by a translation of axes (Figure 7-9). Let us determine the coordinate transformation that relates the systems.

Let the origins of the coordinate systems be O and O'. Let the coordinates of O' in the xy system be $\begin{pmatrix} h \\ k \end{pmatrix}$. Consider the point A, which has representations $\begin{pmatrix} x \\ y \end{pmatrix}$ and $\begin{pmatrix} x' \\ y' \end{pmatrix}$ in the two coordinate systems. Thus $OB = x$, $AB = y$, $O'C = x'$, and $AC = y'$. We know that

$$x = OB = O'C + h = x' + h$$
$$y = AB = AC + k = y' + k$$

Thus

$$\begin{pmatrix} x \\ y \end{pmatrix} = \begin{pmatrix} x' \\ y' \end{pmatrix} + \begin{pmatrix} h \\ k \end{pmatrix}$$

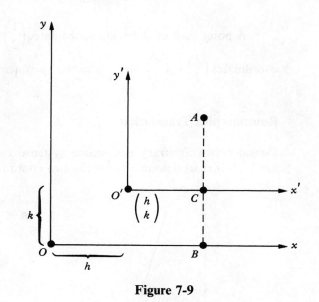

Figure 7-9

✦ *Example 2* ───

Let us discuss the coordinate transformation

$$\begin{pmatrix} x \\ y \end{pmatrix} = \begin{pmatrix} x' \\ y' \end{pmatrix} + \begin{pmatrix} 4 \\ 2 \end{pmatrix}$$

Using the above theory, we see that the origin of the $x'y'$ system relative to the xy system is the point $\begin{pmatrix} 4 \\ 2 \end{pmatrix}$. We get Figure 7-10.

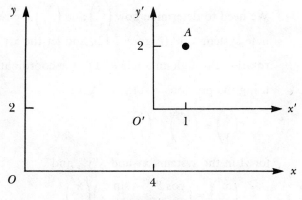

Figure 7-10

A point such as A having coordinates $\begin{pmatrix} 1 \\ 2 \end{pmatrix}$ in the $x'y'$ system would have coordinates $\begin{pmatrix} 1 \\ 2 \end{pmatrix} + \begin{pmatrix} 4 \\ 2 \end{pmatrix}$, or $\begin{pmatrix} 5 \\ 4 \end{pmatrix}$, in the xy system. ✦

Rotation plus Translation

Consider two arbitrary coordinate systems xy and $x'y'$ illustrated in Figure 7-11. Let us determine how the two coordinate systems are related.

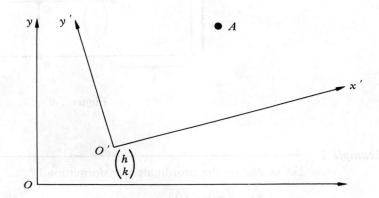

Figure 7-11

The coordinates of the origin O' in the xy system are $\begin{pmatrix} h \\ k \end{pmatrix}$. Let A be an arbitrary point whose representations are $\begin{pmatrix} x \\ y \end{pmatrix}$ and $\begin{pmatrix} x' \\ y' \end{pmatrix}$ in the two systems. We need to determine how $\begin{pmatrix} x \\ y \end{pmatrix}$ and $\begin{pmatrix} x' \\ y' \end{pmatrix}$ are related. Introduce a new coordinate system $x''y''$ (Figure 7-12), and let the $x'y'$ system be obtained from it by rotation through an angle θ. Let the coordinates of A in $x''y''$ be $\begin{pmatrix} x'' \\ y'' \end{pmatrix}$. Then, using the previous results, we have

$$\begin{pmatrix} x \\ y \end{pmatrix} = \begin{pmatrix} x'' \\ y'' \end{pmatrix} + \begin{pmatrix} h \\ k \end{pmatrix}$$

for A in the systems xy and $x''y''$, and

$$\begin{pmatrix} x'' \\ y'' \end{pmatrix} = \begin{pmatrix} \cos \theta & -\sin \theta \\ \sin \theta & \cos \theta \end{pmatrix} \begin{pmatrix} x' \\ y' \end{pmatrix}$$

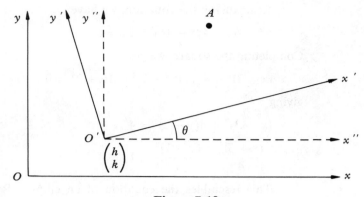

Figure 7-12

for A in the systems $x''y''$ and $x'y'$. Eliminating $\begin{pmatrix} x'' \\ y'' \end{pmatrix}$, we get

$$\begin{pmatrix} x \\ y \end{pmatrix} = \begin{pmatrix} \cos\theta & -\sin\theta \\ \sin\theta & \cos\theta \end{pmatrix} \begin{pmatrix} x' \\ y' \end{pmatrix} + \begin{pmatrix} h \\ k \end{pmatrix}$$

for the general coordinate transformation. Thus the coordinate systems are related by a rotation and a translation.

Here we have examined coordinate transformations between rectangular axes in the plane. Rotations and translations of axes also occur in three-dimensional space. In three-dimensional space, 3×3 orthogonal matrices having determinant 1 define coordinate transformations due to rotations of rectangular axes. The coordinate transformation between two arbitrary rectangular coordinate systems xyz and $x'y'z'$ in three dimensions can be written

$$\mathbf{x} = M\mathbf{x}' + \mathbf{k}$$

where M, an orthogonal matrix having determinant 1, defines the rotation and $\mathbf{k}$, a column vector containing the coordinates of the origin of the $x'y'z'$ system in the xyz system, defines the translation. The column vectors of M give the directions of the new axes, as in two-dimensional space.

✦ *Example 3*
_____

Discuss and sketch the curve

$$x^2 + 4y^2 - 6x - 16y + 21 = 0$$

This is a quadratic equation having no xy term. We do not recognize the equation as being in the form of any of the standard curves, such as a circle, parabola, ellipse, or hyperbola. However, by introducing a second coordinate system, we shall find that it is, in fact, the equation of an ellipse.

Rearranging the equation, we have

$$x^2 - 6x + 4y^2 - 16y + 21 = 0$$

Completing the square, we get

$$(x - 3)^2 - 9 + 4(y - 2)^2 - 16 + 21 = 0$$

giving

$$(x - 3)^2 + 4(y - 2)^2 = 4$$

$$\frac{(x - 3)^2}{4} + \frac{(y - 2)^2}{1} = 1$$

This resembles the equation of an ellipse. Performing the coordinate transformation $x' = x - 3$, $y' = y - 2$, we get the equation into the standard form for an ellipse,

$$\frac{x'^2}{4} + \frac{y'^2}{1} = 1$$

We know now that the curve is an ellipse with a major axis of length 2 and a minor axis of length 1 in the $x'y'$ coordinate system (Figure 7-13).

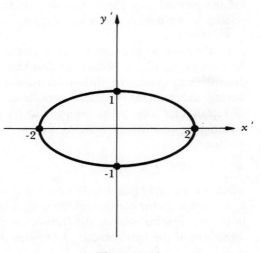

Figure 7-13

We must still locate this coordinate system in the original one. The coordinate transformation can be rewritten in the standard form $x = x' + 3$, $y = y' + 2$. We see that it is a translation of axes with the origin of the $x'y'$ system at the point $\begin{pmatrix} 3 \\ 2 \end{pmatrix}$ of the xy system. In Figure 7-14, we sketch the curve in the original coordinate system.

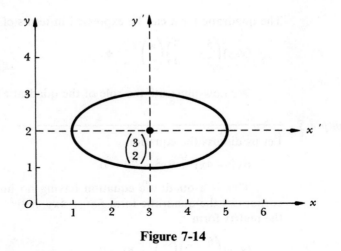

Figure 7-14

In this example,

1. We found the coordinate system that fits the symmetry of the ellipse, the $x'y'$ coordinate system.
2. We analyzed the equation in this system.
3. We interpreted the analysis in the original system.   ✦

An expression of the type

$$ax^2 + bxy + cy^2$$

where a, b, and c are constants, is called a *quadratic form*. This expression, which plays an important role in geometry and in applications, can be expressed in the matrix form

$$(x, y)\begin{pmatrix} a & b/2 \\ b/2 & c \end{pmatrix}\begin{pmatrix} x \\ y \end{pmatrix}$$

The reader can show that these two expressions are identical by multiplying out the matrices.

The symmetric matrix $\begin{pmatrix} a & b/2 \\ b/2 & c \end{pmatrix}$ associated with the quadratic form is called the *matrix of the quadratic form*.

✦ *Example 4*
─────────────────────────────────────

$5x^2 + 6xy - 4y^2$ is a quadratic form.

On comparison with the standard form $ax^2 + bxy + cy^2$, we have that $a = 5$, $b = 6$, and $c = -4$. The matrix of the quadratic form is thus

$$\begin{pmatrix} a & b/2 \\ b/2 & c \end{pmatrix} = \begin{pmatrix} 5 & 3 \\ 3 & -4 \end{pmatrix}$$

The quadratic form can be expressed in terms of matrices:

$$(x, y)\begin{pmatrix} 5 & 3 \\ 3 & -4 \end{pmatrix}\begin{pmatrix} x \\ y \end{pmatrix} \quad \blacklozenge$$

We now illustrate the role of the quadratic form in analyzing equations.

✦ **Example 5**

Let us discuss the equation

$$6x^2 + 4xy + 9y^2 - 20 = 0$$

This is a quadratic equation having no linear terms in x and y. It incorporates the quadratic form $6x^2 + 4xy + 9y^2$. Let us write the equation in the matrix form

$$(x, y)\begin{pmatrix} 6 & 2 \\ 2 & 9 \end{pmatrix}\begin{pmatrix} x \\ y \end{pmatrix} - 20 = 0 \qquad (1)$$

Our aim will be to introduce a new coordinate system in which the equation assumes a simpler, recognizable form. The matrix $\begin{pmatrix} 6 & 2 \\ 2 & 9 \end{pmatrix}$ is symmetric; it can thus be diagonalized through a similarity transformation. We shall diagonalize it into the form having eigenvalues on the main diagonal. Let us determine the eigenvalues:

$$\begin{vmatrix} 6 - \lambda & 2 \\ 2 & 9 - \lambda \end{vmatrix} = 0 \Rightarrow (6 - \lambda)(9 - \lambda) - 4 = 0$$

$$\Rightarrow \lambda^2 - 15\lambda + 50 = 0$$

giving $(\lambda - 10)(\lambda - 5) = 0$. The eigenvalues are 10 and 5.

We now find the corresponding eigenvectors. For $\lambda = 10$,

$$\begin{pmatrix} -4 & 2 \\ 2 & -1 \end{pmatrix}\begin{pmatrix} x \\ y \end{pmatrix} = 0 \Rightarrow 2x - y = 0 \Rightarrow 2x = y$$

The eigenvectors are $r\begin{pmatrix} 1 \\ 2 \end{pmatrix}$.

For $\lambda = 5$,

$$\begin{pmatrix} 1 & 2 \\ 2 & 4 \end{pmatrix}\begin{pmatrix} x \\ y \end{pmatrix} = 0 \Rightarrow x + 2y = 0 \Rightarrow x = -2y$$

The eigenvectors are $s\begin{pmatrix} -2 \\ 1 \end{pmatrix}$.

Normalizing these vectors, we get unit orthogonal eigenvectors $\begin{pmatrix} 1/\sqrt{5} \\ 2/\sqrt{5} \end{pmatrix}$ and $\begin{pmatrix} -2/\sqrt{5} \\ 1/\sqrt{5} \end{pmatrix}$. Write these vectors as the columns of an orthogonal matrix C

having determinant 1:

$$C = \begin{pmatrix} 1/\sqrt{5} & -2/\sqrt{5} \\ 2/\sqrt{5} & 1/\sqrt{5} \end{pmatrix}$$

(The order of selection of the eigenvectors is important, for it decides whether C has determinant $+1$ or -1.)

The matrix $\begin{pmatrix} 6 & 2 \\ 2 & 9 \end{pmatrix}$ can be transformed into the diagonal form $\begin{pmatrix} 10 & 0 \\ 0 & 5 \end{pmatrix}$ using the orthogonal matrix C.

Let us now return to the quadratic equation. We shall rearrange it into a form that suggests a coordinate transformation. Write equation (1) as

$$\mathbf{x}^t A \mathbf{x} - 20 = 0$$

where $\mathbf{x} = (x, y)$ and $A = \begin{pmatrix} 6 & 2 \\ 2 & 9 \end{pmatrix}$. Since C is orthogonal, $C^t C = I$ and this equation can be written

$$\mathbf{x}^t (CC^t) A (CC^t) \mathbf{x} - 20 = 0$$

giving

$$(\mathbf{x}^t C)(C^t A C)(C^t \mathbf{x}) - 20 = 0$$

$$(C^t \mathbf{x})^t \begin{pmatrix} 10 & 0 \\ 0 & 5 \end{pmatrix} (C^t \mathbf{x}) - 20 = 0$$

Introduce a coordinate transformation $\mathbf{x}' = C^t \mathbf{x}$, where $\mathbf{x}' = \begin{pmatrix} x' \\ y' \end{pmatrix}$. In the new coordinate system the equation becomes

$$(x', y') \begin{pmatrix} 10 & 0 \\ 0 & 5 \end{pmatrix} \begin{pmatrix} x' \\ y' \end{pmatrix} - 20 = 0$$

giving

$$10(x')^2 + 5(y')^2 - 20 = 0$$

This equation can be written in standard form as

$$\frac{(x')^2}{2} + \frac{(y')^2}{4} = 1$$

In the $x'y'$ coordinate system, the graph is an ellipse with major axis of length 2 and minor axis of length $\sqrt{2}$. (See Figure 7-15.)

It remains to locate the $x'y'$ coordinate system relative to the original xy system. The coordinate transformation is the rotation

$$\mathbf{x}' = C^t \mathbf{x}$$

or

$$\mathbf{x} = C \mathbf{x}' \qquad \text{since } C^{-1} = C^t, \ C \text{ being orthogonal}$$

$$\begin{pmatrix} x \\ y \end{pmatrix} = \begin{pmatrix} 1/\sqrt{5} & -2/\sqrt{5} \\ 2/\sqrt{5} & 1/\sqrt{5} \end{pmatrix} \begin{pmatrix} x' \\ y' \end{pmatrix}$$

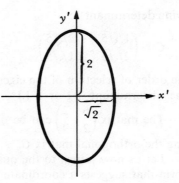

Figure 7-15

The x' axis is in the direction $\begin{pmatrix} 1/\sqrt{5} \\ 2/\sqrt{5} \end{pmatrix}$, and the y' axis is in the direction $\begin{pmatrix} -2/\sqrt{5} \\ 1/\sqrt{5} \end{pmatrix}$. Thus the graph in the xy coordinate system is Figure 7-16.   ✦

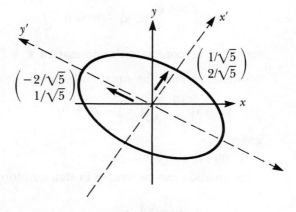

Figure 7-16

In the last two examples, we discussed the graphs of quadratic equations of the types

$$ax^2 + by^2 + cx + dy + e = 0 \text{ (having no } xy \text{ terms)}$$

and

$$ax^2 + bxy + cy^2 + d = 0 \text{ (having no linear terms in } x \text{ and } y)$$

We found that these could be interpreted through a translation of coordinates (in the former) and a rotation of coordinates (in the latter). A suitable coordinate system for discussing a general quadratic equation of the form

$$ax^2 + bxy + cy^2 + dx + ey + f = 0$$

having both an xy term and linear terms in x and y can be obtained by performing first a rotation of axes and then a translation. We shall not pursue this general case, but rather introduce the reader to further techniques and applications involving coordinate transformations.

Coordinate Transformations and Systems of Equations

Another use of coordinate transformations and similarity transformations is in the solution of certain systems of linear equations. Let $A\mathbf{x} = \mathbf{y}$ represent a system of n equations in n unknowns. If C is a nonsingular matrix, let $\mathbf{x} = C\mathbf{x}'$ and $\mathbf{y} = C\mathbf{y}'$. Upon substituting for $\mathbf{x}$ and $\mathbf{y}$, the system of equations becomes $AC\mathbf{x}' = C\mathbf{y}'$; that is, $C^{-1}AC\mathbf{x}' = \mathbf{y}'$. If A is a symmetric matrix, we know that there exists a matrix C such that $C^{-1}AC$ is a diagonal matrix. Using this matrix C, we can easily solve the system $C^{-1}AC\mathbf{x}' = \mathbf{y}'$. $\mathbf{x}$, the required solution, is then obtained from $\mathbf{x} = C\mathbf{x}'$. If the matrix A is not symmetric, similarity transformations can often be used to obtain a simpler system of equations, although the matrix of coefficients of the resulting system will not always be diagonal. Similarity transformations can always be chosen so that the matrix of coefficients has only nonzero elements on its main diagonal and immediately above the diagonal, if we work with complex numbers. Readers who go on to study differential equations will find that this technique is used in solving certain systems of differential equations.

✦ *Example 6*

Solve the system of equations

$$\begin{pmatrix} \frac{3}{2} & -\frac{1}{2} & 0 \\ -\frac{1}{2} & \frac{3}{2} & 0 \\ 0 & 0 & 3 \end{pmatrix} \begin{pmatrix} x \\ y \\ z \end{pmatrix} = \begin{pmatrix} \frac{4}{\sqrt{2}} \\ \frac{2}{\sqrt{2}} \\ 6 \end{pmatrix}$$

using a similarity transformation.

The matrix of coefficients A is symmetric; it can thus be transformed into a diagonal matrix through an orthogonal similarity transformation. This matrix was discussed in Section 7-3, where it was found that its eigenvalues were 1, 2, and 3 with corresponding eigenvectors $r(1, 1, 0)$, $s(-1, 1, 0)$, and $t(0, 0, 1)$.

Normalizing these vectors, we have the transforming orthogonal matrix

$$C = \begin{pmatrix} \dfrac{1}{\sqrt{2}} & -\dfrac{1}{\sqrt{2}} & 0 \\[2mm] \dfrac{1}{\sqrt{2}} & \dfrac{1}{\sqrt{2}} & 0 \\[2mm] 0 & 0 & 1 \end{pmatrix}$$

It will transform the matrix of coefficients into the diagonal form $\begin{pmatrix} 1 & 0 & 0 \\ 0 & 2 & 0 \\ 0 & 0 & 3 \end{pmatrix}$.

In a new coordinate system $x'y'z'$ defined by the rotation $\mathbf{x} = C\mathbf{x}'$, the system of equations becomes $C^{-1} AC\mathbf{x}' = \mathbf{y}'$; that is,

$$\begin{pmatrix} \dfrac{1}{\sqrt{2}} & -\dfrac{1}{\sqrt{2}} & 0 \\[2mm] \dfrac{1}{\sqrt{2}} & \dfrac{1}{\sqrt{2}} & 0 \\[2mm] 0 & 0 & 1 \end{pmatrix}^{-1} \begin{pmatrix} \frac{3}{2} & -\frac{1}{2} & 0 \\ -\frac{1}{2} & \frac{3}{2} & 0 \\ 0 & 0 & 3 \end{pmatrix} \begin{pmatrix} \dfrac{1}{\sqrt{2}} & -\dfrac{1}{\sqrt{2}} & 0 \\[2mm] \dfrac{1}{\sqrt{2}} & \dfrac{1}{\sqrt{2}} & 0 \\[2mm] 0 & 0 & 1 \end{pmatrix} \begin{pmatrix} x' \\ y' \\ z' \end{pmatrix}$$

$$= \begin{pmatrix} \dfrac{1}{\sqrt{2}} & -\dfrac{1}{\sqrt{2}} & 0 \\[2mm] \dfrac{1}{\sqrt{2}} & \dfrac{1}{\sqrt{2}} & 0 \\[2mm] 0 & 0 & 1 \end{pmatrix}^{-1} \begin{pmatrix} \dfrac{4}{\sqrt{2}} \\[2mm] \dfrac{2}{\sqrt{2}} \\[2mm] 6 \end{pmatrix}$$

This equation becomes

$$\begin{pmatrix} 1 & 0 & 0 \\ 0 & 2 & 0 \\ 0 & 0 & 3 \end{pmatrix} \begin{pmatrix} x' \\ y' \\ z' \end{pmatrix} = \begin{pmatrix} \dfrac{1}{\sqrt{2}} & -\dfrac{1}{\sqrt{2}} & 0 \\[2mm] \dfrac{1}{\sqrt{2}} & \dfrac{1}{\sqrt{2}} & 0 \\[2mm] 0 & 0 & 1 \end{pmatrix}^{t} \begin{pmatrix} \dfrac{4}{\sqrt{2}} \\[2mm] \dfrac{2}{\sqrt{2}} \\[2mm] 6 \end{pmatrix}$$

leading to

$$\begin{pmatrix} 1 & 0 & 0 \\ 0 & 2 & 0 \\ 0 & 0 & 3 \end{pmatrix} \begin{pmatrix} x' \\ y' \\ z' \end{pmatrix} = \begin{pmatrix} 3 \\ -1 \\ 6 \end{pmatrix}$$

$$\begin{pmatrix} x' \\ y' \\ z' \end{pmatrix} = \begin{pmatrix} 3 \\ -\frac{1}{2} \\ 2 \end{pmatrix}$$

The solution of the original system is given by $\mathbf{x} = C\mathbf{x}'$,

$$\begin{pmatrix} x \\ y \\ z \end{pmatrix} = \begin{pmatrix} \dfrac{1}{\sqrt{2}} & -\dfrac{1}{\sqrt{2}} & 0 \\ \dfrac{1}{\sqrt{2}} & \dfrac{1}{\sqrt{2}} & 0 \\ 0 & 0 & 1 \end{pmatrix} \begin{pmatrix} x' \\ y' \\ z' \end{pmatrix}$$

or

$$\begin{pmatrix} x \\ y \\ z \end{pmatrix} = \begin{pmatrix} \dfrac{7}{2\sqrt{2}} \\ \dfrac{5}{2\sqrt{2}} \\ 2 \end{pmatrix} \quad \blacklozenge$$

✦ Example 7*

Coordinate transformations are used extensively in computer graphics. In this example, we illustrate their role by discussing how an ellipse is graphed using an Apple microcomputer.

The output from the computer is to a TV screen. The locations of points on the screen are described by the computer relative to the coordinate system in Figure 7-17.

Thus, in the computer's coordinate system, the y axis is inverted and points have x values in the interval $0 \le x \le 279$ and y values in the interval $0 \le y \le 159$. For example, typing the following on the computer keyboard

```
HGR
HCOLOR = 6
HPLOT 39, 20
```

would plot a blue dot in location $x = 39$, $y = 20$. HGR tells the computer to go into high resolution graphics mode; color 6 corresponds to blue. There are 8 colors available. (In practice, dots can only be placed at locations corresponding to integer values of x and y. For example, HPLOT 39.2, 20.7 would lead to the same point on the screen as HPLOT 39, 20; the x and y values are truncated in plotting.)

A more natural coordinate system for the user to work in is the $x'y'$ system in Figure 7-18, with the origin O' at the center of the screen and the positive y' axis up. The coordinate transformation between the xy and $x'y'$ systems is

$$x' = x - 140$$
$$y' = -y + 80$$

The transformation involves a reorientation of axes.

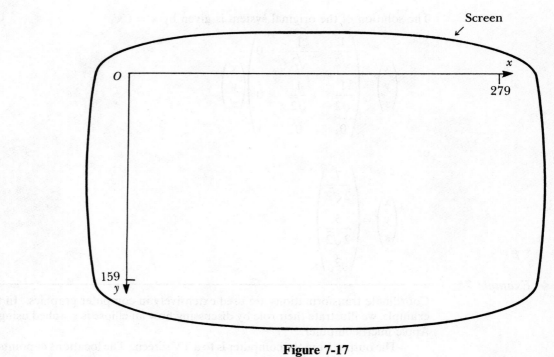

Figure 7-17

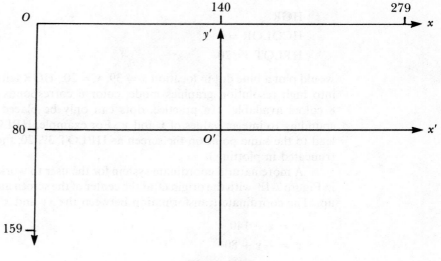

Figure 7-18

The user can discuss geometrical shapes relative to the $x'y'$ coordinate system, then communicate the information to the computer relative to its system, xy, in order to get the points graphed out on the screen. For example, the ellipse

$$\frac{(x')^2}{900} + \frac{(y')^2}{400} = 1$$

has major axis of length 30 and minor axis of length 20. Its location on the screen is shown in Figure 7-19.

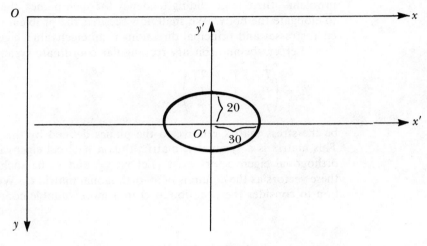

Figure 7-19

To get the computer to actually graph the ellipse in this manner, the user must write the equation in the xy coordinates. The equation in the xy coordinates is

$$\frac{(x-140)^2}{900} + \frac{(-y+80)^2}{400} = 1$$

This equation is rearranged to express y in terms of x and then included in a computer program. The ellipse is then graphed on the screen by the computer. Readers who are interested in the program are referred to Appendix I.

✦

✦ *Example 8**

In this example, we return to the stress analysis of Example 6, Section 2-4. The stress matrix is symmetric—we shall make use of the eigenspace properties of such matrices. We shall also see the significance of using a coordinate transformation that takes us into a special coordinate system that displays physical properties of the situation.

Consider a body subject to external forces. Let O be an arbitrary point in the body. We have seen that associated with every plane through O will be a stress, a force per unit area. This stress is usually discussed in terms of its component perpendicular to the plane, normal stress, and its component parallel to the plane, shearing stress.

Suppose a plane is oriented in such a way that the shearing stress is zero; the resultant stress at the point is then the normal stress. The plane is then called a *principal plane* at the point, the perpendicular direction is called a *principal direction*, and the stress is called a *principal stress* (Figure 7-20).

Determining the principal planes and stresses is an important engineering problem—there is no sliding tendency for such planes, the stress tending only to elongate the body. We shall now see that the problem of determining principal stresses and principal directions is an eigenvalue, eigenvector problem.

Let xyz be an arbitrary rectangular coordinate system at O and let

$$T = \begin{pmatrix} T_x & T_{xy} & T_{xz} \\ T_{yx} & T_y & T_{yz} \\ T_{zx} & T_{zy} & T_z \end{pmatrix}$$

be the stress matrix relative to the planes defined by this coordinate system. This matrix is a symmetric matrix; thus it has real eigenvalues and three unit orthogonal eigenvectors exist. Let v_1, v_2, and v_3 be such eigenvectors. Use these vectors as the columns of an orthogonal matrix C. We are now in a position to consider the situation at O in a more suitable coordinate system. Let

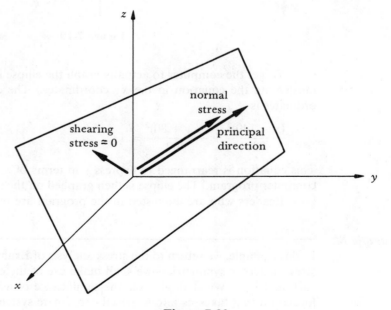

Figure 7-20

$x'y'z'$ be a rectangular coordinate system defined by the vectors $\mathbf{v}_1$, $\mathbf{v}_2$, and $\mathbf{v}_3$ in the sense that Ox' is along $\mathbf{v}_1$, etc. Then the stress matrix at O, relative to the planes of this system of coordinates, can be shown to be (we do not prove this)

$$C^{-1}TC$$

We know that this will be a diagonal matrix, having eigenvalues on the diagonal:

$$\begin{pmatrix} T_{x'} & 0 & 0 \\ 0 & T_{y'} & 0 \\ 0 & 0 & T_{z'} \end{pmatrix}$$

Observe that all shearing stresses are zero. The new coordinate planes are in fact principal planes, and the new coordinate axes give principal directions. In terms of the original coordinate system, the eigenvectors of the stress matrix give principal directions and its eigenvalues principal stresses.

Thus the result that a symmetric matrix can be diagonalized by an orthogonal transformation means here that there always exists an orthogonal set of principal directions for a body subject to external forces.

Let us look at the various possible situations that can arise. There are three possibilities for the roots of the characteristic polynomial of the stress matrix. (σ is usually used in this context for these roots.)

1. If the roots of the characteristic polynomial are distinct, σ_1, σ_2, and σ_3, then these are the principal stresses and there will be three distinct orthogonal eigenvectors giving the corresponding principal directions.
2. If there is a repeated root σ_1 and a single root σ_2 of the characteristic polynomial, then there are just two principal stresses. Corresponding to the repeated root there will be a two-dimensional eigenspace representing a two-dimensional space of principal directions. Corresponding to the single root there will be a principal direction orthogonal to the two-dimensional space.
3. If all three roots of the characteristic polynomial are equal, then there is a single principal stress. All vectors are eigenvectors; thus any direction is a principal direction. This corresponds to a state of so-called hydrostatic stress.

Let us consider a specific stress matrix

$$\begin{pmatrix} 5,000 & 4,000 & 2,000 \\ 4,000 & 5,000 & 2,000 \\ 2,000 & 2,000 & 2,000 \end{pmatrix}$$

where the units are pounds per square inch.

The eigenvalues are 1,000 (repeated) and 10,000. Since 1,000 is a repeated eigenvalue, we have case 2 above.

For $\sigma = 1,000$, we have the two-dimensional eigenspace where eigenvectors are of the form $\begin{pmatrix} -r - \frac{1}{2}s \\ r \end{pmatrix}$.

For $\sigma = 10,000$, we have the one-dimensional eigenspace $t\begin{pmatrix} 2 \\ 2 \\ 1 \end{pmatrix}$.

Thus the principal stresses are 1,000 pounds per square inch with principal directions $\begin{pmatrix} -r - \frac{1}{2}s \\ r \\ s \end{pmatrix}$ and 10,000 pounds per square inch with principal direction $t\begin{pmatrix} 2 \\ 2 \\ 1 \end{pmatrix}$.

We shall not continue any further here with this analysis of the stress matrix; we wish only to show the central role of the concept of eigenvalues and eigenvectors. A further discussion would involve certain invariants that are defined in terms of the coefficients of the characteristic polynomial of the stress matrix. For further reading, the reader is referred to *Plasticity: Theory & Practice*, by Alexander Mendelson, Macmillan, 1968. The book does not use eigenvector and eigenvalue terminology, but the mathematics developed is that of eigenvalues and eigenvectors. ✦

EXERCISES

1. Consider two rectangular coordinate systems xy and $x'y'$ in the plane with a common origin O. The second system is obtained from the first by a rotation about O through an angle θ in a counterclockwise direction. In each case, determine the matrix that defines the coordinate transformation and the coordinates of the point (1, 1) in the new system.
   *(a) $\theta = 90°$ (b) $\theta = 45°$ *(c) $\theta = 60°$

2. Illustrate the coordinate transformations described by the following matrix equations, sketching the relative locations of the axes:

   *(a) $\begin{pmatrix} x \\ y \end{pmatrix} = \begin{pmatrix} x' \\ y' \end{pmatrix} + \begin{pmatrix} 1 \\ 2 \end{pmatrix}$ (b) $\begin{pmatrix} x \\ y \end{pmatrix} = \begin{pmatrix} x' \\ y' \end{pmatrix} + \begin{pmatrix} -1 \\ 3 \end{pmatrix}$

   (c) $\begin{pmatrix} x \\ y \end{pmatrix} = \begin{pmatrix} x' \\ y' \end{pmatrix} + \begin{pmatrix} -2 \\ -3 \end{pmatrix}$ (d) $\begin{pmatrix} x \\ y \end{pmatrix} = \begin{pmatrix} x' \\ y' \end{pmatrix} + \begin{pmatrix} 2 \\ -1 \end{pmatrix}$

   *(e) $\begin{pmatrix} x \\ y \end{pmatrix} = \begin{pmatrix} x' \\ y' \end{pmatrix} + \begin{pmatrix} 0 \\ 2 \end{pmatrix}$ (f) $\begin{pmatrix} x \\ y \end{pmatrix} = \begin{pmatrix} x' \\ y' \end{pmatrix} + \begin{pmatrix} -3 \\ 0 \end{pmatrix}$

(g) $\begin{pmatrix} x \\ y \end{pmatrix} = \begin{pmatrix} 0 & -1 \\ 1 & 0 \end{pmatrix} \begin{pmatrix} x' \\ y' \end{pmatrix}$ **(h)** $\begin{pmatrix} x \\ y \end{pmatrix} = \begin{pmatrix} 0 & 1 \\ -1 & 0 \end{pmatrix} \begin{pmatrix} x' \\ y' \end{pmatrix}$

*(i) $\begin{pmatrix} x \\ y \end{pmatrix} = \begin{pmatrix} 1/\sqrt{2} & -1\sqrt{2} \\ 1/\sqrt{2} & 1/\sqrt{2} \end{pmatrix} \begin{pmatrix} x' \\ y' \end{pmatrix}$

(j) $\begin{pmatrix} x \\ y \end{pmatrix} = \begin{pmatrix} 1/\sqrt{2} & 1/\sqrt{2} \\ -1/\sqrt{2} & 1/\sqrt{2} \end{pmatrix} \begin{pmatrix} x' \\ y' \end{pmatrix}$

(k) $\begin{pmatrix} x \\ y \end{pmatrix} = \begin{pmatrix} \sqrt{3}/2 & -\frac{1}{2} \\ \frac{1}{2} & \sqrt{3}/2 \end{pmatrix} \begin{pmatrix} x' \\ y' \end{pmatrix}$

(l) $\begin{pmatrix} x \\ y \end{pmatrix} = \begin{pmatrix} -\frac{1}{2} & -\sqrt{3}/2 \\ \sqrt{3}/2 & -\frac{1}{2} \end{pmatrix} \begin{pmatrix} x' \\ y' \end{pmatrix}$

(m) $\begin{pmatrix} x \\ y \end{pmatrix} = \begin{pmatrix} -\frac{3}{5} & \frac{4}{5} \\ -\frac{4}{5} & -\frac{3}{5} \end{pmatrix} \begin{pmatrix} x' \\ y' \end{pmatrix}$

(n) $\begin{pmatrix} x \\ y \end{pmatrix} = \begin{pmatrix} 1/\sqrt{2} & -1/\sqrt{2} \\ 1/\sqrt{2} & 1/\sqrt{2} \end{pmatrix} \begin{pmatrix} x' \\ y' \end{pmatrix} + \begin{pmatrix} 2 \\ 4 \end{pmatrix}$

*(o) $\begin{pmatrix} x \\ y \end{pmatrix} = \begin{pmatrix} \sqrt{3}/2 & -\frac{1}{2} \\ \frac{1}{2} & \sqrt{3}/2 \end{pmatrix} \begin{pmatrix} x' \\ y' \end{pmatrix} + \begin{pmatrix} -1 \\ 3 \end{pmatrix}$

(p) $\begin{pmatrix} x \\ y \end{pmatrix} = \begin{pmatrix} 0 & -1 \\ 1 & 0 \end{pmatrix} \begin{pmatrix} x' \\ y' \end{pmatrix} + \begin{pmatrix} -1 \\ -2 \end{pmatrix}$

(q) $\begin{pmatrix} x \\ y \end{pmatrix} = \begin{pmatrix} \frac{3}{5} & \frac{4}{5} \\ -\frac{4}{5} & \frac{3}{5} \end{pmatrix} \begin{pmatrix} x' \\ y' \end{pmatrix} + \begin{pmatrix} 2 \\ -1 \end{pmatrix}$

3. Discuss and sketch the graphs of each of the following equations. Each graph will be an ellipse, hyperbola, or parabola. (Use a coordinate transformation. Observe that each equation has no term in xy.)

   *(a) $x^2 + 4y^2 - 4x + 3 = 0$

   (b) $x^2 - 4y^2 - 4x + 3 = 0$

   (c) $9x^2 + 4y^2 + 18x - 8y - 16 = 0$

   *(d) $x^2 + y^2 - 4x + 6y + 9 = 0$

   *(e) $x^2 - 4y^2 + 2x + 16y - 19 = 0$

   (f) $y^2 - 4y - 4x = 0$

   *(g) $x^2 + 4y^2 - 6x + 16y + 9 = 0$

4. Express each of the following quadratic forms in terms of matrices.

   *(a) $x^2 + 4xy + 2y^2$ **(b)** $3x^2 + 2xy - 4y^2$

   *(c) $7x^2 - 6xy - y^2$ **(d)** $2x^2 + 5xy + 3y^2$

   (e) $-3x^2 - 7xy + 4y^2$

5. Discuss and sketch the graphs of each of the following equations. Each graph will be an ellipse, hyperbola, or pair of straight lines. (Use a coordinate transformation. Observe that each equation has no linear terms in x and y.)

   *(a) $11x^2 + 4xy + 14y^2 - 60 = 0$

   (b) $3x^2 + 2xy + 3y^2 - 12 = 0$

*(c) $x^2 - 6xy + y^2 - 8 = 0$

(d) $4x^2 + 4xy + 4y^2 - 5 = 0$

*(e) $-x^2 - 16xy + 11y^2 - 30 = 0$

(f) $-7x^2 - 18xy + 17y^2 = 0$

*(g) $3x^2 - 10xy + 3y^2 = 0$

6. Solve the following systems of equations using similarity transformations. (The diagonal forms and similarity transformations were determined in the exercises in Section 7-3.)

*(a) $\begin{pmatrix} 1 & 2 \\ 2 & 1 \end{pmatrix} \begin{pmatrix} x \\ y \end{pmatrix} = \begin{pmatrix} 4 \\ 2 \end{pmatrix}$

(b) $\begin{pmatrix} \frac{1}{2} & -\frac{3}{2} & 0 \\ -\frac{3}{2} & \frac{1}{2} & 0 \\ 0 & 0 & -2 \end{pmatrix} \begin{pmatrix} x \\ y \\ z \end{pmatrix} = \begin{pmatrix} 4 \\ 6 \\ 4 \end{pmatrix}$

*(c) $\begin{pmatrix} \frac{3}{2} & -\frac{1}{2} & 0 \\ -\frac{1}{2} & \frac{3}{2} & 0 \\ 0 & 0 & 1 \end{pmatrix} \begin{pmatrix} x \\ y \\ z \end{pmatrix} = \begin{pmatrix} 1 \\ 2 \\ 2 \end{pmatrix}$

(d) $\begin{pmatrix} 1 & -2 & 2 \\ 4 & 5 & -4 \\ 0 & -2 & 3 \end{pmatrix} \begin{pmatrix} x \\ y \\ z \end{pmatrix} = \begin{pmatrix} 1 \\ -1 \\ 1 \end{pmatrix}$

7. Determine the principal stresses and principal directions for each of the following stress matrices:

*(a) $\begin{pmatrix} 1{,}000 & 1{,}000 & 1{,}000 \\ 1{,}000 & 1{,}000 & 1{,}000 \\ 1{,}000 & 1{,}000 & 1{,}000 \end{pmatrix}$ (b) $\begin{pmatrix} 1{,}000 & 0 & 0 \\ 0 & 2{,}000 & 1{,}000 \\ 0 & 1{,}000 & 2{,}000 \end{pmatrix}$

(c) $\begin{pmatrix} 2{,}000 & 0 & 0 \\ 0 & 2{,}000 & 0 \\ 0 & 0 & 2{,}000 \end{pmatrix}$

7-5*   NORMAL MODES OF OSCILLATING SYSTEMS

In this section, we present the analysis of an oscillating system. This analysis involves eigenvalues, eigenvectors, and coordinate transformations.

✦ *Example 1* ——————————————————————————————

Consider a horizontal string AB of length $4a$ and negligible mass loaded with three particles, each of mass m. Let the masses be located at fixed distances a, $2a$, and $3a$ from A. The particles are displaced slightly from their equilibrium position and released. We wish to analyze the subsequent motion, assuming that it all takes place in a vertical plane through A and B.

Let the vertical displacements of the particles at any instant during the subsequent motion be x_1, x_2 and x_3, as illustrated in Figure 7-21. Let T be the tension in the string. Consider the motion of the particle at C. The resultant force on this particle in a vertical direction is $T \sin \alpha - T \sin \theta$. If we assume that the displacements are small, the motion of each particle can be assumed to be vertical and the tension can be assumed to be unaltered throughout the motion. Since the angles α and θ will be small, $\sin \alpha$ can be taken to be equal to $\tan \alpha$ and $\sin \theta$ equal to $\tan \theta$. Thus the resultant vertical force at C is

$$T \tan \alpha - T \tan \theta = T \frac{(x_2 - x_1)}{a} - T \frac{x_1}{a}$$

$$= -\frac{2Tx_1}{a} + \frac{Tx_2}{a}$$

Applying Newton's second law of motion (force = mass × acceleration), we find that the motion of the first particle is described by the equation

$$m\ddot{x}_1 = \frac{2Tx_1}{a} + \frac{Tx_2}{a}$$

Similarly, the motions of the other two particles are described by the equations

$$m\ddot{x}_2 = \frac{Tx_1}{a} - \frac{2Tx_2}{a} + \frac{Tx_3}{a}$$

and

$$m\ddot{x}_3 = \frac{Tx_2}{a} - \frac{2Tx_3}{a}$$

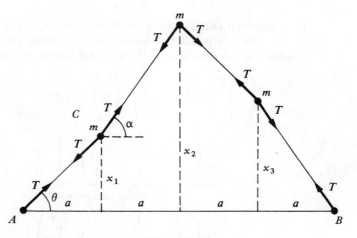

Figure 7-21

These equations can be combined into one matrix equation

$$\begin{pmatrix} \ddot{x}_1 \\ \ddot{x}_2 \\ \ddot{x}_3 \end{pmatrix} = \frac{T}{ma} \begin{pmatrix} -2 & 1 & 0 \\ 1 & -2 & 1 \\ 0 & 1 & -2 \end{pmatrix} \begin{pmatrix} x_1 \\ x_2 \\ x_3 \end{pmatrix}$$

We shall now see how the theory of eigenvalues, eigenvectors, and coordinate transformations enables us to solve this matrix equation elegantly, leading to solutions of the three original equations which describe the motions of the three particles.

The matrix $\begin{pmatrix} -2 & 1 & 0 \\ 1 & -2 & 1 \\ 0 & 1 & -2 \end{pmatrix}$ is symmetric; thus it can be transformed into

a diagonal form through a similarity transformation. A matrix of transformation will have linearly independent eigenvectors of this matrix as column vectors. The eigenvalues of this matrix are -2, $-2 - \sqrt{2}$, and $-2 + \sqrt{2}$. Corresponding eigenvectors are $(1, 0, -1)$, $(1, -\sqrt{2}, 1)$, and $(1, \sqrt{2}, 1)$. Thus a matrix of transformation is

$$\begin{pmatrix} 1 & 1 & 1 \\ 0 & -\sqrt{2} & \sqrt{2} \\ -1 & 1 & 1 \end{pmatrix}$$

and

$$\begin{pmatrix} 1 & 1 & 1 \\ 0 & -\sqrt{2} & \sqrt{2} \\ -1 & 1 & 1 \end{pmatrix}^{-1} \begin{pmatrix} -2 & 1 & 0 \\ 1 & -2 & 1 \\ 0 & 1 & -2 \end{pmatrix} \begin{pmatrix} 1 & 1 & 1 \\ 0 & -\sqrt{2} & \sqrt{2} \\ -1 & 1 & 1 \end{pmatrix}$$

$$= \begin{pmatrix} -2 & 0 & 0 \\ 0 & -2 - \sqrt{2} & 0 \\ 0 & 0 & -2 + \sqrt{2} \end{pmatrix}$$

a diagonal matrix where the eigenvalues are diagonal elements.

We introduce this similarity transformation into the matrix equation that describes the motion by rearranging the equations.

$$\begin{pmatrix} 1 & 1 & 1 \\ 0 & -\sqrt{2} & \sqrt{2} \\ -1 & 1 & 1 \end{pmatrix}^{-1} \begin{pmatrix} \ddot{x}_1 \\ \ddot{x}_2 \\ \ddot{x}_3 \end{pmatrix}$$

$$= \frac{T}{ma} \underbrace{\begin{pmatrix} 1 & 1 & 1 \\ 0 & -\sqrt{2} & \sqrt{2} \\ -1 & 1 & 1 \end{pmatrix}^{-1} \begin{pmatrix} -2 & 1 & 0 \\ 1 & -2 & 1 \\ 0 & 1 & -2 \end{pmatrix} \begin{pmatrix} 1 & 1 & 1 \\ 0 & -\sqrt{2} & \sqrt{2} \\ -1 & 1 & 1 \end{pmatrix}}_{\text{similarity transformation that diagonalizes the matrix}}$$

$$\times \begin{pmatrix} 1 & 1 & 1 \\ 0 & -\sqrt{2} & \sqrt{2} \\ -1 & 1 & 1 \end{pmatrix}^{-1} \begin{pmatrix} x_1 \\ x_2 \\ x_3 \end{pmatrix}$$

This equation is equivalent to the original matrix equation. Thus the motion is described by

$$
\begin{pmatrix} 1 & 1 & 1 \\ 0 & -\sqrt{2} & \sqrt{2} \\ -1 & 1 & 1 \end{pmatrix}^{-1} \begin{pmatrix} \ddot{x}_1 \\ \ddot{x}_2 \\ \ddot{x}_3 \end{pmatrix}
$$

$$
= \frac{T}{ma} \begin{pmatrix} -2 & 0 & 0 \\ 0 & -2-\sqrt{2} & 0 \\ 0 & 0 & -2+\sqrt{2} \end{pmatrix} \begin{pmatrix} 0 & 1 & 1 \\ 0 & -\sqrt{2} & \sqrt{2} \\ -1 & 1 & 1 \end{pmatrix}^{-1} \begin{pmatrix} x_1 \\ x_2 \\ x_3 \end{pmatrix}
$$

We now introduce new coordinates y_1, y_2, and y_3 defined by the transformation:

$$
\begin{pmatrix} y_1 \\ y_2 \\ y_3 \end{pmatrix} = \begin{pmatrix} 1 & 1 & 1 \\ 0 & -\sqrt{2} & \sqrt{2} \\ -1 & 1 & 1 \end{pmatrix}^{-1} \begin{pmatrix} x_1 \\ x_2 \\ x_3 \end{pmatrix}
$$

In this new coordinate system, the matrix equation that describes the motion assumes the particularly simple form

$$
\begin{pmatrix} \ddot{y}_1 \\ \ddot{y}_2 \\ \ddot{y}_3 \end{pmatrix} = \frac{T}{ma} \begin{pmatrix} -2 & 0 & 0 \\ 0 & -2-\sqrt{2} & 0 \\ 0 & 0 & -2+\sqrt{2} \end{pmatrix} \begin{pmatrix} y_1 \\ y_2 \\ y_3 \end{pmatrix}
$$

(Coordinates such as y_1, y_2, y_3 and x_1, x_2, x_3 that in some way describe the configuration of a system are called *generalized coordinates*. They may not be interpretable in terms of rectangular cartesian axes, but since they lead to a description of the configuration, they are coordinates.)

Thus, in terms of the new coordinates y_1, y_2, and y_3, the motion is described by the equations

$$
\ddot{y}_1 = -\frac{2T}{ma} y_1
$$

$$
\ddot{y}_2 = \frac{(-2-\sqrt{2})T}{ma} y_2
$$

$$
\ddot{y}_3 = \frac{(-2+\sqrt{2})T}{ma} y_3
$$

three simple harmonic motions. Solutions of these equations are

$$
y_1 = b_1 \cos\left(\sqrt{\frac{2T}{ma}}\, t + \gamma_1\right)
$$

$$
y_2 = b_2 \cos\left(\sqrt{\frac{(2+\sqrt{2})T}{ma}}\, t + \gamma_2\right)
$$

$$
y_3 = b_3 \cos\left(\sqrt{\frac{(2-\sqrt{2})T}{ma}}\, t + \gamma_3\right)
$$

where $b_1, b_2, b_3, \gamma_1, \gamma_2,$ and γ_3 are constants of integration that depend on the configuration at time $t = 0$. They are not all independent. In standard interpretation of simple harmonic motion, $b_1, b_2,$ and b_3 are amplitudes and $\gamma_1, \gamma_2,$ and γ_3 phases.

These special coordinates $y_1, y_2,$ and y_3 are called the *normal coordinates* of the motion. The general motion in terms of the original coordinates is described by

$$\begin{pmatrix} x_1 \\ x_2 \\ x_3 \end{pmatrix} = \begin{pmatrix} 1 & 1 & 1 \\ 0 & -\sqrt{2} & \sqrt{2} \\ -1 & 1 & 1 \end{pmatrix} \begin{pmatrix} y_1 \\ y_2 \\ y_3 \end{pmatrix}$$

$$= y_1 \begin{pmatrix} 1 \\ 0 \\ -1 \end{pmatrix} + y_2 \begin{pmatrix} 1 \\ -\sqrt{2} \\ 1 \end{pmatrix} + y_3 \begin{pmatrix} 1 \\ \sqrt{2} \\ 1 \end{pmatrix}$$

giving

$$\begin{pmatrix} x_1 \\ x_2 \\ x_3 \end{pmatrix} = b_1 \cos\left(\sqrt{\frac{2T}{ma}} t + \gamma_1\right) \begin{pmatrix} 1 \\ 0 \\ -1 \end{pmatrix}$$

$$+ b_2 \cos\left(\sqrt{\frac{(2+\sqrt{2})T}{ma}} t + \gamma_2\right) \begin{pmatrix} 1 \\ -\sqrt{2} \\ 1 \end{pmatrix}$$

$$+ b_3 \cos\left(\sqrt{\frac{(2-\sqrt{2})T}{ma}} t + \gamma_3\right) \begin{pmatrix} 1 \\ \sqrt{2} \\ 1 \end{pmatrix}$$

The motion can thus be interpreted as a combination of the three motions

$$\begin{pmatrix} x_1 \\ x_2 \\ x_3 \end{pmatrix} = \cos\left(\sqrt{\frac{2T}{ma}} t + \gamma_1\right) \begin{pmatrix} 1 \\ 0 \\ -1 \end{pmatrix}$$

$$\begin{pmatrix} x_1 \\ x_2 \\ x_3 \end{pmatrix} = \cos\left(\sqrt{\frac{(2+\sqrt{2})T}{ma}} t + \gamma_2\right) \begin{pmatrix} 1 \\ -\sqrt{2} \\ 1 \end{pmatrix}$$

$$\begin{pmatrix} x_1 \\ x_2 \\ x_3 \end{pmatrix} = \cos\left(\sqrt{\frac{(2-\sqrt{2})T}{ma}} t + \gamma_3\right) \begin{pmatrix} 1 \\ \sqrt{2} \\ 1 \end{pmatrix}$$

The actual combination will be a linear combination determined by $b_1, b_2,$ and b_3 that will depend on the initial configuration.

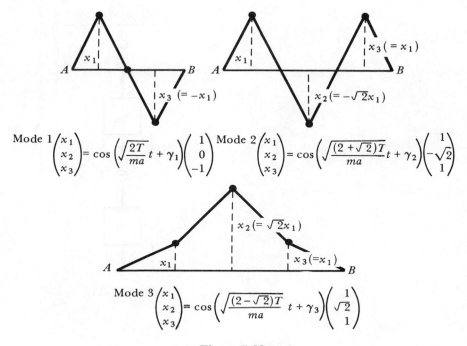

Figure 7-22

Each of these motions is a simple harmonic motion of a particle. These modes, called the *normal modes of oscillation*, are illustrated in Figure 7-22.

                                                                         ✦

EXERCISES

*1. Determine the normal modes of oscillation if the system in Figure 7-23 is displaced slightly from equilibrium.

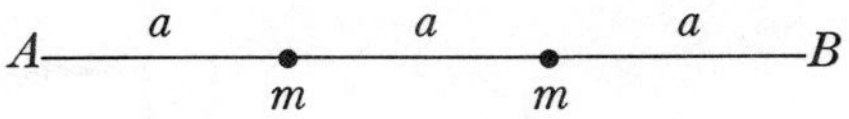

$$A \overset{a}{\rule{2cm}{0.4pt}} \underset{m}{\bullet} \overset{a}{\rule{2cm}{0.4pt}} \underset{m}{\bullet} \overset{a}{\rule{2cm}{0.4pt}} B$$

Figure 7-23

2. The motion of a weight attached to a spring is governed by Hooke's law: tension $=$ $k \times$ extension, where k is a constant for the spring. For oscillations of the system in Figure 7-24, if x_1 and x_2 are the displacements of weights of masses m_1 and m_2 at any instant, the extensions of the two springs are x_1 and $x_2 - x_1$ at that instant. Thus the application of Hooke's law gives the equations

$$m_1 \ddot{x}_1 = -k_1 x_1 + k_2 (x_2 - x_1)$$

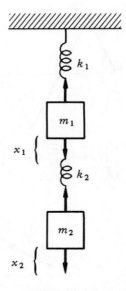

Figure 7-24

and

$$m_2 \ddot{x}_2 = -k_2(x_2 - x_1)$$

The general motion can be analyzed in terms of normal modes in a manner similar to the one used in Example 1. If $m_1 = m_2 = M$, $k_1 = 3$, and $k_2 = 2$, analyze the motion.

*3. Hooke's law also applies to forces in the extended springs of the system in Figure 7-25. The constants of the springs are 1, 2, and 3, as indicated. Applications of the law give the following equations of motion.

$$m\ddot{x}_1 = -x_1 + 2(x_2 - x_1)$$
$$3m\ddot{x}_2 = -2(x_2 - x_1) - 3x_2$$

Analyze this motion in terms of normal modes.

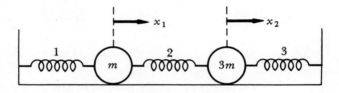

Figure 7-25

8*

8*

$$\overline{\overline{}}$$

Numerical Techniques

$$\overline{\overline{}}$$

In this chapter, we focus on numerical techniques for handling systems of linear equations and for calculating eigenvalues and eigenvectors.

8-1   PIVOTING AND SCALING TECHNIQUES

In this course, we have used the method of Gauss-Jordan elimination to determine the solution to a system of linear equations, and we have seen how practical problems can be solved using these tools. The examples considered involved systems of linear equations with uncomplicated coefficients. The aim of the examples was to enable the reader to understand various techniques without being distracted by arithmetic. However, in practice, the coefficients often arise from measurements taken and can involve numbers of very different magnitudes. Having numbers of very different magnitudes in a system can lead to large errors. Furthermore, when calculations are performed, often on a computer, only a finite number of digits can be carried—rounding off numbers leads to errors. In this section, we illustrate errors that can arise in these ways. In many cases these errors cannot be eliminated entirely, but they can be minimized by two techniques that we shall introduce: *pivoting* and *scaling*.

The method of Gauss-Jordan elimination involves using the first equation to eliminate the first variable from other equations, and so on. In general, at the rth stage, the rth equation is used to eliminate the rth variable from all other equations. However, we are not restricted to using the rth equation to eliminate the rth variable from all other equations. Any later equation that contains the rth variable can be used to accomplish this. The coefficient of the rth variable in that equation can be taken as a *pivot*—we now introduce freedom in selection of pivots. The choice of pivots is very important in obtaining accurate numerical solutions. The following example illustrates how crucial the choice of pivots can be.

✦ *Example 1*

Solve the system

$$10^{-3}x_1 - x_2 = 1$$
$$2x_1 + x_2 = 0$$

assuming that you are working to three significant figures.

Let us solve this system in two ways, using two distinct selections of pivots. The augmented matrix is

$$\begin{pmatrix} 10^{-3} & -1 & 1 \\ 2 & 1 & 0 \end{pmatrix}$$

Approach 1: Let us select the (1, 1) element as pivot:

$$\begin{pmatrix} \boxed{10^{-3}} & -1 & 1 \\ 2 & 1 & 0 \end{pmatrix}$$

$$\underset{(\frac{1}{10^{-3}})R1}{\approx} \begin{pmatrix} \text{①} & -1000 & 1000 \\ 2 & 1 & 0 \end{pmatrix} \underset{R2+(-2)R1}{\approx} \begin{pmatrix} 1 & -1000 & 1000 \\ 0 & \boxed{2000} & -2000 \end{pmatrix}$$

$\uparrow$
This element should have been 2001, *but has been rounded to three significant figures.*

$$\underset{(\frac{1}{2000})R3}{\approx} \begin{pmatrix} 1 & -1000 & 1000 \\ 0 & \text{①} & -1 \end{pmatrix} \underset{R1+(1000)R2}{\approx} \begin{pmatrix} 1 & 0 & 0 \\ 0 & 1 & -1 \end{pmatrix}$$

The solution is $x_1 = 0$, $x_2 = -1$.

Approach 2: Let us now solve the system by initially pivoting on the (2, 1) element.

$$\begin{pmatrix} 10^{-3} & -1 & 1 \\ \text{②} & 1 & 0 \end{pmatrix} \underset{R1 \leftrightarrow R2}{\approx} \begin{pmatrix} \text{②} & 1 & 0 \\ 10^{-3} & -1 & 1 \end{pmatrix}$$

(Note that once the pivot has been selected, we interchange rows, if necessary, to bring it into the customary location for creating zeros in that particular column.)

$$\underset{(\frac{1}{2})R1}{\approx} \begin{pmatrix} \text{①} & \frac{1}{2} & 0 \\ 10^{-3} & -1 & 1 \end{pmatrix} \underset{R2+(-10^{-3})R1}{\approx} \begin{pmatrix} 1 & \frac{1}{2} & 0 \\ 0 & \boxed{-1} & 1 \end{pmatrix}$$

$$\underset{(-1)R2}{\approx} \begin{pmatrix} 1 & \frac{1}{2} & 0 \\ 0 & \text{①} & -1 \end{pmatrix} \underset{R1+(-\frac{1}{2})R2}{\approx} \begin{pmatrix} 1 & 0 & \frac{1}{2} \\ 0 & 1 & -1 \end{pmatrix}$$

The solution obtained by pivoting in this manner is $x_1 = \frac{1}{2}$, $x_2 = -1$.

The exact solution to the system is $x_1 = 1000/2001$, $x_2 = -2000/2001$. The second solution, obtained by pivoting on the (2, 1) element, is a reasonably good approximation; the first solution, obtained by pivoting on the (1, 1) element, is not. ✦

Having seen that the choice of pivots is important, let us now use the above example to decide how pivots can be selected to improve accuracy. Why was the selection of the (2, 1) element as pivot a better choice than the selection of the (1, 1) element?

Let us return to look at the matrix prior to the reduced echelon form in the first approach. The matrix is

$$\begin{pmatrix} 1 & -1000 & 1000 \\ 0 & 1 & -1 \end{pmatrix}$$

The corresponding system of equations is

$$x_1 - 1000x_2 = 1000$$
$$x_2 = -1$$

This first equation becomes

$$x_1 - (1000)(-1) = 1000$$

Since we are working to three significant figures, observe that $x_1 = 4$ satisfies this equation. $x_1 = -3$ also does; in fact, there are many possible values of x_1. When we work to three significant figures, the x_1 term is overpowered by the $(1000)(-1)$ term. The introduction of the large numbers -1000 and 1000 during the first transformation, involving pivoting on the (1, 1) element, led to this inaccuracy in x_1. Observe that in the second approach, pivoting on the (2, 1) element, no such large numbers were introduced. The solution obtained was therefore more accurate.

This discussion illustrates the fact that the introduction of large numbers during the elimination process can cause round-off errors. The large numbers in the first approach resulted from the choice of a small number, 10^{-3}, as pivot; dividing by a small number introduces large numbers. Thus the rule of thumb in selection of pivots is to choose the largest number available as pivot.

Furthermore, the above example illustrates that errors can be severe if there are large differences in magnitude between the numbers involved. The final inaccuracy in the x_1 above resulted from the large difference in magnitude between the coefficient of x_1, namely 1, and the coefficient of x_2, namely 1000. An initial procedure known as *scaling* can often be used to make the elements of the augmented matrix of comparable magnitude, so that one starts off on the best possible footing.

We now give the procedures that are usually used to minimize these causes of error. It should be stressed that these procedures work most of the time and are the ones most widely adopted. There are, however, exceptions— situations where these procedures do not work. We shall not discuss them here.

Pivoting and Scaling Procedures

The scaling procedure is an attempt to make the elements of the augmented matrix as uniform as possible in magnitude. There are two operations used in scaling a system of equations:

1. Multiplying an equation throughout by a nonzero constant.

2. Replacing any variable by a new variable that is a multiple of the original one.

In practice, one attempts to scale the system so that the largest element in each row and column of the augmented matrix is of order unity. There is no automatic method for scaling.

Pivots are then selected, eliminating variables in the natural order x_1, $x_2, \ldots$. At the rth stage, the pivot is generally taken to be the coefficient of x_r, in the rth or later equation, that has the largest absolute value. In terms of elementary matrix transformations, we scan the relevant column from the rth position down for the number that has the largest absolute value. Rows are then interchanged if necessary to bring the pivot to the right location. This procedure ensures that the largest number available is selected as pivot.

We now give an example to illustrate these techniques.

✦ Example 2

Solve the system of equations

$$0.002x_1 + 4x_2 - 2x_3 = 1$$
$$0.001x_1 + 2.0001x_2 + x_3 = 2$$
$$0.001x_1 + 3x_2 + 3x_3 = -1$$

Assume that calculations can be carried out to five significant figures—rounded off after five.

First of all, we scale the system by introducing the new variable $y_1 = 0.001x_1$. All the coefficients in the first column will then be of a magnitude similar to that of the other coefficients. For consistency of notation, let $y_2 = x_2$ and $y_3 = x_3$. We now have the properly scaled system

$$2y_1 + 4y_2 - 2y_3 = 1$$
$$y_1 + 2.0001y_2 + y_3 = 2$$
$$y_1 + 3y_2 + 3y_3 = -1$$

Proceed using the pivoting sequence of transformations:

$$\begin{pmatrix} ② & 4 & -2 & 1 \\ 1 & 2.0001 & 1 & 2 \\ 1 & 3 & 3 & -1 \end{pmatrix} \underset{(\frac{1}{2})R1}{\approx} \begin{pmatrix} ① & 2 & -1 & 0.5 \\ 1 & 2.0001 & 1 & 2 \\ 1 & 3 & 3 & -1 \end{pmatrix}$$

$$\underset{\substack{R2+(-1)R1 \\ R3+(-1)R1}}{\approx} \begin{pmatrix} 1 & 2 & -1 & 0.5 \\ 0 & \boxed{0.0001} & 2 & 1.5 \\ 0 & 1 & 4 & -1.5 \end{pmatrix} \underset{R2\leftrightarrow R3}{\approx} \begin{pmatrix} 1 & 2 & -1 & 0.5 \\ 0 & ① & 4 & -1.5 \\ 0 & 0.0001 & 2 & 1.5 \end{pmatrix}$$

Scan these elements to determine the one with the largest absolute value. It is 1. This element will be the pivot. Interchange rows 2 and 3 to bring the element to the desired location.

$$\underset{\substack{R1+(-2)R2 \\ R3+(-.0001)R2}}{\approx} \begin{pmatrix} 1 & 0 & -9 & 3.5 \\ 0 & 1 & 4 & -1.5 \\ 0 & 0 & 1.9996 & 1.5002 \end{pmatrix} \underset{(\frac{1}{1.9996})R3}{\approx} \begin{pmatrix} 1 & 0 & -9 & 3.5 \\ 0 & 1 & 4 & -1.5 \\ 0 & 0 & \textcircled{1} & 0.75025 \end{pmatrix}$$

This element should have been 1.50015, *but has been rounded off to five significant figures.*

$$\underset{\substack{R1+9R3 \\ R2+(-4)R3}}{\approx} \begin{pmatrix} 1 & 0 & 0 & 10.252 \\ 0 & 1 & 0 & -4.501 \\ 0 & 0 & 1 & 0.75025 \end{pmatrix}$$

The solution is $y_1 = 10.252$, $y_2 = -4.501$, $y_3 = 0.75025$.

In terms of the original variable x, the solution is

$$x_1 = 10,252, \qquad x_2 = -4.501, \qquad x_3 = 0.75025$$

It is interesting to compare this solution with the solution that would have been obtained through the Gauss-Jordan method without the pivoting and scaling refinements. The solution to five significant figures would have been $x_1 = 1100$, $x_2 = -5$, $x_3 = .75025$. Substituting into the original equations, we see that this solution is very unsatisfactory, whereas the solution obtained using the refinements is very accurate. ✦

EXERCISES

Solve the following systems of equations (if possible), using the pivoting procedure and the scaling technique when appropriate.

*1.
$$\begin{aligned} -\ x_2 + 2x_3 &= 4 \\ x_1 + 2x_2 -\ x_3 &= 1 \\ x_1 + 2x_2 + 2x_3 &= 4 \end{aligned}$$

2.
$$\begin{aligned} -x_1 +\ x_2 + 2x_3 &= 8 \qquad \text{(to 3 decimal places)} \\ 2x_1 + 4x_2 -\ x_3 &= 10 \\ -x_1 + 2x_2 + 2x_3 &= 2 \end{aligned}$$

*3.
$$\begin{aligned} -x_1 + 0.002x_2 &= 0 \\ x_1 \qquad\quad + 2x_3 &= -2 \\ 0.001x_2 +\ x_3 &= 1 \end{aligned}$$

4.
$$\begin{aligned} 2x_1 \qquad\qquad +\qquad x_3 &= 1 \\ 0.0001x_1 + 0.0002x_2 + 0.0004x_3 &= 0.0004 \\ x_1 -\qquad 2x_2 -\qquad 3x_3 &= -3 \end{aligned}$$

*5.
$$\begin{aligned} x_2 - 0.1x_3 &= 2 \qquad \text{(to 3 decimal places)} \\ 0.02x_1 + 0.01x_2 \qquad\quad &= 0.01 \\ x_1 -\quad 4x_2 + 0.1x_3 &= 2 \end{aligned}$$

6.
$$\begin{aligned} -\ 0.4x_2 + 0.002x_3 &= 0.2 \\ 100x_1 +\quad x_2 -\ 0.01x_3 &= 1 \\ 200x_1 +\quad 2x_2 -\ 0.02x_3 &= 1 \end{aligned}$$

*7. $\begin{aligned} -x_1 \qquad\qquad + 2x_3 &= 1 \\ 0.1x_1 + 0.001x_2 \qquad\quad &= 0.1 \\ 0.1x_1 + 0.002x_2 \qquad\quad &= 0.2 \end{aligned}$

8. $\begin{aligned} -x_2 + \quad 0.001x_3 &= 6 \\ 0.1x_1 \qquad\quad + \; 0.0002x_3 &= -0.2 \\ 0.01x_1 + 0.01x_2 + 0.00003x_3 &= 0.02 \end{aligned}$

*9. $\begin{aligned} x_1 - \qquad 2x_2 - \qquad x_3 &= 1 \\ x_1 - 0.001x_2 - \qquad x_3 &= 2 \\ x_1 + \qquad 3x_2 - 0.002x_3 &= -1 \end{aligned}$ (to 3 decimal places)

10. Write a computer program for determining a reduced echelon form that incorporates the pivoting technique. Check your answers to Exercises 1–6 using your program.

11. Modify your program in Exercise 10 to get a printout after each transformation, with a statement describing the transformation.

8-2   GAUSSIAN ELIMINATION

We have introduced and used the method of Gauss-Jordan elimination to solve systems of linear equations. We now discuss a second method of solving systems of linear equations, called Gaussian elimination, and then compare the merits of the two methods. The method of Gaussian elimination involves the echelon form of the matrix of coefficients. The echelon form was introduced in Section 4-7; Definition 4-12 is repeated here as a reminder:

Definition 4-12   A matrix is in *echelon form* if

1. Any rows consisting entirely of zeros are grouped at the bottom of the matrix.
2. The first nonzero element of each row is 1. This element is called a *leading* 1.
3. The leading 1 of each row after the first is positioned to the right of the leading 1 of the previous row.

The following matrices are all in echelon form.

$$\begin{pmatrix} 1 & -1 & 2 \\ 0 & 1 & 2 \\ 0 & 0 & 1 \end{pmatrix}, \qquad \begin{pmatrix} 1 & 3 & -1 & 4 \\ 0 & 0 & 1 & 3 \\ 0 & 0 & 0 & 0 \end{pmatrix}, \qquad \begin{pmatrix} 1 & 4 & 6 & 2 & 5 & 2 \\ 0 & 0 & 1 & 2 & 3 & 4 \\ 0 & 0 & 0 & 0 & 1 & 6 \end{pmatrix}$$

The difference between echelon form and reduced echelon form is that the elements above leading ones need not be zeros in the former, whereas they are zeros in the latter.

The method of *Gaussian elimination* involves reducing the augmented matrix of a system of linear equations to echelon form and then using back substitution to solve. In the process of arriving at the echelon form, zeros are created *only below* the pivots. We illustrate the method with the following example.

✦ *Example 1*

Let us solve the following system using Gaussian elimination:

$$x_1 + 2x_2 + 3x_3 + 2x_4 = -1$$
$$-x_1 - 2x_2 - 2x_3 + x_4 = 2$$
$$2x_1 + 4x_2 + 8x_3 + 12x_4 = 4$$

Starting with the augmented matrix, we get

$$\begin{pmatrix} ① & 2 & 3 & 2 & -1 \\ -1 & -2 & -2 & 1 & 2 \\ 2 & 4 & 8 & 12 & 4 \end{pmatrix} \underset{\substack{R2+R1 \\ R3+(-2)R1}}{\approx} \begin{pmatrix} 1 & 2 & 3 & 2 & -1 \\ 0 & 0 & ① & 3 & 1 \\ 0 & 0 & 2 & 8 & 6 \end{pmatrix}$$

At this stage we create zeros *only below the pivot* to get

$$\underset{R3+(-2)R2}{\approx} \begin{pmatrix} 1 & 2 & 3 & 2 & -1 \\ 0 & 0 & 1 & 3 & 1 \\ 0 & 0 & 0 & ② & 4 \end{pmatrix} \underset{(\frac{1}{2})R3}{\approx} \begin{pmatrix} 1 & 2 & 3 & 2 & -1 \\ 0 & 0 & 1 & 3 & 1 \\ 0 & 0 & 0 & ① & 2 \end{pmatrix}$$

This matrix is in echelon form.

The corresponding system of equations is

$$x_1 + 2x_2 + 3x_3 + 2x_4 = -1$$
$$x_3 + 3x_4 = 1$$
$$x_4 = 2$$

This system is now solved by *back substitution.* The value of x_4 is substituted into the second equation to give x_3. x_3 and x_4 are then substituted into the first equation to give x_1.

Substituting for $x_4 = 2$ in the second equation, we get

$$x_3 + 6 = 1$$
$$x_3 = -5$$

Substituting for $x_3 = -5$ and $x_4 = 2$ in the first equation gives

$$x_1 + 2x_2 - 15 + 4 = -1$$
$$x_1 + 2x_2 = 10$$

Expressing the leading variable x_1 in terms of x_2 gives $x_1 = -2x_2 + 10$. The system has many solutions, $x_1 = -2x_2 + 10$, $x_3 = -5$, $x_4 = 2$, with x_2 taking on any value. Letting $x_2 = r$, we can express the many solutions as

$$x_1 = -2r + 10, \qquad x_2 = r, \qquad x_3 = -5, \qquad x_4 = 2 \qquad ✦$$

The back substitution involved in the method of Gaussian elimination can be performed in terms of matrices. The final matrix then becomes the reduced echelon form of the augmented matrix. We now illustrate this complete matrix method for the previous example. The first four steps leading to the echelon form are repeated from the previous example.

✦ *Example 2*

Solve the system

$$x_1 + 2x_2 + 3x_3 + 2x_4 = -1$$
$$-x_1 - 2x_2 - 2x_3 + x_4 = 2$$
$$2x_1 + 4x_2 + 8x_3 + 12x_4 = 4$$

Starting with the augmented matrix, we have

$$\begin{pmatrix} ① & 2 & 3 & 2 & -1 \\ -1 & 2 & -2 & 1 & 2 \\ 2 & 4 & 8 & 12 & 4 \end{pmatrix} \underset{\substack{R2+R1 \\ R3+(-2)R1}}{\approx} \begin{pmatrix} 1 & 2 & 3 & 2 & -1 \\ 0 & 0 & ① & 3 & 1 \\ 0 & 0 & 2 & 8 & 6 \end{pmatrix}$$

$$\underset{R3+(-2)R2}{\approx} \begin{pmatrix} 1 & 2 & 3 & 2 & -1 \\ 0 & 0 & 1 & 3 & 1 \\ 0 & 0 & 0 & ② & 4 \end{pmatrix}$$

$$\underset{(\frac{1}{2})R3}{\approx} \begin{pmatrix} 1 & 2 & 3 & 2 & -1 \\ 0 & 0 & 1 & 3 & 1 \\ 0 & 0 & 0 & ① & 2 \end{pmatrix}$$

We have arrived at an echelon form. This marks the end of the forward elimination of variables from equations. We now commence back substitution in matrix manner.

$$\underset{\substack{R1+(-2)R3 \\ R2+(-3)R3}}{\approx} \begin{pmatrix} 1 & 2 & 3 & 0 & -5 \\ 0 & 0 & 1 & 0 & -5 \\ 0 & 0 & 0 & 1 & 2 \end{pmatrix}$$
↑

Zeros have been created above the leading 1 in row 3. This is equivalent to substituting for x_4 from the third equation in the first two equations.

$$\underset{R1+(-3)R2}{\approx} \begin{pmatrix} 1 & 2 & 0 & 0 & 10 \\ 0 & 0 & 1 & 0 & -5 \\ 0 & 0 & 0 & 1 & 2 \end{pmatrix}$$
↑

A zero has been created above the leading 1 in row 2. This is equivalent to substituting for x_3 from the second equation in the first equation.

This matrix is the reduced echelon form of the augmented matrix. The corresponding system of linear equations is

$$x_1 + 2x_2 = 10$$
$$x_3 = -5$$
$$x_4 = 2$$

Letting $x_2 = r$, we get the same set of solutions as previously:

$$x_1 = -2r + 10, \qquad x_2 = r, \qquad x_3 = -5, \qquad x_4 = 2 \quad ✦$$

Gaussian elimination is more efficient than Gauss-Jordan elimination in that it involves fewer operations of addition and multiplication. It is during the back substitution that Gaussian elimination picks up this advantage. We illustrate the saving of two operations in the above example of Gaussian elimination. Consider the final transformation that brings the matrix to reduced echelon form,

$$\begin{pmatrix} 1 & 2 & 3 & 0 & \boxed{-5} \\ 0 & 0 & 1 & 0 & -5 \\ 0 & 0 & 0 & 1 & 2 \end{pmatrix} \underset{R1 + (-3)R2}{\approx} \begin{pmatrix} 1 & 2 & 0 & 0 & \boxed{10} \\ 0 & 0 & 1 & 0 & -5 \\ 0 & 0 & 0 & 1 & 2 \end{pmatrix}$$

The aim of this transformation is to create a 0 in location $(1, 3)$. The transformation $R1 + (-3)R2$ involves only two arithmetic operations, one of multiplication and one of addition, in changing the circled -5 to 10:

$$-5 + (-3)(-5) = 10$$
$$\quad \uparrow \qquad \uparrow$$

operation   operation
of   of
addition   multiplicaton

Changing the 3 in location $(1, 3)$ to a 0 does not involve any arithmetic operations, as one (or the computer) knows in advance that the element is to be a 0! The 0 in the $(1, 4)$ location remains unchanged; no arithmetic operations are performed on it. It is in this $(1, 4)$ location that Gaussian elimination saves two operations over Gauss-Jordan elimination. No operations take place on the $(1, 4)$ element during the creation of the 0 in the $(1, 3)$ location in Gaussian elimination. However, in Gauss-Jordan elimination, operations have to be performed on the $(1, 4)$ location when the 0 is created in the $(1, 3)$ location. In larger systems, many more operations are saved in Gaussian elimination during the back substitution process. The reduction in the number of operations not only saves time on the computer but also increases the accuracy of the final answer. With each arithmetic operation there is a possibility of truncation error on the computer. Decimal numbers have to be converted to binary form for the computer; very often the conversion involves truncation. With very large systems, the method of Gauss-Jordan elimination involves approximately 50% more arithmetic operations than does Gaussian elimination.

EXERCISES

1. Determine whether or not each of the following matrices is in echelon form.

$$*(a) \begin{pmatrix} 1 & 2 & 1 \\ 0 & 1 & 3 \\ 0 & 0 & 0 \end{pmatrix} \qquad (b) \begin{pmatrix} 1 & 2 & 3 & 4 \\ 0 & 0 & 1 & 0 \\ 0 & 0 & 0 & 1 \end{pmatrix} \qquad *(c) \begin{pmatrix} 1 & 0 & 0 & 3 & 0 \\ 0 & 0 & 1 & 2 & 0 \\ 0 & 0 & 0 & 0 & 1 \end{pmatrix}$$

(d) $\begin{pmatrix} 1 & 2 & 4 & -1 & 2 & 3 \\ 0 & 0 & 1 & 2 & 4 & 6 \\ 0 & 0 & 0 & 1 & 3 & 2 \\ 0 & 0 & 0 & 0 & 1 & 4 \end{pmatrix}$ *(e) $\begin{pmatrix} 1 & 2 & 4 & 6 \\ 0 & 0 & 1 & 2 \\ 0 & 1 & 3 & 3 \\ 0 & 0 & 0 & 1 \end{pmatrix}$

(f) $\begin{pmatrix} 1 & 0 & 0 & 4 & 0 & 2 \\ 0 & 1 & 3 & 2 & 4 & 6 \\ 0 & 0 & 1 & -1 & 3 & 0 \\ 0 & 0 & 0 & 0 & 1 & 2 \end{pmatrix}$ *(g) $\begin{pmatrix} 1 & 3 & 4 & 2 & 3 \\ 0 & 0 & 2 & -1 & 1 \\ 0 & 0 & 0 & 1 & 4 \\ 0 & 0 & 0 & 0 & 6 \end{pmatrix}$

2. Give examples of four 4×5 matrices that are in echelon form but not in reduced echelon form.

In Exercises 3–7, solve each of the systems using Gaussian elimination. (a) Perform the back substitution using equations. (b) Perform the back substitution using matrices.

*3. $\begin{aligned} x_1 + x_2 + x_3 &= 6 \\ x_1 - x_2 + x_3 &= 2 \\ x_1 + 2x_2 + 3x_3 &= 14 \end{aligned}$

4. $\begin{aligned} x_1 - x_2 - x_3 &= 2 \\ x_1 - x_2 + x_3 &= 2 \\ 3x_1 - 2x_2 + x_3 &= 5 \end{aligned}$

*5. $\begin{aligned} x_1 - 2x_2 + 3x_3 &= 1 \\ 3x_1 - 4x_2 + 5x_3 &= 3 \\ 2x_1 - 3x_2 + 4x_3 &= 2 \end{aligned}$

6. $\begin{aligned} x_1 - x_2 + 2x_3 &= 3 \\ 2x_1 - 2x_2 + 5x_3 &= 4 \\ x_1 + 2x_2 - x_3 &= -3 \\ 2x_2 + 2x_3 &= 1 \end{aligned}$

*7. $\begin{aligned} x_1 - x_2 + x_3 + 2x_4 - 2x_5 &= 1 \\ 2x_1 - x_2 - x_3 + 3x_4 - x_5 &= 3 \\ -x_1 - x_2 + 5x_3 \qquad - 4x_5 &= -3 \end{aligned}$

*8. Consider a general system of four equations in five variables. In general, how many arithmetic operations will be saved by using the method of Gaussian elimination rather than Gauss-Jordan elimination to arrive at the reduced echelon form? Determine exactly where operations are saved.

9. Compare Gaussian and Gauss-Jordan elimination for a system of four equations in six variables. Determine where operations are saved during Gaussian elimination.

10. Can a matrix have more than one echelon form? (*Hint*: Consider a 2×3 matrix and arrive at an echelon form using two distinct sequences of transformations.)

11. Suggest a pivoting technique for Gaussian elimination that will reduce round-off errors, based on the pivoting technique introduced in the last section for Gauss-Jordan elimination. Solve the systems in Exercises 3–7 using the pivoting technique.

8-3 ITERATIVE METHODS FOR SOLVING SYSTEMS OF LINEAR EQUATIONS

We have discussed both the Gauss-Jordan and the Gaussian elimination methods for solving systems of linear equations. We now introduce two iterative methods, the Jacobi method and the Gauss-Seidel method. At the close of this section, we shall compare the merits of the Gaussian elimination method (the more efficient of the two elimination methods) and the iterative methods.

Jacobi Method

We introduce the *Jacobi method* by means of an example.

✦ *Example 1*

Solve the system of equations

$$6x + 2y - z = 4$$
$$x + 5y + z = 3 \qquad\qquad (1)$$
$$2x + y + 4z = 27$$

Rewriting the system in the form

$$x = \frac{4 - 2y + z}{6}$$

$$y = \frac{3 - x - z}{5} \qquad\qquad (2)$$

$$z = \frac{27 - 2x - y}{4}$$

we have isolated x in the first equation, y in the second, and z in the third.

Now make a guess at the solution, say $x = 1$, $y = 1$, $z = 1$. The accuracy of the guess affects only the speed with which we get a good approximation. Let us call these values $x^{(0)} = 1$, $y^{(0)} = 1$, $z^{(0)} = 1$. Substitute these values into the right-hand side of system (2) to get values of x, y, and z that we denote $x^{(1)}$, $y^{(1)}$, and $z^{(1)}$:

$$x^{(1)} = \tfrac{1}{2}, \qquad y^{(1)} = \tfrac{1}{5}, \qquad z^{(1)} = 6$$

These values are then substituted again into system (2) to get

$$x^{(2)} = 1.6, \qquad y^{(2)} = -0.7, \qquad z^{(2)} = 6.45$$

The process is repeated to get $x^{(3)}$, $y^{(3)}$, and $z^{(3)}$, and so on. Repeating the iteration will give us a better approximation of the exact solution each time. For this simple system, the solution is easily seen to be $x = 2$, $y = -1$, $z = 6$, so that after the second iteration one has quite a long way to go. ✦

The following theorem gives one set of conditions under which the Jacobi iterative method can be used.

Theorem 8-1 *Consider a system of n linear equations in n variables. If for each row of the matrix of coefficients, the absolute value of the diagonal element is greater than the sum of the absolute values of the other elements in that row, then the system has a unique solution. The Jacobi iterative method will converge to this unique solution, no matter what values are selected for the initial guess.*

Thus the Jacobi method can be used when the diagonal elements *dominate* the rows.

Table 8-1

Row	Absolute Value of Diagonal Element	Sum of the Absolute Values of the Other Elements in That Row
1	6	$\lvert2\rvert + \lvert-1\rvert = 3$
2	5	$\lvert1\rvert + \lvert1\rvert = 2$
3	4	$\lvert2\rvert + \lvert1\rvert = 3$

Let us return to our previous example to see that this condition is indeed satisfied for that system. The matrix of coefficients of that system is

$$\begin{pmatrix} 6 & 2 & -1 \\ 1 & 5 & 1 \\ 2 & 1 & 4 \end{pmatrix}$$

Looking at the elements of each row in turn, we get Table 8-1. Observe that the diagonal elements dominate the rows; the Jacobi method can be used.

In some systems it may be necessary to rearrange the equations before the above condition is satisfied and this method becomes applicable.

The better the initial guess, the sooner one gets a result to within a required degree of accuracy. Note that this method has an advantage: if an error is made at any stage, you merely make a new initial guess at that stage.

There are many theorems similar to the one above that guarantee convergence under varying conditions. Investigations have made available various swiftly convergent methods for various special systems of equations.

Gauss-Seidel Method

The Gauss-Seidel method, a refinement of the above, usually leads to a more rapid convergence. The latest value of each variable is substituted at each stage.[†] This method also works if the diagonal elements dominate the rows. We illustrate this method for the previous system of equations.

As before, let $x^{(0)} = 1$, $y^{(0)} = 1$, $z^{(0)} = 1$ be the initial guess. Substituting the latest value of each variable into (2) every time, we get

$$x^{(1)} = \frac{4 - 2y^{(0)} + z^{(0)}}{6} = \frac{1}{2} = 0.5$$

$$y^{(1)} = \frac{3 - x^{(1)} - z^{(0)}}{5} = \frac{3}{10} = 0.3$$

$$z^{(1)} = \frac{27 - 2x^{(1)} - y^{(1)}}{4} = \frac{25.7}{4} = 6.4250$$

[†] There may be times when this refinement does not lead to more rapid convergence, but in general it is a good strategy to follow. For a discussion see, for example, *Numerical Analysis*, by R. L. Burden, J. D. Faires, and A. C. Reynolds, Prindle, Weber and Schmidt, 1978, page 393.

Thus after one iteration, $x^{(1)} = 0.5$, $y^{(1)} = 0.3$, $z^{(1)} = 6.4250$. Notice that we have used $x^{(1)}$, the most up-to-date value of x, to get $y^{(1)}$, and we have used $x^{(1)}$ and $y^{(1)}$ to get $z^{(1)}$.

Continuing, we have

$$x^{(2)} = \frac{4 - 2y^{(1)} + z^{(1)}}{6} = 1.6375$$

$$y^{(2)} = \frac{3 - x^{(2)} - z^{(1)}}{5} = -1.0125$$

$$z^{(2)} = \frac{27 - 2x^{(2)} - y^{(2)}}{4} = 6.1844$$

Note how, after only two iterations, this set is much closer than the previous set $x^{(2)} = 1.6$, $y^{(2)} = -0.7$, $z^{(2)} = 6.45$ to the exact solution of $x = 2$, $y = -1$, $z = 6$. Tables 8-2 and 8-3 give the results obtained for this particular system by both methods. They illustrate the Gauss-Seidel method's more rapid convergence to the exact solution.

Table 8-2 *Jacobi Method*

Iteration	x	y	z
Initial Guess	1	1	1
1	0.5	0.2	6
2	1.6	−0.7	6.45
3	1.975	−1.01	6.125
4	2.024167	−1.02	6.015
5	2.009167	−1.007833	5.992917
6	2.001431	−1.000417	5.997375

Table 8-3 *Gauss-Seidel Method*

Iteration	x	y	z
Initial Guess	1	1	1
1	0.5	0.3	6.425
2	1.6375	−1.0125	6.184375
3	2.034896	−1.043854	5.993516
4	2.013537	−1.001411	5.993584
5	1.999401	−0.998597	5.999949
6	1.999524	−0.9998945	6.000212

Table 8-4 gives the differences between the solutions obtained in the two methods and the actual solution after six iterations. The Gauss-Seidel method converges much more rapidly.

Table 8-4

	x	y	z
Jacobi Method	0.001431	0.000417	0.002625
Gauss-Seidel Method	0.000476	0.0001055	0.000212

Let us now compare the Gaussian elimination method with the Gauss-Seidel iterative method.

The Gaussian elimination method is finite and leads to a solution for any system of linear equations. The Gauss-Seidel method converges only for special systems of equations; thus it can be used only for such systems.

A second factor of comparison must be the efficiency of the two methods, a function of the number of arithmetic operations (addition, subtraction, multiplication, and division) involved in each method. For a system of n equations in n variables where the solution is unique, Gaussian elimination involves $(4n^3 + 9n^2 - 7n)/6$ arithmetic operations. The Gauss-Seidel method requires $2n^2 - n$ arithmetic operations per iteration. For large values of n, the two methods require, respectively, approximately $2n^3/3$ and $2n^2$ arithmetic operations per iteration. Therefore, if the number of iterations is less than or equal to $n/3$, the iterative method requires fewer arithmetic operations. As a specific example, consider a system of 300 equations in 300 variables. Elimination requires 18,000,000 operations, whereas iteration requires 180,000 operations per iteration. For 100 or fewer iterations the Gauss-Seidel method involves less arithmetic; it is more efficient. It should be stated that the Gaussian elimination method involves movement of data; for example, several rows may need to be interchanged. This is time-consuming and costly on computers. Iterative processes suffer much less from this factor. Thus, even if the number of iterations is more than $n/3$, iteration may require less computer time.

A final factor in the comparison of the two methods is the accuracy of the methods. Round-off errors are minimized in the Gaussian elimination method by using the pivoting technique. However they can still be sizeable. The errors in the Gauss-Seidel method, on the other hand, are only the round-off errors committed in the final iteration—the result of the next-to-last iteration can be interpreted as a very good initial guess! Thus, in general, when the Gauss-Seidel method is applicable, it is more accurate than the Gaussian elimination method. This fact often justifies the use of the Gauss-Seidel method over the Gaussian elimination method even when the total amount of computation time involved is greater.

For more in-depth discussions of iterative methods the reader is referred to *Numerical Methods with Fortran IV Case Studies*, by William S. Dorn and Daniel D. McCracken, John Wiley & Sons, Chap. 4. Interesting and important surveys of this field are to be found in "Solving Linear Equations Can Be Interesting," by George R. Forsythe, *Bulletin of the American Mathematical Society*, Vol. 59, 1953, pp. 299–329, and "On the Solution of Linear Systems by Iteration," by David Young, *Proceedings of Symposia in Applied Mathematics*, Vol. 6, McGraw-Hill Book Company, 1956.

EXERCISES

Determine approximate solutions to the following systems using the Gauss-Seidel iterative method. Work to two decimal places if you are performing the computations by hand. Use the given initial value. [Initial values are given solely so that answers may be checked.]

*1. $4x + y - z = 8$
$\quad 5y + 2z = 6$
$\quad x - y + 4z = 10$
Let $x^{(0)} = 1$, $y^{(0)} = 2$, $z^{(0)} = 3$.

2. $4x - y \quad\quad = 6$
$\quad 2x + 4y - z = 4$
$\quad x - y + 5z = 10$
Let $x^{(0)} = 1$, $y^{(0)} = 2$, $z^{(0)} = 3$.

*3. $5x - y + z = 20$
$\quad 2x + 4y \quad\quad = 30$
$\quad x - y + 4z = 10$
Let $x^{(0)} = 0$, $y^{(0)} = 0$, $z^{(0)} = 0$.

4. $6x + 2y - z = 30$
$\quad -x + 8y + 2z = 20$
$\quad 2x - y + 10z = 40$
Let $x^{(0)} = 5$, $y^{(0)} = 6$, $z^{(0)} = 7$.

*5. $5x - y + 2z = 40$
$\quad 2x + 4y - z = 10$
$\quad -2x + 2y + 10z = 8$
Let $x^{(0)} = 20$, $y^{(0)} = 30$, $z^{(0)} = -40$.

6. $6x - y + z + w = 20$
$\quad x + 8y + 2z \quad\quad = 30$
$\quad -x + y + 6z + 2w = 40$
$\quad 2x \quad - 3z + 10w = 10$
Let $x^{(0)} = y^{(0)} = z^{(0)} = w^{(0)} = 0$.

7. Apply the Gauss-Seidel method to the following system to observe that the values of $x^{(n)}$, $y^{(n)}$, $z^{(n)}$ do not converge. The method does not lead to the solution here.

$5x + 4y - z = 8$
$\quad x - y + z = 4$
$2x + y + 2z = 1$
Let $x^{(0)} = y^{(0)} = z^{(0)} = 0$.

8. Write a computer program for the Gauss-Seidel method. Check your answers to the exercises above. [*Hint*: Suppose you use 10 iterations. List values of **x** as components of a vector $(x_1, \ldots, x_{10})$. For this vector, use the dimension statement DIM X (10). **x** is a one-dimensional array—you do not need two subscripts to handle it.]

8-4 EIGENVECTORS BY ITERATION; APPLICATIONS IN GEOGRAPHY

Numerical techniques exist for evaluating certain eigenvalues and eigenvectors of various types of matrices. Here we present an iterative method called the *power method*. It can be used for determining the eigenvalue with the largest absolute value (if it exists) and a corresponding eigenvector for certain matrices. Such an eigenvalue is called the *dominant eigenvalue* of the matrix. In many applications one is only interested in the dominant eigenvalue. The reader will see such an application in geography.

✦ *Example 1*

Let A be a square matrix with eigenvalues -5, -2, 1, and 3. Then -5 is the dominant eigenvalue, since $|-5| > |-2|$, $|-5| > |1|$, and $|-5| > |3|$.

Let B be a square matrix with eigenvalues -4, -2, 1, and 4. There is no dominant eigenvalue, since $|-4| = |4|$.   ✦

The power method is based on the following theorem.

Theorem 8-2   *Let A be an $n \times n$ matrix having n linearly independent eigenvectors and a dominant eigenvalue. Let $\mathbf{x}_0$ be an arbitrarily chosen initial vector such that $A\mathbf{x}_0$ exists. The sequence of vectors*

$$\mathbf{x}_1 = A\mathbf{x}_0, x_2 = A\mathbf{x}_1, \mathbf{x}_3 = A\mathbf{x}_2, \ldots, \mathbf{x}_k = A\mathbf{x}_{k-1}, \ldots$$

as k becomes larger, will approach an eigenvector for λ_1, the dominant eigenvalue, if $\mathbf{x}_0$ has a nonzero component in the direction of an eigenvector for λ_1.

Proof: Let the eigenvalues of A be $\lambda_1, \ldots, \lambda_n$ with corresponding linearly independent eigenvectors $\mathbf{z}_1, \ldots, \mathbf{z}_n$. It is assumed that the initial guess $\mathbf{x}_0$ has a nonzero component in the direction of $\mathbf{z}_1$. Thus $\mathbf{x}_0$ can be expressed

$$\mathbf{x}_0 = a_1\mathbf{z}_1 + \cdots + a_n\mathbf{z}_n$$

where $a_1 \neq 0$.

$$\begin{aligned}
\mathbf{x}_k = A\mathbf{x}_{k-1} = A^2\mathbf{x}_{k-2} &= \cdots = A^k\mathbf{x}_0 \\
&= A^k[a_1\mathbf{z}_1 + \cdots + a_n\mathbf{z}_n] \\
&= [a_1A^k\mathbf{z}_1 + \cdots + a_nA^k\mathbf{z}_n] \\
&= [a_1(\lambda_1)^k\mathbf{z}_1 + \cdots + a_n(\lambda_n)^k\mathbf{z}_n] \\
&= (\lambda_1)^k\left[a_1\mathbf{z}_1 + a_2\left(\frac{\lambda_2}{\lambda_1}\right)^k\mathbf{z}_2 + \cdots + a_n\left(\frac{\lambda_n}{\lambda_1}\right)^k\mathbf{z}_n\right]
\end{aligned}$$

As k increases, since $|\lambda_j/\lambda_1| < 1$ for $j \geq 2$, $(\lambda_j/\lambda_1)^k$ will approach 0, and the vector $\mathbf{x}_k$ will approach the direction of the vector $\mathbf{z}_1$, an eigenvector for the dominant eigenvalue λ_1.

We have seen that symmetric $n \times n$ matrices have n linearly independent eigenvectors, satisfying one requirement of this theorem. The method can thus be tried for a symmetric matrix.

It remains to find an approximation to the dominant eigenvalue. Let λ be an eigenvalue of A and $\mathbf{x}$ a corresponding eigenvector. Then

$$\frac{\mathbf{x} \cdot A\mathbf{x}}{\mathbf{x} \cdot \mathbf{x}} = \frac{\mathbf{x} \cdot \lambda\mathbf{x}}{\mathbf{x} \cdot \mathbf{x}} = \frac{\lambda\mathbf{x} \cdot \mathbf{x}}{\mathbf{x} \cdot \mathbf{x}} = \lambda$$

This result is used in practice to find an approximation to the dominant eigenvalue. The expression $\dfrac{\mathbf{x}_i \cdot A\mathbf{x}_i}{\mathbf{x}_i \cdot \mathbf{x}_i}$ is computed concurrently with each approximation $\mathbf{x}_i$ to the eigenvector, resulting in a sequence of scalars that approaches the dominant eigenvalue. The method can be continued until successive approximations to the eigenvalue are within the required accuracy. $\Big($The ratio

$\dfrac{\mathbf{x} \cdot A\mathbf{x}}{\mathbf{x} \cdot \mathbf{x}}$ is called the *Rayleigh quotient.*$\Big)$

There is one modification that is carried out in practice. The components of the vectors $\mathbf{x}_1, \mathbf{x}_2, \ldots$ may, as the method now stands, become very large, causing significant round-off errors to occur. This problem is overcome by dividing each component of $\mathbf{x}_i$ by the absolute value of its largest component and then using this vector (a vector in the same direction as $\mathbf{x}_i$) in the following iteration.

✦ *Example 2*

Consider the matrix $A = \begin{pmatrix} 5 & 4 & 2 \\ 4 & 5 & 2 \\ 2 & 2 & 2 \end{pmatrix}$ of Section 7-1. We already know that a dominant eigenvalue exists and is 10 and that the corresponding eigenvectors are of the form $c \begin{pmatrix} 2 \\ 2 \\ 1 \end{pmatrix}$. Let us illustrate the method for this known result.

Let $\mathbf{x}_0 = \begin{pmatrix} -4 \\ 2 \\ 6 \end{pmatrix}$—an arbitrarily selected vector. We get Table 8.5.

Table 8-5

Iteration	$A\mathbf{x}$	Adjusted Vector	$(\mathbf{x} \cdot A\mathbf{x})/(\mathbf{x} \cdot \mathbf{x})$
1	$A\mathbf{x}_0 = \begin{pmatrix} 0 \\ 6 \\ 8 \end{pmatrix}$	$\mathbf{x}_1 = \begin{pmatrix} 0 \\ 0.75 \\ 1 \end{pmatrix} \left[= \dfrac{1}{8} A\mathbf{x}_0 \right]$	5
2	$A\mathbf{x}_1 = \begin{pmatrix} 5 \\ 5.75 \\ 3.5 \end{pmatrix}$	$\mathbf{x}_2 = \begin{pmatrix} 0.869565 \\ 1 \\ 0.608696 \end{pmatrix} \left[= \dfrac{1}{5.75} A\mathbf{x}_1 \right]$	9.88889
3	$A\mathbf{x}_2 = \begin{pmatrix} 9.56522 \\ 9.69565 \\ 4.95652 \end{pmatrix}$	$\mathbf{x}_3 = \begin{pmatrix} 0.986547 \\ 1 \\ 0.511211 \end{pmatrix}$	9.99888
4	$A\mathbf{x}_3 = \begin{pmatrix} 9.95516 \\ 9.96861 \\ 4.99552 \end{pmatrix}$	$\mathbf{x}_4 = \begin{pmatrix} 0.998651 \\ 1 \\ 0.501125 \end{pmatrix}$	9.99999

Thus, after 4 iterations, an approximation to the dominant eigenvalue is 9.99999 and an approximation to a corresponding eigenvector is $\begin{pmatrix} 0.998651 \\ 1 \\ 0.501125 \end{pmatrix}$.   ✦

This method has an advantage similar to that of the Gauss-Seidel iterative method discussed earlier. Any error in computation only means that a new arbitrary vector has been introduced at that stage. The method is very accurate in that the only round-off errors that occur are those arising from the matrix multiplication carried out during the final iteration and the final computation of $\mathbf{x} \cdot A\mathbf{x}/\mathbf{x} \cdot \mathbf{x}$.

If A is symmetric, a technique called *deflation* can be used for determining further eigenvalues and corresponding eigenvectors. The method is based on the following theorem, which we state without proof.

Theorem 8-3   *Let A be a symmetric matrix with eigenvalues $\lambda_1, \lambda_2, \ldots, \lambda_n$, where $\lambda_1, \lambda_2, \ldots, \lambda_n$ are in order according to absolute values, λ_1 being the largest. Let $\mathbf{x}$ be a unit eigenvector for λ_1, in column form. Then*

(1) *The matrix $B = A - \lambda_1 \mathbf{x}\mathbf{x}^t$ has eigenvalues $0, \lambda_2, \ldots, \lambda_n$. The matrix B is symmetric.*

(2) *If $\mathbf{y}$ is an eigenvector of B corresponding to one of the eigenvalues $\lambda_2, \ldots, \lambda_n$, it is also an eigenvector of A corresponding to the same eigenvalue.*

One determines λ_1 and $\mathbf{x}$ (or good approximations) using the previous technique. B is then found. Note that, according to the theorem, λ_2 will be the dominant eigenvalue of B (if one exists); hence, on applying the iterative method to B, one gets λ_2 and a corresponding eigenvector which is also an eigenvector of λ_2 for A. In this manner, all the eigenvalues of A and a corresponding eigenvector in each case can be found. The disadvantage of the method is that the eigenvalues become increasingly inaccurate through compounding of errors. If at any stage λ_i is not dominant, the method breaks down at that stage.

Readers who are interested in further numerical techniques for determining eigenvalues and eigenvectors will find articles and references in *Methods for Digital Computers*, Vols. 1 and 2, John Wiley & Sons, 1968, and in *Numerical Methods*, by Germund Dahlquist and Åke Björck, Section 5-8, Prentice-Hall, 1974.

Eigenvectors and Eigenvalues in Geography

Geographers use eigenvectors within the context of graph theory to analyze networks. The most common types of networks looked at are roads, airways, sea routes, navigable rivers, and canals. Here we discuss the concepts involved in terms of a road network.

Consider a network of n cities linked by two-way roads. The cities can be represented as vertices of a digraph and the roads as edges. Let us assume that it is possible to travel along roads from any city to another; the digraph is then said to be *connected*. Geographers are interested in the "accessibility" of each city. The *Gould index* is used as a measure of accessibility. This index is obtained from an eigenvector corresponding to the dominant eigenvalue of the $n \times n$ adjacency matrix of the digraph that describes the network.

The network consists of two-way roads. Thus if there is an edge from vertex i of the digraph to vertex j, there will also be an edge from vertex j to vertex i. The adjacency matrix A will therefore be symmetric. As mentioned previously, this means that the matrix has n linearly independent eigenvectors. Furthermore, there is a theorem called the Perron-Frobenius theorem which implies that the adjacency matrix of such a connected digraph does have a dominant eigenvalue and that the corresponding eigenspace is one-dimensional, having a basis vector consisting of all positive components. The conditions of Theorem 8-2 are met. This theorem is used in practice to compute the dominant eigenvalue and a corresponding eigenvector. The components of this eigenvector lead to the accessibilities of the cities corresponding to the vertices of the digraph.

Let us now justify this use of the components of the eigenvector.[†] A highly accessible vertex should have a large number of paths to other vertices. We know that the elements of A^n give the number of paths of length n between vertices. The sum of the elements in row i of A^n would give the number of n-paths out of vertex i. Let **e** be the column vector having n components all of which are 1. Then A^n**e** will be a column vector, the ith component of which is the sum of the elements in row i of A^n. Thus the components of the vector A^n**e** will give the numbers of n-paths out of the various vertices. A satisfies the conditions of Theorem 8-2, and since **e** is made up of ones it has a nonzero component in the direction of the eigenvector of the dominant eigenvalue of A. Thus as n increases, A^n**e** approaches an eigenvector for the dominant eigenvalue of A. As n increases, A^n**e** accounts for longer and longer paths out of each vertex; thus the ratios of the components of the eigenvector, which is the limit of this sequence of vectors, will give the relative accessibilities of the vertices.

We now illustrate the method for the following simple transportation network.

✦ *Example 3*
───

Consider the transportation network in Figure 8-1. Each edge is two-way.
The adjacency matrix is

$$A = \begin{pmatrix} 0 & 1 & 0 & 0 \\ 1 & 0 & 1 & 1 \\ 0 & 1 & 0 & 1 \\ 0 & 1 & 1 & 0 \end{pmatrix}$$

[†] This justification of the accessibility index is adapted from Philip D. Straffin, Jr., "Linear Algebra in Geography: Eigenvectors of Networks," *Mathematics Magazine*, Vol. 53, No. 5, November, 1980, pages 269–276. This excellent article has a discussion of the Perron-Frobenius theorem and references to applications ranging from the growth of the São Paolo economy to an analysis of the U.S. interstate road system.

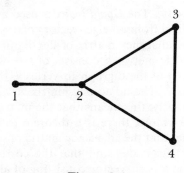

Figure 8-1

The dominant eigenvalue and corresponding eigenvector are found to be $\lambda = 2.17$ and $(0.44, 1, 0.84, 0.84)$, in adjusted form. The components of the eigenvector give the relative accessibilities of the vertices. Thus the second component being the largest implies that vertex 2 is the most accessible in the network. Component one being smallest implies that vertex 1 is the least accessible. Vertices 3 and 4 have equal accessibility. We can see that, for this simple network, these results agree with observation of the network. The convention adopted by geographers is to divide the eigenvector by the sum of its elements, the components of the resulting vector being the *Gould indices of accessibility* of the various vertices. Thus we get

$$\frac{1}{(0.44 + 1 + 0.84 + 0.84)} (0.44, 1, 0.84, 0.84) = (0.14, 0.32, 0.27, 0.27)$$

The indices of the various vertices are shown in Table 8-6.

Table 8-6

Vertex	Gould Index
1	0.14
2	0.32
3	0.27
4	0.27

vertex with →
highest
accessibility (pointing to vertex 2)

✦ *Example 4*

This example illustrates a network that has many vertices. The graph in Figure 8-2 represents the major river trade routes in central Russia in the twelfth and thirteenth centuries. The vertices 1–39 represent various centers of population; Moscow is 35, Kiev 4, Novgorod 1, Smolensk 3, to mention a few.

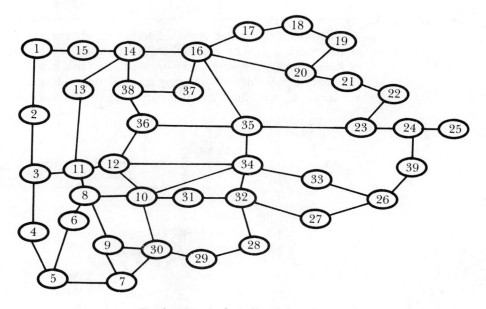

Trade routes of medieval Russia

Figure 8-2

Techniques of linear algebra have been used by Pitts[†] to review claims by geographers that Moscow eventually assumed the dominant position in central Russia because of its strategic location on medieval trade routes.

The accessibility index for each city was found. This involved finding the dominant eigenvalue and corresponding eigenvector for a 39×39 matrix, using the computer. We quote the seven highest indices in Table 8-7, giving the seven most accessible cities. Observe that Moscow ranks sixth. The conclusion arrived at in this analysis was that other sociological and political factors must have been important in Moscow's rise.

Table 8-7

	City		*Gould Accessibility Index*
city with →	Kozelsk	(10)	0.0837
highest	Kolomna	(34)	0.0788
accessibility	Vyazma	(12)	0.0722
	Bryansk	(8)	0.0547
	Mtsensk	(30)	0.0493
	Moscow	(35)	0.0490
	Dorogobusch	(11)	0.0477

[†] "A Graph Theoretical Approach to Historical Geography," by F. R. Pitts, *Professional Geographer*, **17**, 1965, pages 15–20.

✦ *Example 5*
_____

The dominant eigenvalue of the adjacency matrix that describes a transportation network is used by geographers as a measure of the overall connectedness of the network.[†] Networks are compared using this index. Table 8-8 gives the dominant eigenvalues of adjacency matrices that represent the internal airline networks of eight countries. It is surprising to find Sweden low down on the list.

Table 8-8

Country	Dominant Eigenvalue
USA	6
France	5.243
United Kingdom	4.610
India	4.590
Canada	4.511
U.S.S.R	3.855
Sweden	3.301
Turkey	2.903

✦

EXERCISES

Using the iterative method in this section, determine the dominant eigenvalue and a corresponding eigenvector for each of the following matrices.

*1. $\begin{pmatrix} 1 & 7 & -7 \\ -1 & 3 & -1 \\ -1 & -5 & 7 \end{pmatrix}$
  2. $\begin{pmatrix} 9 & 4 & -4 \\ -1 & 1 & 1 \\ 7 & 4 & -2 \end{pmatrix}$

*3. $\begin{pmatrix} 13 & 7 & -7 \\ -2 & -1 & 2 \\ 12 & 7 & -6 \end{pmatrix}$
  4. $\begin{pmatrix} 17 & 8 & -8 \\ 0 & 1 & 0 \\ 16 & 8 & -7 \end{pmatrix}$

*5. $\begin{pmatrix} 1 & 1 & -1 & 1 \\ 1 & 1 & 1 & -1 \\ -3 & 3 & 3 & 3 \\ -3 & 3 & 1 & 5 \end{pmatrix}$
  6. $\begin{pmatrix} 3 & -1 & 1 & -1 \\ -3 & 1 & -3 & 3 \\ -3 & 3 & 7 & 3 \\ -1 & 5 & 7 & 3 \end{pmatrix}$

*7. $\begin{pmatrix} 84 & 5 & -5 & 5 \\ 1 & 0 & 1 & -1 \\ -1 & 1 & 0 & 1 \\ 3 & 5 & -5 & 6 \end{pmatrix}$
  8. $\begin{pmatrix} 4.5 & 5.5 & -5.5 & 5.5 \\ 1.5 & 0.5 & 1.5 & -1.5 \\ -1 & 1 & 0 & 1 \\ 3 & 5 & -6 & 7 \end{pmatrix}$

[†] *Applications of Graph Theory*, edited by Robin J. Wilson and Lowell W. Beineke, Academic Press, Chap. 10, 1979.

Determine *all* eigenvalues and a corresponding eigenvector in each case for the following *symmetric* matrices, using the iterative method of this section.

*9. $\begin{pmatrix} 5 & 4 & 2 \\ 4 & 5 & 2 \\ 2 & 2 & 2 \end{pmatrix}$
10. $\begin{pmatrix} 4 & 0 & 2 \\ 0 & 4 & 0 \\ 2 & 0 & 4 \end{pmatrix}$

*11. $\begin{pmatrix} 7 & 0 & 5 \\ 0 & 2 & 0 \\ 5 & 0 & 7 \end{pmatrix}$
12. $\begin{pmatrix} 4 & 0 & 0 & 6 \\ 0 & 2 & -4 & 0 \\ 0 & -4 & 2 & 0 \\ 6 & 0 & 0 & 4 \end{pmatrix}$

*13. $\begin{pmatrix} 5 & 4 & 1 & 1 \\ 4 & 5 & 1 & 1 \\ 1 & 1 & 4 & 2 \\ 1 & 1 & 2 & 4 \end{pmatrix}$

14. Compute the accessibility index for each vertex in the following graphs. The iterative method should be used in each case. Use three iterations with the given initial vectors. Initial vectors have been given so that answers can be compared. Comment on the selections of the initial vectors.

*(a) initial vector (1, 2, 2, 1)

(b) initial vector (1, 1, 3, 3, 1, 1)

(c) initial vector (2, 1, 1, 1, 1)

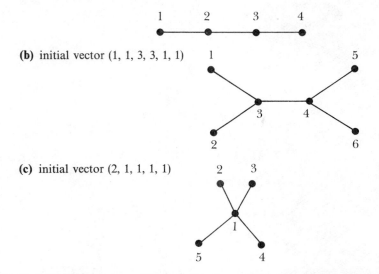

15. Compute the dominant eigenvalue and a corresponding eigenvector for each of the following matrices by going through the following steps: draw the graph represented by the matrix; arrive at an eigenvector by inspection of the graph; and use the equation $A\mathbf{x} = \mathbf{x}$ to compute the eigenvalue.

(a) $\begin{pmatrix} 0 & 1 & 1 \\ 1 & 0 & 1 \\ 1 & 1 & 0 \end{pmatrix}$
(b) $\begin{pmatrix} 0 & 1 & 0 & 1 \\ 1 & 0 & 1 & 0 \\ 0 & 1 & 0 & 1 \\ 1 & 0 & 1 & 0 \end{pmatrix}$

9*

Linear Programming

We complete this course with an introduction to linear programming. Linear programming is a mathematical method that is widely used in industry and in government. It is based on the concepts of linear algebra developed in this course.

Historically, linear programming was first developed and applied in 1947 by George B. Dantzig, Marshall Wood, and their associates of the U.S. Department of the Air Force; the early applications of linear programming were thus in the military field. However, the emphasis in applications has now moved to the general industrial area. Linear programming today is concerned with the efficient use or allocation of limited resources to meet desired objectives. To illustrate the current importance of linear programming, we point out that the 1975 Nobel Prize in Economic Science was awarded to two scientists, Professors Leonid Kantorovich of the Soviet Union and Tjalling C. Koopmans of the United States, for their "contributions to the theory of optimum allocation of resources."

Kantorovich, born in 1912, a winner of the Stalin and Lenin prizes, is presently with the Mathematics Institute of the Sibirskoje Otdelenie Akademi Nauk, Adademgodorske, in Novosibirsk. He is considered the leading representative of the school in Soviet economic research. Koopmans, who was born in the Netherlands in 1913, is currently with the Cowles Foundation for Research in Economics at Yale University in New Haven, Connecticut. Both economists worked independently on the problem of optimum allocation of scarce resources. Kantorovich showed how linear programming can be used to improve economic planning in Russia. He analyzed efficiency conditions for an economy as a whole, demonstrating the connection between the allocation of resources and the price system. An important element in his analysis was the demonstration of how the decentralizing of decisions in a planned economy is dependent on the existence of a rational price system. According to Professor

Assar Lindbeck of the International Economics Prize Committee, Kantorovich's work changed the views on economic planning in the Soviet Union.

Koopmans developed his linear programming theory while working with planning of optimal transportation of ships back and forth across the Atlantic during World War II. Koopmans's work gave new ways of interpreting the relationships between inputs and outputs of a production process. His methods are used to clarify the correspondence between efficiency in production and the existence of a price system.

We now mention a few examples to illustrate the industrial economic benefits of linear programming. In the mid-1950s, linear programming techniques were developed to guide the blending of gasoline by Exxon Corporation. As a result, the company began saving 2 to 3% of the cost of the blending operations. At about the same time, other industries, notably those involved in paper products, food distribution, agriculture, steel, and metalworking, began to adopt linear programming. A paper manufacturer estimated that it increased its profits by $15 million in a single year by employing linear programming to determine its assortment of products.

9-1 LINEAR PROGRAMMING: A GEOMETRICAL INTRODUCTION

We introduce the concepts involved in linear programming by means of a sample problem.

A company makes two products on separate production lines X and Y. It commands a labor force that is equivalent to 900 hours per week, and it has $2800 outlay weekly on running costs. It takes 5 hours and 2 hours to produce a single item on lines X and Y, respectively. The cost of producing a single item on X is $8 and on Y is $10. The aim of the company is to maximize its profits. If the profit on each item produced on line X is $3 and that on each item produced on line Y is $2, how should the scheduling be arranged to lead to maximum profit?

In this problem there are two restrictions: time and funds available. The aim of the company is to maximize its profits under these restrictions. We can solve the problem in three stages:

1. Constructing the mathematical model; that is, using mathematics to describe the situation.
2. Illustrating the mathematics by means of a graph.
3. Using the graph to determine the solution.

Let us discuss each of these steps in detail.

The Mathematical Model

Let x items be produced on production line X and y items on line Y. The weekly profit on line X is $3x$, and on line Y it is $2y$. The total weekly profit is thus $(3x + 2y)$.

Let us examine the time constraint. The total time involved in producing x items on line X at the rate of 5 hours per item is $5x$ hours. The time it takes to manufacture y items on line Y at the rate of 2 hours per item is $2y$ hours. Hence, the total time involved for both production lines is $5x + 2y$ hours. Since the number of hours available is 900, we must ensure that $5x + 2y$ is less than or equal to 900. We write this mathematically as

$$5x + 2y \leq 900$$

Next we examine the cost involved. The cost of producing x items on line X at $8 an item is $8x$. The cost for y items on line Y at $10 an item is $10y$. Hence, the total cost is $8x + 10y$. Since the outlay is $2800, we must have

$$8x + 10y \leq 2800$$

Finally, we observe that x and y cannot be negative; it would be meaningless for these quantities to be negative, because they represent the number of items produced on the production lines. Thus, there are additional restrictions:

$$x \geq 0 \quad \text{and} \quad y \geq 0$$

The problem reduces to determining the values of x and y that maximize $3x + 2y$ under the following constraints:

$$5x + 2y \leq 900$$
$$8x + 10y \leq 2800$$

with $x \geq 0$ and $y \geq 0$. The function $f = 3x + 2y$ is called the *objective function* of the linear programming problem.

Graphical Representation of the Constraints

The above system of constraints is called a *system of linear inequalities*. In the case of two linear equations in x and y, we saw how we could represent the equations graphically and how the solutions, pairs (x, y) that satisfied both equations simultaneously, were the points of intersection. Here we shall show how the above system of inequalities can be represented graphically and how the solutions—all the points (x, y) that satisfy all the inequalities—are points that make up a region of the plane.

Consider the inequality $5x + 2y \leq 900$. Certain points (x, y) satisfy this inequality; others do not. The line $5x + 2y = 900$ divides the plane into three parts: the part above the line, the line itself, and the part below the line. All points on the line and on one side of the line will satisfy the condition; all points on the other side will not. Thus, to find the region of the plane that satisfies this condition, one draws the line and examines whether or not a point on one side satisfies the condition. Consider the origin $(0, 0)$. This point is not on the line $5x + 2y = 900$; it is below it. The point satisfies the condition $5x + 2y \leq 900$,

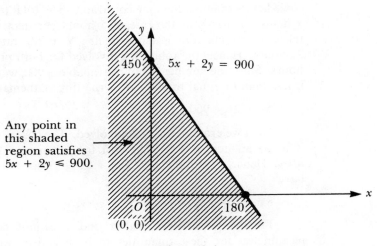

Any point in this shaded region satisfies $5x + 2y \leq 900$.

Figure 9-1

because $(5 \times 0) + (2 \times 0) \leq 900$. Hence $(0, 0)$ is on the side that satisfies the inequality. Therefore the points that satisfy this inequality are on or below the line in Figure 9-1.

Similarly, the points that satisfy the inequality $8x + 10y \leq 2800$ are on and below the line $8x + 10y = 2800$. The condition $x \geq 0$ is satisfied by all points on and to the right of the y axis (the line $x = 0$). The condition $y \geq 0$ is satisfied by all points on and above the x axis (the line $y = 0$). The region that is common to all of these regions is the region that satisfies all the inequalities simultaneously. It is the shaded region $ABCO$ in Figure 9-2.

The points in the shaded region $ABCO$ make up the *set of solutions* to the system of inequalities. The *boundaries* of the solution set are the line segments

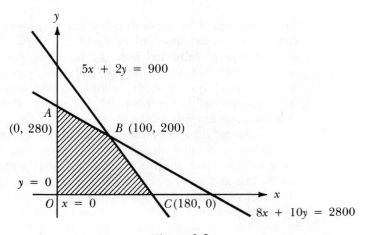

Figure 9-2

AB, BC, CO, and OA. The points A, B, C, and O are the *vertices* of the set. These points are found in the usual way by determining the points of intersection of the lines.

We now find the vertices. Vertex A is the intersection of the lines $x = 0$ and $8x + 10y = 2800$. Substituting for $x = 0$ into the equation gives $y = 280$. Thus, A is the point $(0, 280)$.

Vertex B is the intersection of the line $8x + 10y = 2800$ and the line $5x + 2y = 900$. Solving the pair of equations

$$8x + 10y = 2800$$
$$5x + 2y = 900$$

gives $x = 100$, $y = 200$. Thus, B is the point $(100, 200)$.

Vertex C is the intersection of the lines $y = 0$, $5x + 2y = 900$. Substituting for $y = 0$ into the equation gives $x = 180$. Thus, C is the point $(180, 0)$, and O is, of course, the origin $(0, 0)$.

Each point in $ABCO$ satisfies all the inequalities; $3x + 2y$ has a value at each of these points. Every one of these points has a chance of being a solution to the problem—that is, a point that leads to a maximum value of $3x + 2y$ under the given constraints. We call such a region, satisfying all the constraints of a linear programming problem, the *feasible region* of the linear programming problem, and a point in the region a *feasible solution*. Among these points will be one point or possibly many points that give a maximum value to the objective function; such a point is called an *optimal solution*.

Determining the Optimal Solution

It would be an endless task to examine the value of the objective function $3x + 2y$ at all feasible solutions. General problems of this type have been analyzed, and it has been found that the maximum will occur at a vertex, or at the points that make up one side of $ABCO$ (in which case the maximum value of $3x + 2y$ is attained at all the points that make up one side). Hence, we have only to examine the values of $f = 3x + 2y$ at the vertices A, B, C, and O:

At A, $x = 0$ and $y = 280$. Thus, $f = 3(0) + 2(280) = 560$.

At B, $x = 100$ and $y = 200$. Thus, $f = 3(100) + 2(200) = 700$.

At C, $x = 180$ and $y = 0$. Thus, $f = 3(180) + 2(0) = 540$.

At O, $x = 0$ and $y = 0$. Thus, $f = 0$.

Thus, the maximum value of $3x + 2y$—namely, 700—occurs at B, when $x = 100$ and $y = 200$. The interpretation of these results is that the maximum weekly profit that can be achieved under the given constraints is \$700. This profit will be achieved when production line X turns out 100 items and line Y turns out 200 items.

The optimal solution to a linear programming problem need not always be unique. Example 1 will illustrate this.

✦ *Example 1*

Determine the maximum value of the function $f = 8x + 2y$ subject to the following constraints:

$$4x + y \leq 32$$
$$4x + 3y \leq 48$$

with $x \geq 0$ and $y \geq 0$.

The constraints are represented graphically by the shaded region in Figure 9-3. The vertices of the region are found to be $A(0, 16)$, $B(6, 8)$, $C(8, 0)$, and $O(0, 0)$. We now determine the values of $8x + 2y$ at each of these vertices.

At A, $8x + 2y = 32$
At B, $= 64$
At C, $= 64$
At O, $= 0$

Thus, the maximum value of $8x + 2y$ is 64, and it occurs at two vertices, B and C. When this happens in a linear programming problem, the objective function will have the same value at every point of the edge joining the two

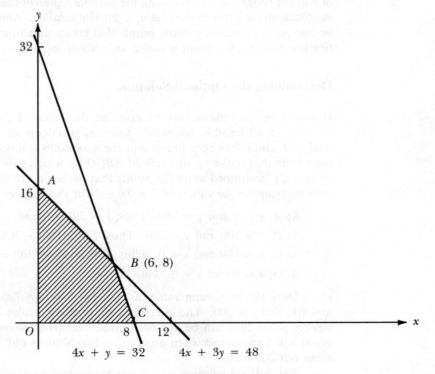

Figure 9-3

vertices. Thus, $8x + 2y$ has a maximum of 64 subject to these constraints. This value occurs at all points along BC. The maximum value occurs at all points along the line $4x + y = 32$ from the point $(6, 8)$ to the point $(8, 0)$.

To illustrate this, let us consider the point $D(7, 4)$. This point lies on the line segment BC. The value of $8x + 2y$ at D is $8(7) + 2(4) = 64$.

If x and y correspond to numbers of items produced and f corresponds to profit, in a situation similar to the one previously discussed, then there is a certain flexibility in the production schedule that will lead to maximum profit. Any production schedule corresponding to points along line segment BC will lead to maximum profit. The deciding factor can be some other factor that has not been brought into the mathematical model.   ✦

It is possible that the constraints in a linear programming problem define an unbounded feasible region and that the objective function is unbounded, having no maximum. The linear programming problem then has no solution. We illustrate this possibility with the following example.

✦ *Example 2*
───────────────────────────────────────

Determine the maximum value of the function $x + 4y$ subject to the following constraints:

$$-4x + y \leq 2$$
$$2x - y \leq 1$$

with $x \geq 0$ and $y \geq 0$.

The constraints are represented by the shaded region in Figure 9-4. Let us show that the objective function has no maximum value within this region.

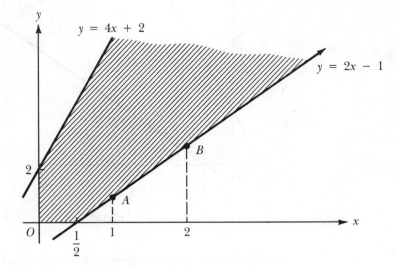

Figure 9-4

Consider the values of f at the points along the line $y = 2x - 1$. Along these points,

$$f = x + 4y = x + 4(2x - 1)$$
$$= 9x - 4$$

Thus, at A, having x value 1, f is 5. At B, having x value 2, f is 14. As we move in the direction indicated along this line, the x values of the points increase, and f increases without limit. The objective function has no maximum value within this region; it increases without limit in the direction indicated.

If this situation arises in practice (when one expects a solution), it may be that a constraint has been omitted. For example, suppose the constraint $x + y \leq 7$ has been omitted in the above case, the correct problem being to maximize $x + 4y$ subject to the following constraints:

$$-4x + y \leq 2$$
$$2x - y \leq 1$$
$$x + y \leq 7$$

with $x \geq 0$ and $y \geq 0$.

The feasible region is the shaded area in Figure 9-5. On examining $f = x + 4y$ at each of the vertices, we can now see that f does have a maximum of 25 at the vertex $(1, 6)$.

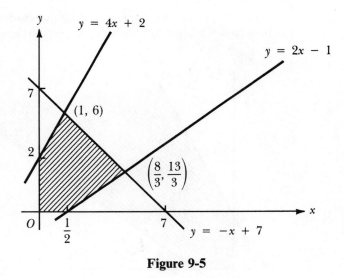

Figure 9-5   ✦

The fact that the feasible region is unbounded does not in itself imply that the objective function has no maximum. Consider the following example.

✦ Example 3

Maximize the function $3y$ subject to the following constraints:

$$-x + 2y \leq 4$$
$$y \leq 3$$

with $x \geq 0$ and $y \geq 0$. Refer to Figure 9-6.

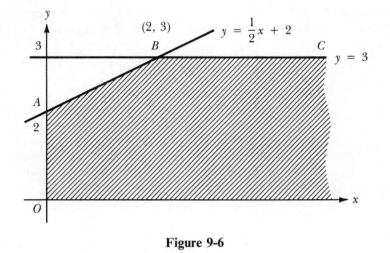

Figure 9-6

The feasible region is unbounded. At O, the value of the function $3y$ is 0; at A, it is 6; at B, it is 9. In fact, the value of the function is 9 at all points along the line BC. The function does have a maximum value of 9; any point along the line BC is an optimal solution.   ✦

Minimum Value of a Function

A problem that involves determining a minimum value of an objective function f can be solved by looking for the maximum value of $-f$, the negative of f, over the same region. Over the given region, the value of f at any point will be the negative of that of $-f$ at that point. *We shall show that the minimum value of f occurs at the point(s) of maximum value of $-f$ and is the negative of that maximum value.*

Denote the value of f at the point A by $f(A)$. If f has a minimum value at A, in the region, then

$$f(A) \leq f(B)$$

where B is any other point in the region. Multiplying both sides by -1, we have

$$-f(A) \geq -f(B)$$

implying that A is a point of maximum value of $-f$. The converse can also be seen to hold—namely, that if A is a point of maximum value of $-f$, then it is a point of minimum value of f. Thus the minimum value of f and maximum value of $-f$ occur at the same point(s).

It can be seen that the minimum value of f, or $f(A)$, is the negative of the maximum value of $-f$, or $-f(A)$.

✦ *Example 4*
—————————————————————————————

Determine the minimum value of the function $2x - 3y$ under the constraints

$$2y + \ x \le 10$$
$$y + 2x \le 11$$

with $x \ge 0$ and $y \ge 0$.

The vertices of the relevant region are $A(0, 5)$, $B(4, 3)$, $C(\frac{11}{2}, 0)$, and $O(0, 0)$. (See Figure 9-7.) To determine the minimum value of $2x - 3y$ over this region, we find the location of the maximum value of $-(2x - 3y)$ in the region. The values of $-(2x - 3y)$ at the vertices are as follows: 15 at A, 1 at B, -11 at C, 0 at O. Thus $-(2x - 3y)$ has a maximum of 15 at $x = 0$, $y = 5$. The minimum of $(2x - 3y)$ will be -15 at $x = 0$, $y = 5$.

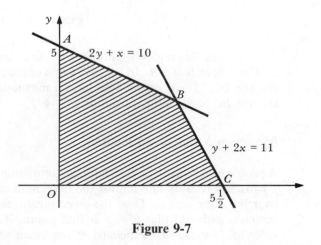

Figure 9-7   ✦

Notice that each feasible region we have discussed is such that the whole of the segment of a straight line joining any two points within the region lies within that region. Such a region is called *convex*. A theorem states that the region satisfying the set of inequalities that represent the constraints in a linear programming problem is always convex. See Figure 9-8.

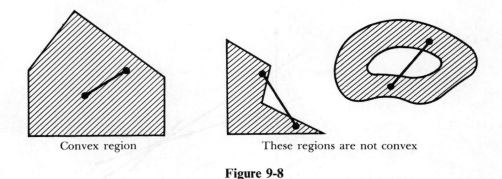

Convex region   These regions are not convex

Figure 9-8

We now give a geometrical reason for why we expect the maximum value of the objective function to occur either at a vertex or along a side of the feasible region for a linear programming problem in two variables. In Figure 9-9, let the feasible region be $ABCO$ and the objective function be $f = ax + by$; f has a value at each point in the region $ABCO$. Let $P(x_P, y_P)$ be an arbitrary point in $ABCO$. Let f have value f_P at P. Thus, $ax_P + by_P = f_P$. This implies that P lies on the line $ax + by = f_P$. The y intercept of this line is f_P/b (see Figure 9-9).

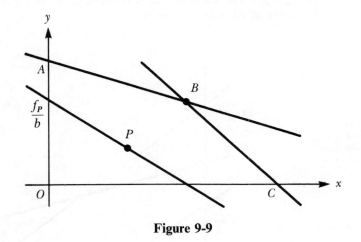

Figure 9-9

Consider the vertex B. In a similar manner, B lies on the line $ax + by = f_B$, a parallel line with y intercept f_B/b. It can be seen in Figure 9-10 that $f_B/b > f_P/b$, implying that $f_B > f_P$. Thus, the maximum value of f occurs at B, a vertex.

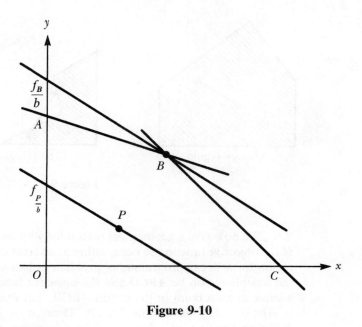

Figure 9-10

If the lines are parallel to a side of the feasible region, there will be many optimal solutions, as shown in Figure 9-11. Here $f_B/b = f_C/b$, implying that $f_B = f_C$. The value of f at B is the same as its value at C; f will have the same value at all points along BC.

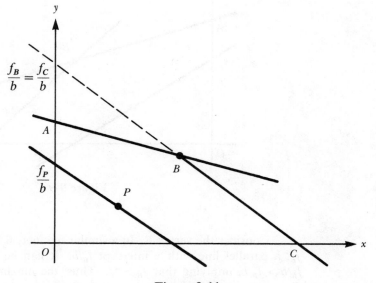

Figure 9-11

Examples of Linear Programming Problems

We complete this section by listing examples of problems that are usually solved using linear programming techniques.

1. *The production-scheduling problem.* A company manufactures a product on a production line. The rate of production can be varied, and the expected sales pattern over a period is known. It is expensive to vary the rate of production, the cost being greater the greater the change. Thus, it is unusual to manufacture the exact amount in each period to satisfy the demand during that period, as the changeover cost would be prohibitive. Furthermore, it is expensive to store the product. Linear programming can be used to determine the production schedule that will minimize the sum of the changeover costs and storage costs.

2. *The transportation problem.* In the transportation problem, one is concerned with a product that must be transported from a number of origins to a number of destinations. The origins could be production plants, and the destinations could be distribution warehouses. The quantity of the product available at each origin and that required at each destination are known. The objective is to determine the amount that should be shipped from each origin to each destination to minimize the total shipping costs.

    The H. J. Heinz Co., for example, which manufactures tomato ketchup in half a dozen factories and distributes the product to about 70 warehouses, has saved several thousand dollars semiannually in freight costs by applying linear programming to determine the optimal pattern of shipments.

3. *The diet problem.* This problem gets its name from the fact that it was first applied to determining economical human diets. In its industrial form it consists of determining the most economical mixture of raw materials that will result in a product with a desired chemical formula.

    Problems of this type occur in the petroleum refinery industry, and linear programming methods have proved useful in solving them. Crude oil products have to be blended into several grades of gasoline, each of which must satisfy certain quality specifications.

    The food industry uses linear programming in manufacturing products that have certain ingredients conforming to certain specifications. Armour Co., for example, uses linear programming in determining specifications for a processed cheese spread which contains 10 ingredients under 7 specifications. The specifications are represented by inequalities under which the cost of ingredients is minimized.

    Since the market prices of various materials change periodically, the most economical mix changes accordingly. Fortunately, this repetitive characteristic occurs in many linear programming problems. We say "fortunately" because the typical problem is usually so large that the cost of the analysis necessary to develop a linear programming model is greater than the savings resulting from a single application. However, if the problem (except for minor changes) must be reexamined periodically, the changes in the mathematical model each time are slight and the cost per solution becomes worthwhile. All practical applications are so involved that the use of a computer becomes mandatory.

EXERCISES

Graph the linear inequalities in Exercises 1–16, indicating in each case the region that satisfies the inequalities.

*1. $2x + y \leq 50$
$4x + 5y \leq 160$
$x \geq 0$
$y \geq 0$

2. $2x + y \leq 60$
$2x + 3y \leq 120$
$x \geq 0$
$y \geq 0$

*3. $4x + 3y \leq 72$
$4x + 9y \leq 144$
$x \geq 2$
$y \geq 0$

4. $4x - 3y \geq 60$
$x + y \leq 10$
$-x + 2y \geq -50$

Solve the following linear programming problems:

*5. Maximize $2x + y$
subject to
$4x + y \leq 36$
$4x + 3y \leq 60$
$x \geq 0$
$y \geq 0$

6. Maximize $x - 4y$
subject to
$x + 2y \leq 4$
$x + 6y \leq 8$
$x \geq 0$
$y \geq 0$

*7. Maximize $2x + y$
subject to
$-3x + y \leq 4$
$x - y \leq 2$
$x \geq 0$
$y \geq 0$

8. Maximize $4x + 2y$
subject to
$x + 3y \leq 15$
$2x + y \leq 10$
$x \geq 0$
$y \geq 0$

*9. Maximize $3x + y$
subject to
$2x - 3y \geq 10$
$x \leq 8$
$x \geq 0$
$y \geq 0$

10. Maximize $x + 5y$
subject to
$x + y \leq 10$
$2x + y \geq 10$
$x + 2y \geq 10$

*11. Maximize $x + 2y$
subject to
$x \geq -2$
$x - y \geq -4$
$x + 2y \leq 6$
$2x + y \leq 6$

12. Maximize $-8x + 10y$
subject to
$x \geq -20$
$x \leq 5$
$y \geq 0$
$4x + 3y \leq 40$
$-4x + 5y \leq 120$

*13. Minimize $-3x - y$
subject to
$x + y \leq 150$
$4x + y \leq 450$
$x \geq 0$
$y \geq 0$

14. Minimize $-2x + y$
subject to
$2x + y \leq 440$
$4x + y \leq 680$
$x \geq 0$
$y \geq 0$

*15. Minimize $-3x - 2y$
subject to
$2x + 3y \leq 260$
$6x + y \leq 300$
$x \geq 0$
$y \geq 0$

16. Minimize $-x + 2y$
subject to
$x + 2y \leq 4$
$x + 4y \leq 6$
$x \geq 0$
$y \geq 0$

*17. A company manufactures two products, X and Y. Two machines, I and II, are needed to manufacture each product. It takes 3 minutes on each machine to produce an X. To produce a Y takes 1 minute on machine I and 2 minutes on machine II. The total time available on machine I is 3000 minutes and on machine II is 4500 minutes. The company realizes a profit of $15 on each X and $7 on each Y. How should the manufacturing of the products be arranged to obtain the largest possible total profit?

18. A company manufactures two types of hand calculators, model C1 and model C2. It takes 1 hour and 4 hours in labor time to manufacture the C1 and the C2, respectively. The cost of manufacturing the C1 is $30, and that of manufacturing the C2 is $20. The company has 1600 hours per week available in labor and $18,000 in running costs. The profit on the C1 model is $10, and on the C2 model it is $8. What should the weekly production schedule be to ensure maximum total profit?

*19. A refrigerator company has two plants, at town X and town Y. Its refrigerators are sold in a certain town Z. It takes 20 hours (packing, transportation, and so on) to transport a refrigerator from X to Z and 10 hours from Y to Z. It costs $60 to transport each refrigerator from X to Z, and $10 per refrigerator from Y to Z. There are 1200 hours available (for packing, transportation, and so on), and $2400 has been budgeted for transportation costs. The profit on each refrigerator manufactured at X is $40 and on each manufactured at Y is $20 (the plant at X is newer and more efficient than that at Y). How should the company allocate the transportation of refrigerators so as to maximize total profits?

20. A company makes a single product on two separate production lines X and Y. It commands a labor force that is equivalent to 1000 hours per week, and it has $3000 outlay weekly on running costs. It takes 1 hour and 4 hours to produce a single item on X and Y, respectively. The cost of producing a single item on X is $5 and on Y is $4. The aim of the company is maximum productivity under these constraints. How can this be achieved?

*21. The maximum daily production of an oil refinery is 1400 barrels. The refinery can produce two types of oil: gasoline for automobiles and heating oil for domestic purposes. The production cost per barrel is $6 for gasoline and $8 for heating oil. The daily production budget is $9600. The profit per barrel is $3.50 on gasoline and $4 on heating oil. What is the maximum total profit that can be realized daily, and what quantities of each type of oil are then produced?

22. The organizer of a day conference has $2000 available for distribution as expense money to participants. The participants fall into two categories: those having $30 expenses and those having $10 expenses. Facilities are available to host 100 participants at the conference. Describe the ways that 100 participants can be invited to the conference.

*23. A manufacturer makes two types of fertilizer, X and Y, using chemicals A and B. Fertilizer X is made up of 80% of chemical A and 20% of chemical B. Fertilizer Y is made up of 60% of chemical A and 40% of chemical B. The manufacturer requires at least 30 tons of X and at least 50 tons of Y, and has available 100 tons of A and 50 tons of B. The aim is to make as much fertilizer as possible. What are the quantities of X and Y that the firm must make to accomplish this?

24. A city has $600,000 to purchase cars. Two models, the Arrow and the Gazelle, are under consideration, costing $4000 and $5000, respectively. The estimated annual maintenance cost on the Arrow is $400 and on the Gazelle is $300. The city will allocate $40,000 for the total annual maintenance of these cars. The Arrow

gets 24 miles per gallon and the Gazelle 20 miles per gallon. The city wants to maximize "the gasoline efficiency number" of this group of cars. For x Arrows and y Gazelles, this number would be $24x + 20y$. How many of each model should be purchased?

*25. A car dealer imports foreign cars by way of two ports of entry, A and B. A total of 120 cars are needed at city C and 180 at city D. There are 100 cars available at A and 200 at B. It takes 2 hours and 6 hours to transport a car from A and B, respectively, to C. It takes 4 hours and 3 hours to transport a car from A and B, respectively, to D. The dealer has 1030 hours available in driver time to move the cars. The aim is to supply as many of the cars for C as possible from B, because drivers will not be available in the future for this route. How can this aim be achieved?

26. A tailor has 80 square yards of cotton material and 120 square yards of woolen material. A suit requires 2 square yards of cotton and 1 square yard of wool, and a dress requires 1 square yard of cotton and 3 square yards of wool. How many of each garment should the tailor make to maximize income if a suit and a dress each sell for $20? What is the maximum income?

*27. A school district is buying new buses. It has a choice of two kinds. The Torro costs $18,000 and holds 25 passengers, while the Sprite costs $22,000 and holds 30 passengers. $572,000 has been budgeted for the new buses. A maximum of 30 drivers will be available to drive the buses. At least 17 Sprite buses must be ordered because of the desirability of having a certain number of large-capacity buses. How many of each type should be purchased to carry a maximum number of students?

28. A company is buying lockers. It has narrowed down the choice to two kinds, X and Y. X has a volume of 36 cubic feet, while Y has a volume of 44 cubic feet. X occupies an area of 6 square feet and costs $54, while Y occupies an area of 8 square feet and costs $60. A total of 256 square feet of floor space is available and $2100 in funds for the purchase of the cabinets. At least 20 of the large lockers are wanted. In order to maximize volume, how many of each should be purchased?

*29. A farmer has to decide how many acres of a 40-acre plot are to be devoted to growing strawberries and how many to growing tomatoes. There will be 300 hours of labor available for the picking. It takes 8 hours to pick an acre of strawberries and 6 hours to pick an acre of tomatoes. The per-acre profit is $700 for the strawberries compared to $600 for the tomatoes. How many acres of each should be grown in order to maximize total profit?

9-2   THE SIMPLEX METHOD

The graphical method of solving a linear programming problem has its limitations. The method demonstrated for two variables can be extended to linear programming problems involving three variables, the feasible region being a convex subset of three-dimensional space. However, for problems involving more than three variables, the geometrical approach becomes impractical. We now introduce the reader to the *simplex method*, an algebraic method that can be used for any number of variables. Furthermore, the simplex method has the advantage of being readily programmable for implementation on the computer. Today "packages" of computer programs based on the simplex method are offered commercially by software companies. Industrial customers often pay sizable monthly fees for the use of these programs. Exxon, for example, currently applies linear programming to the schedule of drilling operations, the allocation of crude oil among refineries, the setting of refinery operating conditions, the distribution of products, and the planning of business strategy. It is estimated that linear programming accounts for from 5 to 10% of the company's total computing load. The simplex method involves reformulating the constraints in terms of linear equations and then using elementary matrix transformations in a manner very similar to that of Gauss-Jordan elimination to arrive at the solution.

We introduce the simplex method by using it to obtain the solution for the first model of the previous section. In that model we wished to determine values of x and y that maximized the objective function $3x + 2y$ subject to the following constraints:

$$5x + 2y \leq 900$$
$$8x + 10y \leq 2800$$

with $x \geq 0$ and $y \geq 0$.

The simplex method involves reformulating these constraints in terms of a system of linear equations by introducing additional variables called *slack variables*. Let us examine the first constraint, $5x + 2y \leq 900$. For each pair of values of x and y that satisfy this condition, there will exist a value of a nonnegative variable u such that

$$5x + 2y + u = 900$$

The value of u will be the number that must be added to $5x + 2y$ to bring it up to 900. Similarly, for each pair of values of x and y that satisfy the second constraint, $8x + 10y \leq 2800$, there will exist a value of a nonnegative variable v such that $8x + 10y + v = 2800$.

Thus, the constraints in the linear programming problem may be represented by the following system:

$$5x + 2y + u = 900$$
$$8x + 10y + v = 2800$$

with $x \geq 0$, $y \geq 0$, $u \geq 0$, and $v \geq 0$. The variables u and v are called *slack variables* because they make up the slack in the original inequalities.

Finally, the objective function $f = 3x + 2y$ may be written in the form

$$-3x - 2y + f = 0$$

The entire problem becomes that of determining the solution to the following system of equations:

$$5x + 2y + u \qquad\qquad = 900$$
$$8x + 10y \qquad + v \qquad = 2800$$
$$-3x - 2y \qquad\qquad + f = \quad 0$$

such that f is as large as possible with $x \geq 0$, $y \geq 0$, $u \geq 0$, and $v \geq 0$.

We have thus reformulated the problem in terms of a system of linear equations under certain constraints. The system of equations, consisting of three equations in the five variables x, y, u, v, and f, will have many solutions in the region defined by $x \geq 0$, $y \geq 0$, $u \geq 0$, $v \geq 0$. Any such solution is a *feasible solution*. A solution that maximizes f is an *optimal solution*; this is the solution in which we are interested.

We determine the optimal solution by using elementary matrix transformations in a very definite sequence. The method of Gauss-Jordan elimination discussed previously involved creating zeros in a special systematic order. The element that is used to create the zeros in any column when elementary matrix transformations are involved is called the *pivot* for that column. The manner of choosing pivots in the simplex method differs from that of Gauss-Jordan elimination. Thus, zeros are created in columns in an order different from that used in Gauss-Jordan elimination.

The augmented matrix of this system, called the *initial simplex tableau*, is

$$\begin{pmatrix} 5 & 2 & 1 & 0 & 0 & 900 \\ 8 & 10 & 0 & 1 & 0 & 2800 \\ -3 & -2 & 0 & 0 & 1 & 0 \end{pmatrix}$$

We now proceed to obtain a *final simplex tableau* by following the steps below:

1. Locate the negative entry in the last row, other than the last element, that is largest in magnitude. If two or more entries share this property, any one of these can be selected arbitrarily. If all such entries are nonnegative, the tableau is in final form.

2. Divide each positive element in the column defined by this negative entry into the corresponding element of the last column.

3. Select as pivot the divisor that yields the smallest quotient.

4. Use this pivot to create a 1 in its location and zeros elsewhere in this column. Create zeros by adding suitable multiples of the pivot row to the other rows.

5. Repeat this sequence until all such negative elements have been eliminated from the last row.

This sequence of transformations ensures that the constraints $x \geq 0, \ldots,$ $v \geq 0$ are not violated at any stage and that we obtain, in an efficient manner, an augmented matrix that leads to the solution. For this particular tableau we get the following:

pivot,
since
$\frac{900}{5} < \frac{2800}{8}$
$$\begin{pmatrix} ⑤ & 2 & 1 & 0 & 0 & 900 \\ 8 & 10 & 0 & 1 & 0 & 2800 \\ -3 & -2 & 0 & 0 & 1 & 0 \end{pmatrix}$$

$\cong$
$(\frac{1}{5})R1$
$$\begin{pmatrix} ① & \frac{2}{5} & \frac{1}{5} & 0 & 0 & 180 \\ 8 & 10 & 0 & 1 & 0 & 2800 \\ -3 & -2 & 0 & 0 & 1 & 0 \end{pmatrix}$$

$\cong$
$R2-(8)R1$
$R3+(3)R1$
$$\begin{pmatrix} 1 & \frac{2}{5} & \frac{1}{5} & 0 & 0 & 180 \\ 0 & ㊞\frac{34}{5} & -\frac{8}{5} & 1 & 0 & 1360 \\ 0 & -\frac{4}{5} & \frac{3}{5} & 0 & 1 & 540 \end{pmatrix}$$

pivot

$\cong$
$(\frac{5}{34})R2$
$$\begin{pmatrix} 1 & \frac{2}{5} & \frac{1}{5} & 0 & 0 & 180 \\ 0 & ① & -\frac{4}{17} & \frac{5}{34} & 0 & 200 \\ 0 & -\frac{4}{5} & \frac{3}{5} & 0 & 1 & 540 \end{pmatrix}$$

$\cong$
$R2-(\frac{2}{5})R2$
$R3+(\frac{4}{5})R2$
$$\begin{pmatrix} 1 & 0 & \frac{25}{85} & -\frac{1}{17} & 0 & 100 \\ 0 & 1 & -\frac{4}{17} & \frac{5}{34} & 0 & 200 \\ 0 & 0 & \frac{7}{17} & \frac{2}{17} & 1 & 700 \end{pmatrix}$$

This final tableau is equivalent to the following system of equations:

$$x + \quad \frac{25}{85}u - \frac{1}{17}v \qquad = 100$$
$$y - \frac{4}{17}u + \frac{5}{34}v \qquad = 200$$
$$\frac{7}{17}u + \frac{2}{17}v + f = 700$$

The solutions to this system are, of course, identical to those of the original system. Since $u \geq 0$ and $v \geq 0$, the last equation tells us that f is a maximum of 700 when $u = v = 0$. On substituting these values back into the system, we have that $x = 100$ and $y = 200$. Thus, the optimal solution is $x = 100$, $y = 200$, and $f = 700$.

In terms of our original maximal problem, this solution means that the maximum value of the function $3x + 2y$ is 700, and it occurs when $x = 100$ and $y = 200$. (The element in the last row and last column of the final tableau is always the maximum value of f.)

In practice, the next to last column of the simplex, $\begin{pmatrix} 0 \\ 0 \\ 1 \end{pmatrix}$, is usually omitted, as it does not change during the process. Its elimination reduces the computation involved when the method is programmed for the computer. We illustrate this in the following example.

✦ *Example 1*

Determine the maximum value of the function $3x + 5y + 8z$ subject to the following constraints:

$$x + y + z \le 100$$
$$3x + 2y + 4z \le 200$$
$$x + 2y \le 150$$

with $x \ge 0$, $y \ge 0$, and $z \ge 0$.

The corresponding system of equations is

$$x + y + z + u = 100$$
$$3x + 2y + 4z + v = 200$$
$$x + 2y + w = 150$$
$$-3x - 5y - 8z + f = 0$$

with $x \ge 0$, $y \ge 0$, $z \ge 0$, $u \ge 0$, $v \ge 0$, and $w \ge 0$.

Thus, the tableaux are as follows:

$$pivot\negthickspace\negthickspace\longrightarrow\begin{pmatrix} 1 & 1 & 1 & 1 & 0 & 0 & 0 & 100 \\ 3 & 2 & ④ & 0 & 1 & 0 & 0 & 200 \\ 1 & 2 & 0 & 0 & 0 & 1 & 0 & 150 \\ -3 & -5 & -8 & 0 & 0 & 0 & 1 & 0 \end{pmatrix}$$

$$\begin{array}{c} \\ \text{Select} \\ \text{this} \\ \text{column.} \end{array} \qquad \begin{array}{c} \\ \text{Leave} \\ \text{this} \\ \text{column} \\ \text{out.} \end{array}$$

$$\underset{(\frac{1}{4})R2}{\cong}\begin{pmatrix} 1 & 1 & 1 & 1 & 0 & 0 & 100 \\ \frac{3}{4} & \frac{1}{2} & ① & 0 & \frac{1}{4} & 0 & 50 \\ 1 & 2 & 0 & 0 & 0 & 1 & 150 \\ -3 & -5 & -8 & 0 & 0 & 0 & 0 \end{pmatrix}$$

$$\underset{\substack{R1-R2 \\ R4+(8)R2}}{\cong}\begin{pmatrix} \frac{1}{4} & \frac{1}{2} & 0 & 1 & -\frac{1}{4} & 0 & 50 \\ \frac{3}{4} & \frac{1}{2} & 1 & 0 & \frac{1}{4} & 0 & 50 \\ 1 & ② & 0 & 0 & 0 & 1 & 150 \\ 3 & -1 & 0 & 0 & 2 & 0 & 400 \end{pmatrix}$$

$$\underset{(\frac{1}{2})R3}{\cong} \begin{pmatrix} \frac{1}{4} & \frac{1}{2} & 0 & 1 & -\frac{1}{4} & 0 & 50 \\ \frac{3}{4} & \frac{1}{2} & 1 & 0 & \frac{1}{4} & 0 & 50 \\ \frac{1}{2} & ① & 0 & 0 & 0 & \frac{1}{2} & 75 \\ 3 & -1 & 0 & 0 & 2 & 0 & 400 \end{pmatrix}$$

$$\underset{\substack{R1-(\frac{1}{2})R3 \\ R2-(\frac{1}{2})R3 \\ R4+R3}}{\cong} \begin{pmatrix} 0 & 0 & 0 & 1 & -\frac{1}{4} & -\frac{1}{4} & 12\frac{1}{2} \\ \frac{1}{2} & 0 & 1 & 0 & \frac{1}{4} & -\frac{1}{4} & 12\frac{1}{2} \\ \frac{1}{2} & 1 & 0 & 0 & 0 & \frac{1}{2} & 75 \\ 3\frac{1}{2} & 0 & 0 & 0 & 2 & \frac{1}{2} & 475 \end{pmatrix} \begin{array}{l} \\ \\ \\ \textit{maximum value} \\ \textit{of the function} \end{array}$$

This is the final tableau. It represents the following system of equations:

$$
\begin{aligned}
u - \tfrac{1}{4}v - \tfrac{1}{4}w &= 12\tfrac{1}{2} \\
\tfrac{1}{2}x \quad + z \quad + \tfrac{1}{4}v - \tfrac{1}{4}w &= 12\tfrac{1}{2} \\
\tfrac{1}{2}x + y \qquad\quad + \tfrac{1}{2}w &= 75 \\
3\tfrac{1}{2}x \qquad\quad + 2v + \tfrac{1}{2}w + f &= 475
\end{aligned}
$$

(The f in the final equation comes from the column that was omitted.) Because $x \geq 0$, $v \geq 0$, and $w \geq 0$, the final equation gives a maximum value of 475 for f when $x = v = w = 0$.

On substituting these values into the system, we get $y = 75$, $z = 12\frac{1}{2}$, $u = 12\frac{1}{2}$. Thus, the maximum value of the function is 475, and it occurs when $x = 0$, $y = 75$, and $z = 12\frac{1}{2}$.   ✦

We now demonstrate the simplex method for a nonunique solution. (We use Example 1 of the previous section.)

✦ **Example 2** _____

Let us determine the maximum value of $8x + 2y$ subject to the constraints

$$
\begin{aligned}
4x + y &\leq 32 \\
4x + 3y &\leq 48
\end{aligned}
$$

with $x \geq 0$ and $y \geq 0$.

The simplex tableaux are as follows:

$$\begin{pmatrix} ④ & 1 & 1 & 0 & 32 \\ 4 & 3 & 0 & 1 & 48 \\ -8 & -2 & 0 & 0 & 0 \end{pmatrix} \cong \begin{pmatrix} ① & \frac{1}{4} & \frac{1}{4} & 0 & 8 \\ 4 & 3 & 0 & 1 & 48 \\ -8 & -2 & 0 & 0 & 0 \end{pmatrix}$$

$$\cong \begin{pmatrix} 1 & \frac{1}{4} & \frac{1}{4} & 0 & 8 \\ 0 & 2 & -1 & 1 & 16 \\ 0 & 0 & 2 & 0 & 64 \end{pmatrix}$$

The final tableau gives the following system of equations:

$$x + \tfrac{1}{4}y + \tfrac{1}{4}u \qquad\qquad = 8$$
$$2y - u + v \qquad = 16$$
$$2u \quad + f = 64$$

with $x \geq 0$, $y \geq 0$, $u \geq 0$, and $v \geq 0$.

The last equation gives that the maximum value of f is 64 when u is zero. If we let u be zero, the other equations become

$$x + \tfrac{1}{4}y \qquad = 8$$
$$2y + v = 16$$

with $x \geq 0$, $y \geq 0$, and $v \geq 0$.

Any values of x and corresponding values of y satisfying these conditions are optimal solutions. The second equation tells us that the largest value y can take under these conditions is 8, when $v = 0$. The corresponding value of x from the first equation is then 6. The second equation also tells us that the smallest value y can take is 0, when $v = 16$. The first equation gives the corresponding x to be 8.

Note also that the first equation can be written $4x + y = 32$.

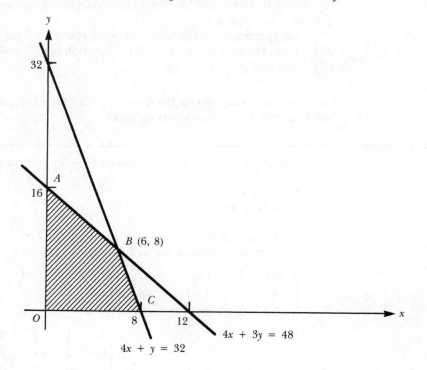

Figure 9-12

Thus, the maximum value of $8x + 2y$ is 64. This is attained at any point of the line $4x + y = 32$ lying between $B(6, 8)$ and $C(8, 0)$ (Figure 9-12). This result agrees with that of Example 1, where the geometric approach was used.
♦

We now demonstrate how the simplex method handles a linear programming problem that has no solution, the feasible region being unbounded. In the following example, we consider the problem that was discussed in Example 2 of the previous section.

✦ *Example 3*

Maximize the function $x + 4y$ subject to the following constraints:

$$-4x + y \le 2$$
$$2x - y \le 1$$

with $x \ge 0$ and $y \ge 0$.

The simplex tableaux are as follows:

$$\begin{pmatrix} -4 & ① & 1 & 0 & 2 \\ 2 & -1 & 0 & 1 & 1 \\ -1 & -4 & 0 & 0 & 0 \end{pmatrix} \cong \begin{pmatrix} -4 & ① & 1 & 0 & 2 \\ -2 & 0 & 1 & 1 & 3 \\ -1 & -4 & 0 & 0 & 0 \end{pmatrix}$$

$$\cong \begin{pmatrix} -4 & 1 & 1 & 0 & 2 \\ -2 & 0 & 1 & 1 & 3 \\ -17 & 0 & 4 & 0 & 8 \end{pmatrix}$$

It is impossible to proceed further, as no pivot exists in the first column because of the negative signs. The simplex method reveals that no maximum exists.

It can be shown that whenever one arrives at a simplex tableau that contains a column, other than the last column, having a negative last entry and all nonpositive elements above it, the feasible region is unbounded and the objective function is unbounded. ♦

We now summarize the simplex method, as it has been introduced here. Since a general linear programming problem can involve many variables, we formulate our results using the notation $x_1, \ldots, x_n$ for variables.

A *standard linear programming problem* is that of maximizing a function $f = c_1 x_1 + \cdots + c_n x_n$ in n variables subject to m constraints:

$$a_{11} x_1 + \cdots + a_{1n} x_n \le b_1$$
$$\vdots$$
$$a_{m1} x_1 + \cdots + a_{mn} x_n \le b_m$$

with $x_1 \ge 0, \ldots, x_n \ge 0$, where $b_1, \ldots, b_m$ are all nonnegative.

The simplex method involves converting the first m inequalities into equations by introducing m slack variables $x_{n+1}, \ldots, x_{n+m}$. The problem then becomes that of maximizing f subject to the following:

$$
\begin{aligned}
a_{11}x_1 + \cdots + a_{1n}x_n + x_{n+1} &&&= b_1 \\
a_{21}x_1 + \cdots + a_{2n}x_n \quad\quad + x_{n+2} &&&= b_2 \\
\vdots \\
a_{m1}x_1 + \cdots + a_{mn}x_n \quad\quad\quad\quad + x_{n+m} &&= b_m \\
-cx_1 - \cdots - \quad c_n x_n \quad\quad\quad\quad\quad\quad + f &&= 0
\end{aligned}
$$

with $x_1 \geq 0, \ldots, x_n \geq 0, x_{n+1} \geq 0, \ldots, x_{n+m} \geq 0$.

The augmented matrix of this system of equations is called the *initial simplex tableau*. One obtains a final simplex tableau by performing a sequence of elementary matrix transformations, the sequence being selected according to the following rules:

1. Locate the negative entry in the last row, other than the last element, that is largest in magnitude. If two or more entries share this property, any one of these can be selected arbitrarily. If all such entries are nonnegative, the tableau is in final form.
2. Divide each positive element in the column defined by this negative entry into the corresponding element of the last column.
3. Select as pivot the divisor that yields the smallest quotient.
4. Use this pivot to create a 1 in its own location and zeros elsewhere in this column. Create zeros by adding suitable multiples of the pivot row to the other rows.
5. Repeat this sequence until all such negative elements have been eliminated from the last row.

The final tableau leads to the solution or gives the information that no solution exists.

It is most important to realize the *limitations* on the method as stated above. In the above form, it can be applied to a standard linear programming problem: that of maximizing a function in which the first set of inequalities is of the type $\leq$, all b_i are nonnegative, and all variables are nonnegative. The simplex method can be applied to other classes of linear programming problems; for example, when constraints are a mixture of $\leq$ and $\geq$ types, or some variables are not restricted to being nonnegative. In such problems the use of slack variables in setting up the initial tableau differs from the above. Once the initial tableau has been constructed, however, the sequence of transformations is performed according to the above rules to determine a final tableau. Interested readers will find a discussion of these other classes of linear programming problems in *Mathematics with Applications in the Management, Natural, and Social Sciences,*" by Gareth Williams, Allyn and Bacon, 1981, Section 4-5.

In the class of standard linear programming problems, there are two *complications* that can arise:

1. There can be a tie between two or more negative entries in the last row for largest in magnitude. In such a case, any of these entries can be selected to provide the pivot column.
2. There can be a tie for pivot. When the divisions are made into the corresponding elements of the last column, two or more positive elements in the relevant column may yield equal divisors. Such a case is called *degeneracy* and will occur, for example, if some of the b_i are zero. If this occurs, any of these tied elements can be selected as pivot. This technique works for all but a few examples which have been artificially constructed.

Here we have introduced the reader to the simplex method. Dantzig's discovery of the simplex method ranks high among the achievements of twentieth-century applied mathematics.

EXERCISES

Use the simplex method to maximize (if possible) the following functions under the given restrictions. (Exercises 1 and 2 are Exercises 5 and 6 of the previous section; check your answers.)

*1. Maximize $2x + y$
subject to
$$4x + y \leq 36$$
$$4x + 3y \leq 60$$
$$x \geq 0$$
$$y \geq 0$$

2. Maximize $x - 4y$
subject to
$$x + 2y \leq 4$$
$$x + 6y \leq 8$$
$$x \geq 0$$
$$y \geq 0$$

*3. Maximize $4x + 6y$
subject to
$$x + 3y \leq 6$$
$$3x + y \leq 8$$
$$x \geq 0$$
$$y \geq 0$$

4. Maximize $2x + 3y$
subject to
$$-4x + 2y \leq 7$$
$$2x - 2y \leq 5$$
$$x \geq 0$$
$$y \geq 0$$

*5. Maximize $10x + 5y$
subject to
$$x + y \leq 180$$
$$3x + 2y \leq 480$$
$$x \geq 0$$
$$y \geq 0$$

6. Maximize $x + 2y + z$
subject to
$$3x + y + z \leq 3$$
$$x - 10y - 4z \leq 20$$
$$x \geq 0$$
$$y \geq 0$$
$$z \geq 0$$

*7. Maximize $100x + 200y + 50z$
subject to
$$5x + 5y + 10z \leq 1000$$
$$10x + 8y + 5z \leq 2000$$
$$10x + 5y \leq 500$$
$$x \geq 0$$
$$y \geq 0$$
$$z \geq 0$$

8. Maximize $2x_1 + 4x_2 + x_3$
subject to
$$-x_1 + 2x_2 + 3x_3 \leq 6$$
$$-x_1 + 4x_2 + 5x_3 \leq 5$$
$$-x_1 + 5x_2 + 7x_3 \leq 7$$
$$x_1 \geq 0$$
$$x_2 \geq 0$$
$$x_3 \geq 0$$

*9. Maximize $2x_1 + x_2 + x_3$
subject to
$$x_1 + 2x_2 + 4x_3 \leq 20$$
$$2x_1 + 4x_2 + 4x_3 \leq 60$$
$$3x_1 + 4x_2 + x_3 \leq 90$$
$$x_1 \geq 0$$
$$x_2 \geq 0$$
$$x_3 \geq 0$$

10. Maximize $x_1 + 2x_2 + 4x_3$
subject to
$$8x_1 + 5x_2 - 4x_3 \leq 30$$
$$-2x_1 + 6x_2 + x_3 \leq 5$$
$$-2x_1 + 2x_2 + x_3 \leq 15$$
$$x_1 \geq 0$$
$$x_2 \geq 0$$
$$x_3 \geq 0$$

*11. Maximize $x_1 + 2x_2 + 4x_3 - x_4$
subject to
$$5x_1 + 4x_3 + 6x_4 \leq 20$$
$$4x_1 + 2x_2 + 2x_3 + 8x_4 \leq 40$$
$$x_1 \geq 0$$
$$x_2 \geq 0$$
$$x_3 \geq 0$$
$$x_4 \geq 0$$

12. Maximize $x_1 + 2x_2 - x_3 + 3x_4$
subject to
$$2x_1 + 4x_2 + 5x_3 + 6x_4 \leq 24$$
$$4x_1 + 4x_2 + 2x_3 + 2x_4 \leq 4$$
$$x_1 \geq 0$$
$$x_2 \geq 0$$
$$x_3 \geq 0$$
$$x_4 \geq 0$$

*13. A manufacturing company uses three machines I, II, and III to produce three products X, Y, and Z. It takes 2, 4, and 0 minutes of time on each of the machines, respectively, to manufacture a single X. It takes 3, 0, and 6 minutes on each of the machines to produce a single Y. To manufacture a Z involves 1, 2, and 3 minutes on each of the machines. The total time available on each machine per day is 6 hours. The profits are $10, $8, and $12 on each of X, Y, and Z, respectively. How should the company allocate the production times on the machines in order to maximize total profit?

14. An industrial furniture company manufactures desks, cabinets, and chairs. These items involve metal, wood, and plastic. The following table represents in convenient units the amounts that go into each product and the profit on each item.

	Metal	Wood	Plastic	Profit
Desk	3	4	2	$16
Cabinet	6	1	1	$12
Chair	1	2	2	$6

If the company has available 800 units of metal, 400 units of wood, and 100 units of plastic, how should it allocate these resources in order to maximize total profit?

*15. A washing machine manufacturing company produces its machines at three factories A, B, and C. The washing machines are sold in a certain city P. It costs $10, $20, and $40 to transport each washing machine, on an average, from A, B, and C, respectively, to P. It has been estimated that it involves 6, 4, and 2 hours of packing and transportation time to get a washing machine from A, B, and C, respectively, to P. There is $6000 budgeted weekly for transportation of the washing machines to P and a total of 4000 hours of labor available. How should the company schedule its weekly transportation arrangements in order to maximize total profit if the profit on each machine from A is $12, on each from B is $20, and on each from C is $16?

16. A manufacturer makes three lines of tents: the Alpine, the Cub, and the Aspen. They are all made out of the same material; the Alpine requires 30 square yards, the Cub 15 square yards, and the Aspen 60 square yards. The manufacturing cost of

the Alpine is $20, the Cub $12, and the Aspen $32. The material is available in amounts of 7800 square yards weekly, and the weekly working budget of the company is $8320. If the profit is $8 on each Alpine, $4 on each Cub, and $12 on each Aspen, what should the weekly production schedule be to maximize total profit?

*17. A furniture company finishes two kinds of tables X and Y. There are three steps in the finishing process: sanding, staining, and varnishing. The various times in minutes of each of these processes for the two tables are as follows:

	Sanding	Staining	Varnishing
X	10	8	4
Y	5	4	8

The three types of equipment needed for sanding, staining, and varnishing are each available for 5 hours per day. Each type of equipment can handle only one table at a time. The profit on each X table is $8 and on each Y table is $4. How many of each should be finished daily to maximize total profit?

In Exercises 18–21, minimize the objective function using the simplex method.

*18. Minimize $f = -x_1 + 2x_2$
subject to
$$x_1 + 2x_2 \leq 4$$
$$x_1 + 4x_2 \leq 6$$
$$x_1 \geq 0$$
$$x_2 \geq 0$$

19. Minimize $f = -2x_1 + x_2$
subject to
$$2x_1 + 2x_2 \leq 8$$
$$x_1 - x_2 \leq 2$$
$$x_1 \geq 0$$
$$x_2 \geq 0$$

*20. Minimize $f = 2x_1 + x_2 - x_3$
subject to
$$x_1 + 2x_2 - 2x_3 \leq 20$$
$$2x_1 + x_2 \leq 10$$
$$x_1 + 3x_2 + 4x_3 \leq 15$$
$$x_1 \geq 0$$
$$x_2 \geq 0$$
$$x_3 \geq 0$$

21. Minimize $f = x_1 - 2x_2 + 4x_3$
subject to
$$x_1 - x_2 + 3x_3 \leq 4$$
$$2x_1 + 2x_2 - 3x_3 \leq 6$$
$$-x_1 + 2x_2 + 3x_3 \leq 2$$
$$x_1 \geq 0$$
$$x_2 \geq 0$$
$$x_3 \geq 0$$

9-3 GEOMETRICAL EXPLANATION OF THE SIMPLEX METHOD

We now give an explanation, by means of an example, of the sequence of transformations used in the simplex method. Let us maximize the function $f = 2x + 3y$ subject to the following constraints:

$$x + 2y \leq 8$$
$$3x + 2y \leq 12$$

with $x \geq 0$ and $y \geq 0$.

The region of interest is *ABCO* in Figure 9-13. Recall that any point in the region *ABCO* is called a *feasible solution*; any such point satisfies the constraints. A feasible solution that gives a maximum value for f is called an *optimal solution*.

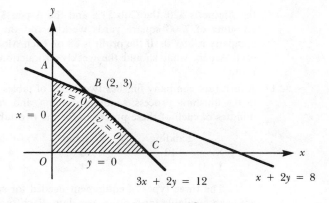

Figure 9-13

We reformulate the constraints using slack variables u and v:

$$\begin{aligned}
x + 2y + u \qquad\qquad &= 8 \\
3x + 2y \qquad + v \qquad &= 12 \\
-2x - 3y \qquad\qquad + f &= 0
\end{aligned} \tag{1}$$

with $x \geq 0$, $y \geq 0$, $u \geq 0$, and $v \geq 0$.

Observe that along AB, $u = 0$, since AB is a segment of the line $x + 2y = 8$. Along BC, $v = 0$, since BC is a segment of the line $3x + 2y = 12$. Furthermore, on OA, $x = 0$; and on OC, $y = 0$. Therefore, the boundaries of the region of interest are such that one variable is 0 along each boundary.

Let us now look at the vertices in terms of the four variables x, y, u, and v. Vertex O is the point $x = 0$, $y = 0$, $u = 8$, $v = 12$. (One obtains the u and v values of O by letting $x = 0$ and $y = 0$ in the constraints.) Vertex A lies on OA and AB. Thus, at A, $x = 0$ and $u = 0$. Substituting these values into the constraints gives $y = 4$ and $v = 4$. Thus A is the point $x = 0$, $y = 4$, $u = 0$, $v = 4$.

Vertex B lies on AB and BC. At B, $x = 2$, $y = 3$, $u = 0$, $v = 0$. Vertex C lies on OC and BC. At C, $y = 0$ and $v = 0$. From the constraints we get that $x = 4$ and $u = 4$. Vertex C is the point $x = 4$, $y = 0$, $u = 4$, $v = 0$.

Observe that at each vertex, certain variables are nonzero and others are zero. The variables that are not zero are called *basic variables* for that vertex;

Table 9-1

Vertex	Coordinates	Basic Variables $(\neq 0)$	Nonbasic Variables $(=0)$
O	$x = 0, y = 0, u = 8, v = 12$	u, v	x, y
A	$x = 0, y = 4, u = 0, v = 4$	y, v	x, u
B	$x = 2, y = 3, u = 0, v = 0$	x, y	u, v
C	$x = 4, y = 0, u = 4, v = 0$	x, u	y, v

the remaining variables are called *nonbasic variables.* We summarize the results thus far with Table 9-1.

The method that we shall develop starts at a vertex feasible solution, O in our case, and then proceeds through a sequence of adjoining vertices, each one giving an increased value of f, until an optimal solution is reached. The initial simplex tableau corresponds (in a way to be explained later) to the situation at the initial feasible solution; further tableaux represent the pictures at other vertices. We must examine each vertex in turn to see if it is an optimal solution. If it is not optimal, we must decide which neighboring vertex to move to next. Our aims are first to develop the tools to carry out this procedure geometrically and then to translate the geometrical concepts into analogous algebraic ones. The advantage of carrying out the procedure algebraically is that algebraic procedures lend themselves to the use of the computer. We shall find that the algebraic procedures that result are those demonstrated previously in the simplex method.

Let us start at O, a vertex feasible solution. We demonstrate whether or not it is necessary to move from O to an adjacent vertex. There are two vertices adjacent to O: A and C. In moving along OC, we find that $y = 0$ and that f increases by $2(f = 2x + 3y)$ for each unit increase in x. On the other hand, in moving along OA, we find that $x = 0$ and that f increases by 3 for each unit increase in y. Because of the larger rate of increase in f along OA, this path is selected; A becomes the next feasible solution to be examined. The value of f at A is 12. This is indeed an improvement on the value of f at O.

We are now confronted with determining whether or not A is optimal. In examining O, we expressed f at that point in terms of the nonbasic variables x and y: $f = 2x + 3y$. This representation led to the decision to move along the path OA to A. At A, we carry out a similar analysis by expressing f in terms of the nonbasic variables x and u at this point. To do this, we substitute the value for y from the first constraint in system (1) into $f = 2x + 3y$. We have from the first constraint that $y = 4 - x/2 - u/2$. Thus, in terms of x and u,

$$f = 2x + 3\left(4 - \frac{x}{2} - \frac{u}{2}\right)$$

$$= 12 + \frac{x}{2} - \frac{3u}{2}$$

In moving along AB from A, we find that $u = 0$ and that f increases by $\frac{1}{2}$ for every unit increase in x. In moving along AO from A, we find that $x = 0$ and that f decreases by $\frac{3}{2}$ for every unit increase in u. (This decrease is to be expected, of course, as we originally moved from O to A in order to increase f.) Thus, because of the associated increase in f, we move from A to B. The value of f at B is 13, an increase over its value at A.

Having arrived at B, we are confronted with the question of whether or not this is an optimal solution. The nonbasic variables at B are u and v. We express f in terms of these variables. From the original form of restrictions (1), we get

$$x = 2 + \frac{u}{2} - \frac{v}{2} \quad \text{and} \quad y = 3 - \frac{3u}{4} + \frac{v}{4}$$

Thus,

$$f = 2x + 3y$$

$$= 2\left(2 + \frac{u}{2} - \frac{v}{2}\right) + 3\left(3 - \frac{3u}{4} + \frac{v}{4}\right)$$

$$= 13 - \frac{5u}{4} - \frac{v}{4}$$

In moving along BC from B, we find that $v = 0$ and that f decreases by $\frac{5}{4}$ for every unit increase in u. In moving along BA from B, we find that $u = 0$ and that f decreases by $\frac{1}{4}$ for every unit increase in v. Thus, f has a maximum value at B. This maximum value is 13.

We have developed the geometrical concepts. We now translate them into algebraic form. The initial feasible solution was O. The basic variables at O are u and v, having values 8 and 12, respectively. The value of f at O is 0. Let us write an initial tableau (in a more complete form than previously) to reflect this:

Coefficients of

$$\begin{array}{cccc}
x & y & u & v \\
\end{array}$$

$$\begin{pmatrix}
1 & 2 & 1 & 0 & 8 \\
3 & 2 & 0 & 1 & 12 \\
-2 & -3 & 0 & 0 & 0
\end{pmatrix} \begin{matrix} \leftarrow u \\ \leftarrow v \\ \leftarrow f \end{matrix} \Big\} \text{basic variables}$$

The last column gives the values of the basic variables and f at O. Knowing the basic variables, we also know the nonbasic variables. This is the tableau associated with O.

Our geometric discussion told us that O was not an optimal solution and took us to A. The basic variables at O are u and v; at A, the basic variables are y and v. In going from O to A, we find that y replaces u as a basic variable. We call u the *departing basic variable* and y the *entering basic variable*. In moving from O to A, we found that y, the entering variable, corresponded to the largest rate of increase in f, 3. In terms of the above tableau, this is reflected by y being the column corresponding to the negative entry in the last row that is largest in magnitude, -3. This rule enables us to select the entering basic variable (if one exists) from any tableau.

The next step is to determine the departing variable for the tableau. In going from O to A, we found that the departing variable was u. The first two rows of the tableau correspond to the following equations:

$$x + 2y + u \quad\quad = 8$$
$$3x + 2y \quad\quad + v = 12$$

In going along OA, we find that $x =$ zero. Thus, the equations may be rewritten as

$$u = 8 - 2y$$
$$v = 12 - 2y$$

Since $u \geq 0$ and $v \geq 0$, the maximum value to which y is allowed to increase is 4, when $u = 0$ and $v = 4$ (at the point A). If y goes beyond 4, u will become negative. Thus, u becomes a nonbasic variable. Looking at the above two equations, we see that the reason u arrives at 0 before v is that $\frac{8}{2} < \frac{12}{2}$. In terms of the tableau, this corresponds to dividing the elements in the entering variable column into the corresponding elements of the last column. The row containing the numbers that give the smallest result gives the departing variable. Thus,

$$
\begin{array}{cccc}
x & y & u & v \\
\end{array}
$$

$$
\left(
\begin{array}{cccc}
1 & 2 & 1 & 0 \\
3 & 2 & 0 & 1 \\
-2 & -3 & 0 & 0 \\
\end{array}
\right.
\left.
\begin{array}{c}
8 \\
12 \\
0 \\
\end{array}
\right)
\begin{array}{l}
u \leftarrow departing \\
v \quad\ variable \\
f \\
\end{array}
$$

$$\uparrow$$
entering
variable

We now have a method for selecting the entering and departing variables in any tableau.

We must next decide how to transform the above tableau into the tableau that corresponds to the vertex having the new basic variables y and v. We want the new tableau to have the value of the new basic variables and f, at the new vertex, as its last column. The initial constraints, from which the initial tableau was derived, were as follows:

$$x + 2y + u \quad\quad\quad = 8$$
$$3x + 2y \quad\ + v \quad\quad = 12$$
$$-2x - 3y \quad\quad\ + f = 0$$

Observe that the basic variables u and v appear in a single equation each, the coefficient being 1 in each case. It is this fact that causes u and v to assume the values 8 and 12, respectively, at O, when we let the nonbasic variables x and y become zero. Furthermore, the fact that the last equation involves f and only the nonbasic variables causes f to assume the value 0 on the right-hand side of the equation. These are the characteristics that we must attempt to obtain for the tableau that represents A, in terms of the basic variables y and v of A. We achieve this by selecting as pivot the element that lies in the entering variable column and departing variable row, creating a 1 in its location and zeros elsewhere in this column. (Create zeros by adding suitable multiples of the pivot row to other rows.) Thus, the sequence of tableaux becomes the following:

$$
\begin{array}{cccc}
x & y & u & v
\end{array}
$$

$$
\begin{array}{c}
\text{pivot} \\
\downarrow
\end{array}
$$

$$
\begin{pmatrix}
1 & ② & 1 & 0 & 8 \\
3 & 2 & 0 & 1 & 12 \\
-2 & -3 & 0 & 0 & 0
\end{pmatrix}
\begin{matrix} u \leftarrow \\ v \\ f \end{matrix}
\;\cong\;
\begin{pmatrix}
\tfrac{1}{2} & ① & \tfrac{1}{2} & 0 & 4 \\
3 & 2 & 0 & 1 & 12 \\
-2 & -3 & 0 & 0 & 0
\end{pmatrix}
\begin{matrix} \leftarrow \\ \\ \end{matrix}
$$

$$
\uparrow \qquad\qquad\qquad\qquad\qquad \uparrow
$$

$$
\begin{array}{cccc}
x & y & u & v
\end{array}
$$

$$
\cong
\begin{pmatrix}
\tfrac{1}{2} & 1 & \tfrac{1}{2} & 0 & 4 \\
2 & 0 & -1 & 1 & 4 \\
-\tfrac{1}{2} & 0 & \tfrac{3}{2} & 0 & 12
\end{pmatrix}
\begin{matrix} y \\ v \end{matrix} \left.\begin{matrix}\ \\ \ \end{matrix}\right\} \begin{matrix}\text{new basic} \\ \text{variables}\end{matrix} \\
\quad f \quad \text{new value of } f
$$

Let us verify that this tableau does indeed correspond to a system of equations that gives $y = 4$, $v = 4$, and $f = 12$ in the above order, if we take y and v as basic variables. The corresponding equations are as follows:

$$
\begin{aligned}
\tfrac{1}{2}x + y + \tfrac{1}{2}u & = 4 \\
2x \quad - u + v & = 4 \\
-\tfrac{1}{2}x \quad + \tfrac{3}{2}u + f & = 12
\end{aligned}
$$

Taking x and u as nonbasic variables, we make them zero, and we do indeed get $y = 4$, $v = 4$, $f = 12$ as claimed. This is the tableau for the vertex A.

The analysis is now repeated for this tableau. Because of the negative entry, $-\tfrac{1}{2}$, in the last row, this value 12 for f is not an optimal solution. The new entering variable is x because of the negative sign in this column; the departing variable is v, since $\tfrac{4}{2} < 4/\tfrac{1}{2}$:

$$
\begin{array}{cccc}
x & y & u & v
\end{array}
$$

$$
\text{pivot} \rightarrow
\begin{pmatrix}
\tfrac{1}{2} & 1 & \tfrac{1}{2} & 0 & 4 \\
② & 0 & -1 & 1 & 4 \\
-\tfrac{1}{2} & 0 & \tfrac{3}{2} & 0 & 12
\end{pmatrix}
\begin{matrix} y \\ v \leftarrow \text{departing} \\ f \quad\ \text{variable} \end{matrix}
$$

$$
\uparrow
$$
$$
\begin{matrix}\text{entering} \\ \text{variable}\end{matrix}
$$

The entering variable, x, will replace the v in the second row. The sequence of transformations is as follows:

$$
\begin{array}{cccc}
x & y & u & v \\
\end{array}
$$

$$
\begin{pmatrix}
\frac{1}{2} & 1 & \frac{1}{2} & 0 & 4 \\
\textcircled{2} & 0 & -1 & 1 & 4 \\
-\frac{1}{2} & 0 & \frac{3}{2} & 0 & 12 \\
\end{pmatrix}
\begin{array}{c} y \\ v \leftarrow \\ f \end{array}
\cong
\begin{pmatrix}
\frac{1}{2} & 1 & \frac{1}{2} & 0 & 4 \\
\textcircled{1} & 0 & -\frac{1}{2} & \frac{1}{2} & 2 \\
-\frac{1}{2} & 0 & \frac{3}{2} & 0 & 12 \\
\end{pmatrix} \leftarrow
$$

$$
\begin{array}{cccc}
x & y & u & v \\
\end{array}
$$

$$
\cong
\begin{pmatrix}
0 & 1 & \frac{3}{4} & -\frac{1}{4} & 3 \\
1 & 0 & -\frac{1}{2} & \frac{1}{2} & 2 \\
0 & 0 & \frac{5}{4} & \frac{1}{4} & 13 \\
\end{pmatrix}
\begin{array}{c} y \\ x \\ f \end{array}
$$

This is the final tableau. The basic variables are x and y, assuming the values 2 and 3, respectively. This is the tableau for the vertex B. The value of f is 13; it is the maximum value possible under the given restrictions.

✦ Example 1

Let us return to Example 2 of Section 9-2, the linear programming problem that had a nonunique solution. We now analyze it in terms of entering and departing variables in order to see how the final tableau can be interpreted for such a problem.

The objective function is $8x + 2y$, and the constraints are

$$
4x + y \le 32
$$
$$
4x + 3y \le 48
$$

with $x \ge 0$ and $y \ge 0$. These lead to the introduction of the slack variables u and v:

$$
\begin{aligned}
4x + y + u \phantom{{}+v} &= 32 \\
4x + 3y \phantom{{}+u} + v &= 48 \\
-8x - 2y \phantom{{}+u+v} + f &= 0
\end{aligned}
$$

with $x \ge 0$, $y \ge 0$, $u \ge 0$, and $v \ge 0$.

The simplex tableaux are

$$
\begin{array}{cccc}
x & y & u & v \\
\end{array}
$$

$$
\begin{pmatrix}
\textcircled{4} & 1 & 1 & 0 & 32 \\
4 & 3 & 0 & 1 & 48 \\
-8 & -2 & 0 & 0 & 0 \\
\end{pmatrix}
\begin{array}{l} u \leftarrow \textit{departing} \\ \textit{variable} \\ f \end{array}
\cong
\begin{pmatrix}
\textcircled{1} & \frac{1}{4} & \frac{1}{4} & 0 & 8 \\
4 & 3 & 0 & 1 & 48 \\
-8 & -2 & 0 & 0 & 0 \\
\end{pmatrix}
\begin{array}{l} u \leftarrow \\ v \\ f \end{array}
$$

entering
variable

$$
\begin{array}{cccc}
x & y & u & v \\
\end{array}
$$

$$
\cong
\begin{pmatrix}
1 & \frac{1}{4} & \frac{1}{4} & 0 & 8 \\
0 & 2 & -1 & 1 & 16 \\
0 & 0 & 2 & 0 & 64 \\
\end{pmatrix}
\begin{array}{l} x \\ v \\ f \end{array}
$$

This is the *final tableau*. It leads to a maximum value of 64 for f. It occurs at $x = 8$, $v = 16$. Since y is a nonbasic variable, $y = 0$. Thus, we have obtained the optimal solution $x = 8$, $y = 0$ (the point C in Figure 9-12), with $f = 64$. When the simplex method is applied to a problem that contains many solutions, it stops as soon as it finds one optimal solution, as shown here. We now show how to extend the tableaux algorithm to find the other solutions.

Observe that in the final tableau, the coefficient of y, a nonbasic variable, is 0 in the row for f. Each coefficient of a nonbasic variable in this row indicates the rate at which f increases as that variable is increased. Thus, making y an entering variable neither increases nor decreases f. We use y as an entering variable; the departing variable is then v. Therefore, we get the following sequence of tableaux:

$$
\begin{array}{cccc}
x & y & u & v \\
\end{array}
\begin{pmatrix}
1 & \frac{1}{4} & \frac{1}{4} & 0 & 8 \\
0 & ② & -1 & 1 & 16 \\
0 & 0 & 2 & 0 & 64
\end{pmatrix}
\begin{array}{l}
x \\
v \leftarrow \textit{departing} \\
f \quad \textit{variable}
\end{array}
\cong
\begin{array}{cccc}
x & y & u & v \\
\end{array}
\begin{pmatrix}
0 & \frac{1}{4} & \frac{1}{4} & 0 & 8 \\
0 & ① & -\frac{1}{2} & \frac{1}{2} & 8 \\
0 & 0 & 2 & 0 & 64
\end{pmatrix}
\begin{array}{l}
x \\
v \leftarrow \\
f
\end{array}
$$

entering variable

$$
\cong
\begin{array}{cccc}
x & y & u & v \\
\end{array}
\begin{pmatrix}
0 & 0 & \frac{3}{8} & -\frac{1}{8} & 6 \\
0 & 1 & -\frac{1}{2} & \frac{1}{2} & 8 \\
0 & 0 & 2 & 0 & 64
\end{pmatrix}
\begin{array}{l}
x \\
y \\
f
\end{array}
$$

This tableau leads to the optimal solution $x = 6$, $y = 8$ (the point B in Figure 9-3), with f having a maximum value of 64. All points on the line between the two optimal solutions obtained using the tableaux—namely, $(8, 0)$ and $(6, 8)$—would also be optimal solutions.   ✦

We now discuss the geometrical interpretation of the simplex method for the problem having no solution, the objective function being unbounded. Let us return to Example 3 of the previous section.

✦ *Example 2*

Maximize the function $x + 4y$ subject to the following constraints:

$$-4x + y \le 2$$
$$2x - y \le 1$$

with $x \ge 0$ and $y \ge 0$.

The simplex tableaux are as follows:

$$
\begin{array}{cccc}
x & y & u & v \\
\end{array}
$$

$$
\left(\begin{array}{rrrrr}
-4 & ① & 1 & 0 & 2 \\
-2 & -1 & 0 & 1 & 1 \\
-1 & -4 & 0 & 0 & 0
\end{array}\right)\begin{array}{c} u \leftarrow \\ v \\ f \end{array}
\cong
\left(\begin{array}{rrrrr}
-4 & ① & 1 & 0 & 2 \\
-2 & 0 & 1 & 1 & 3 \\
-1 & -4 & 0 & 0 & 0
\end{array}\right)
$$

$$
\begin{array}{cccc}
x & y & u & v \\
\end{array}
$$

$$
\cong
\left(\begin{array}{rrrrr}
-4 & ① & 1 & 0 & 2 \\
-2 & 0 & 1 & 1 & 3 \\
-17 & 0 & 4 & 0 & 8
\end{array}\right)\begin{array}{c} y \\ v \\ f \end{array}
$$

Because of the negative signs in the pivot column, the first column, one cannot choose a pivot; there is no departing variable. What is the geometrical interpretation? The associated system of equalities is

$$
\begin{array}{rl}
-4x + y + u & = 2 \\
-2x + u + v & = 3 \\
-17x + 4u + f & = 8
\end{array}
$$

The entering variable should be x (that is, the variable that is currently 0 and should be increased). It can be seen that x can be increased indefinitely, if y and v (the basic variables) and f are increased. There is no limit to f.

     Recall that selecting the column with the negative entry in the last row that is largest in magnitude to be the pivot column is a convenience that enables us to obtain a final tableau efficiently. At any stage, any nonbasic variable can be selected as an entering variable. In a manner similar to the above, if a tableau appears having a column containing a negative last entry and all nonpositive elements above it, selecting the variable corresponding to that column as entering variable enables the objective function to increase indefinitely; there is no maximum to the objective function.   ✦

EXERCISES

Solve the following linear programming problems by the simplex method. Determine the basic and nonbasic variables and the entering and departing variables for each tableau. Determine the optimal solution and the maximum value of the objective function directly from the final tableau. (These problems were given in the previous set of exercises. Use the tableaux that you have already derived, and check your previous answers.)

*1. Maximize $2x + y$
     subject to
         $4x + y \leq 36$
         $4x + 3y \leq 60$
         $x \geq 0$
         $y \geq 0$
     (Exercise 1, Section 9-2)

2. Maximize $x - 4y$
     subject to
         $x + 2y \leq 4$
         $x + 6y \leq 8$
         $x \geq 0$
         $y \geq 0$
     (Exercise 2, Section 9-2)

*3. Maximize $x + 2y + z$
subject to
$$3x + y + z \leq 3$$
$$x - 10y - 4z \leq 20$$
$$x \geq 0$$
$$y \geq 0$$
$$z \geq 0$$

(Exercise 6, Section 9-2)

*5. Maximize $2x_1 + x_2 + x_3$
subject to
$$x_1 + 2x_2 + 4x_3 \leq 20$$
$$2x_1 + 4x_2 + 4x_3 \leq 60$$
$$3x_1 + 4x_2 + x_3 \leq 90$$
$$x_1 \geq 0$$
$$x_2 \geq 0$$
$$x_3 \geq 0$$

(Exercise 9, Section 9-2)

4. Maximize $100x + 200y + 50z$
subject to
$$5x + 5y + 10z \leq 1000$$
$$10x + 8y + 5z \leq 2000$$
$$10x + 5y \leq 500$$
$$x \geq 0$$
$$y \geq 0$$
$$z \geq 0$$

(Exercise 7, Section 9-2)

6. Maximize $x_1 + 2x_2 + 4x_3 - x_4$
subject to
$$5x_1 + 4x_3 + 6x_4 \leq 20$$
$$4x_1 + 2x_2 + 2x_3 + 8x_4 \leq 40$$
$$x_1 \geq 0$$
$$x_2 \geq 0$$
$$x_3 \geq 0$$
$$x_4 \geq 0$$

(Exercise 11, Section 9-2)

Computing

These appendixes are intended for readers who will be using the computer in the course. BASIC is the language used. Programs are related to sections in the main text where methods have been developed.

APPENDIX A   MATRICES (SECTION 1-2)

There are various techniques for reading in and printing out matrices. In this section, we discuss the approaches used in later programs.

MAT READ and MAT PRINT Statements*

Many computer systems have built-in matrix functions that enable matrices to be read in and printed out as arrays.† The following program reads in and then prints out the matrix

$$\begin{pmatrix} 1 & 2 & 3 \\ 4 & 0 & -4 \end{pmatrix}$$

Statement 5 is an example of a dimension statement. Here it tells the computer to expect an array, called A, consisting of 2 rows and 3 columns.

5 DIM A(2,3)

---

† Some computer systems do not have these built-in matrix commands. Users who do not have these commands available can modify the programs given in the text by using techniques given in Appendix E.

Statement 10 commands the computer to read in the matrix A. It will scan the remainder of the program until it gets to the data in statement 25, then read in these data as a matrix A.

```
10 MAT READ A
```

The computer will print out whatever you enclose between quotation marks. This statement is not essential but makes the printout clearer.

```
15 PRINT "THE MATRIX IS"
```

Tells the computer to print out A.

```
20 MAT PRINT A
```

Note the way the data are typed in; the first row of A is followed by the second row.

```
25 DATA 1,2,3,4,0, −4
```

The last statement must always be END.

```
30 END
```

You tell the computer to execute the program by typing RUN.

```
RUN
```

The output. In this text, boldface type will be used to indicate output.

THE MATRIX IS
1 2 3
4 0 −4

The dimension function in statement 5 in the above program had constant arguments 2 and 3. Many systems allow variables as arguments, such as DIM A(M,N). For such a system, the above program could be modified as follows:

A is m × n. m and n are read in from statement 22.

```
2 READ M,N
5 DIM A(M,N)
10 MAT READ A
15 PRINT "THE MATRIX IS"
```

```
20 MAT PRINT A
22 DATA 2,3
25 DATA 1,2,3,4,0,−4
30 END
```

This program is more versatile than the original one in that the DATA statements are the only ones that need to be modified to accommodate matrices of various sizes.

MAT INPUT Statement

The MAT INPUT statement enables the user to feed in the desired matrix while the program is running, rather than through DATA statements as in the case of MAT READ.

Very often it is a matter of personal preference which of the two MAT statements one uses. If the data set is large, it is usually preferable to use READ. INPUT, on the other hand, allows the user to supply data after seeing partial results.

The MAT INPUT statement causes the computer to request a 2 × 3 matrix. The user types in the elements of the 2 × 3 matrix. The computer then executes the program with this matrix as data. Some systems may require input thus:
?1,2,3
?4,0,−4
The semicolon in the MAT PRINT statement will lead to a more compressed output.

```
5 DIM A(2,3)
10 MAT INPUT A
15 PRINT "THE MATRIX IS"
20 MAT PRINT A;
25 END

RUN

?1,2,3,4,0,−4
```

THE MATRIX IS

1	2	3
4	0	−4

TAB Function for Matrix Output

If the output matrix has elements consisting of various significant figures, the MAT PRINT statement may not, on some systems, lead to a tidy output—the columns of the matrix may not be properly aligned. When such an output is expected, one can use the following method to print out the matrix:

The TAB(X) function causes the computer to "tab" to column X using the current value of X and there print the current value of a_{ij}.

```
5 READ M,N
10 MAT READ A(M,N)
15 PRINT "THE MATRIX IS"
20 FOR I = 1 TO M
25 FOR J = 1 TO N
30 PRINT TAB (J*8−7);A(I,J);
35 NEXT J
40 PRINT
45 NEXT I
50 DATA 2,3
55 DATA 1.2,2.3456,6,2.7314,7,2
60 END

RUN
```

THE MATRIX IS

1.2	**2.3456**	**6**
2.7314	**7**	**2**

EXERCISES

The following programs are to be run on the computer. What will the output be in each case?

*1.
```
5 DIM A(2,4)
10 MAT READ A
15 PRINT "THE MATRIX A IS"
20 MAT PRINT A;
25 DATA 0,1,−2,4,−3,6,7,8
30 END
```

2.
```
5 DIM A(4,1)
10 MAT READ A
15 PRINT "A ="
20 MAT PRINT A;
25 DATA −1,2,3,4
30 END
```

*3.
```
5 DIM A(1,3)
10 MAT READ A
15 MAT PRINT A;
20 DATA −1,2,3
40 END

5 DIM A(2,3)
20 DATA −1,2,3,4,5,1
```

4. Write programs to print out the following matrices.

(a) $\begin{pmatrix} 1 & 2 \\ 3 & 4 \end{pmatrix}$

(b) $\begin{pmatrix} 1 \\ 2 \\ 3 \end{pmatrix}$

(c) $(-1 \quad 0.3 \quad 4 \quad 6)$

(d) $\begin{pmatrix} 1.32 & -76 & 2 \\ 0 & 1 & 4 \end{pmatrix}$

(e) $\begin{pmatrix} 1 & 2 & -3.6 \\ 7.2 & 8 & 9 \\ 28 & 307 & 21 \end{pmatrix}$

5. Write a program to print out the matrix

$$\begin{pmatrix} 1 & 2 & 3 \\ 4 & 5 & 6 \end{pmatrix}$$

By updating statements in this program, print out each of the following matrices in turn.

(a) $\begin{pmatrix} 1 & 2.3 & 4 \\ 5 & 0 & 7 \end{pmatrix}$ **(b)** $\begin{pmatrix} 1 & 2 \\ 3 & 4 \\ -1 & 2 \end{pmatrix}$ **(c)** $\begin{pmatrix} 1 & 3 & 5 & 6 \\ 7 & 1 & 2 & 4 \end{pmatrix}$

The following programs contain errors. Find them and correct them. Run each program to investigate if and in what manner your system diagnoses each error. Then correct your error.

*6. 5 DIM A(2,3)
   10 MAT READ A
   15 PRINT "THE MATRIX IS"
   20 MAT PRINT A;
   25 DATA 1,2,3,4,5
   30 END

7. 5 DIM A(2,2
   10 MAT READ A
   15 MAT PRINT A;
   20 DATA 1,2,5,7
   25 END

*8. 5 DIM A(1,4)
   10 PRINT "THE MATRIX IS"
   15 MAT PRINT A;
   20 DATA 5, −1,3.2,6
   25 END

9. 5 DIM A(2,2)
   10 DATA 1,2,3,4
   15 MAT PRINT A
   20 MAT READ A
   25 END

10. The statement PRINT A(I,J) prints the specified element of a matrix. For example, PRINT A(2,3) prints out the element in the second row and third column of *A*.

Write out a program that reads in the matrix $A = \begin{pmatrix} 1 & 2 & 3 \\ 4 & 5 & 6 \end{pmatrix}$ and prints out $A(2, 3)$.

11. Let $A = \begin{pmatrix} 1 & 2 & 4 & 5 \\ 0 & 1 & -1 & 2 \end{pmatrix}$. Write a program to print out $A(2, 2)$ and $A(2, 4)$ on a single line.

APPENDIX B MULTIPLICATION OF A MATRIX BY A SCALAR (SECTION 2-1)

The following is a program written to perform the scalar multiplication

$$3\begin{pmatrix} 1 & 2 & 3 \\ -3 & 4 & 8 \end{pmatrix}.$$

5 PRINT "THE SCALAR MULTIPLE IS"

We tell the computer that it will be handling two 2 × 3 matrices, called A and B.

10 DIM A(2,3),B(2,3)

We call the scalar c and the matrix that is read in A.

```
15 READ C
20 MAT READ A
```

The matrix A is multiplied by the scalar c. The resulting matrix is called B. (C) indicates that c is a scalar.

```
25 MAT B = (C)*A
```

; can be used to compress the matrix output.

```
30 MAT PRINT B;
```

All the data could have been placed on one line. The computer would read the first number in the data as c, the remaining numbers as making up A.

```
35 DATA 3
40 DATA 1,2,3,−3,4,8
```

```
45 END

RUN
```

THE SCALAR MULTIPLE IS

3	6	9
−9	12	24

EXERCISES

The following programs are run. What will the output be for each program?

*1.
```
5 DIM A(2,2),B(2,2)
10 READ C
15 MAT READ A
20 MAT B = (C)*A
25 MAT PRINT B;
30 DATA 2
35 DATA 1,−1,2,3
40 END
```

2.
```
5 DIM A(1,2),B(1,2)
10 READ C
15 MAT READ A
20 MAT B = (C)*A
25 PRINT "THE SCALAR MULTIPLE IS"
30 MAT PRINT B;
35 DATA −1,3,2
40 END
```

3. Write programs to perform the following multiplications:

(a) $2\begin{pmatrix} 1 & -2 \\ 3 & 4 \end{pmatrix}$ **(b)** $-3\begin{pmatrix} 2 & -4 & 6 \\ 3.2 & 78 & 1 \end{pmatrix}$ **(c)** $4\begin{pmatrix} 1 \\ 2 \\ -3 \\ 4 \end{pmatrix}$

4. Write a program to execute $4\begin{pmatrix} 1 & 2 \\ 3 & 4 \end{pmatrix}$. Update your program to perform the following operations in turn:

(a) $3\begin{pmatrix} -1 & 0 \\ 2 & 4 \end{pmatrix}$ **(b)** $4\begin{pmatrix} 1 & 2 \\ 3 & 4 \\ 5 & 6 \end{pmatrix}$ **(c)** $-3(1 \quad 2 \quad 4 \quad 1)$

5. Modify the program for scalar multiplication given in this section to contain input statements.

*6. Write a program that multiplies every element in the third row of the matrix

$$A = \begin{pmatrix} 1 & 2 & 3 \\ 0 & -1 & 2 \\ 2 & 1 & -1 \\ 1 & 2 & -3 \end{pmatrix}$$

by 3 to give a matrix B. All the other rows of B are identical to the corresponding rows of A. Multiplying a row of a matrix by a nonzero scalar is an elementary matrix transformation. This program illustrates how the computer can be used to carry out this type of transformation.

7. Write a program that interchanges the third and fifth rows of the matrix

$$A = \begin{pmatrix} 1 & 3 & -1 & 4 \\ 2 & 0 & -1 & 4 \\ 5 & 3 & 2 & -2 \\ 0 & 4 & 1 & 2 \\ 3 & -1 & 2 & 0 \end{pmatrix}$$

Observe that interchanging two rows of a matrix is another elementary matrix transformation.

8. Let

$$A = \begin{pmatrix} 3 & -1 & 2 & 4 \\ 0 & 2 & 2 & 1 \\ 1 & 3 & 4 & 5 \end{pmatrix}$$

(a) Use your program from Exercise 6 to divide the second row of A by 2.

(b) Use your program from Exercise 7 to interchange the first and third rows of A.

APPENDIX C   ADDITION OF MATRICES (SECTION 2-1)

We will add the matrices

$$A = \begin{pmatrix} 1 & 3 & 7 \\ 9 & -9 & 0.8 \\ 0 & 5 & -6 \\ 0.5 & -3 & 6 \end{pmatrix} \quad \text{and} \quad B = \begin{pmatrix} 0 & 9 & -7 \\ 9 & 4 & 34 \\ 7 & -6 & 8 \\ 1 & 0 & 8 \end{pmatrix}$$

5 PRINT "THE SUM IS"

We tell the computer to expect to be working with three 4 × 3 matrices A, B, and C.

10 DIM A(4,3),B(4,3),C(4,3)

The computer reads in the matrices A and B from the data statements and calls their sum C.

15 MAT READ A
20 MAT READ B
25 MAT C = A + B

30 MAT PRINT C;
35 DATA 1,3,7,9,−9,.8,0,5,−6,.5,−3,6
40 DATA 0,9,−7,9,4,34,7,−6,8,1,0,8
45 END

RUN

THE SUM IS

1	12	0
18	−5	34.8
7	−1	2
1.5	−3	14

EXERCISES

The following programs are run. Give the output that will be obtained for each program.

*1. 5 DIM X(2,1),Y(2,1),Z(2,1)
    10 MAT READ X
    15 MAT READ Y
    20 MAT Z = X + Y
    25 MAT PRINT Z;
    30 DATA 1,2,3,4
    35 END

2. 5 PRINT "C, THE SUM OF A AND B, IS"
10 DIM A(2,3),B(2,3),C(2,3)
15 MAT READ A
20 MAT READ B
25 MAT C = A + B
30 MAT PRINT C;
35 DATA $-1,2,3,4,-3,5,7,2,0,1,-3,2$
40 END

3. Write programs to perform the following matrix additions (if possible) on the computer:

(a) $\begin{pmatrix} 1 & 2 \\ 3 & 4 \end{pmatrix} + \begin{pmatrix} 0 & 1 \\ 4 & 6 \end{pmatrix}$ (b) $\begin{pmatrix} 1 & 2 & -3 \\ 4 & 1 & 2 \end{pmatrix} + \begin{pmatrix} 0 & 1 & 2 \\ -1 & 3 & 4 \end{pmatrix}$

(c) $\begin{pmatrix} 1 \\ 2 \end{pmatrix} + (3 \quad 4)$ (Investigate to see how the computer tells you that this addition is not possible.)

(d) $\begin{pmatrix} 1 & 3 \\ 2 & 4 \end{pmatrix} + \begin{pmatrix} 3 & 4 \\ -1 & 2 \end{pmatrix} + \begin{pmatrix} 4 & 6 \\ -1 & 3 \end{pmatrix}$

(e) $(1 \quad 2 \quad 3 \quad 4) + (0 \quad -1 \quad 4 \quad 6) + (7 \quad 3 \quad 1 \quad 2)$

4. Write programs to evaluate the following on the computer:

(a) $3\begin{pmatrix} 1 & 2 \\ 3 & 4 \end{pmatrix} + \begin{pmatrix} 1 & 0 \\ -1 & 2 \end{pmatrix}$ (b) $4\begin{pmatrix} 3 & 1 \\ 4 & 1 \end{pmatrix} + 5\begin{pmatrix} 1 & 3 \\ 3 & -1 \end{pmatrix}$

(c) $-1(1 \quad 3 \quad 4) + 2(3 \quad 5 \quad -1) - 4(2 \quad 4 \quad 6)$

5. Modify the given program for adding matrices to use input statements.

6. Write a program to determine a matrix B whose second row is the second row of

$$A = \begin{pmatrix} 1 & 2 & 3 & -1 \\ 1 & 1 & -1 & 2 \\ 2 & 0 & -2 & 4 \end{pmatrix}$$ minus the first row of A. The first and third rows of B

are identical to those of A. Note that this creates from matrix A a matrix B which has a zero in its (2, 1) location. Write a program to determine a matrix C whose third row is the third row of B minus twice the first row of B. The first and second rows of C are identical to those of B. This creates from B a matrix C which has a 0 in its (3, 1) location. In this manner we can use the computer to perform Gauss-Jordan elimination. In Appendix F this type of program is incorporated into a larger program that performs Gauss-Jordan elimination.

7. Let A be a 3 × 6 matrix and let B be the 3 × 3 matrix whose columns are the last three columns of A. Write a program that will print out B given A. Test your program on

$$A = \begin{pmatrix} 1 & 2 & 3 & 4 & 1 & 0 \\ 0 & 1 & 2 & -1 & 0 & 4 \\ 3 & 1 & -2 & 4 & 1 & -3 \end{pmatrix}$$

B will be the matrix

$$B = \begin{pmatrix} 4 & 1 & 0 \\ -1 & 0 & 4 \\ 4 & 1 & -3 \end{pmatrix}$$

We shall use this program later as a subroutine in a program for determining the inverse of a matrix.

8. Let *A* be a 5 × 4 matrix. Write a program that will scan the last four elements of the second column to determine the element with the largest absolute value and then interchange the row containing this element with the second row. Test your program on the matrix

$$A = \begin{pmatrix} 1 & 2 & -1 & 3 \\ 0 & 2 & 0 & 4 \\ 0 & 3 & 2 & 1 \\ 0 & -4 & 0 & 1 \\ 0 & 2 & 1 & 1 \end{pmatrix}$$

APPENDIX D   MULTIPLICATION OF MATRICES (SECTION 2-2)

We will perform the following matrix multiplication using the computer:

$$\begin{pmatrix} 4 & 7 \\ -6 & -89 \\ 0 & 2 \end{pmatrix} \begin{pmatrix} 4 & -1 \\ 7 & 3 \end{pmatrix}$$

5 PRINT "THE PRODUCT IS"

We are going to call the above matrices A and B. Their product C will be a 3 × 2 matrix. Here we see the need to be able to predict the shape of the product matrix. (On some systems the C(3, 2) can be omitted.)

10 DIM A(3,2),B(2,2),C(3,2)

15 MAT READ A
20 MAT READ B

This is the statement for multiplying the matrices A and B and naming their product C.

25 MAT C = A∗B

```
30 MAT PRINT C;
35 DATA 4,7,−6,−89,0,2
40 DATA 4,−1,7,3
45 END

RUN
```

THE PRODUCT IS

```
  65     17
−647   −261
  14      6
```

EXERCISES

Give the output that would be obtained on running each of the following programs.

*1.
```
 5 DIM A(2,1),B(1,3),C(2,3)
10 MAT READ A
15 MAT READ B
20 MAT C = A ∗ B
25 MAT PRINT C;
30 DATA 1,0,−1,2,3
35 END
```

2.
```
 5 DIM A(2,2),B(2,1),C(2,1)
10 MAT READ A
15 MAT READ B
20 MAT C = A ∗ B
25 PRINT "THE PRODUCT MATRIX IS"
30 MAT PRINT C;
35 DATA 1,2 0,−1
40 DATA 0,1
45 END
```

3. If $A = \begin{pmatrix} 1 & 2 & -1 \\ 3 & 1 & 0 \\ 4 & 1 & 1 \end{pmatrix}$, $B = \begin{pmatrix} 2 & 0 \\ -1 & 7 \\ -4 & 5 \end{pmatrix}$, $C = \begin{pmatrix} 2 & 3 & 1 \\ -1 & 4 & 2 \\ -5 & 1 & 3 \end{pmatrix}$, write programs to determine

(a) AB **(b)** AC **(c)** CA **(d)** BA

(e) A^2 **(f)** ACB **(g)** A^2BC

if these products exist.

4. Write a program to evaluate the following:

(a) $3\begin{pmatrix} 1 & 2 & 4 \\ -1 & 3 & 2 \end{pmatrix}\begin{pmatrix} 1 & 0 \\ 2 & 1 \\ 3 & 4 \end{pmatrix} + 2\begin{pmatrix} 1 & 2 \\ 3 & 1 \end{pmatrix} + 4\begin{pmatrix} 2 & 0 \\ 1 & 3 \end{pmatrix}$

(b) $-1\begin{pmatrix} 2 & 3 \\ 4 & 1 \end{pmatrix}\begin{pmatrix} 1 & 3 \\ 2 & 4 \end{pmatrix}\begin{pmatrix} 1 & 0 \\ 1 & 1 \end{pmatrix}$

(c) $2(1 \quad 2) + 3(1 \quad 4) + 5(1 \quad 3)\begin{pmatrix} 1 & 2 \\ 3 & 4 \end{pmatrix}$

5. Write a program that can be used to compute powers of a given square matrix. Use your program to compute the first 20 powers of the matrix $\begin{pmatrix} 1 & 0.2 \\ 3 & -4 \end{pmatrix}$.

6. Write a program that can be used to determine the transpose of a given matrix. Your computer system may have a built-in transpose function TRN(A). Investigate. (Transpose is defined in Section 2-4.) Use your program to determine the transpose of the matrix

$$\begin{pmatrix} 1 & 2 \\ 3 & 4 \\ -1 & 2 \end{pmatrix}$$

7. Write a program to determine the trace of a matrix. Use your program to determine the traces of the following matrices. (Trace is defined in Section 2-4.)

(a) $\begin{pmatrix} 1 & 2 & 0 \\ 0 & -1 & 3 \\ 4 & 1 & 2 \end{pmatrix}$ (b) $\begin{pmatrix} -1 & 2 & 3 & 1 \\ 4 & 2 & 1 & 3 \\ 1 & 2 & 4 & 1 \\ 2 & 1 & 3 & 4 \end{pmatrix}$

APPENDIX E   PROGRAMS FOR SYSTEMS WITHOUT BUILT-IN MATRIX COMMANDS

Some computer systems do not have built-in matrix functions such as MAT READ and MAT PRINT. Microcomputers do not, for example, currently have these functions. This section is intended for the users of such systems. We illustrate how matrices can be handled on these systems.

MAT READ/WRITE

```
10 REM PROGRAM TO ILLUSTRATE
20 REM HOW TO READ IN AND
30 REM PRINT OUT A MATRIX
50 REM ***********************

60 REM READ IN SIZE OF MATRIX
70 READ M,N
80 DIM A(M,N)
90 REM READ ELEMENTS OF MATRIX
100 FOR I = 1 TO M
110 FOR J = 1 TO N
120 READ A(I,J)
130 NEXT J
140 NEXT I
150 REM ***********************

160 REM PRINT OUT THE MATRIX
170 PRINT
```

```
180 PRINT "THE MATRIX IS"
190 PRINT
200 FOR I = 1 TO M
210 FOR J = 1 TO N
220 PRINT TAB(J*8 − 7);A(I,J);
230 NEXT J
240 PRINT
250 PRINT
260 NEXT I
270 REM **********************

280 REM SIZE OF MATRIX
290 DATA 2,3
300 REM ELEMENTS OF MATRIX
310 DATA 1,2,3,4,0, − 4
320 END

RUN
```

THE MATRIX IS
```
  1    2    3
  4    0   −4
```

Two loops are *nested*. Statements 90 through 140 feed in the matrix, the inside loop scanning each row, the outside loop then leading to the next row. These statements can be used to read in a matrix on a system that does not have MAT READ.

Statements 200 through 260 cause the elements to be printed out in matrix form.

The TAB(X) function causes the computer to "tab" to column X for the current value of X and there print the current value of a_{ij}. These statements can be used to print out a matrix on a system that does not have MAT PRINT.

Programs in the text will be written using matrix commands. Users who do not have these functions available can modify the programs by using programs provided in this section.

MAT INPUT/WRITE

```
10 REM PROGRAM TO ILLUSTRATE
20 REM HOW TO INPUT A MATRIX
30 REM AS PROGRAM IS RUNNING
50 REM **********************

60 PRINT
70 PRINT "GIVE SIZE M,N OF MATRIX"
80 INPUT M,N
90 DIM A(M,N)
100 PRINT
110 PRINT "GIVE ELEMENTS OF MATRIX"
```

```
120 PRINT "(PRESS RETURN AFTER EACH ELEMENT)"
130 FOR I = 1 TO M
140 FOR J = 1 TO N
150 INPUT A(I,J)
160 NEXT J
170 NEXT I
180 REM **********************
```
```
190 REM PRINT OUT THE MATRIX
200 PRINT
210 PRINT "THE MATRIX IS"
220 PRINT
230 FOR I = 1 TO M
240 FOR J = 1 TO N
250 PRINT TAB(J*8 − 7);A(I,J);
260 NEXT J
270 PRINT
280 PRINT
290 NEXT I
300 REM **********************
```
```
310 END
```
```
RUN
```

GIVE SIZE M,N OF MATRIX
?2,3

GIVE ELEMENTS OF MATRIX
(PRESS RETURN AFTER EACH ELEMENT)
?1
?2
?3
?4
?5
?6

THE MATRIX IS

1	2	3
4	5	6

MAT SCALAR MULT

```
10 REM SCALAR MULT OF A MATRIX
30 REM **********************
```
```
40 REM READ IN THE SCALAR
50 READ C
60 REM READ SIZE OF MATRIX
70 READ M,N
80 DIM A(M,N)
```

```
90 REM READ ELEMENTS OF MATRIX
100 FOR I = 1 TO M
110 FOR J = 1 TO N
120 READ A(I,J)
130 LET A(1,J) = (C)*A(I,J)
140 NEXT J
150 NEXT I
160 REM ***********************
170 REM PRINT OUT SCALAR MULT
180 PRINT
190 PRINT "THE SCALAR MULTIPLE IS"
200 PRINT
210 FOR I = 1 TO M
220 FOR J = 1 TO N
230 PRINT TAB(J*8−7);A(I,J);
240 NEXT J
250 PRINT
260 PRINT
270 NEXT I
280 REM ***********************

290 REM THE SCALAR
300 DATA 3
310 REM SIZE OF MATRIX
320 DATA 2,3
330 REM ELEMENTS OF MATRIX
340 DATA 1,2,3,−3,4,8
350 END

RUN
```

THE SCALAR MULTIPLE IS

```
  3     6     9
 −9    12    24
```

MAT ADD

```
10 REM ADDITION OF MATRICES
30 REM ***********************

40 REM READ SIZE OF MATRICES
50 READ M,N
60 DIM A(M,N),B(M,N),C(M,N)
70 REM READ ELEMENTS OF A
80 FOR I = 1 TO M
90 FOR J = 1 TO N
100 READ A(I,J)
110 NEXT J
120 NEXT I
130 REM ***********************
```

```
140 REM READ B, ADD B TO A
150 FOR I = 1 TO M
160 FOR J = 1 TO N
170 READ B(I,J)
180 LET C(I,J) = A(I,J) + B(I,J)

190 NEXT J
200 NEXT I
210 REM ***********************

220 REM PRINT OUT SUM C
230 PRINT
240 PRINT "THE SUM IS"
250 PRINT
260 FOR I = 1 TO M
270 FOR J = 1 TO N
280 PRINT TAB(J*8 − 7);C(I,J);
290 NEXT J
300 PRINT
310 PRINT
320 NEXT I
330 REM ***********************

340 REM SIZE OF MATRICES
350 DATA 2,3
360 REM MATRIX A
370 DATA 1,2,3, − 1,2,4
380 REM MATRIX B
390 DATA 0, − 1,4,2,3,1
400 END

RUN

THE SUM IS
    1    1    7
    1    5    5
```

MAT PROD

```
10 REM PRODUCT OF MATRICES
30 REM ***********************

40 REM READ SIZE OF MATRIX A
50 READ M,N
60 DIM A(M,N)
70 REM READ ELEMENTS OF A
80 FOR I = 1 TO M
90 FOR J = 1 TO N
100 READ A(I,J)
110 NEXT J
120 NEXT I
130 REM ***********************
```

```
140 REM READ SIZE OF MATRIX B
150 READ P,Q
155 IF N< >P PRINT "PRODUCT DOES NOT EXIST": GOTO 550
160 DIM B(P,Q)
170 REM READ ELEMENTS OF B
180 FOR I = 1 TO P
190 FOR J = 1 TO Q
200 READ B(I,J)
210 NEXT J
220 NEXT I
230 REM *********************

240 REM MULT A BY B TO GET C
250 DIM C(M,Q)
260 FOR I = 1 TO M
270 FOR J = 1 TO Q
280 LET C(I,J) = 0
290 FOR K = 1 TO N
300 LET C(I,J) = C(I,J)+A(I,K)*B(K,J)
310 NEXT K
320 NEXT J
330 NEXT I
340 REM *********************

350 REM PRINT OUT PRODUCT C
360 PRINT
370 PRINT "THE PRODUCT MATRIX IS"
380 PRINT
390 FOR I = 1 TO M
400 FOR J = 1 TO Q
410 PRINT TAB(J*8−7);C(I,J);
420 NEXT J
430 PRINT
440 PRINT
450 NEXT I
460 REM *********************

470 REM SIZE OF A
480 DATA 3,2
490 REM ELEMENTS OF A
500 DATA 4,7,−6,−89,0,2
510 REM SIZE OF B
520 DATA 2,2
530 REM ELEMENTS OF B
540 DATA 4,−1,7,3
550 END

RUN

THE PRODUCT MATRIX IS
  65     17
−647   −261
  14      6
```

MAT POWERS

```
10 REM POWERS OF A MATRIX
20 REM USING DATA STATEMENTS
40 REM **********************

50 PRINT : PRINT : PRINT
60 PRINT "PRESS RETURN TO CONTINUE"
70 PRINT : PRINT
80 REM READ NO OF POWERS
90 READ D
100 REM READ SIZE OF MATRIX
110 READ N
120 DIM A(N,N),B(N,N)
130 REM READ GIVEN MATRIX A
140 REM GIVE B INITIAL VALUE A
150 FOR I = 1 TO N
160 FOR J = 1 TO N
170 READ A(I,J)
180 LET B(I,J) = A(I,J)
190 NEXT J
200 NEXT I
210 REM **********************

220 REM COMPUTES NEXT POWER
230 FOR P = 2 TO D
240 FOR I = 1 TO N
250 FOR J = 1 TO N
260 LET C(I,J) = 0
270 FOR K = 1 TO N
280 LET C(I,J) = C(I,J) + A(I,K)*B(K,J)
290 NEXT K
300 NEXT J
310 NEXT I
320 REM **********************

330 REM PRINTS THE MATRIX AND
340 REM UPDATES B TO COMPUTE
350 REM THE NEXT POWER
360 PRINT "A↑";P;" ="
370 PRINT
380 FOR I = 1 TO N
390 FOR J = 1 TO N
400 PRINT TAB(J*13 − 12);C(I,J);
410 LET B(I,J) = C(I,J)
420 NEXT J
430 PRINT : PRINT
450 NEXT I
460 PRINT : PRINT
480 INPUT " ";D$
490 NEXT P
500 REM **********************
```

```
510 REM NO OF POWERS
520 DATA 4
530 REM SIZE OF MATRIX
540 DATA 2
550 REM ELEMENTS OF MATRIX
560 DATA 1,.2,3, − 4
570 END

RUN
```

PRESS RETURN TO CONTINUE
A↑2 =
  1.6   −.6
−9   16.6
A↑3 =
−.2   2.72
 40.8   −68.2
A↑4 =
  7.96   −10.92
−163.8   280.96

APPENDIX F   REDUCED ECHELON FORM OF A MATRIX (SECTIONS 1-3, 1-4)

Determine the reduced echelon form of $\begin{pmatrix} 1 & 1 & 1 & 3 \\ 2 & 3 & 1 & 5 \\ 1 & -1 & -2 & -5 \end{pmatrix}$.

```
10 READ N,M
20 DIM A(N,M)
30 MAT READ A
40 PRINT "THE ORIGINAL MATRIX IS"
50 MAT PRINT A
60 PRINT
70 LET L = 0
80 FOR K = 1 TO N
90     LET L = L +1
100    IF L > M THEN GOTO 370
110    IF A(K,L) < > 0 THEN GOTO 230
120    IF K = N THEN GOTO 380
130    FOR I = K +1 TO N
140       IF A(I,L) = 0 THEN GOTO 210
150       FOR J = L TO M
160          LET B = A(K,J)
170          LET A(K,J) = A(I,J)
180          LET A(I,J) = B
190       NEXT J
200       GOTO 230
210    NEXT I
```

These statements scan the lth column from a_{kl} down to determine a nonzero element. The row containing this element is made into the kth row by interchanging rows. If such a nonzero element does not exist, we start scanning the following column.

Divide the kth row by a_{kl} to make the first nonzero element in this row unity.

```
220   GOTO 90
230   IF A(K,L) = 1 THEN GOTO 280
240   LET Y = A(K,L)
250   FOR P = L TO M
260      LET A(K,P) = A(K,P)/Y
270   NEXT P
```

These statements make the elements above and below the first nonzero element in the kth row zero.

```
280   FOR I = 1 TO N
290      IF A(I,L) = 0 THEN GOTO 350
300      IF I = K THEN GOTO 350
310      LET Z = A(I,L)
320      FOR J = L TO M
330         LET A(I,J) = A(I,J) − Z∗A(K,J)
340      NEXT J
350   NEXT I
360 NEXT K
370 PRINT "THE REDUCED ECHELON FORM IS"
380 MAT PRINT A
```

Kind of matrix.
Matrix elements.

```
390 DATA 3,4
400 DATA 1,1,1,3,1, − 1, − 2, − 5,2,3,1,5
410 END

RUN
```

THE ORIGINAL MATRIX IS

```
 1    1    1    3
 1   −1   −2   −5
 2    3    1    5
```

THE REDUCED ECHELON FORM IS

```
 1    0    0    0
 0    1    0    1
 0    0    1    2
```

EXERCISES

1. Use the above program to determine the reduced echelon forms for the following matrices:

(a) the augmented matrix of the system of equations in Example 3, Section 1-3:

$$\begin{pmatrix} 4 & 8 & -12 & 44 \\ 3 & 6 & -8 & 32 \\ -2 & -1 & 0 & -7 \end{pmatrix}$$

(b) the augmented matrix of the system of equations in Example 2, Section 1-4:

$$\begin{pmatrix} 2 & -4 & 12 & -10 & 58 \\ -1 & 2 & -3 & 2 & -14 \\ 2 & -4 & 9 & -6 & 44 \end{pmatrix}$$

(c) the augmented matrix of the system of equations in Example 3, Section 1-4:

$$\begin{pmatrix} 1 & 2 & -1 & 3 & 1 & 2 \\ 2 & 4 & -2 & 6 & 3 & 6 \\ -1 & -2 & 1 & -1 & 3 & 4 \end{pmatrix}$$

(d) the augmented matrix of the system of equations in Example 4, Section 1-4:

$$\begin{pmatrix} 1 & -1 & 2 & 3 \\ 2 & -2 & 5 & 4 \\ 1 & 2 & -1 & -3 \\ 0 & 2 & 2 & 1 \end{pmatrix}$$

2. Check your answers to Exercises 4–17, Section 1-4, using the above program.

3. Modify the above program to get a printout after each elementary matrix transformation with a statement describing the transformation, e.g., ROW 3 − (2) ROW 1. The technique of inserting statements to obtain printouts at various locations in a program can be used to analyze programs that are not running correctly—that is, to debug programs.

4. Modify the above program to give the echelon form of a matrix that is used in the method of Gaussian elimination (Section 8-2).

5. Check your answers to Exercises 3–7, Section 8-2, using the program of Exercise 4.

6. Let $A = \begin{pmatrix} 2 & 1 & 3 \\ 1 & 2 & 3 \\ -1 & 3 & 1 \end{pmatrix}$. Write a program that performs the following operations on A, printing out each matrix in the sequence and stating the operations that have been performed:

$$\begin{pmatrix} 2 & 1 & 3 \\ 1 & 2 & 3 \\ -1 & 3 & 1 \end{pmatrix} \underset{R1 \leftrightarrow R2}{\approx} \begin{pmatrix} 1 & 2 & 3 \\ 2 & 1 & 3 \\ -1 & 3 & 1 \end{pmatrix}$$

$$\underset{\substack{R2-(2)R1 \\ R3+R1}}{\approx} \begin{pmatrix} 1 & 2 & 3 \\ 0 & -3 & -3 \\ 0 & 5 & 4 \end{pmatrix}$$

APPENDIX G INVERSE OF A MATRIX USING GAUSS-JORDAN ELIMINATION (SECTION 2-5)

Determine the inverse of the matrix $\begin{pmatrix} 0 & 3 & 3 \\ 1 & 2 & 3 \\ 1 & 4 & 6 \end{pmatrix}$.

```
10 READ N
20 DIM B(N,N),A(N,2*N)
30 MAT READ B
40 PRINT "THE ORGINAL MATRIX IS"
50 MAT PRINT B
```

```
60 LET M = 2*N
70 FOR I = 1 TO N
80   FOR J = 1 TO M
90     IF J > N THEN GOTO 120
100    LET A(I,J) = B(I,J)
110    GOTO 160
120    IF J = N+I THEN GOTO 150
130    LET A(I,J) = 0
140    GOTO 160
150    LET A(I,J) = 1
160  NEXT J
170 NEXT I
180 FOR K = 1 TO N
190  IF A(K, K) < > 0 THEN GOTO 310
200  IF  K = N THEN GOTO 540
210  FOR I = K+1 TO N
220    IF A(I,K) = 0 THEN GOTO 290
230    FOR J = K TO M
240      LET B = A(K,J)
250      LET A(K,J) = A(I,J)
260      LET A(I,J) = B
270    NEXT J
280    GOTO 310
290  NEXT I
300  GOTO 540
310  IF A(K,K) = 1 THEN GOTO 360
320  LET Y = A(K,K)
330  FOR P = K TO M
340    LET A(K,P) = A(K,P)/Y
350  NEXT P
360  FOR I = 1 TO N
370    IF I = K THEN GOTO 430
380    IF A(I,K) = 0 THEN GOTO 430
390    LET Z = A(I,K)
400    FOR J = K TO M
410      LET A(I,J) = A(I,J) - Z*A(K,J)
420    NEXT J
430  NEXT I
440 NEXT K
450 FOR P = 1 TO N
460   FOR Q = N+1 TO M
470     LET R = Q-N
480     LET B(P,R) = A(P,Q)
490   NEXT Q
500 NEXT P
510 PRINT "THE INVERSE IS"
520 MAT PRINT B
530 GOTO 570
540 PRINT "THE INVERSE DOES NOT EXIST"
550 DATA 3
```

```
560 DATA 0,3,3,1,2,3,1,4,6
570 END

RUN
```

THE ORIGINAL MATRIX IS

0	3	3
1	2	3
1	4	6

THE INVERSE IS

0	2	−1
1	1	−1
−.66666667	−9	1

EXERCISES

1. Modify the above program to get a printout after each transformation with a statement describing the transformation.

2. Find out whether your system has a built-in matrix inverse function,

$$\text{MAT B} = \text{INV(A)}$$

If so, use this function in a program to determine the inverse of the matrix $\begin{pmatrix} 1 & 2 \\ 3 & 4 \end{pmatrix}$.

APPENDIX H   EVALUATION OF DETERMINANTS (SECTION 3-3)

Evaluate $\begin{vmatrix} 1 & 2 & 3 & 1 \\ 2 & 4 & 3 & 1 \\ 1 & 3 & 4 & 2 \\ 2 & 5 & 6 & 4 \end{vmatrix}$.

S will be used as a counter to determine the number of row interchanges.

```
10 READ N
20 DIM A(N,N)
30 MAT READ A
40 PRINT "THE MATRIX IS"
50 MAT PRINT A
60 LET S = 0
70 FOR K = 1 TO N−1
80    IF A(K,K) < >0 THEN GOTO 210
90    FOR I = K+1 TO N
100      IF A(I,K) = 0 THEN GOTO 180
110      FOR J = K TO N
120        LET B = A(K,J)
130        LET A(K,J) = A(I,J)
140        LET A(I,J) = B
```

```
150      NEXT J
160      LET S = S + 1
170      GOTO 210
180    NEXT I
190    PRINT "THE DETERMINANT IS ZERO"
200    GOTO 360
210    FOR I = K + 1 TO N
220      IF A(I,K) = 0 THEN GOTO 270
230      LET Z = A(I,K)/A(K,K)
240      FOR J = K TO N
250        LET A(I,J) = A(I,J) − Z*A(K,J)
260      NEXT J
270    NEXT I
280 NEXT K
290 LET D = 1
300 FOR P = 1 TO N
310    LET D = D*A(P,P)
320 NEXT P
330 LET D = D*((−1)↑S)
340 PRINT "THE DETERMINANT IS"; D
350 DATA 4
360 DATA 1,2,3,1,2,4,3,1,1,3,4,2,2,5,6,4
370 END

RUN

THE MATRIX IS
  1    2    3    1
  2    4    3    1
  1    3    4    2
  2    5    6    4
THE DETERMINANT IS 4
```

EXERCISES

1. Modify the above program to get a printout after each transformation with a statement describing the transformation.

2. Check your answers to the exercises of Section 3-3 using the computer.

APPENDIX I GRAPHING AN ELLIPSE ON AN APPLE MICROCOMPUTER (EXAMPLE 7, SECTION 7-4)

The following program can be used to graph the ellipse

$$\frac{(x')^2}{900} + \frac{(y')^2}{400} = 1$$

of Example 7, Section 7-4. The coordinate system $x'y'$ is set up with its origin O' at the center of the screen, as discussed in Example 7.

Draw the x and y axes.

Draw the x' and y' axes.

Various ellipses can be graphed.

Solutions of the equation of the ellipse for y' in terms of x'.

```
5 HGR
10 HCOLOR = 6

15 HPLOT 279,0 TO 0,0 TO 0,159

20 HPLOT 4,80 TO 279,80
25 HPLOT 140,4 TO 140,159

30 INPUT "GIVE A, B"; A,B

35 FOR X = 140 − A TO 140 + A

40 LET Y1 = − B*SQR(1 − (X − 140)↑2/A↑2) + 80
45 LET Y2 = B*SQR(1 − (X − 140)↑2/A↑2) + 80

50 HPLOT X,Y1
55 HPLOT X,Y2
60 NEXT X
65 END

RUN
GIVE A,B 30,20
```

The result is shown in Figure A-1.

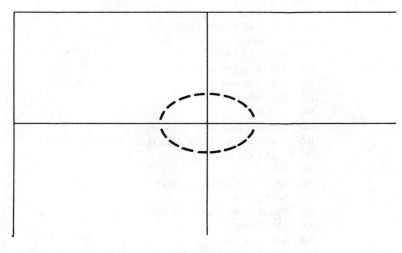

Figure A-1

EXERCISES

1. Explain the gaps in the graph surrounding the x axis in the above output.
2. The above program contains a loop involving x. Modify the program, replacing the loop involving x by one involving y. Explain the gaps that appear in the graph.

3. Write a program that contains loops involving both *x* and *y*, leading to a more complete graph for an ellipse.

4. Write a program that graphs a solid ellipse.

5. Modify the given program for graphing the ellipse to draw line segments between consecutive points of output.

APPENDIX J   EIGENVALUES AND EIGENVECTORS (SECTION 8-3)

Determine the dominant eigenvalue and a corresponding eigenvector of the matrix $\begin{pmatrix} 5 & 4 & 2 \\ 4 & 5 & 2 \\ 2 & 2 & 2 \end{pmatrix}$.

Let us use five iterations, using the initial vector $(-4, 2, 6)$.

```
10 READ M
20 DIM A(M,M),X(M,1),Y(M,1),Z(M,1)
30 PRINT "GIVE NUMBER OF ITERATIONS"
40 INPUT N
50 PRINT "GIVE INITIAL VECTOR"
60 MAT INPUT X
70 MAT READ A
80 PRINT "THE INITIAL VECTOR IS"
90 MAT PRINT X
100 FOR I = 1 TO N
110    PRINT "ITERATION", I
120    MAT Y = A*X
130    MAT PRINT Y
140    LET K = 1
150    FOR J = 2 TO 3
160       IF ABS(Y(K,1)) > = ABS(Y(J,1)) THEN GOTO 180
170       LET K = J
180    NEXT J
190    MAT Y = (1/ABS(Y(K,1)))*Y
200    MAT X = Y
210    PRINT "THE ADJUSTED VECTOR IS"
220    MAT PRINT X
230    MAT Z = A*X
240    LET S = 0
250    LET T = 0
260    FOR L = 1 TO M
270       LET S = S+X(L,1)*Z(L,1)
280       LET T = T+X(L,1)*X(L,1)
290    NEXT L
300    LET E = S/T
310    PRINT "APPROX EIGENVALUE"; E
```

```
320    PRINT
330    INPUT "PRESS RETURN TO CONTINUE"; A$
340    PRINT
350 NEXT I
360 DATA 3
370 DATA 5,4,2,4,5,2,2,2,2
380 END
```

RUN

GIVE NUMBER OF ITERATIONS
  ? 5
GIVE INITIAL VECTOR
  ? −4, 2, 6
THE INITIAL VECTOR IS
−4
 2
 6
ITERATION 1
 0
 6
 8
THE ADJUSTED VECTOR IS
 0
 .75
 1
APPROX EIGENVALUE 5
PRESS RETURN TO CONTINUE
                ⋮
ITERATION 4
 9. 955157
 9. 9686099
 4. 9955157
THE ADJUSTED VECTOR IS
 . 99865047
 1
 . 50112461
APPROX EIGENVALUE 9. 9999888
PRESS RETURN TO CONTINUE
ITERATION 5
 9. 9955016
 9. 9968511
 4. 9995502
THE ADJUSTED VECTOR IS
 . 999865
 1
 . 5001125
APPROX EIGENVALUE 9. 9999999
PRESS RETURN TO CONTINUE

EXERCISES

1. Use this program to answer Exercises 1–8 of Section 8-3.
2. Extend the program to determine further eigenvalues, and a corresponding eigenvector in each case, for a symmetric matrix (Theorem 8-3, Section 8-3). Use your program to check your answers to Exercises 9–13, Section 8-3.
3. Determine the dominant eigenvalue and a corresponding eigenvector for each of the following matrices:

(a) $\begin{pmatrix} 7.2 & -0.6 & -2.4 & 1.8 \\ 3.8 & 5.6 & 0.4 & 4.2 \\ -4.4 & -0.8 & 3.8 & -3.6 \\ -1.4 & 2.2 & 0.8 & 2.4 \end{pmatrix}$

(b) $\begin{pmatrix} 9.6 & 3.2 & 0.8 & 2.4 \\ 6 & 5 & -1 & 5 \\ -5.2 & 0.6 & 7.4 & -5.8 \\ -2 & -1 & -3 & 3 \end{pmatrix}$

(c) $\begin{pmatrix} 6.4 & 8.8 & 3.2 & 5.6 \\ 9.2 & -2.6 & -5.4 & 3.8 \\ -6.8 & -0.6 & 4.6 & -6.2 \\ 0.4 & -6.2 & -5.8 & -1.4 \end{pmatrix}$

(d) $\begin{pmatrix} -2366 & 1470 & 525 & -318 \\ -3498 & 2170 & 783 & -474 \\ -2166 & 1362 & 445 & -270 \\ -1908 & 1200 & 390 & -236 \end{pmatrix}$

(e) $\begin{pmatrix} 1990 & -1200 & -495 & 290 \\ 2786 & -1680 & -693 & 406 \\ 2358 & -1422 & -585 & 342 \\ 1940 & -1170 & -480 & 280 \end{pmatrix}$

APPENDIX K   SIMPLEX METHOD (SECTION 9-2)

The initial tableau is

$$\begin{pmatrix} 1 & 1 & 1 & 1 & 0 & 0 & 100 \\ 3 & 2 & 4 & 0 & 1 & 0 & 200 \\ 1 & 2 & 0 & 0 & 0 & 1 & 150 \\ -3 & -5 & -8 & 0 & 0 & 0 & 0 \end{pmatrix}$$

Determine the final tableau.

```
10 READ N, M
20 DIM A(N,M)
30 MAT READ A
40 PRINT "THE INITIAL TABLEAU IS"
50 GOSUB 500
60 LET T = 0
70 FOR K = 1 TO M−1
80    IF T−A(N,K) < = 0 THEN GOTO 110
90    LET T = A(N,K)
100   LET J = K
110 NEXT K
120 IF T = 0 THEN GOTO 410
130 FOR I = 1 TO N−1
140   IF A(I,J) < = 0 THEN GOTO 160
150   GOTO 180
160 NEXT I
170 GOTO 450
180 LET Q = 10↑10
190 FOR I = 1 TO N−1
200   IF A(I,J) < = 0 THEN GOTO 250
210   LET B = A(I,M)/A(I,J)
220   IF B > = Q THEN GOTO 250
230   LET Q = B
240   LET S = I
250 NEXT I
260 IF Q = 10↑10 THEN GOTO 480
270 IF A(S,J) = 1 THEN GOTO 320
280 LET Y = A(S,J)
290 FOR P = 1 TO M
300   LET A(S,P) = A(S,P)/Y
310 NEXT P
320 FOR I = 1 TO N
330   IF I = S THEN GOTO 390
340   IF A(I,J) = 0 THEN GOTO 390
350   LET Z = A(I,J)
360   FOR R = 1 TO M
370     LET A(I,R) = A(I,R)−Z*A(S,R)
380   NEXT R
390 NEXT I
400 GOTO 60
410 PRINT
420 PRINT "THE FINAL TABLEAU IS"
430 GOSUB 500
440 GOTO 620
450 PRINT
460 PRINT "REGION UNBOUNDED AND OBJECTIVE FUNCTION UNBOUNDED"
470 GOTO 620
480 PRINT "PROGRAM CANNOT BE USED, ELEMENTS GET TOO LARGE"
490 GOTO 620
```

```
500 FOR U = 1 TO N
510    FOR V = 1 TO M
520       PRINT TAB(V*8 − 7);A(U,V);
530    NEXT V
540    PRINT
550 NEXT U
560 RETURN
570 DATA 4,7
580 DATA 1,1,1,1,0,0,100
590 DATA 3,2,4,0,1,0,200
600 DATA 1,2,0,0,0,1,150
610 DATA  −3, −5, −8,0,0,0,0
620 END

RUN
```

THE INITIAL TABLEAU IS

1	1	1	1	8	0	100
3	2	4	0	1	0	200
1	2	0	0	0	1	150
−3	−5	−8	0	0	0	0

THE FINAL TABLEAU IS

0	0	0	1	−.25	−.25	12.5
.5	0	1	0	.25	−.25	12.5
.5	1	0	0	0	.5	75
3.5	0	0	0	2	.5	475

The output matrix is likely to be large here. We use the subroutine consisting of statements 500 to 550 to print out the matrix. The output can be controlled using the TAB function.

EXERCISES

1. Modify the above program to get a printout after each transformation with a statement describing the transformation.
2. Check you answers to the exercises of Section 9-2 using the computer.

Answers to Selected Exercises

CHAPTER 1

SECTION 1-1

1. $x = \frac{3}{2}, y = -\frac{1}{2}$ **3.** No solutions   **5.** $x = 11, y = -4$ **7.** $x = \frac{6}{7}, y = \frac{1}{7}$
9. $x - y = -1, 2x - y = 0$ **12.** e.g., $4x - y = 1, 8x - 2y = 2$
14. e.g., $2x + y = 1, x + \frac{1}{2}y = \frac{1}{2}$
17. e.g., $-x + y = 2, -2x + y = 1, x + y = 4$; three lines through point $(1, 3)$
19. (a) $d \neq 2$ **(b)** $d = 2$

SECTION 1-2

1. (a) 2×2 **(b)** 3×2 **(c)** 1×3 **(d)** 2×5 **3.** $1, 7, 1, 8$

5. $\begin{pmatrix} 1 & 0 & 0 & 0 \\ 0 & 1 & 0 & 0 \\ 0 & 0 & 1 & 0 \\ 0 & 0 & 0 & 1 \end{pmatrix}$

6. (a) $\begin{pmatrix} 1 & -1 \\ 2 & 1 \end{pmatrix}, \begin{pmatrix} 1 & -1 & 1 \\ 2 & 1 & 3 \end{pmatrix}$ **(c)** $\begin{pmatrix} 0 & 1 & -1 \\ 1 & 2 & -4 \\ 2 & -1 & 0 \end{pmatrix}, \begin{pmatrix} 0 & 1 & -1 & 5 \\ 1 & 2 & -4 & 2 \\ 2 & -1 & 0 & 4 \end{pmatrix}$

(e) $\begin{pmatrix} 7 & -1 & 1 & 2 \\ 0 & 1 & -8 & -1 \\ 1 & 0 & 7 & 1 \end{pmatrix}, \begin{pmatrix} 7 & -1 & 1 & 2 & 4 \\ 0 & 1 & -8 & -1 & -1 \\ 1 & 0 & 7 & 1 & -2 \end{pmatrix}$

7. (a) $\begin{aligned} 2x_1 + x_2 &= 5 \\ 3x_1 - 2x_2 &= 3 \end{aligned}$ **(c)** $\begin{aligned} 3x_1 + 5x_2 - 4x_3 &= 1 \\ -x_1 + 2x_2 + 2x_3 &= 2 \\ 2x_1 + 3x_2 &= 3 \end{aligned}$

8. (a) $\begin{aligned} x_1 + 4x_2 + 3x_3 + 4x_4 &= -1 \\ x_2 - 8x_3 + 9x_4 &= 0 \\ x_4 &= 4 \end{aligned}$ **(c)** $\begin{aligned} x_1 + 7x_2 + 5x_3 &= -8 \\ x_2 + 8x_3 &= 2 \end{aligned}$

9.
$$\begin{array}{l} x_1 \quad\quad = 2 \\ \quad x_2 \quad = 1; \\ \quad\quad x_3 = -3 \end{array} \begin{pmatrix} 1 & 0 & 0 \\ 0 & 1 & 0 \\ 0 & 0 & 1 \end{pmatrix}; x_1 = 2, x_2 = 1, x_3 = -3$$

SECTION 1-3

1. (a) $\begin{pmatrix} 1 & 2 & -2 & 1 \\ 1 & 2 & -1 & 3 \\ -1 & 0 & 4 & 2 \end{pmatrix}$ **(c)** $\begin{pmatrix} 1 & 0 & -1 & -3 \\ 0 & 1 & 2 & 1 \\ 0 & 0 & 13 & 1 \end{pmatrix}$

(e) $\begin{pmatrix} 1 & -2 & 0 & 3 \\ 0 & -5 & 3 & 10 \\ 0 & 7 & 2 & -3 \end{pmatrix}$

2. (a) Divide row 2 by a pivot to get a 1 in the (2, 2) location.
(c) Switch rows to get a pivot for row 2 in the (2, 2) location.
(e) Create zeros above the 1 in the (3, 3) location.

3. $x_1 = 2, x_2 = 5$ **5.** $x_1 = 4, x_2 = -3, x_3 = -1$ **7.** $x_1 = 3, x_2 = 0, x_3 = 2$
9. $x_1 = 0, x_2 = 4, x_3 = 2$ **11.** $x_1 = 2, x_2 = 3, x_3 = -1$
13. $x_1 = 6, x_2 = -3, x_3 = 2$ **15.** $x_1 = 1, x_2 = 2, x_3 = 3, x_4 = 4$
17. $x_1 = -1, x_2 = -2, x_3 = 1, x_4 = 2$

SECTION 1-4

1. (a) Yes   **(c)** No; leading nonzero element in row 2 is not 1.
(e) No; leading 1 in row 3 is not to right of leading 1 in row 2.
(g) No; elements above leading 1 in row 3 are not zero.
(i) No; leading nonzero element in row 2 is not 1.

2. (a) $\begin{pmatrix} 1 & 0 & 1 & 4 \\ 0 & 1 & 2 & 3 \\ 0 & 0 & ③ & 6 \end{pmatrix}$ **(c)** $\rightarrow \begin{pmatrix} 1 & 3 & 4 & -1 \\ 0 & 0 & 6 & 2 \\ 0 & 4 & 3 & 4 \end{pmatrix} \underset{R2 \leftrightarrow R3}{\approx} \begin{pmatrix} 1 & 3 & 4 & -1 \\ 0 & ④ & 3 & 4 \\ 0 & 0 & 6 & 2 \end{pmatrix}$

(e) $\rightarrow \begin{pmatrix} 1 & 4 & 2 & 4 & 5 \\ 0 & 0 & 0 & 0 & 6 \\ 0 & 0 & 0 & 2 & 3 \\ 0 & 0 & 3 & 1 & 2 \end{pmatrix} \underset{R2 \leftrightarrow R4}{\approx} \begin{pmatrix} 1 & 4 & 2 & 4 & 5 \\ 0 & 0 & ③ & 1 & 2 \\ 0 & 0 & 0 & 2 & 3 \\ 0 & 0 & 0 & 0 & 6 \end{pmatrix}$

3. (a) $(2, -2r + 3, r, -1)$ **(c)** $(-2r - 2s - 1, -4r - 3s + 2, r, -4s + 1, s)$
(e) $(-2r + 2, 3r + 1, r, 3)$ **(g)** $(-2r - s + 4, r, -3s - 1, s, 3, 2)$
4. $(-2r + 3, r, 2, 1)$ **7.** $(2, 1, 3)$ **9.** No solution
11. $(-2r - 3s + t + 3, r, s, -2t - 1, -t + 2, t, 1)$ **13.** $(-1, 2, 3, 1, -2r + 4, r)$
15. $(-2r + 3s - t + 3w, r, s, -2t - w, t, w)$ **17.** $(2, -2r - 1, r, 3)$

SECTION 1-5

1. $I_1 = 12, I_2 = I_4 = I_5 = 4, I_3 = 8$ **3.** 1, 1, 0   **5.** 1, 3, 4
7. (a) 3, 2, 1   **(b)** $\frac{17}{5}, \frac{22}{5}, 1$. No current through AB when voltage is 16.
8. Minimum flow along AB is 150. Two distinct flows might be $x_1 = 200, x_2 = 50,$
$x_3 = 50, x_4 = 100; x_1 = 220, x_2 = 30, x_3 = 70, x_4 = 80.$

CHAPTER 2

SECTION 2-1

1. (a) $2A = \begin{pmatrix} 2 & 4 \\ 6 & 0 \end{pmatrix}$, $3B = \begin{pmatrix} -3 & 6 \\ 3 & 3 \end{pmatrix}$, $-2C = \begin{pmatrix} 0 & -2 \\ 2 & -8 \end{pmatrix}$

(b) $A + B = \begin{pmatrix} 0 & 4 \\ 4 & 1 \end{pmatrix}$, $B + A = \begin{pmatrix} 0 & 4 \\ 4 & 1 \end{pmatrix}$, $A + C = \begin{pmatrix} 1 & 3 \\ 2 & 4 \end{pmatrix}$, $B + C = \begin{pmatrix} -1 & 3 \\ 0 & 5 \end{pmatrix}$

(c) $A + 2B = \begin{pmatrix} -1 & 6 \\ 5 & 2 \end{pmatrix}$, $3A + C = \begin{pmatrix} 3 & 7 \\ 8 & 4 \end{pmatrix}$, $2A + B - C = \begin{pmatrix} 1 & 5 \\ 8 & -3 \end{pmatrix}$

3. $a_{11} = 1$, $a_{24} = -1$, $a_{34} = 3$, $a_{25} = 5$ **5.** $c_{12} = 5$, $c_{21} = 2$, $c_{23} = 9$

7. (a) $a(B + C) = a[(b_{ij}) + (c_{ij})] = a(b_{ij} + c_{ij}) = (ab_{ij} + ac_{ij}) = (ab_{ij}) + (ac_{ij})$
$= a(b_{ij}) + a(c_{ij}) = aB + aC$

SECTION 2-2

1. $AB = \begin{pmatrix} 5 \\ 23 \end{pmatrix}$, $AC = \begin{pmatrix} 2 & 1 \\ 4 & 9 \end{pmatrix}$, $AD = \begin{pmatrix} 2 & 4 & 1 \\ 8 & 18 & 1 \end{pmatrix}$, BA does not exist, DB does not exist, BD does not exist, $A^2 = \begin{pmatrix} 2 & 3 \\ 6 & 11 \end{pmatrix}$, $C^3 = \begin{pmatrix} -7 & 21 \\ 14 & 7 \end{pmatrix}$, $AC + CA = \begin{pmatrix} 8 & 9 \\ 6 & 14 \end{pmatrix}$.

3. (a) $AB + BA = \begin{pmatrix} 8 & 12 \\ 8 & 16 \end{pmatrix}$ **(b)** $AC + BC = \begin{pmatrix} 11 \\ 19 \end{pmatrix}$

(c) $AD - 3(BD) = \begin{pmatrix} 3 & -2 & -27 \\ -4 & -7 & 7 \end{pmatrix}$ **(d)** $3(DC) + (4A)C$ does not exist.

(e) $AC + 2(BD)$ does not exist.

4. (a) 7   **(b)** -21

6. (a) $AB + C$ does not exist.   **(c)** 4×2 **(e)** $2(EB) + DA$ does not exist.

9. $AC = BC$ does not imply that $A = B$. (Hint: Use the result of Exercise 8.)

10. (a) $B = \begin{pmatrix} -1 & -2 \\ 0 & 3 \\ 4 & 1 \end{pmatrix}$ or $\begin{pmatrix} -1 & -2 \\ 0 & 3 \\ 4 & 1 \end{pmatrix}$. AB is 2×2 or 2×1.

11. (a) $AB = \begin{pmatrix} A_1 B \\ \vdots \\ A_n B \end{pmatrix}$ **(c)** $AB = \begin{pmatrix} A_1 B^1 & \cdots & A_1 B^n \\ \vdots & & \\ A_n B^1 & \cdots & A_n B^n \end{pmatrix}$, the usual product

SECTION 2-3

1. $AB = \begin{pmatrix} 17 & 7 \\ -2 & -4 \end{pmatrix}$, $BA = \begin{pmatrix} -2 & 6 \\ 4 & 15 \end{pmatrix}$ **3.** $AB = BA = \begin{pmatrix} 4 & 4 \\ -2 & 2 \end{pmatrix}$

5. $A(BC) = (AB)C = \begin{pmatrix} 36 & 50 \\ 34 & 50 \end{pmatrix}$ **6. (a)** Does not exist   **(c)** 3×6 **(e)** 4×4

8. (a) $a_{41}b_{13} + a_{42}b_{23} + a_{43}b_{33} + a_{44}b_{43}$, element in row 4 and column 3 of the product AB. Matrix A has 4 columns, B has 4 rows.

(c) $a_{31}b_{11} + a_{32}b_{21} + a_{33}b_{31}$, element in row 3 and column 1 of the product AB. A has 3 columns, B has 3 rows.

(e) $p_{11}q_{14} + p_{12}q_{24} + p_{13}q_{34}$, element in row 1 and column 4 of the product PQ. P has 3 columns, Q has 3 rows.

9. (a) $\sum_{k=1}^{5} a_{1k}b_{k1}$ **(c)** $\sum_{k=1}^{5} a_{4k}b_{k5}$ **14. (a)** $\sum_{k=1}^{6} p_{2k}q_{k3}$ **(c)** $\sum_{k=1}^{6} p_{1k}q_{k7}$

15. (a) $\begin{pmatrix} 1 & -R-R_1 \\ -1/R_2 & 1 + (R/R_2) + (R_1/R_2) \end{pmatrix}$ **18. (a)** $\begin{pmatrix} 1 & 0 \\ -\frac{1}{3} & 1 \end{pmatrix}$, $V_2 = 7$, $i_2 = 3$

SECTION 2-4

1. (a) $\begin{pmatrix} -1 & 2 \\ 2 & -3 \end{pmatrix}$, symmetric   **(c)** $\begin{pmatrix} 3 & 2 \\ -1 & 4 \end{pmatrix}$, not symmetric

(e) $\begin{pmatrix} 2 & 4 & 7 \\ -1 & 5 & 8 \\ 3 & 6 & 9 \end{pmatrix}$, not symmetric   **(g)** $\begin{pmatrix} 1 & -1 & 3 \\ -1 & 2 & 0 \\ 3 & 0 & 4 \end{pmatrix}$, symmetric

(i) $(1 \quad 2 \quad 3)$, not symmetric

2. (a) $\begin{pmatrix} 1 & 2 & 4 \\ 2 & 6 & 5 \\ 4 & 5 & 2 \end{pmatrix}$ **4. (a)** 4   **(d)** 3

7. (c) $(A + B + C)^t = (A + (B + C))^t = A^t + (B + C)^t = A^t + B^t + C^t$

8. (a) $\text{tr}(cA) = \text{tr}(ca_{ij}) = \sum ca_{ii} = c \sum a_{ii} = c\text{tr}(A)$

10. (a) Let $A = A^t$, $B = B^t$, $(A + B)^t = A^t + B^t = A + B$

13. (a) $(A + A^t)^t = A^t + (A^t)^t = A^t + A = A + A^t$

17. (a) $A = \begin{pmatrix} 0 & 1 & 0 & 0 \\ 0 & 0 & 1 & 0 \\ 0 & 0 & 0 & 1 \\ 0 & 0 & 0 & 0 \end{pmatrix}$ **18. (a)** a_{24}

20. (a) Graves 2-3-1, pottery 1-2, no information.

(c) Graves 2-3-1, pottery 1-3$\begin{smallmatrix} 2 \\ 4 \end{smallmatrix}$, 2 and 4 are contemporary.

SECTION 2-5

1. (a) $\begin{pmatrix} 1 & 3 \\ 2 & -1 \end{pmatrix}\begin{pmatrix} x_1 \\ x_2 \end{pmatrix} = \begin{pmatrix} 5 \\ 6 \end{pmatrix}$ **2.** $x_1 = 1, x_2 = 1; x_1 = -2, x_2 = 3; x_1 = -1, x_2 = ?$

4. $x_1 = 1, x_2 = 2, x_3 = 3; x_1 = -1, x_2 = 2, x_3 = 0; x_1 = 0, x_2 = 1, x_3 = 2$

6. $x_1 = -2r - 1, x_2 = -3r + 3, x_3 = r; x_1 = -2r + 4, x_2 = -3r - 1, x_3 = r$

8. $\begin{pmatrix} 1 & 0 \\ -2 & 1 \end{pmatrix}$ **10.** $\begin{pmatrix} \frac{1}{2} & -3 \\ \frac{1}{2} & 0 \end{pmatrix}$ **12.** $\begin{pmatrix} \frac{5}{6} & \frac{2}{3} & -2 \\ \frac{1}{3} & \frac{2}{3} & -1 \\ -\frac{1}{6} & -\frac{1}{3} & 1 \end{pmatrix}$

14. Inverse does not exist.   **16.** $\begin{pmatrix} \frac{3}{2} & -\frac{1}{2} & \frac{1}{2} \\ \frac{3}{4} & -\frac{1}{4} & -\frac{1}{4} \\ 2 & -1 & 0 \end{pmatrix}$ **18.** Inverse does not exist.

19. $x_1 = 5$, $x_2 = 0$ **21.** $x_1 = \frac{1}{4}$, $x_2 = -\frac{3}{4}$, $x_3 = \frac{5}{4}$

24. $57, 44, 26, 24, 17, 13, -1, 6$ **26.** PEACE   **31.** $\begin{pmatrix} \frac{3}{2} & -\frac{1}{2} \\ -2 & 1 \end{pmatrix}$

SECTION 2-6

1. (a) 0.25   **(c)** Electrical industry   **(e)** Steel industry

3. $\begin{pmatrix} 16 \\ 21 \end{pmatrix}$, $\begin{pmatrix} 28 \\ 18 \end{pmatrix}$, $\begin{pmatrix} 40 \\ 30 \end{pmatrix}$ **5.** $\begin{pmatrix} 15 \\ 24 \\ 32 \end{pmatrix}$, $\begin{pmatrix} 15 \\ 32 \\ 56 \end{pmatrix}$, $\begin{pmatrix} 30 \\ 56 \\ 48 \end{pmatrix}$ **7.** $\begin{pmatrix} 2.4 \\ 5 \end{pmatrix}$ **9.** $\begin{pmatrix} 4 \\ 0.9 \\ 0.65 \end{pmatrix}$

SECTION 2-7

1. (a) Stochastic   **(b)** Not stochastic   **(c)** Stochastic
3. (a) 0.02   **(b)** 0.05   **(c)** Low-density residential   **(d)** Auto commercial
5. (a) 0.078   **(b)** 0.028525   **8. (a)** 0.2   **(b)** (14,000, 16,000), (19,760, 10,240)
14. $P = \begin{pmatrix} 0.8 & 0.2 \\ 0.4 & 0.6 \end{pmatrix}$, $X_4 = (40{,}000, 50{,}000) \begin{pmatrix} 0.8 & 0.2 \\ 0.4 & 0.6 \end{pmatrix}^4 = (59{,}488, 30{,}512)$

SECTION 2-8

1. (a) $\begin{pmatrix} 0 & 1 & 0 \\ 0 & 0 & 1 \\ 1 & 0 & 0 \end{pmatrix}$ **(b)** $\begin{pmatrix} 0 & 0 & 1 & 0 \\ 0 & 1 & 0 & 1 \\ 1 & 1 & 0 & 0 \\ 0 & 0 & 0 & 0 \end{pmatrix}$ **(c)** $\begin{pmatrix} 0 & 0 & 0 & 0 & 0 \\ 1 & 0 & 0 & 0 & 0 \\ 1 & 0 & 0 & 0 & 0 \\ 1 & 0 & 0 & 0 & 0 \\ 1 & 0 & 0 & 0 & 0 \end{pmatrix}$

2.   **4.**

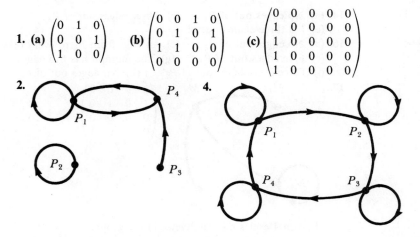

8.

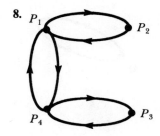

(a) 2 links, path $P_1 \rightarrow P_4 \rightarrow P_3$
(b) 3 links, path $P_2 \rightarrow P_1 \rightarrow P_4 \rightarrow P_3$

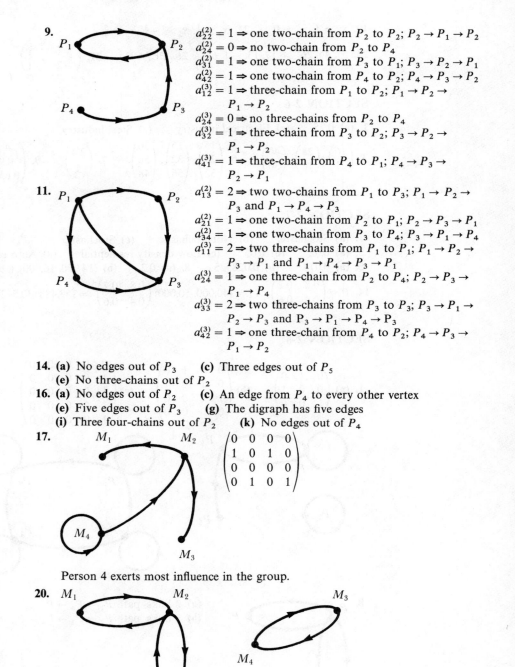

9. $a_{22}^{(2)} = 1 \Rightarrow$ one two-chain from P_2 to P_2; $P_2 \rightarrow P_1 \rightarrow P_2$
$a_{24}^{(2)} = 0 \Rightarrow$ no two-chain from P_2 to P_4
$a_{31}^{(2)} = 1 \Rightarrow$ one two-chain from P_3 to P_1; $P_3 \rightarrow P_2 \rightarrow P_1$
$a_{42}^{(2)} = 1 \Rightarrow$ one two-chain from P_4 to P_2; $P_4 \rightarrow P_3 \rightarrow P_2$
$a_{12}^{(3)} = 1 \Rightarrow$ three-chain from P_1 to P_2; $P_1 \rightarrow P_2 \rightarrow$
$\qquad P_1 \rightarrow P_2$
$a_{24}^{(3)} = 0 \Rightarrow$ no three-chains from P_2 to P_4
$a_{32}^{(3)} = 1 \Rightarrow$ three-chain from P_3 to P_2; $P_3 \rightarrow P_2 \rightarrow$
$\qquad P_1 \rightarrow P_2$
$a_{41}^{(3)} = 1 \Rightarrow$ three-chain from P_4 to P_1; $P_4 \rightarrow P_3 \rightarrow$
$\qquad P_2 \rightarrow P_1$

11. $a_{13}^{(2)} = 2 \Rightarrow$ two two-chains from P_1 to P_3; $P_1 \rightarrow P_2 \rightarrow$
$\qquad P_3$ and $P_1 \rightarrow P_4 \rightarrow P_3$
$a_{21}^{(2)} = 1 \Rightarrow$ one two-chain from P_2 to P_1; $P_2 \rightarrow P_3 \rightarrow P_1$
$a_{34}^{(2)} = 1 \Rightarrow$ one two-chain from P_3 to P_4; $P_3 \rightarrow P_1 \rightarrow P_4$
$a_{11}^{(3)} = 2 \Rightarrow$ two three-chains from P_1 to P_1; $P_1 \rightarrow P_2 \rightarrow$
$\qquad P_3 \rightarrow P_1$ and $P_1 \rightarrow P_4 \rightarrow P_3 \rightarrow P_1$
$a_{24}^{(3)} = 1 \Rightarrow$ one three-chain from P_2 to P_4; $P_2 \rightarrow P_3 \rightarrow$
$\qquad P_1 \rightarrow P_4$
$a_{33}^{(3)} = 2 \Rightarrow$ two three-chains from P_3 to P_3; $P_3 \rightarrow P_1 \rightarrow$
$\qquad P_2 \rightarrow P_3$ and $P_3 \rightarrow P_1 \rightarrow P_4 \rightarrow P_3$
$a_{42}^{(3)} = 1 \Rightarrow$ one three-chain from P_4 to P_2; $P_4 \rightarrow P_3 \rightarrow$
$\qquad P_1 \rightarrow P_2$

14. (a) No edges out of P_3 **(c)** Three edges out of P_5
(e) No three-chains out of P_2
16. (a) No edges out of P_2 **(c)** An edge from P_4 to every other vertex
(e) Five edges out of P_3 **(g)** The digraph has five edges
(i) Three four-chains out of P_2 **(k)** No edges out of P_4
17.

$$\begin{pmatrix} 0 & 0 & 0 & 0 \\ 1 & 0 & 1 & 0 \\ 0 & 0 & 0 & 0 \\ 0 & 1 & 0 & 1 \end{pmatrix}$$

Person 4 exerts most influence in the group.

20.

CHAPTER 3

SECTION 3-1

1. (a) 1   **(c)** 7   **(e)** 12

2. (a) Singular   **(c)** Nonsingular   **(e)** Singular

3. (a) $\sum_{j=1}^{5} (-1)^{i+j} a_{ij} A_{ij}$ **4. (a)** 0   **(b)** 19   **(c)** 14

5. (a) 3   **(b)** 1   **(c)** -3 **(d)** -1 **(e)** 6

7. $a_{13} A_{13} - a_{23} A_{23} + a_{33} A_{33} = \sum_{i=1}^{3} (-1)^{i+3} a_{i3} A_{i3}$

SECTION 3-2

1. $|A| = -1, |B| = 1$ **2.** $|A| = -4, |B| = 4$ **4. (a)** -5 **(b)** 5

(c) 0   **(d)** 16   **6.** Second result is correct, 4.   **8.** 1, 1, 0

SECTION 3-3

1. -4 **3.** -20 **5.** -9 **7.** 0

SECTION 3-4

1. $x_1 = 2, x_2 = 3$ **3.** $x_1 = 2, x_2 = 0, x_3 = -2$ **5.** $x_1 = \frac{8}{3}, x_2 = 1, x_3 = -\frac{2}{3}$

7. $\begin{pmatrix} -0.2 & 0.4 \\ 0.3 & -0.1 \end{pmatrix}$ **9.** No inverse   **11.** $\begin{pmatrix} \frac{7}{3} & -3 & -\frac{1}{3} \\ -\frac{8}{3} & 3 & \frac{2}{3} \\ \frac{4}{3} & -1 & -\frac{1}{3} \end{pmatrix}$

13. No inverse   **15.** $\lambda = 7, (r, r); \lambda = -4, (r, -5r/6)$

CHAPTER 4

SECTION 4-1

1.

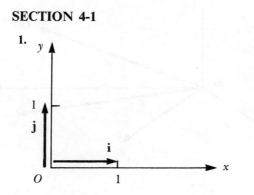

3. (a)

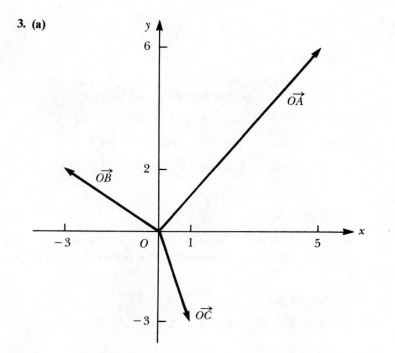

4. (a)

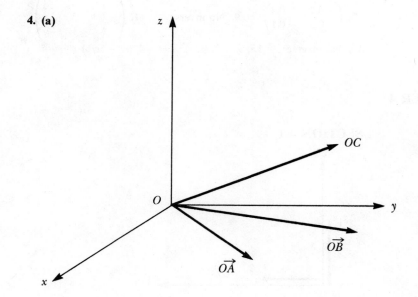

5. (a) $3(1, 4) = (3, 12)$

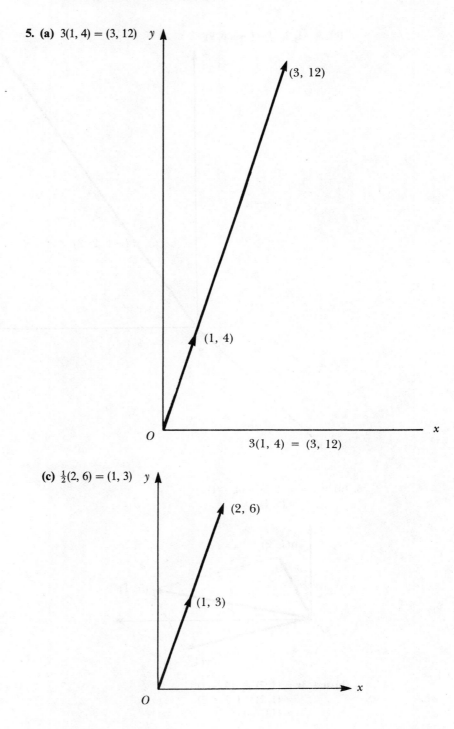

$3(1, 4) = (3, 12)$

(c) $\frac{1}{2}(2, 6) = (1, 3)$

(e) $3(-1, 2, 3) = (-3, 6, 9)$

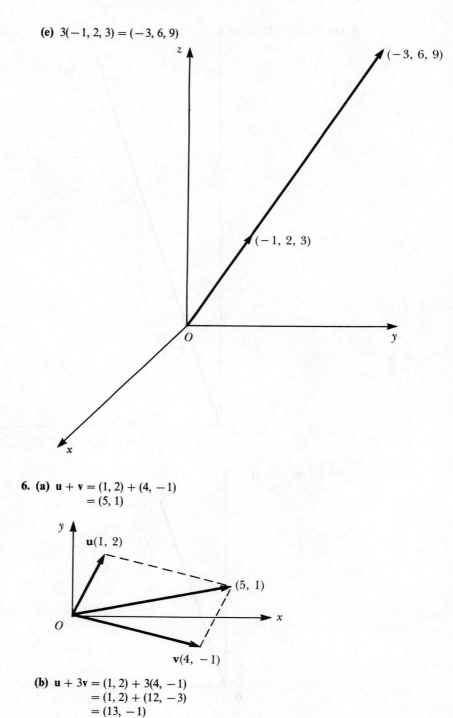

6. (a) $\mathbf{u} + \mathbf{v} = (1, 2) + (4, -1)$
$\qquad\qquad = (5, 1)$

(b) $\mathbf{u} + 3\mathbf{v} = (1, 2) + 3(4, -1)$
$\qquad\qquad = (1, 2) + (12, -3)$
$\qquad\qquad = (13, -1)$

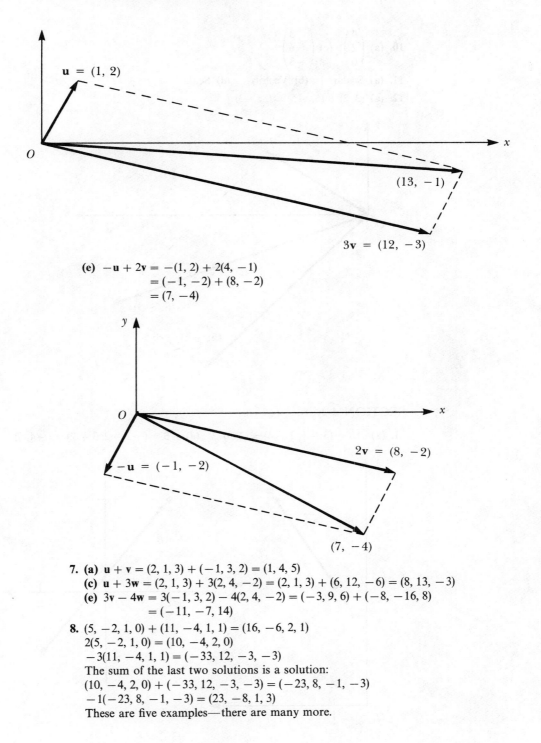

u = (1, 2)

(13, −1)

3**v** = (12, −3)

(e) −**u** + 2**v** = −(1, 2) + 2(4, −1)
   = (−1, −2) + (8, −2)
   = (7, −4)

−**u** = (−1, −2)

2**v** = (8, −2)

(7, −4)

7. **(a)** **u** + **v** = (2, 1, 3) + (−1, 3, 2) = (1, 4, 5)
   (c) **u** + 3**w** = (2, 1, 3) + 3(2, 4, −2) = (2, 1, 3) + (6, 12, −6) = (8, 13, −3)
   (e) 3**v** − 4**w** = 3(−1, 3, 2) − 4(2, 4, −2) = (−3, 9, 6) + (−8, −16, 8)
        = (−11, −7, 14)

8. (5, −2, 1, 0) + (11, −4, 1, 1) = (16, −6, 2, 1)
   2(5, −2, 1, 0) = (10, −4, 2, 0)
   −3(11, −4, 1, 1) = (−33, 12, −3, −3)
   The sum of the last two solutions is a solution:
   (10, −4, 2, 0) + (−33, 12, −3, −3) = (−23, 8, −1, −3)
   −1(−23, 8, −1, −3) = (23, −8, 1, 3)
   These are five examples—there are many more.

10. (a) $\begin{pmatrix} 4 \\ 2 \\ 0 \end{pmatrix}$ **(c)** $\begin{pmatrix} 3 \\ 6 \\ -3 \end{pmatrix}$

11. (a) Scalar   **(b)** Vector   **(d)** Scalar

12. (a) $(3, 2) + (5, -5) = (8, -3)$

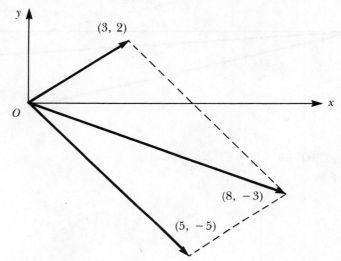

SECTION 4-2

1. (a) $\overrightarrow{AB} = (3 - 1, 7 - 4) = (2, 3)$ **(d)** $\overrightarrow{AB} = (-2 - 2, 3 + 2) = (-4, 5)$

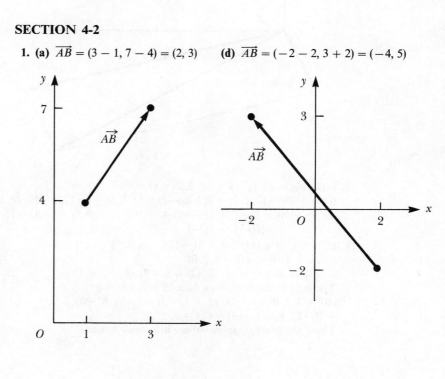

2.

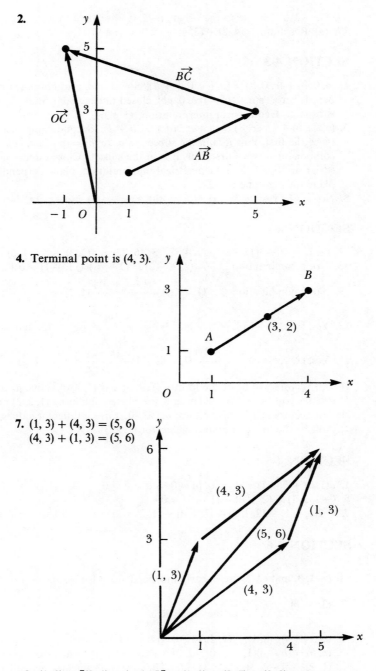

4. Terminal point is (4, 3).

7. (1, 3) + (4, 3) = (5, 6)
   (4, 3) + (1, 3) = (5, 6)

8. (1, 2) + [(3, 1) + (−1, 6)] = (1, 2) + (2, 7) = (3, 9)
   [(1, 2) + (3, 1)] + (−1, 6) = (4, 3) + (−1, 6) = (3, 9)
12. $\overrightarrow{OC} = \overrightarrow{OA} + \overrightarrow{AB} + \overrightarrow{BC}$ = (1, 2) + (4, 2) + (−1, −3) = (4, 1)

14. $(1, -2, 6)$ relative to the origin, $(-2, 0, 7)$ relative to A **16. (a)** $(a, b) = (1, 5)$
17. (a) Resultant $= (4, 2) + (3, 4) + (7, 3) = (14, 9)$

SECTION 4-3

1. $(a, 0, b) + (c, 0, d) = (a + c, 0, b + d)$; closed under addition, since second component is zero. $p(a, 0, b) = (pa, 0, pb)$; closed under scalar multiplication. Therefore subspace. Geometrical interpretation, xz plane.
3. $(a, 2a, b) + (c, 2c, d) = (a + c, 2(a + c), b + d)$. Second component, $2(a + c)$, is twice the first; thus it is in the subset. $p(a, 2a, b) = (pa, 2pa, pb)$. Again second component is twice first; thus it is in the subset. Closed under addition and scalar multiplication. Geometrical interpretation, plane perpendicular to xy plane through line $y = 2x$.
5. No   **7.** Yes   **9. (a)** No   **(b)** Yes   **(c)** No   **11.** No

SECTION 4-4

1. $(-1, 7) = -3(1, -1) + (2, 4)$ **3.** $(6, 22) = 4(2, 3) + 2(-1, 5)$
5. Not a combination   **7.** $(2, 2, -2) = (2 - a_3)(1, 1, -1) - a_3(2, 1, 3) + a_3(4, 3, 1)$
9. Not a combination   **11.** $(x, y) = \dfrac{(-x + 3y)}{5}(1, 2) + \dfrac{(2x - y)}{5}(3, 1)$
13. Yes; $(x, y) = -y(3, -1) + 0(2, 3) + \dfrac{(x + 3y)}{4}(4, 0)$ **15.** Do not span $\mathbf{R}^3$
17. Yes; $(x, y, z) = \dfrac{(3x + z - 2y)}{4}(1, -1, -1) + \dfrac{(x - 2y + 3z)}{4}(0, 1, 2) - \dfrac{(z - x - 2y)}{4}(1, 2, 1)$
18. All vectors will be of the form $a_1(1, 2, 3) + a_2(1, 2, 0)$. Give a_1 and a_2 any values you like. The subspace is the plane containing $(1, 2, 3)$ and $(1, 2, 0)$.
20. All vectors will be of the form $a(1, 2, 3)$. Give a different values. The subspace will be the line containing the vector $(1, 2, 3)$.

SECTION 4-5

1. $a(1, 0) + b(0, 1) = 0 \Rightarrow (a, b) = 0 \Rightarrow a = b = 0$ **3.** $3(1, -2) - (3, -6) = 0$
5. $(-1, 3, 1) + (2, 1, 0) - (1, 4, 1) = 0$
7. $-2(1, 1, 1) - (2, 1, 0) + (3, 1, 4) + (1, 2, -2) = 0$ **10.** $\mathbf{v} = a(2, 1, 0) + b(-1, 2, 0)$

SECTION 4-6

1. $(-1, 3)$ and $(2, 2)$ are linearly independent. Also $(a, b) = \dfrac{(b - a)}{4}(-1, 3) + \dfrac{(3a + b)}{4}(2, 2)$. Thus they span $\mathbf{R}^2$.
3. e.g., $\{(1, 0), (0, 1)\}, \{(1, 2), (0, 3)\}, \{(4, 5), (0, 1)\}$
5. $(a, b) = (b - a)(1, 2) + 0(-2, 3) + (2a - b)(1, 1)$. Thus they span $\mathbf{R}^2$.
    $(1, 2) - \frac{1}{5}(-2, 3) - \frac{7}{5}(1, 1) = 0$. Thus they are linearly dependent; e.g.,
    $(3, 4) = \frac{17}{7}(1, 2) - \frac{2}{7}(-2, 3) + 0(1, 1)$ and $(3, 4) = (1, 2) + 0(-2, 3) + 2(1, 1)$.
7. $(1, 2, 0)$ lies in the plane defined by $(1, 0, 0)$ and $(0, 1, 0)$. All vectors lie in the xy plane.

9. $(1, 2, 0)$ and $(2, 3, 0)$ are linearly independent. Thus they form a basis for this subspace. It is of dimension 2. It is the xy plane.

11. The subspace consists of all vectors of the type $a(1, -1, 2)$. The subspace is the line defined by $(1, -1, 2)$.

13. $(1, 2, -1) = 2(1, 3, 1) - (1, 4, 3)$

15. $(1, 2, -1, 3) = (1, 2, 1, 0) + 2(1, -2, 3, 1) + (-2, 4, -8, 1)$

17. e.g., $(1, 0, 3), (0, -1, 0)$

19. e.g., $(1, -1, 3, -1), (0, 1, 2, 0), (0, 0, 1, -1)$

SECTION 4-7

1. (a) 2   **(b)** 2   **(c)** 3   **(d)** 1   **3.** 2   **5.** 3   **7.** 3   **9.** 3

11. $\{(1, \frac{1}{2}, \frac{3}{2}, 0, 2), (0, 1, \frac{9}{5}, \frac{2}{5}, \frac{4}{5})\}$

13. $\{(1, -2, 0, 1, 3), (0, 1, 0, \frac{3}{2}, \frac{7}{4}), (0, 0, 2, -9, -6)\}$

18. (a) Row rank is 3; column rank is the row rank of the transpose of the matrix.

$$\begin{pmatrix} 1 & 2 & 3 & 5 \\ 0 & 1 & 2 & 3 \\ 0 & 0 & 1 & 2 \\ 0 & 0 & 0 & 0 \end{pmatrix}^t = \begin{pmatrix} 1 & 0 & 0 & 0 \\ 2 & 1 & 0 & 0 \\ 3 & 2 & 1 & 0 \\ 5 & 3 & 2 & 0 \end{pmatrix} \approx \cdots \approx \begin{pmatrix} 1 & 0 & 0 & 0 \\ 0 & 1 & 0 & 0 \\ 0 & 0 & 1 & 0 \\ 0 & 0 & 0 & 0 \end{pmatrix}$$

Row rank of the transpose is 3. Thus column rank is 3.

SECTION 4-8

1. 6   **3.** $\sqrt{5}, 5$ **5.** $(1/\sqrt{13})(2, 3), (1/\sqrt{5})(-1, 2)$

9. $45°$ **11.** $\cos\theta = \sqrt{2/5}, \theta = 50.77°$ **14.** $b(-2, 1)$ for any b

20. (a) 5   **(c)** $\sqrt{69} = 8.3066$ **(e)** $\sqrt{48} = 6.9282$

22. (a) Vector   **(c)** No sense   **(e)** No sense   **(g)** No sense   **(i)** Scalar   **(k)** Vector

SECTION 4-9

1. $12/\sqrt{29}, \frac{12}{29}(2, 5)$ **3.** $\sqrt{5}, (1, 2, 0)$

5. $0, (0, 0, 0, 0)$; vectors are orthogonal   **8.** The vector $(1/\sqrt{2}, b)$ for any b

11. $(a, b, c, d) \cdot (1, 2, -1, 1) = 0; a = -2b + c - d$; thus $(-2b + c - d, b, c, d)$ for any b, c, d

15. (a) $3/\sqrt{2}$ **(b)** 1   **(c)** 2

17. (a) $10/\sqrt{3}, 5, 5/\sqrt{3}$ **(b)** $10/\sqrt{2}, 10/\sqrt{2}, 20$ **(c)** $5\sqrt{2}, 5\sqrt{2}, 0$

SECTION 4-10

1. (a) $(8, -7, 2)$ **(c)** $(3, 10, 23)$ **(e)** -8 **5.** $\frac{1}{2}\sqrt{45}$ **6.** 4

8. (a) $(25, -3, -7)$ **(b)** $(-9, 6, -1)$

SECTION 4-11

1. (a) $1(x - 1) + 1(y + 2) + 1(z - 4); x + y + z - 3 = 0$
    (c) Both forms are $x + 2y + 3z = 0$

2. (a) $6x + y + 2z - 10 = 0$ **(c)** $3x - 2y + 1 = 0$

3. Normals are parallel; thus planes are parallel.   **5.** $2x - 3y + z + 7 = 0$

6. (a) $x = 1 - t, y = 2 + 2t, z = 3 + 4t$, where $-\infty < t < \infty$; $\dfrac{x - 1}{-1} = \dfrac{y - 2}{2} = \dfrac{z - 3}{4}$

(c) $x = -2t$, $y = -3t$, $z = 5t$, where $-\infty < t < \infty$; $\dfrac{x}{-2} = \dfrac{y}{-3} = \dfrac{z}{5}$

7. $x = 1 + 2t$, $y = 2 + 3t$, $z = -4 + t$, where $-\infty < t < \infty$

9. $x = 4 + 2t$, $y = -1 - t$, $z = 3 + 4t$, where $-\infty < t < \infty$

11. Show that P_1, P_2, and P_3 are colinear.

13. Show that every point on line satisfies the equation of the plane.

15. $x = 5 + (3b + 2c)t/2$, $y = -1 + bt$, $z = 2 + ct$ for any values of b and c

CHAPTER 5

SECTION 5-1

3. (a) $(f + g)(x) = x^2 + x + 1$, $(2f)(x) = 2x + 4$, $(3g)(x) = 3x^2 - 3$

(b) $(f + g)(x) = 4$, $(3f)(x) = 6x$, $(-g)(x) = 2x - 4$

5. (a) Yes   **(b)** No, the sum of the two discontinuous functions can be continuous.

7. (a) Subspace   **(b)** Subspace

10. No, not closed under addition or multiplication.

12. (a) Subspace   **(b)** Not a subspace   **(c)** Subspace

14. (a) Yes, $\begin{pmatrix} 5 & 7 \\ 5 & -10 \end{pmatrix} = 2\begin{pmatrix} 1 & 2 \\ 3 & -4 \end{pmatrix} - \begin{pmatrix} 0 & 3 \\ 1 & 2 \end{pmatrix} + 3\begin{pmatrix} 1 & 2 \\ 0 & 0 \end{pmatrix}$ **(c)** No

15. (a) Yes, $3x^2 + 2x + 9 = 3(x^2 + 1) + 2(x + 3)$

(c) Yes, $x^2 + 4x + 5 = -2(x^2 + x - 1) + 3(x^2 + 2x + 1)$

16. (a) Yes, $(2x^2 + 1) - (x^2 + 4x) - (x^2 - 4x + 1) = 0$ **(c)** No   **(e)** No

17. (a) $1, x, x^2, x^3$; dimension is 4.   **(c)** $\begin{pmatrix} 1 & 0 \\ 0 & 0 \end{pmatrix}, \begin{pmatrix} 0 & 0 \\ 0 & 1 \end{pmatrix}$; dimension is 2.

18. (a) Yes, $x + 5 = -(x + 1) + 2(x + 3)$

(c) e.g., $(g + h)(x) = 3x^2 + 3x + 2$, $2g(x) = 4x^2 + 6$, $-h(x) = -x^2 - 3x + 1$

19. (a) No, these vectors are linearly dependent; $f(x) + g(x) - h(x) = 0$

(c) No; they are linearly dependent; $\begin{pmatrix} 1 & 2 \\ 0 & 1 \end{pmatrix} - \begin{pmatrix} 3 & 4 \\ 1 & 1 \end{pmatrix} + 2\begin{pmatrix} 1 & 2 \\ 1 & 1 \end{pmatrix} - \begin{pmatrix} 0 & 2 \\ 1 & 2 \end{pmatrix} = 0$

21. (a) $\begin{vmatrix} 1 & x & x^2 \\ 0 & 1 & 2x \\ 0 & 0 & 2 \end{vmatrix} = 2 \neq$ zero function; thus linearly independent.

(c) $\begin{vmatrix} \sin x & \cos x & x \sin x \\ \cos x & -\sin x & \sin x + x \cos x \\ -\sin x & -\cos x & 2\cos x - x \sin x \end{vmatrix} = -2\cos x \neq$ zero function

SECTION 5-2

3. $\langle \mathbf{u}, \mathbf{u} \rangle = 2x_1^2 - x_2^2$. Let $\mathbf{u} = (x_1, x_2) = (1, \sqrt{2})$. Then $\langle \mathbf{u}, \mathbf{u} \rangle = 2(1)^2 - (\sqrt{2})^2 = 0$. But $\mathbf{u} \neq \mathbf{0}$. Thus the positive definite axiom is not satisfied.

6. $\langle af, g \rangle = \int_0^1 af(x)g(x)\, dx = a\int_0^1 f(x)g(x)\, dx = a\langle f, g \rangle$

$\langle f, f \rangle = \int_0^1 [f(x)]^2 dx \geq 0$; $\langle f, f \rangle = 0$ if and only if $f = \mathbf{0}$ **8.** $\sqrt{7}$, $7/\sqrt{5}$

10. $\sqrt{15}/4$ **13.** $\sqrt{76/3}$ **15.** Magnitude of each matrix is 1.

18. e.g., $\begin{pmatrix} -9 & 1 \\ 1 & 1 \end{pmatrix}$ **20.** $\sqrt{10}$ **22. (a)** $2, 3, \sqrt{13}$ **(c)** $\sqrt{13}$

SECTION 5-3

3. 0   **5.** 12 years, 0.8 speed of light   **7.** 8.2098 centuries

CHAPTER 6

SECTION 6-1

1. (a) $\begin{pmatrix} 1/\sqrt{2} & -1\sqrt{2} \\ 1/\sqrt{2} & 1/\sqrt{2} \end{pmatrix}$; image of $\begin{pmatrix} 2 \\ 1 \end{pmatrix}$ is $\begin{pmatrix} 1/\sqrt{2} \\ 3/\sqrt{2} \end{pmatrix}$.

(c) $\begin{pmatrix} -1 & 0 \\ 0 & -1 \end{pmatrix}$; image of $\begin{pmatrix} 2 \\ 1 \end{pmatrix}$ is $\begin{pmatrix} -2 \\ -1 \end{pmatrix}$.

3. $\begin{pmatrix} 1 & 0 \\ 0 & -1 \end{pmatrix}$ **7.** $\begin{pmatrix} c\cos\theta & -c\sin\theta \\ c\sin\theta & c\cos\theta \end{pmatrix}$

8. (a) $\overset{O}{\begin{pmatrix} 0 \\ 0 \end{pmatrix}} \mapsto \overset{O}{\begin{pmatrix} 0 \\ 0 \end{pmatrix}}, \overset{A}{\begin{pmatrix} 1 \\ 0 \end{pmatrix}} \mapsto \overset{A'}{\begin{pmatrix} 0 \\ 1 \end{pmatrix}}, \overset{B}{\begin{pmatrix} 1 \\ 1 \end{pmatrix}} \mapsto \overset{B'}{\begin{pmatrix} -1 \\ 1 \end{pmatrix}}, \overset{C}{\begin{pmatrix} 0 \\ 1 \end{pmatrix}} \mapsto \overset{C'}{\begin{pmatrix} -1 \\ 0 \end{pmatrix}}$

(c) $\overset{O}{\begin{pmatrix} 0 \\ 0 \end{pmatrix}} \mapsto \overset{O}{\begin{pmatrix} 0 \\ 0 \end{pmatrix}}, \overset{A}{\begin{pmatrix} 1 \\ 0 \end{pmatrix}} \mapsto \overset{A'}{\begin{pmatrix} 3 \\ 1 \end{pmatrix}}, \overset{B}{\begin{pmatrix} 1 \\ 1 \end{pmatrix}} \mapsto \overset{B'}{\begin{pmatrix} 3 \\ 5 \end{pmatrix}}, \overset{C}{\begin{pmatrix} 0 \\ 1 \end{pmatrix}} \mapsto \overset{C'}{\begin{pmatrix} 0 \\ 4 \end{pmatrix}}$

$$
\begin{array}{cccccccc}
O & O & A & A' & B & B' & C & C' \\
\end{array}
$$

(e) $\begin{pmatrix} 0 \\ 0 \end{pmatrix} \mapsto \begin{pmatrix} 0 \\ 0 \end{pmatrix}, \begin{pmatrix} 1 \\ 0 \end{pmatrix} \mapsto \begin{pmatrix} -2 \\ 0 \end{pmatrix}, \begin{pmatrix} 1 \\ 1 \end{pmatrix} \mapsto \begin{pmatrix} -5 \\ 4 \end{pmatrix}, \begin{pmatrix} 0 \\ 1 \end{pmatrix} \mapsto \begin{pmatrix} -3 \\ 4 \end{pmatrix}$

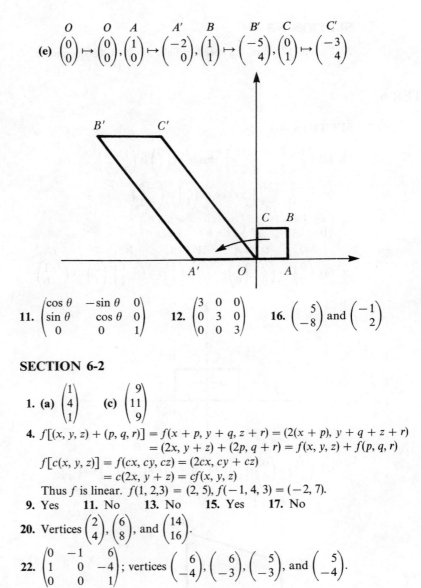

11. $\begin{pmatrix} \cos\theta & -\sin\theta & 0 \\ \sin\theta & \cos\theta & 0 \\ 0 & 0 & 1 \end{pmatrix}$ **12.** $\begin{pmatrix} 3 & 0 & 0 \\ 0 & 3 & 0 \\ 0 & 0 & 3 \end{pmatrix}$ **16.** $\begin{pmatrix} 5 \\ -8 \end{pmatrix}$ and $\begin{pmatrix} -1 \\ 2 \end{pmatrix}$

SECTION 6-2

1. (a) $\begin{pmatrix} 1 \\ 4 \\ 1 \end{pmatrix}$ **(c)** $\begin{pmatrix} 9 \\ 11 \\ 9 \end{pmatrix}$

4. $f[(x, y, z) + (p, q, r)] = f(x + p, y + q, z + r) = (2(x + p), y + q + z + r)$
$\qquad\qquad\qquad\qquad\qquad = (2x, y + z) + (2p, q + r) = f(x, y, z) + f(p, q, r)$
$\quad f[c(x, y, z)] = f(cx, cy, cz) = (2cx, cy + cz)$
$\qquad\qquad\qquad = c(2x, y + z) = cf(x, y, z)$
Thus f is linear. $f(1, 2, 3) = (2, 5), f(-1, 4, 3) = (-2, 7)$.

9. Yes   **11.** No   **13.** No   **15.** Yes   **17.** No

20. Vertices $\begin{pmatrix} 2 \\ 4 \end{pmatrix}, \begin{pmatrix} 6 \\ 8 \end{pmatrix}$, and $\begin{pmatrix} 14 \\ 16 \end{pmatrix}$.

22. $\begin{pmatrix} 0 & -1 & 6 \\ 1 & 0 & -4 \\ 0 & 0 & 1 \end{pmatrix}$; vertices $\begin{pmatrix} 6 \\ -4 \end{pmatrix}, \begin{pmatrix} 6 \\ -3 \end{pmatrix}, \begin{pmatrix} 5 \\ -3 \end{pmatrix}$, and $\begin{pmatrix} 5 \\ -4 \end{pmatrix}$.

24. Yes, $\begin{pmatrix} 1 & 0 & -h \\ 0 & 1 & -k \\ 0 & 0 & 1 \end{pmatrix}$. Note also that determinant = 1, matrix is nonsingular, inverse exists.

SECTION 6-3

1. (a) Kernel is $\left\{ \begin{pmatrix} 0 \\ 0 \end{pmatrix} \right\}$, range is $\mathbf{R}^2$ **(c)** Kernel is $\left\{ \begin{pmatrix} -2r \\ r \end{pmatrix} \right\}$, range is $\left\{ \begin{pmatrix} s \\ 2s \end{pmatrix} \right\}$

(e) Kernel is $\left\{\begin{pmatrix} r \\ -2r \\ r \end{pmatrix}\right\}$, range is $\mathbf{R}^2$ **(g)** Kernel is x axis, range is $\left\{\begin{pmatrix} r \\ 2r \\ s \end{pmatrix}\right\}$

3. (a) Kernel is yz plane, range is x axis
   (c) Kernel is plane $(x, y, -x - y)$, range is $\mathbf{R}$

9.

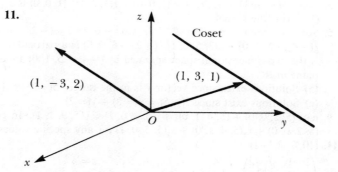

11.

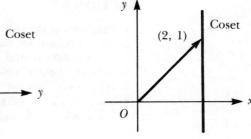

13. Coset $(1, 2, 0) + a(0, 0, 1)$

15. Coset $(2, 1) + a(0, 1)$

SECTION 6-4

1. Solutions are $(r, r, 1) = (0, 0, 1) + r(1, 1, 0)$. $(0, 0, 1)$ is a particular solution.
   $\{r(1, 1, 0)\}$ is the kernel of the mapping defined by the matrix of coefficients.

It is the set of solutions to the corresponding system of homogeneous equations.

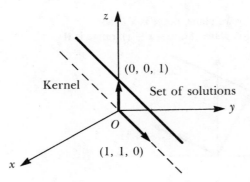

3. Solutions are $(1 + r, 2r, r) = (1, 0, 0) + r(1, 2, 1)$. $(1, 0, 0)$ is a particular solution. $(1, 2, 1)$ is the kernel.

5. Solutions are $(-4r + 3s - 2, -5r + 5s - 6, r, s) = (-2, -6, 0, 0) + r(-4, -5, 1, 0) + s(3, 5, 0, 1)$. $(-2, -6, 0, 0)$ is a particular solution. The kernel is the two-dimensional space spanned by $(-4, -5, 1, 0)$ and $(3, 5, 0, 1)$. This is a plane in $\mathbf{R}^4$.

7. **(a)** Solutions exist, since vector **y** is in the range. $z(=2) = x(=1) + y(=1)$
   (c) Solutions exist since $z(=5) = x(=3) + y(=2)$

9. e.g., $(5, 2, 0, 0) + (5, 4, 1, 0) + (5, 3, 0, 1) = (15, 9, 1, 1)$. In general, $(5, 2, 0, 0) + r_1(5, 4, 1, 0) + s_1(5, 3, 0, 1)$ for any specific values of r_1 and s_1.

11. $(10, 7, 2, -1)$

13. $\begin{pmatrix} 2 & 1 & 0 \\ 3 & 3 & 1 \\ 0 & 1 & 1 \end{pmatrix} \begin{pmatrix} 2 \\ 0 \\ 3 \end{pmatrix} = \begin{pmatrix} 4 \\ 9 \\ 3 \end{pmatrix}$; thus $\begin{aligned} 2x_1 + x_2 &= 4 \\ 3x_1 + 3x_2 + x_3 &= 9 \\ x_2 + x_3 &= 3 \end{aligned}$

SECTION 6-5

1. Functions of the type $a_1 x + a_0$; linear functions

3. x^2, x, 1 is basis for the range; dimension is 3. x^4, x^3, x^2, x, 1 is basis for the domain; dimension is 5. The kernel is the set of linear functions; dimension is 2.

5. $(D^2 + D - 2)e^x = e^x + e^x - 2e^x = 0$; $(D^2 + D - 2)e^{-2x} = 4e^{-2x} - 2e^{-2x} - 2e^{-2x} = 0$

7. **(a)** $y = c_1 e^{-4x} + c_2 e^{2x}$ **(b)** e^{-4x}, e^{2x} 9. $y = c_1 e^{-2x} + c_2 x e^{-2x} + \frac{3}{4}$

11. $y = c_1 + c_2 e^{3x} - 8x/3$ 13. $y = c_1 e^x \cos \sqrt{5}x + c_2 e^x \sin \sqrt{5}x + 1$

15. $y = c_1 \cos 2x + c_2 \sin 2x - (\sin 3x)/5$ 17. $(D - 4)$ 18. $(D^2 - D)$

SECTION 6-6

1. **(a)** One-to-one   **(b)** Onto   **(c)** Isomorphism
3. **(a)** Not one-to-one   **(b)** Not onto   **(c)** Not an isomorphism
5. **(a)** One-to-one   **(b)** Onto   **(c)** Isomorphism
7. **(a)** Not one-to-one   **(b)** Not onto   **(c)** Not an isomorphism
9. **(a)** One-to-one   **(b)** Onto   **(c)** Isomorphism

11. (a) Not one-to-one   **(b)** Not onto   **(c)** Not an isomorphism
13. (a) One-to-one   **(b)** Onto   **(c)** Isomorphism
15. (a) One-to-one   **(b)** Not onto   **(c)** Not an isomorphism
17. (a) One-to-one   **(b)** Not onto   **(c)** Not an isomorphism
19. (a) One-to-one   **(b)** Onto   **(c)** Isomorphism
21. (a) One-to-one   **(b)** Onto   **(c)** Isomorphism
23. (a) Not one-to-one   **(b)** Onto   **(c)** Not an isomorphism
25. (a) Not one-to-one   **(b)** Onto   **(c)** Not an isomorphism
27. $f(ax^3 + bx^2 + cx + d) = (a, b, c, d)$, $f(3x^3 + 2x^2 + 2x + 2) = (3, 2, 1, 2)$
29. $f(ax^2 + bx + c) = (a, b, c - b)$, $f(x^2 + 2x + 6) = (1, 2, 4)$

31. $f\begin{pmatrix} a & b \\ c & d \end{pmatrix} = (a, b, c, d)$, $f\begin{pmatrix} 1 & 2 \\ 3 & 4 \end{pmatrix} = (1, 2, 3, 4)$

34. (a) $f(a, a + b, a - b) = f[a(1, 1, 1) + b(0, 1, -1)]$
$= a(1, 0) + b(0, 1) = (a, b)$

(c) $f(a + b, a, b) = f[a(1, 1, 0) + b(1, 0, 1)]$ **(e)** $f\begin{pmatrix} a & 0 \\ 0 & b \end{pmatrix} = (a, b)$
$= a(1, 1, 1, 1) + b(0, 1, 2, 3)$
$= (a, a + b, a + 2b, a + 3b)$

37. $\begin{pmatrix} b_1 \\ b_2 \\ b_3 \end{pmatrix} = \begin{pmatrix} \frac{1}{3} & 0 & 0 \\ 0 & 1 & 0 \\ 0 & \frac{1}{4} & \frac{1}{4} \end{pmatrix} \begin{pmatrix} a_1 \\ a_2 \\ a_3 \end{pmatrix}$

$(1, 4, 3), (2, 0, 1), (0, 8, 5), (1, 4, 2)$

39. $\begin{pmatrix} a \\ b \\ c \\ d \end{pmatrix} = \begin{pmatrix} \frac{1}{2} & 0 & 0 & 0 \\ 0 & -\frac{2}{3} & \frac{1}{3} & 0 \\ 0 & \frac{1}{6} & \frac{1}{6} & 0 \\ 0 & 0 & 0 & \frac{1}{3} \end{pmatrix} \begin{pmatrix} p \\ q \\ r \\ s \end{pmatrix}$

$(1, 0, \frac{3}{2}, 1), (-1, -4, \frac{5}{2}, 2), (0, -5, \frac{7}{2}, 0)$

41. $\begin{pmatrix} a_1 \\ a_2 \\ a_3 \end{pmatrix} = \begin{pmatrix} -\frac{1}{3} & \frac{2}{3} & 0 \\ \frac{2}{3} & -\frac{1}{3} & 0 \\ 0 & 0 & \frac{1}{4} \end{pmatrix} \begin{pmatrix} b_1 \\ b_2 \\ b_3 \end{pmatrix}$

$(3, 0, 1), (4, -2, 2), (-2, 4, 0), (-2, 7, -1)$

SECTION 6-7

1. (a) $\begin{pmatrix} 1 & 0 & 0 \\ 0 & 0 & 1 \end{pmatrix}$ **(b)** $\begin{pmatrix} 1 & 1 & 0 \\ 2 & -1 & 0 \end{pmatrix}$

3. $\begin{pmatrix} 0 & 3 & 0 \\ 2 & 0 & 0 \\ 0 & 0 & -1 \end{pmatrix}, \begin{pmatrix} 6 \\ 2 \\ -3 \end{pmatrix}$ **5.** $\begin{pmatrix} 1 & 3 & 1 & 2 \\ 1 & -2 & 2 & 1 \\ 0 & 1 & -1 & -2 \end{pmatrix}, 12v_1 - 2v_2 - 3v_3$

7. $\begin{pmatrix} 0 & 2 & 0 \\ 0 & 0 & 3 \\ 0 & 0 & 0 \end{pmatrix}$ **10. (a)** $\begin{pmatrix} 0 & 1 & 1 \\ 0 & 1 & -1 \\ 0 & 0 & 0 \end{pmatrix}$ **(c)** $\begin{pmatrix} 0 & 0 & 1 \\ 0 & 1 & 0 \\ 1 & 0 & 0 \end{pmatrix}$ **11.** $\begin{pmatrix} 1 & 0 & 0 \\ 0 & 0 & 1 \end{pmatrix}$

14. $\begin{pmatrix} 1 & 1 \\ -1 & 0 \end{pmatrix}, \begin{pmatrix} \frac{1}{2} & -\frac{1}{2} \\ \frac{3}{2} & \frac{1}{2} \end{pmatrix}$ **16.** $\begin{pmatrix} 1 & -1 \\ 1 & 1 \end{pmatrix}, \begin{pmatrix} 1 & 1 \\ -1 & 1 \end{pmatrix}$

CHAPTER 7

SECTION 7-1

1. $\lambda = 1, r\begin{pmatrix} 1 \\ -1 \end{pmatrix}$; $\lambda = 6, r\begin{pmatrix} 4 \\ 1 \end{pmatrix}$

3. $\lambda = -1, r\begin{pmatrix} 1 \\ -1 \\ 1 \end{pmatrix}$; $\lambda = 2, r\begin{pmatrix} 0 \\ 1 \\ 1 \end{pmatrix}$; $\lambda = 1, r\begin{pmatrix} 1 \\ 0 \\ 1 \end{pmatrix}$

5. $\lambda = 3, r\begin{pmatrix} 0 \\ 1 \\ 1 \end{pmatrix}$; $\lambda = 1, 1, \begin{pmatrix} r \\ s \\ r \end{pmatrix}$; orthogonal basis $\left\{ \dfrac{1}{\sqrt{2}}\begin{pmatrix} 1 \\ 0 \\ 1 \end{pmatrix}, \begin{pmatrix} 0 \\ 1 \\ 0 \end{pmatrix} \right\}$

7. $\lambda = 1, r\begin{pmatrix} 1 \\ -1 \\ 1 \end{pmatrix}$; $\lambda = 2, r\begin{pmatrix} 0 \\ 1 \\ 1 \end{pmatrix}$; $\lambda = 8, r\begin{pmatrix} 1 \\ 0 \\ 1 \end{pmatrix}$

9. $\lambda = 4, r\begin{pmatrix} 0 \\ -1 \\ 0 \\ 1 \end{pmatrix}$; $\lambda = 6, r\begin{pmatrix} 1 \\ 0 \\ 0 \\ 1 \end{pmatrix}$; $\lambda = 2, 2, \begin{pmatrix} r \\ -r \\ s \\ s \end{pmatrix}$; orthogonal basis $\left\{ \dfrac{1}{\sqrt{2}}\begin{pmatrix} 1 \\ -1 \\ 0 \\ 0 \end{pmatrix}, \dfrac{1}{\sqrt{2}}\begin{pmatrix} 0 \\ 0 \\ 1 \\ 1 \end{pmatrix} \right\}$

11. $\lambda = 1, 2, 3 + 5/\sqrt{2}$ **12.** $\lambda = 3$; $\mathbf{R}^2$ is eigenspace.

14. $\begin{vmatrix} 1 - \lambda & 1 \\ -2 & -1 - \lambda \end{vmatrix} = 0 \Rightarrow \lambda^2 + 1 = 0$; no real roots

SECTION 7-2

2. 132,603.34 metro; 66,301.67 nonmetro

5. (a) 0.506   **(b)** $\begin{pmatrix} 0.4 & 0.6 \\ 0.4 & 0.6 \end{pmatrix}$ **(c)** 0.123   **(d)** $(p_{22})^{m-1} p_{21} (p_{11})^{n-1} p_{12}$

7. Distribution of 3:3:2 in rooms 1, 2, 3, respectively; $p = \frac{3}{8}$ **9.** A 44.4%; B 55.6%

11. $\mathbf{x} = (93\frac{3}{4}, 6\frac{1}{4})$; $93\frac{3}{4}\%$ will want ranch style, $6\frac{1}{4}\%$ will want split level.

SECTION 7-3

1. (a) $\begin{pmatrix} 7 & 13 \\ -2 & -3 \end{pmatrix}$ **(c)** $\begin{pmatrix} 21 & -31 \\ 14 & -21 \end{pmatrix}$ **(e)** $\begin{pmatrix} 10 & 24 & 39 \\ 5 & 16 & 24 \\ -\frac{13}{3} & -\frac{37}{3} & -19 \end{pmatrix}$

(g) $-\frac{1}{8}\begin{pmatrix} -32 & -22 & -2 \\ -8 & 7 & 5 \\ 16 & 3 & -23 \end{pmatrix}$

2. (a) $C = \begin{pmatrix} 1 & 4 \\ -1 & 1 \end{pmatrix}, \begin{pmatrix} 1 & 0 \\ 0 & 6 \end{pmatrix}$ **(c)** $C = \begin{pmatrix} 1 & 1 \\ 2 & 1 \end{pmatrix}, \begin{pmatrix} 2 & 0 \\ 0 & 3 \end{pmatrix}$

(e) $C = \begin{pmatrix} 1 & 0 & 1 \\ -1 & 1 & 0 \\ 1 & 1 & 1 \end{pmatrix}, \begin{pmatrix} 1 & 0 & 0 \\ 0 & 2 & 0 \\ 0 & 0 & 8 \end{pmatrix}$ **(g)** $C = \begin{pmatrix} 1 & 0 & 0 \\ 0 & 1 & 1 \\ 1 & 0 & 1 \end{pmatrix}, \begin{pmatrix} 1 & 0 & 0 \\ 0 & 1 & 0 \\ 0 & 0 & 3 \end{pmatrix}$

(i) $C = \begin{pmatrix} 0 & 1 & 1 \\ 1 & 0 & -1 \\ 1 & 1 & 1 \end{pmatrix}, \begin{pmatrix} 1 & 0 & 0 \\ 0 & 3 & 0 \\ 0 & 0 & 5 \end{pmatrix}$

3. (a) $C = \begin{pmatrix} 1/\sqrt{2} & 1/\sqrt{2} \\ -1/\sqrt{2} & 1/\sqrt{2} \end{pmatrix}, \begin{pmatrix} -1 & 0 \\ 0 & 3 \end{pmatrix}$ **(c)** $C = \begin{pmatrix} 1/\sqrt{2} & 1/\sqrt{2} \\ -1/\sqrt{2} & 1/\sqrt{2} \end{pmatrix}, \begin{pmatrix} 2 & 0 \\ 0 & 4 \end{pmatrix}$

(e) $C = \begin{pmatrix} 0 & 1/\sqrt{2} & 1/\sqrt{2} \\ 0 & 1/\sqrt{2} & -1/\sqrt{2} \\ 1 & 0 & 0 \end{pmatrix}, \begin{pmatrix} -2 & 0 & 0 \\ 0 & -1 & 0 \\ 0 & 0 & 2 \end{pmatrix}$

(g) $C = \begin{pmatrix} 1/\sqrt{2} & 0 & 1/\sqrt{2} \\ -1/\sqrt{2} & 0 & 1/\sqrt{2} \\ 0 & 1 & 0 \end{pmatrix}, \begin{pmatrix} -2 & 0 & 0 \\ 0 & 1 & 0 \\ 0 & 0 & 2 \end{pmatrix}$

SECTION 7-4

1. (a) $\begin{pmatrix} 0 & -1 \\ 1 & 0 \end{pmatrix}, \begin{pmatrix} 1 \\ -1 \end{pmatrix}$ **(c)** $\begin{pmatrix} \frac{1}{2} & -\sqrt{3}/2 \\ \sqrt{3}/2 & \frac{1}{2} \end{pmatrix}, \begin{pmatrix} \dfrac{1-\sqrt{3}}{2} \\ \dfrac{1+\sqrt{3}}{2} \end{pmatrix}$

2. (a)

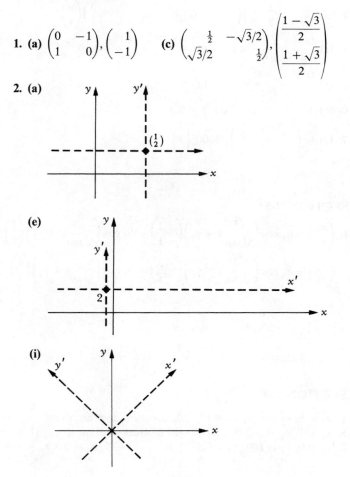

(o)

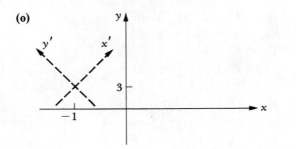

3. (a) $\dfrac{(x')^2}{1} + \dfrac{(y')^2}{(\frac{1}{2})^2} = 1$ **(d)** $(x')^2 + (y')^2 = 2^2$ **(e)** $\dfrac{(x')^2}{4} - \dfrac{(y')^2}{1} = 1$

   (g) $\dfrac{(x')^2}{16} + \dfrac{(y')^2}{4} = 1$

4. (a) $(x \quad y)\begin{pmatrix} 1 & 2 \\ 2 & 2 \end{pmatrix}\begin{pmatrix} x \\ y \end{pmatrix}$ **(c)** $(x \quad y)\begin{pmatrix} 7 & -3 \\ -3 & -1 \end{pmatrix}\begin{pmatrix} x \\ y \end{pmatrix}$

5. (a) $\dfrac{(x')^2}{4} + \dfrac{(y')^2}{6} = 1$ **(c)** $\dfrac{-(x')^2}{4} + \dfrac{(y')^2}{2} = 1$

   (e) $\dfrac{(x')^2}{2} - \dfrac{(y')^2}{6} = 1$ **(g)** $x' = \pm 2y'$

6. (a) $x = 0,\ y = 2$ **(c)** $x = \frac{5}{4},\ y = \frac{7}{4},\ z = 2$

7. (a) $0,\ \begin{pmatrix} -x_2 & -x_3 \\ x_2 \\ x_3 \end{pmatrix};\ 3000,\ \begin{pmatrix} 1 \\ 1 \\ 1 \end{pmatrix}$

SECTION 7-5

1. $\begin{pmatrix} x_1 \\ x_2 \end{pmatrix} = b_1 \cos\left(\sqrt{\dfrac{3T}{ma}}\, t + \gamma_1\right)\begin{pmatrix} 1 \\ -1 \end{pmatrix} + b_2 \cos\left(\sqrt{\dfrac{T}{ma}}\, t + \gamma_2\right)\begin{pmatrix} 1 \\ 1 \end{pmatrix}$

3. $\begin{pmatrix} x_1 \\ x_2 \end{pmatrix} = b_1 \cos\left(\sqrt{\dfrac{11}{3m}}\, t + \gamma_1\right)\begin{pmatrix} -3 \\ 1 \end{pmatrix} + b_2 \cos\left(\sqrt{\dfrac{1}{m}}\, t + \gamma_2\right)\begin{pmatrix} 1 \\ 1 \end{pmatrix}$

CHAPTER 8

SECTION 8-1

1. $x_1 = 6,\ x_2 = -2,\ x_3 = 1$ **3.** Solution does not exist.
5. $x_1 = 1,\ x_2 = -0.999,\ x_3 = -30$ (correct solution is $x_1 = 1,\ x_2 = -1,\ x_3 = -30$)
7. $x_1 = 0,\ x_2 = 100,\ x_3 = 0.5$ **9.** $x_1 = -2.5,\ x_2 = 0.5,\ x_3 = -4.511$

SECTION 8-2

1. (a) Yes   **(c)** Yes   **(e)** No   **(g)** No   **3.** $x_1 = 1, x_2 = 2, x_3 = 3$
5. $x_1 = r + 1, x_2 = 2r, x_3 = r$
7. $x_1 = 2r - s - t + 2, x_2 = 3r + s - 3t + 1, x_3 = r, x_4 = s, x_5 = t; r, s,$ and t are parameters that can take on all values.
8. Six arithmetic operations are saved: two saved in each of (1, 3) and (1, 4) locations on creation of 0 in (1, 2) location, two saved in (2, 4) location on creation of 0 in (2, 3) location.

SECTION 8-3

1. $x = 2.4, y = 0.4, z = 2$ **3.** $x = 4.51, y = 5.25, z = 2.69$
5. $x = 7, y = -0.43, z = 2.29$

SECTION 8-4

1. $\lambda = 8, r \begin{pmatrix} 1 \\ 0 \\ -1 \end{pmatrix}$ **3.** $\lambda = 6, r \begin{pmatrix} 1 \\ 0 \\ 1 \end{pmatrix}$ **5.** $\lambda = 6, r \begin{pmatrix} 0 \\ 0 \\ 1 \\ 1 \end{pmatrix}$

7. $\lambda = 84.311, r \begin{pmatrix} 1 \\ 0.0112559 \\ -0.0112559 \\ 0.039746 \end{pmatrix}$ **9.** $\lambda = 10, r \begin{pmatrix} 2 \\ 2 \\ 1 \end{pmatrix}; \lambda = 1, r \begin{pmatrix} -1 \\ 1 \\ 0 \end{pmatrix}$

11. $\lambda = 12, r \begin{pmatrix} 1 \\ 0 \\ 1 \end{pmatrix}; \lambda = 2, 2, \begin{pmatrix} r \\ s \\ -r \end{pmatrix}$

13. $\lambda = 10, r \begin{pmatrix} 2 \\ 2 \\ 1 \\ 1 \end{pmatrix}; \lambda = 5, r \begin{pmatrix} 1 \\ 1 \\ -2 \\ -2 \end{pmatrix}; \lambda = 2, r \begin{pmatrix} 0 \\ 0 \\ 1 \\ -1 \end{pmatrix}; \lambda = 1, r \begin{pmatrix} 1 \\ -1 \\ 0 \\ 0 \end{pmatrix}$

14. (a)

Iteration	$A\mathbf{x}$	Adjusted Vector
1	(2, 3, 3, 2)	$(\frac{2}{3}, 1, 1, \frac{2}{3})$
2	$(1, 1\frac{2}{3}, 1\frac{2}{3}, 1)$	(0.6, 1, 1, 0.6)
3	(1, 1.6, 1.6, 1)	

$$\left(\frac{1}{5.2}\right)(1, 1.6, 1.6, 1) = (0.1923, 0.3077, 0.3077, 0.1923)$$

CHAPTER 9

SECTION 9-1

1.

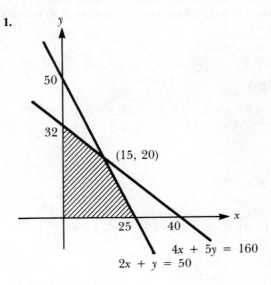

3.

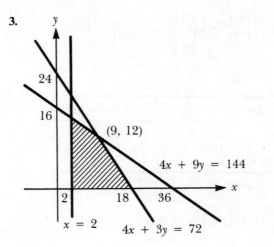

5. $2x + y$ is a maximum of 24 at $A(6, 12)$.

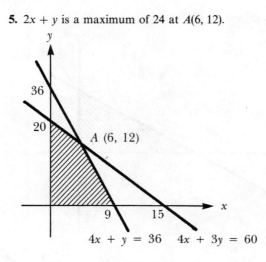

7. The feasible region is unbounded, and there is no maximum.

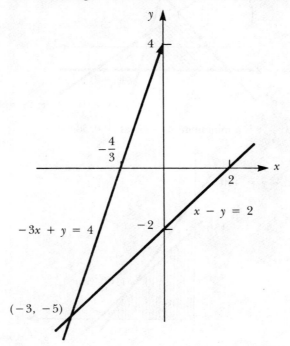

9. $3x + y$ is a maximum of 26 at $(8, 2)$.

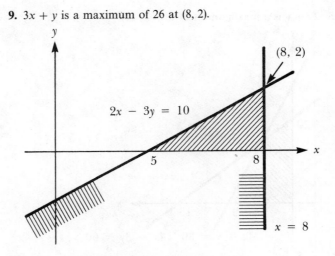

11. $x + 2y$ is a maximum of 6 along BC.

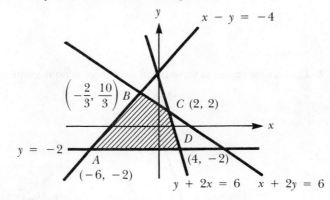

13. $-3x - y$ is a minimum of -350 at $(100, 50)$.

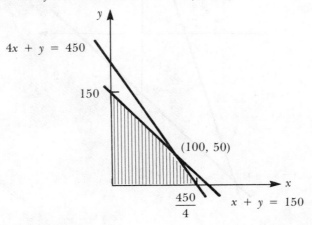

15. $-3x - 2y$ is a minimum of -240 at (40, 60).

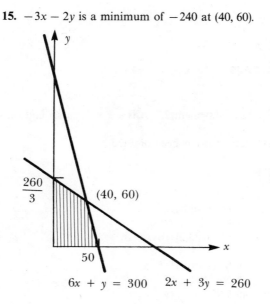

$\dfrac{260}{3}$ (40, 60)

50

x

$6x + y = 300$ $2x + 3y = 260$

17. $15x + 7y$ is a maximum of 18,000 at $x = 500$, $y = 1500$.
19. $40x + 20y$ is a maximum of 2400 along line $20x + 10y = 1200$ between (0, 120) and (30, 60).
21. Let x and y be barrels of oil for automobiles and heating, respectively. $3.5x + 2y$ is a maximum of 5200 when $x = 800$, $y = 600$.
23. $x + y$ is a maximum of 150 when $x = 50$, $y = 100$.
25. Let distribution of cars be x from A to C, y from B to C, p from A to D, q from B to D. Maximum y is 54 when $q = 146$, $x = 66$, $p = 34$.
27. 11 Torros and 17 Sprites   **29.** 30 acres of strawberries and 10 acres of tomatoes

SECTION 9-2

1. 24 at $x = 6$, $y = 12$ **3.** 16.5 at $x = 2.25$, $y = 1.25$
5. 1600 at $x = 160$, $y = 0$ **7.** 22,500 at $x = 0$, $y = 100$, $z = 50$
9. 40 at $x_1 = 20$, $x_2 = x_3 = 0$ **11.** 50 at $x_1 = 0$, $x_2 = 15$, $x_3 = 5$, $x_4 = 0$
13. Maximum profit of \$1740 when $x = 30$, $y = 0$, $z = 120$
15. Maximum profit of \$7200 when 600 transported from A to P and none from B or C
17. Maximum profit is \$240 daily when $x + \dfrac{1}{2}y = 30$; for example, $x = 10$, $y = 40$.
18. -4 at $x_1 = 4$, $x_2 = 0$ **20.** -3.75 at $x_1 = x_2 = 0$, $x_3 = 3.75$

SECTION 9-3

1. 24 at $x = 6$, $y = 12$ **3.** 6 at $x = 0$, $y = 3$ **5.** 40 at $x_1 = 20$, $x_2 = x_3 = 0$

APPENDIXES

APPENDIX A

1. THE MATRIX A IS   **3.** −1   2   3
    0   1 −2   4   4   5   1
 −3   6   7   8

6. Insufficient data in statement 25. *A* is a 2 × 3 matrix and thus should have six elements.

8. MAT READ statement has been omitted.

APPENDIX B

1. 2 −2   **6.** 5 DIM A(4,3)
   4   6   10 MAT READ A
                15 FOR J = 1 TO 4
                20 LET A(3,J) = 3*A(3,J)
                25 NEXT J
                30 MAT PRINT A;
                35 DATA 1,2,3,0, −1,2,2,1, −1,1,2, −3
                40 END

APPENDIX C

1. 4
  6

APPENDIX D

1. −1   2   3
   0   0   0

Index